The Chi-squared Distribution

The Chi-squared Distribution

H. O. LANCASTER

Professor of Mathematical Statistics
University of Sydney, Australia

John Wiley & Sons, Inc. New York   London   Sydney   Toronto

To J. O. Irwin,
Statistician and Editor

Preface

This monograph gives an exposition of the theory and application of the Pearson χ^2 as well as some related topics. Although the Pearson χ^2 is quite commonly used in statistical practice, there has been as yet no detailed and comprehensive account of it.

Chapter I gives an historical introduction with special attention to the works of K. Pearson and R. A. Fisher. Chapters II through IV build the necessary mathematical techniques. Chapter II is devoted to the distribution of the Γ-variable and of quadratic forms in jointly normal variables. Emphasis here is on the theoretical, continuous distributions. Some brief mention is made of the characterization theorems, which show that some desirable properties of test functions, such as stochastic independence from other random variables, can only hold in normal systems.

In Chapter III, the relations between discrete and continuous distributions are discussed in some detail. The possibility of a good approximation by a continuous random variable to a discrete random variable is examined numerically, since analytic methods are not always adequate or available. Criteria for a satisfactory approximation are stated and a compromise solution is given for the determination of the probability to be assigned to a sample when the distribution is discrete.

We do not draw, as many authors do, a sharp line between variables assuming a continuum of values and those assuming a finite or denumerable set of values. In many cases of practical interest, this distinction breaks down and the class frequencies are merely the number of metrical observations in certain intervals.

In Chapter IV, some special properties of orthogonal matrices are studied and orthonormal functions are defined as elements of vectors or functions on a finite set of points. Orthonormal functions are then defined on general statistical distributions to give such sets of functions as the Hermite polynomials on the normal distribution.

Chapter V gives a number of different proofs of the distribution of χ^2 in discrete distributions. Several proofs are obvious generalizations of the normal approximation to the binomial distribution. Preference is given to a proof based on the multivariate form of the central limit theorem. Some of these proofs are of historical interest and others are used commonly. At this point, it is possible to consider some practical examples without the development of further theory. This has the advantage of introducing not too early the difficult problems that arise from the fitting of parameters estimated from the data.

In Chapter VI, a general discussion is given of classes of distributions dependent on certain unknown parameters. The likelihood ratio or, in measure theoretical terms, the Radon-Nikodym derivative, is expressed as a series of orthonormal functions. In some cases, it is possible to choose the set of orthonormal functions so as to contain the standardized form of the likelihood ratio as a member. Dependence in multivariate systems is examined and some theorems on stochastic independence in multivariate distribution are proved, which give a basis for the later tests of independence in contingency tables. For many years, notions of dependence had been based on the notion of correlation in normal systems in which correlations or covariances between pairs of variables determined the form of the joint distribution. In this chapter, this single parameter is replaced by a, possibly infinite, matrix of parameters, the correlations between members of complete orthonormal sets of functions defined on the marginal distributions.

In Chapter VII, the distribution of the non-central χ^2 is described. Methods of computing parameters of non-centrality, making use of the canonical forms of Chapter VI, are given. Relations between the theory of χ^2 and sufficiency are derived. The criteria of efficient χ^2 tests are also given.

Chapter VIII considers the distribution of the Pearson χ^2 when parameters have been estimated from the data. With sufficient statistics, the asymptotic theory is relatively easy. However, no proof can be simple and short in the general case. Fisher's original proof and an alternative derivation of the joint distribution of linear forms in the cell frequencies are expressed. It is shown that an orthonormal set of linear forms can be chosen in such a way that one subset retains the orthonormal property with variations in the fitted parameter and the complementary set is identically zero if maximum likelihood estimates are used. Fitting parameters thus corresponds to the deletion of certain linear orthonormal forms in the cell frequencies and χ^2 is the sum of squares of the remaining linear forms.

Chapter IX deals with problems of inference as they arise in the interpretation and application of the χ^2 test. It is in no sense a summary of a general theory of inference. Special attention is paid to the problems of discreteness, multiple comparisons, and the χ^2 interpretations of high and low values of χ^2. It is shown that there is a close relation between the likelihood ratio test and the χ^2 test. In fact, the two tests only differ in the order in which the normal approximation and the calculation of the likelihood ratio are carried out. The examples and exercises in Chapters V, VIII, and IX cover many situations likely to be met in applications, other than in contingency tables.

In Chapter X, some topics in multivariate normal theory are briefly discussed—partial and multiple correlation, biserial η, the correlation ratio, the tetrachoric and polychoric correlations, Hotelling's canonical correlation and its generalization.

Chapter XI deals with the analysis of contingency tables of two dimensions. Special attention is paid to canonical forms of distribution and to a description of the models of the sampling. Combinatorial and asymptotic methods of testing the tables are described. Parameters of non-centrality are obtained and hence the powers of the tests can, in principle at any rate, be determined.

Chapter XII deals with the controversial treatment of contingency tables of higher dimensions. The passage from two to three dimensions cannot be lightly dismissed as a straightforward generalization as is often stated. Definitions of interactions are then expressed and the favored definitions, related to the generalized coefficients of correlation, are shown to correspond to familiar ideas in regression theory. In this treatment of the contingency tables of higher dimensions, close analogies with the analysis of variance in factorial experiments are evident. Indeed, the aim of the author has always been to bring out these analogies and devise new tests with their aid.

An extensive bibliography and author and subject indexes conclude the monograph.

We have used Chapter II and Sections 1–3 of Chapter IV in an advanced undergraduate course on the analysis of variance. All of Chapter III and most of Chapters V, VII, IX, and XI are also appropriate at that level. The material of Chapter IV, Section 4 and of Chapter VI is more difficult and the discussion needs to be supplemented from works on the general theory of orthogonal functions. The exercises of these Chapters are designed partly as a guide to further reading. Most of Chapter X will be best studied at the level of our final Bachelor of Science Honours year and the same applies to Chapters XI and XII.

The aims of the monograph have been to give:
  (i)   an historical survey of the theory of χ^2;
 (ii)   a discussion of the problems of approximation to empirical discrete distributions by theoretical continuous distributions;
(iii)   an exposition of a systematic use of the theory of orthogonal transformations and functions;
 (iv)   an exposition of the theory of dependence;
  (v)   a cover of the problems of inference;
 (vi)   a treatment of the theory of contingency tables;
(vii)   an extensive classified bibliography;

(viii)   examples of the application of the theory.

It is hoped that these aims have been realised.

I am indebted to the Literary Executor of the late Sir Ronald A. Fisher, F.R.S., and to Oliver & Boyd Ltd., Edinburgh, for their permission to reprint quotations from the book *Statistical Methods for Research Workers*, 13th edition, 1963, in my Section I.4, Quotation 20.

I am indebted to the Biometrika Trustees for permission to reproduce extracts from the articles of K. Pearson and other authors which have appeared in *Biometrika*.

I should like to acknowledge the assistance received in discussions with G. S. Watson, J. N. Darroch, J. O. Irwin, and my departmental colleagues. H. Mulhall and G. K. Eagleson have given detailed help in the revision of the manuscript. For careful typing and other assistance, I must especially thank Mrs. Janet Fish, Mrs. Elsie Adler, Mrs. L. Ferrier, and Miss Rhonda Yee of this department.

H. O. LANCASTER

March 1969
University of Sydney

Contents

The Chi-squared Distribution

Historical Survey of χ^2

1. FORERUNNERS OF THE PEARSON χ^2

The study of games of chance by Pascal, Fermat, Huygens, Cardano and others had led to a great number of particular results. Some generality was first attained when de Moivre expressed many results in approximate form by the formula now known after James Stirling. De Moivre in the second supplement to the *Miscellanea Analytica* of 1733 and in the later English translation of 1758 had obtained an approximation to the terms of the binomial distribution by areas under the normal curve. This result enables binomial terms to be expressed approximately in exponential form and related to ordinates of the normal curve. Sums of adjacent terms can then be expressed as areas under the normal curve by equating Riemann sums to integrals. De Moivre was thus the first to determine areas under the normal curve by quadrature. Due appreciation of this work is found in Todhunter (1865), in Pearson (1924c) and in Walker (1929).

The work of de Moivre and Laplace established the asymptotic normality of the standard variable,

$$(1.1) \qquad \chi = (m - Np)/(Npq)^{\frac{1}{2}},$$

where m is the observed number of successes in N independent trials with constant probability, p, of success at each trial. Squaring the expressions (1.1) yields

$$(1.2) \qquad \begin{aligned} \chi^2 &= (m - Np)^2/(Npq) \\ &= (m - Np)^2/(Np) + (N - m - Nq)^2/(Nq). \end{aligned}$$

The expression on the right is in a form that was to be generalized by Pearson (1900a) to

$$(1.3) \qquad \chi^2 = \sum (a_i - Np_i)^2/(Np_i)$$

where a_i is the observed number in the ith cell of a multinomial distribution. Since the χ of (1.2) is standardized and asymptotically normal, χ^2 of (1.2) is asymptotically distributed as the square of a standardized normal variable.

Bienaymé (1858) obtained the distribution of the sum of squares of m independently distributed standardized normal variables in the gamma function form. In particular, he evaluated the integral

(1.4) $$\mathscr{P}(U^2 < \gamma^2) = 2\int_0^{\gamma} u^{m-1}e^{-u^2}\,du/\Gamma(\tfrac{1}{2}m),$$

and gave numerical values for P for odd and even m, such as were to appear later in Pearson (1900a). This work of Bienaymé does not appear to have been known to Helmert (1875b) who also derived (1.4) and further gave a useful orthogonal transformation. Laplace's work on the normal distribution was extended by Poisson, Bienaymé and Todhunter. Poisson (1827 and 1832) gave the distribution of the standardized sum of independently, but not necessarily identically, distributed variables. Bienaymé (1838) gave formulae, which suggest that he was proposing to extend the normal approximation to the multinomial distribution. Bienaymé (1852, reprinted 1868) obtains a version of the multivariate central limit theorem and obtains the χ^2 distribution for a sum of standardized squares.

A special theme, the distribution of a linear form in the class frequencies of a multinomial distribution, was considered by Bienaymé (1838) and later by Sheppard (1898a). The solution of this problem is closely related to the characterization of the joint normal distribution by the property that every linear form in the marginal variables is normally distributed. Fréchet (1951) has defined a multivariate normal distribution by this property. This is a useful generalisation as the joint distribution of random variables possessing a singular covariance matrix may be required. The new definition also permits an easy extension of the central limit theorem to the multivariate case.

The joint multivariate normal distribution is of fundamental importance in the study of χ^2. Seal (1967) gives most credit to Bravais (1846), Schols (1875) and Edgeworth (1892) for its development. Before passing on to an appreciation of K. Pearson's work, we might mention other workers active in related fields towards the end of the nineteenth century.

Sheppard (1898a) considered possible tests of goodness of fit for the multinomial distribution. He derived expressions for the moments of the class totals and of linear combinations of the class totals. He gave the correlations between the class totals and the expectations for one class conditional on a given value for another class. As a test of goodness of fit, he proposed to work out the value of the difference of the observed frequency from the expected frequency for each cell of a contingency table and to see how often it exceeded its probable error. In a good fit, this would happen less often than in a bad one. Sheppard (1898a and b) also considered the fourfold table as a dichotomy of a bivariate normal distribution—a special case of a theme, the tetrachoric expansion developed in greater detail by Pearson in his later papers. Looking back on these early papers, Sheppard (1929) comments that he had not appreciated the importance of the correlations between cells in the multinomial or contingency table. Sheppard's lack of success in reaching a satisfactory general solution was due partly to the awkward form of the variance-covariance matrix in the general two-way contingency table. The variance-covariance matrix of the multinomial distribution is more tractable and so it was Pearson (1900a) who obtained the solution, providing the widely applicable test of goodness of fit. Sheppard's (1903) table of the normal integral has been widely used as a basis for the tabulation of χ^2. W. Lexis (1877) had been independently trying to establish norms of behaviour of repeated binary trials with varying probabilities. This was completed by Pearson's discovery of χ^2, as R. A. Fisher has pointed out. Lexis may be regarded as the forerunner of A. A. Tchouproff and V. I. Romanovskii.

G. U. Yule reconsidered the theory of normal correlation, including partial correlation, and the more general concepts of association. When R. A. Fisher began to modify the Pearson theory, Yule was active in popularising the new ideas on degrees of freedom.

2. THE CONTRIBUTIONS OF K. PEARSON

In the series, *Mathematical Contributions to the Theory of Evolution*, Karl Pearson introduced a number of theoretical statistical distributions, which were new to statistics, and among which the Type III is, after an appropriate choice of scale and origin, the distribution of χ^2 or alternatively the gamma-distribution. Given any particular set of empirical data, it became necessary to distinguish those distributions which fitted it closely from those which did not. Pearson realised that the normal curve had too often been accepted uncritically as fitting empirical data.

Pearson had been much concerned with generalizing the univariate normal distribution to the general normal correlation; so that, it appeared natural for him to provide a normal approximation to the multinomial distribution, just as de Moivre and Laplace had given normal approximations to the binomial distribution. The symbol, χ^2, was first introduced by Pearson (1896), where it is written in place of $\mathbf{x}^T\mathbf{R}^{-1}\mathbf{x}$ for brevity.

Pearson's contributions to statistical theory were numerous but, perhaps, the greatest of them was the χ^2 test of goodness of fit, which has remained one of the most useful of all statistical tests. Pearson (1900a) states "the object of this paper is to investigate a criterion of the probability on any theory of an observed system of errors, and to apply it to the determination of goodness of fit in the case of frequency curves". He begins with a preliminary lemma, which is more conveniently expressed in matrix notation. Let us define

$$(2.1) \qquad\qquad \chi^2 = \mathbf{Y}^T\mathbf{V}^{-1}\mathbf{Y}$$

where $\mathbf{Y}$ is a vector, whose elements are random variables and $\mathbf{V}$ is a positive definite variance-covariance matrix of size, n. Then Pearson's lemma states that there is a linear transformation, $\mathbf{Y} \to \mathbf{Z}$, such that

$$(2.2) \qquad \mathbf{Y}^T \mathbf{V}^{-1} \mathbf{Y} = \mathbf{Z}^T \mathbf{Z} = \sum_1^n Z_i^2$$

Further, if the joint frequency distribution is given by

$$(2.3) \qquad f(y_1, y_2, \ldots, y_n) = \text{constant} \exp -\tfrac{1}{2} \mathbf{y}^T \mathbf{V}^{-1} \mathbf{y}$$

then $\mathbf{Y}^T \mathbf{V}^{-1} \mathbf{Y} = \chi^2$ is distributed as the sum of squares of n independently distributed standardized normal variables.

Pearson's proof of this important lemma is given in geometrical language but amounts to saying that there exists a linear, possibly non-orthogonal transformation on the Y's to variables, $Z_1, Z_2, \ldots, Z_n$, so that the distribution is spherically symmetrical. As a matter of fact, we have that the joint distribution of the $\{Z_i\}$ is given by

$$(2.4) \qquad g(z_1, z_2, \ldots, z_n) = \text{constant} \exp -\tfrac{1}{2} \mathbf{z}^T \mathbf{z}.$$

It is thus evident that the $\{Z_i\}$ form a mutually independent set of standardized normal variables. Pearson refers to χ as the *ray* in this paper and as the *generalised probable error* in Pearson and Lee (1908).

Pearson (1900a) next defines for the multinomial distribution $(p_1 + p_2 + \ldots + p_{n+1})^N$, observed and expected numbers, m_i' and m_i, and differences, $e_i = m_i' - m_i$, which he calls the errors.

$$(2.5) \qquad e_i = m_i' - m_i, \qquad i = 1, 2, \ldots, (n+1); \sum_1^{n+1} e_i = 0;$$

$$(2.6) \qquad \text{var } e_i = Np_i(1 - p_i) = N(1 - m_i/N)/(m_i/N) = \sigma_i^2$$

$$(2.7) \qquad \text{cov } (e_i, e_j) = -m_i m_j/N = -Np_i p_j = \sigma_i \sigma_j r_{ij}.$$

Owing to the identity (2.5), Pearson considers only the first n of the e_i to be "variables". By assuming that he will always be dealing with large numbers, Pearson then considers the e_i to be approximately distributed as normal variables with an $n \times n$ covariance matrix, $\mathbf{V}$, given by (2.6) and (2.7). It is of interest to note that he assumed that normal variables have a joint normal distribution. However, counter-examples to this proposition are easily found. But, having made the assumption of joint normality, Pearson (1900a) gives the joint distribution of the first n of the set, $\{e_i\}$, by an equation of the form (2.3); he has yet to give $\chi^2 = \mathbf{e}^T \mathbf{V}^{-1} \mathbf{e}$ in a form convenient for computation. This takes up a good deal of the paper. Later authors have pointed out that once he had derived the inverse of $\mathbf{V}$, he could have dispensed with the steps which gave the method of derivation. It can in fact be verified that

(2.8) $$\mathbf{V}^{-1} = N^{-1}(p_i^{-1}\delta_{ij} + p_{n+1}^{-1}),$$

by multiplication of the matrix on the right of (2.8) by $\mathbf{V}$. Pearson (1916a) himself felt that this part of the proof might be simplified and determined the elements of $\mathbf{V}^{-1}$ by an application of the theory of partial correlation, but the proof is not as simple as the method of verification; nor is the treatment quite rigorous, especially in the derivation of the correlation of e_i with e_j, given e_k, where i, j and k are three pairwise different indices. In the same year, Pearson (1916b) adopted a suggestion of H. E. Soper to simplify the proof. A supernumerary class with an expectation, very large in comparison to those of the other cells, is introduced; the errors of the first $(n + 1)$ cells have then zero correlation and variances approximately equal to Np_i. The matrix $\mathbf{V}$, say, for the first $(n + 1)$ variables would thus be diagonal and its reciprocal readily computed. This device has become well known after the paper of Fisher (1922a), who treats the multinomial distribution as a conditional distribution arising from a multivariate Poisson distribution. Later authors have noted that Pearson (1900a) did not give a rigorous proof of (2.7) but its truth was surely well known to him. Pearson (1916c) once again dealt with the evaluation of the covariances but rather casually. Pearson (1916a) gives the relation,

(2.9) $$\sigma_{\chi^2}^2 = \{2(q - 1) + q/H + q(q - 2)/N\},$$

where H is the harmonic mean of the expected cell frequencies, q is the number of cells in the multinomial.

Even in the initial paper of Pearson (1900a), an awkward problem, nowadays considered under the heading of the estimation of parameters, arises. We may wish not to test only the simple hypothesis but also a composite hypothesis. As is well known, this problem of estimating parameters was only solved after Fisher (1922a and b) began his investigations.

Pearson (1900b) derived the tetrachoric expansion of the bivariate normal distribution in collaboration with Bramley-Moore. Pearson (1904) also studied contingency tables, with special emphasis on the possibility of the theoretical or underlying distribution being the joint normal distribution. He defined contingency and several numerical measures of it, among them the χ^2 of the contingency table. He obtained the asymptotic result

(2.10) $$r = \pm\phi(1 + \phi^2)^{-\frac{1}{2}}, \qquad \phi^2 = \chi^2/N$$

whereby r can be estimated from ϕ^2. The r defined by (2.10) is his "first coefficient of contingency." If the distribution were the joint normal, r

would approximate for large N to the theoretical parameter of the distribution, the coefficient of correlation. He made much of the point that χ^2 is unaffected by a reordering of the marginal classes, and this enabled him to think in terms of a metrical character for the marginal classifications if it had not already been given. Alternatively it enabled him to dispense with any knowledge of the marginal distributions if he considered only a test of independence of the two marginal variables. For multivariate normal distributions he obtained a generalization of (2.10).

3. THE CONTRIBUTIONS OF R. A. FISHER

R. A. Fisher made many contributions to the theory and application of χ^2, as will be clear from the numerous references to his work later in the text and exercises. We may briefly note some of his early interests in order to see how he was so well fitted to clear up some inconsistencies and obscurities in the theory of χ^2. Fisher had devised tests of hypothesis in regression analysis, in which the joint normal distribution arises naturally and in which sums of squares are quadratic forms of normal variables, or at least are approximately such. Moreover, the values of certain parameters may have to be estimated from the data and allowance has to be made for this in the tests of goodness of fit. Fisher had read Pearson's *Mathematical Contributions to the Theory of Evolution* and was thus familiar with the work on normal correlation and with Pearson's systems of curves. In this field, Fisher (1915) gave the distribution of the correlation coefficient, as estimated from samples. In this work, he showed how powerful the ideas of the geometry of n-dimensional space were in obtaining exact sampling distributions in variables related to the normal; he also used orthogonal matrices for convenient changes of variables, which is equivalent to rotations in the sample spaces considered. Among other contributions arising from these ideas, Fisher (1922b) obtained the distribution of the correlation ratio, Pearson's η^2. Fisher (1928b and 1935b) gave the distribution of the "central" or theoretical χ^2 and (1928c) gave the distribution of non-central χ^2. On the applied side, Fisher, Thornton and Mackenzie (1922) gave extensive applications of the Pearson χ^2 to the test of consistency of parallel bacterial counts.

Fisher (1922c) gave new theoretical foundations to mathematical statistics, introducing such new ideas as sufficiency of estimators and efficiency of statistical tests. These new ideas were applied in Fisher (1922a, 1923 and 1924) to the distribution of χ^2 in contingency tables and frequency distributions. Fisher (1924) gave the first proof of the asymptotic distribution of χ^2 in the general case when parameters are estimated from the data; this proof is fundamental to every proof given by later authors. Some special cases, the fourfold table, with sufficient statistics had already been treated in

Fisher (1922a). Fisher made so many contributions which are used later in this monograph, that it is unnecessary to detail them in this historical section. Leading ideas include the partition of χ^2 in which the total χ^2 is resolved into components, giving comparisons of interest. Numerous examples of these partitions are available in his *Statistical Methods*. The present theory of two-dimensional contingency tables owes much to him, whether the correct number of degrees of freedom or the combinatorial theory is considered. Fisher also suggested a criterion of second order interaction in the contingency tables of higher dimensions, which was worked out by Bartlett (1935) for the $2 \times 2 \times 2$ tables. He also devised the alternative method of partitioning the overall χ^2 in the higher tables by implication, although it remained for Haldane (1939) to interpret the higher order interactions. Another fruitful idea was the application of Hotelling's canonical analysis to contingency tables as quoted in Maung (1942).

In this historical introduction, we do not mention any further individual contributions to the theory. Many of them are mentioned in the text and we hope that few have been omitted from the bibliography.

We end this chapter with some quotations from the works of Fisher, Pearson and others, which may help to recall the atmosphere of the times.

4. SOME QUOTATIONS OF HISTORICAL INTEREST

1. The object of this paper is to investigate a criterion of the probability on any theory of an observed system of errors, and to apply it to the determination of goodness of fit in the case of frequency curves.

(1) Preliminary Proposition. Let $x_1, x_2, \ldots, x_n$ be a system of deviations from the means of n variables with standard deviations $\sigma_1, \sigma_2, \ldots, \sigma_n$ and with correlations $r_{12}, r_{13}, r_{23}, \ldots, r_{n-1,n}$.

Then the frequency surface is given by

$$Z = Z_0 e^{-\frac{1}{2}\left\{\mathscr{S}_1\left(\frac{R_{pp}}{R}\frac{x_p{}^2}{\sigma_p{}^2}\right)+2\mathscr{S}_2\left(\frac{R_{pq}}{R}\frac{x_p}{\sigma_p}\frac{x_p}{\sigma_p}\right)\right\}}$$

where R is the determinant

$$\begin{vmatrix} 1 & r_{12} & r_{13} & \cdots & r_{1n} \\ r_{21} & 1 & r_{23} & \cdots & r_{2n} \\ r_{31} & r_{32} & 1 & \cdots & r_{3n} \\ \cdot & \cdot & \cdot & \cdot & \cdot \\ \cdot & \cdot & \cdot & \cdot & \cdot \\ r_{n1} & r_{n2} & r_{n3} & & 1 \end{vmatrix}$$

and R_{pp}, R_{pq} the minors obtained by striking out the pth row and p^{th} column and the pth row and q^{th} column. $\mathscr{S}_1$ is the sum for every value of p, and $\mathscr{S}_2$ for every pair of values of p and q.

Now let

$$\chi^2 = \mathscr{S}_1\left(\frac{R_{pp}}{R}\frac{x_p^2}{\sigma_p^2}\right) + 2\mathscr{S}_2\left(\frac{R_{pq}}{R}\frac{x_p x_q}{\sigma_p \sigma_q}\right) \quad . \quad . \quad . \quad . \quad \text{(ii.)}$$

Then, $\chi^2 = $ constant, is the equation to a generalized "ellipsoid," all over the surface of which the frequency of the system of errors or deviations $x_1, x_2, \ldots, x_n$ is constant. The values which χ must be given to cover the whole of space are from 0 to ∞. Now suppose the "ellipsoid" referred to its principal axes, and then by squeezing reduced to a sphere, $X_1, X_2, \ldots, X_n$ being now the coordinates; then the chances of a system of errors with as great or greater frequency than that denoted by χ is given by

$$P = \frac{\left[\displaystyle\iiint\cdots e^{-\frac{1}{2}\chi^2}\,dX_1\,dX_2 \ldots\,dX_n\right]_\chi^\infty}{\left[\displaystyle\iiint\cdots e^{-\frac{1}{2}\chi^2}\,dX_1\,dX_2 \ldots\,dX_n\right]_0^\infty}$$

the numerator being an n-fold integral from ellipsoid χ to the ellipsoid ∞, and the denominator an n-fold integral from the ellipsoid 0 to the ellipsoid ∞. A common constant factor divides out. Now suppose a transformation of coordinates to generalized polar coordinates, in which χ may be treated as the ray, then the numerator and denominator will have common integral factors really representing the generalized "solid angles" and having identical limits. Thus we shall reduce our result to

$$P = \frac{\displaystyle\int_\chi^\infty e^{-\frac{1}{2}\chi^2}\chi^{n-1}\,d\chi}{\displaystyle\int_0^\infty e^{-\frac{1}{2}\chi^2}\chi^{n-1}\,d\chi} \quad . \quad . \quad . \quad . \quad \text{(iii.)}$$

This is the measure of the probability of a complex system of n errors occurring with a frequency as great or greater than that of the observed system. (Pearson 1900a).

●

2. Let

$$\chi^2 = \frac{1}{R}\{\mathscr{S}(R_{pp}x_p/\sigma_p^2) + 2\mathscr{S}'R_{pq}x_p x_q/\sigma_p\sigma_q\}$$

then $\chi^2 = $ constant, is an equation of an ellipsoid in n-fold space. If transferred to its principal axes, it will take the form

$$\chi^2 = \overset{n}{\underset{1}{\mathscr{S}}}\, X_p^2/\Sigma_p^2$$

Further $z \, dx_1 \, dx_2 \ldots dx_n$ will equal

$$\frac{N}{(2\pi)^{\frac{1}{2}n}} \frac{1}{\Sigma_1 \Sigma_2 \ldots \Sigma_n} e^{-\frac{1}{2}\chi^2} \, dX_1 \, dX_2 \ldots dX_n$$

representing the frequency due to n independent variables. Pearson and Lee (1908).

•

3. Let $\mathbf{X}$ be a vector of n centred and jointly normal variables, and suppose that $\mathscr{E}\mathbf{XX}^T$ is not positive definite but of rank $m < n$. Let $\mathbf{U}$ be any vector containing the maximum number of elements such that $\mathscr{E}\mathbf{UU}^T = \mathbf{V}$ is positive definite, then $\mathbf{U}^T\mathbf{V}^{-1}\mathbf{U}$ is invariant for all choices of $\mathbf{U}$ and the density function of $\mathbf{U}$ is $A(2\pi)^{-\frac{1}{2}m} \exp{-\frac{1}{2}\mathbf{u}^T\mathbf{V}^{-1}\mathbf{u}}$, where A is a constant for any particular choice of the $m \times n$ matrix, $\mathbf{B}$, and $\mathbf{U} = \mathbf{BX}$. This is a version of a theorem which is closely related to the so-called Pearson's lemma.

[Bienaymé, 1852; Todhunter, 1869; Pearson, 1900a; Sheppard, 1929.]

•

4. Suppose instead of a single correlation table we have a multiple correlation system. Such a system is well illustrated by the cabinet at Scotland Yard, which contains the measurements of habitual criminals on the old system of body measurements, now discarded in favour of a finger-print index. We have in this case a division of the cabinet into 3 compartments, which mark a threefold division of long, medium, and short head lengths. Each of these vertical divisions is then subdivided horizontally into three divisions giving the corresponding divisions for head breadth; each of these head-breadth divisions has three drawers for large, moderate, and small face breadths. Each drawer is sub-divided into three sections for three finger groups, and these again into compartments for cubit groups, and so on. If this be carried out for the seven characters dealt with, we should have ultimately 3^7 sub-groups forming a multiple correlation system of the 7th order. We may ask what is the mean square contingency of such a system and to what extent does it diverge from an independent probability system? Of course, for an ideal anthropometric index system the divergence should be very slight. (Pearson 1904.)

•

5. And here we will at once emphasise the fundamental difference between Mr. Yule and ourselves. Mr. Yule, as we will indicate later, does not stop to discuss whether his attributes are really continuous or are discrete, or hide under discrete terminology true continuous variates. We see under such class-indices as 'death' or 'recovery', 'employment' or 'non-employment' of the mother, only measures of continuous variates—which of course are not *a priori* and necessary Gaussian. Mr. Greenwood in the discussion on Mr.

Yule's paper referred to the jibe* about persons being 'dead' or 'not dead' and questioned whether Mr. Yule was correct in treating the variate behind the class-index as discrete and not continuous. In the original paper of Pearson "strength to resist smallpox when incurred" was stated to be the variate, and all the evidence that has been produced since indicates its continuity; in precisely the same way vaccination and non-vaccination represent degrees of immunity in a continuous variate of which area of vaccination as indicated by extent of cicatrix and period since vaccination are contributory quantitative factors. Again "employment or non-employment of the mother" are not taken by us as signifying the presence or absence of a mere discrete attribute—for example whether she works in a factory or not—but as a class-index indicating that employed women, who have not only their home work but factory labour also, have on the whole more physical exertion to endure than those who are simply housewives. We are really seeking how far the continuous variate physical exertion of women affects infant welfare, and this is not a discrete variate any more than survival or death of infant is a discrete variate, when you view them merely as class-indices of physical fitness to survive in the child. In other words, for the great bulk of attributes, to which Mr. Yule without analysis of their nature applies association, we should assert continuous variation. We hold therefore that in the main we are applying fourfold or other class divisions to continuous variates. Mr. Yule thinks he has freed himself from all consideration of what the nature of this continuity may be; we consider his brief wholly fallacious. You cannot free yourself from some assumption as to the nature of the distribution when you are dealing with the association of attributes. And in ignorance of what the true distribution may be, what assumption will help you to the most probable result? On the basis of a very large experience of frequency curves and surfaces we have no hesitation in saying that up to the present time no distribution has been proposed which roundly represents experience so effectively as the Gaussian frequency. One of the present writers has indicated over and over again how it fails, and he has measured the significance of its failure, but has always recognised that he must put against this the large percentage of cases in which it gives reasonable results, close enough for all practical purposes. (Pearson and Heron, 1913.)

•

6. The following deduction of the value of χ^2 is a variant from my *Phil. Mag.* proof. I owe the suggestion of it to Mr. H. E. Soper, although I have

---

* "We are considering", writes Mr. Yule, "simply the performance as against the non-performance of the operation of vaccination. Similarly all those who have died of small-pox are equally dead: no one of them is more dead or less dead than another, and the dead are quite distinct from the survivors".

deviated somewhat from his track. Let an indefinitely large population N consist of the classes $C_0, C_1, \ldots, C_l$ in the quantities $n_0, n_1, n_2, \ldots, n_l$ respectively. Then $p_s = n_s/N =$ chance of drawing a member of the class C_s, and the standard deviation of the distribution of frequency in samples of M drawn from the population will in this class C_s be as above

$$\sigma_s = \sqrt{\{Mp_s(1 - p_s)\}} \quad \cdot \quad \cdot \quad \cdot \quad \cdot \quad \cdot \quad \text{(xiii)}^{\text{bis}}.$$

Further, the mean of samples for this class will be Mp_s by (xii).

In the next place the correlation between deviations from the means in classes C_s and C_t will be in our present notation,

$$\sigma_s \sigma_{s'} r_{ss'} = -Mn_s n_{s'}/N^2 \quad \cdot \quad \cdot \quad \cdot \quad \cdot \quad \cdot \quad \text{(xiv)}^{\text{bis}},$$

or by (xiii)^{bis}

$$r_{ss'} = -\frac{\sqrt{(p_s p_{s'})}}{\sqrt{(1 - p_s)}\sqrt{(1 - p_{s'})}} \quad \cdot \quad \cdot \quad \cdot \quad \cdot \quad \text{(xv)}.$$

The distribution of frequency of the different classes in these samples of M will be given by the terms of the multinomial

$$(p_0 C_0 + p_1 C_1 + p_2 C_2 + \ldots + p_l C_l)^M \quad \cdot \quad \cdot \quad \cdot \quad \text{(xvi)}$$

$$\cdot \quad \cdot \quad \cdot$$

Now it is clear that the factor

$$(p_{1'} C_1 + p_{2'} C_2 + \ldots + p_{l'} C_l)^m$$

gives the frequency distribution of samples of m drawn from a population of indefinitely large size of which the proportions of the classes $C_1, C_2, \ldots, C_l$ are $p_{1'}, p_{2'}, \ldots, p_{l'}$ and in which no class C_0 occurs. But by (xx) these proportions are the same as in original population which contains C_0.

Hence if we take samples of M from an indefinitely large population with classes, $C_0, C_1, C_2, \ldots, C_l$, those that contain m of the classes $C_1, C_2, \ldots, C_l$ will be distributed in the same proportions as if we had extracted m from an indefinitely large population consisting only of those l classes in the same proportions.

Now thus far the nature of the class C_0 is at our choice. In the original population N it appears with the total frequency n_0. Let $n_0 + n' = N$, and suppose n_0 is indefinitely greater than n', then p_0 will be indefinitely greater than $p_1, p_2, \ldots, p_n$. It follows that p_s if s be not zero is very small compared to unity, because

$$p_0 + p_1 + p_2 + \ldots + p_n = 1.$$

Hence in such a system from (xiii)^{bis}, if s be not zero,

$$\sigma_s = \sqrt{(Mp_s)} \quad \cdot \quad \cdot \quad \cdot \quad \cdot \quad \cdot \quad \cdot \quad \text{(xxi)},$$

and from (xv) to the same degree of approximation $r_{st} = 0$. That is to say, if p_0 be large in taking samples of size M from an indefinitely large population, there will be no correlation in deviations in the frequency of the classes $C_1, \ldots, C_l$.

.   .   .

From this we deduce that the probability of a sample which gives $\chi^2 = \chi_0^2$ with q linear conditions must be obtained from

$$P = \frac{\int_{\chi_0^2 - K^2}^{\infty} e^{-\frac{1}{2}\chi^2} \chi^{n-q-1} \, d\chi}{\int_0^{\infty} e^{-\frac{1}{2}\chi^2} \chi^{n-q-1} \, d\chi} \qquad . \quad . \quad . \quad . \tag{xxxvi}$$

We must therefore in order to find P enter the Tables of "Goodness of Fit" with $\chi^2 = \chi_0^2 - K^2$ and with $n' = n - q + 1$.

The reader who is familiar with the theory of the multiple total and partial correlation coefficients will note how closely analogous the formulae (xxxiii) to (xxxv) are to results in that theory. The fundamental determinant of the correlation may be made to agree with Δ if we merely write $r_{ss'} = \cos(ss')$. In fact both theories really reduce to the discussion of the formulae of spherical trigonometry in multiple space. (Pearson, 1916b.)

●

7. "In a paper published in 1898, one object of which was to obtain formulae for testing the hypothesis of normal distribution or normal correlation, I met (A) [i.e., the fitting of parameters] so far as these particular cases were concerned, by using the true s.d. of the discrepancy in place of the s.d. of the error. But the method was not really satisfactory as I did not realise the importance of (B) [the correlation between cells of a frequency distribution]. (Page 117 of Sheppard, 1929.)

●

8. This short paper, with all its juvenile inadequacies, yet did something to break the ice. Any reader who feels exasperated by its tentative and piecemeal character should remember that it had to find its way to publication past critics who, in the first place, could not believe that Pearson's work stood in need of correction, and who, if this had to be admitted, were sure that they themselves had corrected it.

The writer's point of view was that certain inconsistencies were manifest in Pearson's uses of the χ^2 test, and certain other writers, notably Yule and Greenwood (1915) and Bowley (1920), had felt that something was wrong, without discovering what it was.

In the writer's thought, though not very explicitly in this paper, the mathematical distribution given by tables of χ^2 was that of the sum of n squares of variates normally and independently distributed about zero with unit variance. χ^2 in fact was the square of the distance of a random point from the centre of a homogeneous normal distribution in n dimensions. The number of dimensions, however, would be reduced by unity for every restriction upon deviations between expectation and observation, and it appeared that the inconsistencies in the literature could be straightened out if account were taken of the true number of degrees of freedom in which observation and expectation might in reality differ.

In the examples chosen the complete equivalence of different rational approaches, once the question of degrees of freedom is rectified, was still somewhat obscured by minor inconsistencies in the method of computation. These arise from the fact that the mathematical distribution is only realised in the limit for large samples, where non-linear restrictions tend to linearity.

The treatment, in a footnote, of the multinomial distribution as a section of a multiple Poisson distribution of independent variates shows the most direct demonstration of the mathematical distribution tabled, as the sampling distribution of χ^2 defined as a measure of discrepancy between observation and hypothesis. This analytical artifice has, I hope, a lasting value.

The algebraic re-examination of Pearson's (1900) proof that the fitting of constants did not affect the distribution of χ^2 was set out in 1924 (Paper 8), in which the effects of inefficient fitting also are worked out. (Comment by R. A. Fisher on his paper *J. Roy. Statist. Soc.* 85, 87–94, in Fisher, 1950*a*.)

●

9. The interesting series of experiments on the distribution of χ^2, reported by Dr. Brownlee [1] affords an opportunity, not only of clearing up such doubts as still remain as to the necessity of entering Elderton's table with a corrected, or reduced, value of n', but also of bringing the conditions under which χ^2 affords a measure of goodness of fit into relation with the general theory of statistical estimation.

If x is the frequency of observations in any compartment of a frequency distribution, and if m is the expectation in that compartment, Pearson introduced ([2] 1900) the statistic

$$\chi^2 = \mathscr{S}\left\{\frac{(x - m)^2}{m}\right\}$$

as a measure of the discrepancy between observation and expectation. He succeeded in calculating the distribution of χ^2, when the values of x were the frequencies in random samples from an infinite population in which the frequencies were proportional to m, and showed that the distribution of χ^2

depends, in the limit when the samples are large, only on the number of classes, n', into which the samples were divided. In the same paper Pearson considered the possibility that when the values of m are not *a priori* expectations, but are themselves calculated from the observed values, the distribution of χ^2 might be modified by this procedure. He concluded that this was not so, and applied the test without correction to several examples in which the expectations in the several classes had been calculated from the distribution in the sample.

In 1922 [3] I was able to show, in the case of contingency tables, for which the margins of the expected table are reconstructed from those of the observed table, that the distribution of χ^2 was given exactly by Pearson's formula if we take for n', not the number of classes in the table, but one more than the number of degrees of freedom in which the expected table might differ from the values observed. The number of degrees of freedom is the number of frequencies which may be given arbitrary values without conflicting with the condition that the marginal totals are already specified. Thus, for a contingency table with two varities having r rows and c columns, should be equated, not to cr, but to $1 + (c - 1)(r - 1)$. In the same paper I expressed the opinion that the same reasoning should be applied to testing the goodness of fit of frequency curves, but that some discrepancy would arise if the grouping used in calculating the theoretical distribution were different from that employed in testing the goodness of fit. (Fisher, 1924.)

●

10. The following quotations are from Fisher's *Statistical Methods*.

It is of interest to note that the measure of dispersion, Q, introduced by the German economist Lexis, is, if accurately calculated, equivalent to χ^2/n of our notation. In the many references in English to the method of Lexis, it has not, I believe, been noted that the discovery of the distribution of χ^2 in reality completed the method of Lexis. If it were desired to use Lexis' notation, our table could be transformed into a table of Q merely by dividing each entry by n. Para. 20.

When a number of quite independent tests of significance have been made, it sometimes happens that although few or none can be claimed individually as significant, yet the aggregate gives an impression that the probabilities are on the whole lower than would often have been obtained by chance. It is sometimes desired, taking account only of these probabilities, and not of the detailed composition of the data from which they are derived, which may be of very different kinds, to obtain a single test of the significance of the aggregate, based on the product of the probabilities individually observed.

The circumstance that the sum of a number of values of χ^2 is itself distributed in the χ^2 distribution with the appropriate number of degrees of freedom, may be made the basis of such a test. For in the particular case when $n = 2$, the natural logarithm of the probability is equal to $-\frac{1}{2}\chi^2$. If therefore we take the natural logarithm of a probability, change its sign and double it, we have the equivalent value of χ^2 for 2 degrees of freedom. Any number of such values may be added together, to give a composite test, using the Table of χ^2 to examine the significance of the result. Para. 21.1.

Just as values of χ^2 may be aggregated together to make a more comprehensive test, so in some cases it is possible to separate the contributions to χ^2 made by the individual degrees of freedom, and so to test the separate components of a discrepancy. Para. 22.

•

11. Several reasons have contributed to the prolonged neglect into which the study of statistics, in its theoretical aspects, has fallen. In spite of the immense amount of fruitful labour which has been expended in its practical applications, the basic principles of this organ of science are still in a state of obscurity, and it cannot be denied that, during the recent development of practical methods, fundamental problems have been ignored and fundamental paradoxes left unresolved. This anomalous state of statistical science is strikingly exemplified by a recent paper (1) entitled "The Fundamental Problem of Practical Statistics," in which one of the most eminent of modern statisticians presents what purports to be a general proof of BAYES' postulate, a proof which, in the opinion of a second statistician of equal eminence, "seems to rest upon a very peculiar—not to say hardly supposable—relation." (2)

Leaving aside the specific question here cited, to which we shall recur, the obscurity which envelops the theoretical bases of statistical methods may perhaps be ascribed to two considerations. In the first place, it appears to be widely thought, or rather felt, that in a subject in which all results are liable to greater or smaller errors, precise definition of ideas or concepts is, if not impossible, at least not a practical necessity. In the second place, it has happened that in statistics a purely verbal confusion has hindered the distinct formulation of statistical problems; for it is customary to apply the same name, *mean, standard deviation, correlation coefficient*, etc., both to the true value which we should like to know, but can only estimate, and to the particular value at which we happen to arrive by our methods of estimation; so also in applying the term probable error, writers sometimes would appear to suggest that the former quantity, and not merely the latter, is subject to error.

It is this last confusion, in the writer's opinion, more than any other, which has led to the survival to the present day of the fundamental paradox of inverse probability, which like an impenetrable jungle arrests progress towards precision of statistical concepts. The criticisms of BOOLE, VENN, and CHRYSTAL have done something towards banishing the method, at least from the elementary text-books of Algebra; but though we may agree wholly with CHRYSTAL that inverse probability is a mistake (perhaps the only mistake to which the mathematical world has so deeply committed itself), there yet remains the feeling that such a mistake would not have captivated the minds of LAPLACE and POISSON if there has been nothing in it but error. (Fisher, 1922c.)

●

12. "Of 10 cases treated by Lister's method, 7 did well and 3 suffered from blood-poisoning; of 14 cases treated with ordinary dressings, 9 did well and 5 had blood-poisoning; what are the odds that the success of Lister's method was due to chance."

(See Winsor (1948) for comments by D. MacAlister, A. MacFarlane, Whitworth and others. MacAlister cited a solution by Liebermeister (1877). This problem, with an essentially correct solution by Liebermeister, is a good example of the haphazard development of statistical theory in the 19th Century.)

CHAPTER II

Distribution Theory

1. THE GAMMA VARIABLE

A random variable U is a Γ-variable with parameter n if its frequency distribution is specified by the probability element

$$(1.1) \qquad dF(u) = [\Gamma(n)]^{-1}u^{n-1}e^{-u}\,du, \qquad n > 0, 0 \leqslant u < \infty.$$

We show that the Γ-variable is reproduced under convolution:

Theorem 1.1. *If U and V are independently distributed Γ-variables with parameters m and n respectively, then $U + V$ is a Γ-variable with parameter, $m + n$. Conversely, a Γ-variable, W, with parameter $(m + n)$ can always be expressed as the sum of two independently distributed Γ-variables, U and V, with parameters m and n respectively.*

Proof. The joint frequency function of U and V is specified by the probability element

$$(1.2) \qquad f(u, v)\,du\,dv = [\Gamma(m)\Gamma(n)]^{-1}e^{-(u+v)}u^{m-1}v^{n-1}\,du\,dv,$$
$$0 \leqslant u < \infty, \qquad 0 \leqslant v < \infty.$$

After the transformation,

$$(1.3) \qquad \begin{aligned} u &= w\cos^2\theta \\ v &= w\sin^2\theta, \end{aligned}$$

for which the Jacobian is $2w\cos\theta\sin\theta$, the joint distribution of the random variables W, θ, is specified by

$$(1.4)$$
$$g(w, \theta)\,dw\,d\theta = \frac{1}{\Gamma(m+n)}w^{m+n-1}e^{-w}\,dw\,2\,\frac{\Gamma(m+n)}{\Gamma(m)\Gamma(n)}\cos^{2m-1}\theta\sin^{2n-1}\theta\,d\theta,$$
$$0 \leqslant w < \infty, \qquad 0 \leqslant \theta \leqslant \tfrac{1}{2}\pi.$$

Thus W and θ are independent and W is a Γ-variable with parameter, $(m + n)$. The converse follows by retracing the steps of the direct proof.

Given a Γ-variable W, an independent variable θ can be introduced giving the joint frequency function as specified in (1.4).

The theorem and its converse also follow from the properties of moment generating functions or characteristic functions. The moment generating function of the Γ-variable with parameter n is:

$$(1.5) \qquad \frac{1}{\Gamma(n)} \int_0^\infty u^{n-1} e^{-u(1-t)} \, du = (1-t)^{-n}$$

and its characteristic function is $(1 - it)^{-n}$. The moment generating function of the sum of two independent Γ-variables with parameters, m and n is therefore, $(1 - t)^{-(m+n)}$.

Lemma 1.1. *Half the square of a standard normal variable is a Γ-variable with parameter, $\frac{1}{2}$.*

Proof. Let X be normally distributed with frequency function

$$(1.6) \qquad f(x) = \frac{1}{\sqrt{(2\pi)}} e^{-\frac{1}{2}x^2}, \qquad -\infty < x < \infty,$$

then the transformation $z = \frac{1}{2}x^2$ yields the frequency function,

$$(1.7) \qquad g(z) = \frac{1}{\sqrt{\pi}} z^{-\frac{1}{2}} e^{-z} = \frac{1}{\Gamma(\frac{1}{2})} z^{-\frac{1}{2}} e^{-z}, \qquad 0 \leqslant z < \infty.$$

Theorem 1.2. *Half the sum of the squares of n independent standard normal variables is a Γ-variable with parameter, $\frac{1}{2}n$.*

Proof. The theorem follows immediately from Theorem 1.1 and Lemma 1.1.

Corollary. *The random variable Z, defined as the sum of the squares of n standard normal variables has the frequency function*

$$(1.8) \qquad 2^{-\frac{1}{2}n} [\Gamma(\tfrac{1}{2}n)]^{-1} z^{\frac{1}{2}n-1} e^{-\frac{1}{2}z}, \qquad 0 \leqslant z < \infty.$$

Finally, the converse of Theorem 1.1 enables us to state that a random variable having the distribution (1.8) can be expressed as the sum of the squares of n independent standard normal variables.

2. THE χ^2 VARIABLE

We say that Z has the χ^2 distribution if it can be exhibited as a sum of squares,

$$(2.1) \qquad Z = U_1^2 + U_2^2 + \ldots + U_n^2,$$

where each U_i is a standard normal variable and the U_i form a mutually independent set, so that their joint distribution has the density function,

$$(2.2) \qquad (2\pi)^{-\frac{1}{2}n} e^{-\frac{1}{2}\sum_1^n u_i^2}, \qquad -\infty < u_i < \infty.$$

Theorem 2.1. *The distribution function of Z, the sum of squares of n mutually independent standard normal variables, is given by*

$$(2.3) \qquad Pr(Z \leqslant z) = 2^{-\frac{1}{2}n}[\Gamma(\tfrac{1}{2}n)]^{-1} \int_0^z x^{\frac{1}{2}n-1} e^{-\frac{1}{2}x} \, dx.$$

Note. The proof given here is classical. Some derivations of this result use geometrical ideas, taking advantage of the "spherical symmetry" of the joint distribution. Essentially, these arguments are statements of the analytical steps given here in geometric language, as is pointed out, for example, by Satterthwaite (1942) and Fog (1948).

Proof. From (2.2), the distribution function of Z is given by

$$(2.3') \qquad Pr(Z \leqslant z) = (2\pi)^{-\frac{1}{2}n} \int \cdots \int_{\Sigma u_i^2 \leqslant z} e^{-\frac{1}{2}\Sigma u_i^2} \, du_1 \ldots du_n,$$

where $Z = \sum_1^n U_i^2$. A transformation is now made to polar co-ordinates.

$$(2.4) \qquad \begin{aligned}
u_1 &= z^{\frac{1}{2}} \cos \theta_1 \\
u_2 &= z^{\frac{1}{2}} \sin \theta_1 \cos \theta_2 \\
u_3 &= z^{\frac{1}{2}} \sin \theta_1 \sin \theta_2 \cos \theta_3 \\
&\;\cdot \quad \cdot \quad \cdot \quad \cdot \quad \cdot \quad \cdot \quad \cdot \\
u_{n-1} &= z^{\frac{1}{2}} \sin \theta_1 \sin \theta_2 \ldots \sin \theta_{n-2} \cos \theta_{n-1} \\
u_n &= z^{\frac{1}{2}} \sin \theta_1 \sin \theta_2 \ldots \sin \theta_{n-2} \sin \theta_{n-1}.
\end{aligned}$$

In the region $\sum_1^n u_i^2 \leqslant z$, the angular co-ordinates $\theta_1, \ldots, \theta_{n-2}$ vary between 0 and π, θ_{n-1} varies between 0 and 2π.

To evaluate the Jacobian

$$\frac{\delta(u_1, \ldots, u_n)}{\delta(z, \theta_1, \ldots, \theta_{n-1})}$$

we note that it has the form $\frac{1}{2}z^{\frac{1}{2}n-1} \Delta_n$ where Δ_n is an n-rowed determinant involving the angular co-ordinates. If this determinant is expanded with respect to the elements of its last column the following recurrence relation is obtained:

$$(2.5) \qquad \Delta_n = \sin \theta_1 \sin \theta_2 \ldots \sin \theta_{n-2} \, \Delta_{n-1}$$

and since $\Delta_2 = 1$, it follows that:

$$(2.6) \qquad \Delta_n = \sin^{n-2} \theta_1 \sin^{n-3} \theta_2 \ldots \sin \theta_{n-2}.$$

Alternatively the elements of the last two rows of the determinant can be replaced by $\mathbf{a}^T \cos \theta_{n-1} + \mathbf{b}^T \sin \theta_{n-1}$, and $-\mathbf{a}^T \sin \theta_{n-1} + \mathbf{b}^T \cos \theta_{n-1}$

where $\mathbf{a}^T$ and $\mathbf{b}^T$ are the $(n-1)^{\text{th}}$ and n^{th} rows respectively. Whence,

$$(2.7) \quad Pr(Z \leqslant z) = \frac{1}{2(2\pi)^{\frac{1}{2}n}} \int_0^z z^{\frac{1}{2}n-1} e^{-\frac{1}{2}z}\, dz \times$$

$$\times \int_0^{2\pi} d\theta_{n-1} \int_0^\pi \sin_{n-2} d\theta_{n-2} \dots \int_0^\pi \sin^{n-2}\theta_1\, d\theta_1.$$

Using the result,

$$(2.8) \qquad\qquad \int_0^\pi \sin^p \theta\, d\theta = \tfrac{1}{2} B(\tfrac{1}{2}p + \tfrac{1}{2}, \tfrac{1}{2}),$$

we finally obtain

$$(2.9) \qquad\qquad Pr(Z \leqslant z) = 2^{-\frac{1}{2}n}[\Gamma(\tfrac{1}{2}n)]^{-1} \int_0^z z^{\frac{1}{2}n-1} e^{-\frac{1}{2}z}\, dz.$$

3. SOME PROPERTIES OF THE χ^2 DISTRIBUTION

The χ^2 distribution has an additive property, *i.e.* it is reproduced under convolution; the sum of two independent χ^2 variables with m and n degrees of freedom, respectively, is a χ^2 variable with $m+n$ degrees of freedom. The moment generating function of χ^2 with 1 degree of freedom is

$$(3.1) \qquad M(t) = (2\pi)^{-\frac{1}{2}} \int \exp tz^2 \exp -\tfrac{1}{2}z^2\, dz = (1-2t)^{-\frac{1}{2}}$$

and from the analysis given in Sections 1 and 2, the moment generating function for χ^2 with n degrees of freedom is $(1-2t)^{-\frac{1}{2}n}$. The characteristic function is obtained by replacing t by it in the moment generating function. From the moment generating function (3.1) we obtain the cumulant generating function:

$$(3.2) \qquad\qquad K(t) = \log M(t) = -\tfrac{1}{2}n \log(1-2t).$$

Moments and cumulants of all orders exist and we have:

$$(3.3) \qquad\qquad \kappa_r = n2^{r-1}(r-1)!$$

In particular, the mean is n and the lower central moments are:

$$(3.4) \qquad\qquad \begin{aligned} \mu_2 &= 2n \\ \mu_3 &= 8n \\ \mu_4 &= 48n + 12n^2. \end{aligned}$$

The coefficients of skewness and kurtosis are given by

$$(3.5) \qquad \gamma_1 = \mu_3 \mu_2^{\frac{3}{2}} = 2^{\frac{3}{2}} n^{-\frac{1}{2}}; \qquad \gamma_2 = \mu_4 \mu_2^{-2} - 3 = 12n^{-1}.$$

The values of the mean and variance also follow from the definition of χ^2 as the sum of squares of independent standardized normal variables.

From the central limit theorem the random variable,

$$(3.6) \qquad W_n = (\chi^2 - n)/(2n)^{\frac{1}{2}},$$

is asymptotically normal. A more accurate approximation is due to R. A. Fisher, who showed that

$$(3.7) \qquad Y_n = (2\chi^2)^{\frac{1}{2}} - (2n - 1)^{\frac{1}{2}},$$

is approximately standard normal.

4. TABLES OF THE χ^2 DISTRIBUTION

Bienaymé (1858) and Pearson (1900a) derived expressions for the "tail" probability that χ_n^2 should exceed a given value x^2. For n odd, this probability is given by

$$(4.1) \quad P = \sqrt{\frac{2}{\pi}}\left[\int_x^\infty e^{-\frac{1}{2}u^2}\,du +\right.$$

$$\left. + e^{-\frac{1}{2}x^2}\left(\frac{x}{1} + \frac{x^3}{1.3} + \frac{x^5}{1.3.5} + \ldots + \frac{x^{n-2}}{1.3\ldots(n-2)}\right)\right].$$

For n even,

$$(4.2) \qquad P = e^{-\frac{1}{2}x^2}\left(1 + \frac{x^2}{2} + \frac{x^4}{2.4} + \ldots + \frac{x^{n-2}}{2.4.6\ldots(n-2)}\right),$$

which is the sum of a set of Poisson probabilities with parameter, $\frac{1}{2}x^2$.

K. Pearson (1922a) gave tables of the incomplete Gamma-function in the form,

$$(4.3) \qquad I(u, p) = \frac{1}{\Gamma(p + 1)}\int_0^{u\sqrt{(p+1)}} v^p e^{-v}\,dv$$

for certain values of p and had noted the relation between the probability, P, of (4.1) and (4.2) and $I(u, p)$, namely,

$$(4.4) \quad I(u, p) = 1 - P, \quad \text{with } n = 2(p + 1) \text{ and } \tfrac{1}{2}x^2 = u\sqrt{(p + 1)}.$$

Tables of the normal distribution constructed by Sheppard (1903 and 1939) and tables of the Poisson distribution constructed by Soper (1914) have been the basis of many later computations of P in (4.1) and (4.2). In Pearson's (1900a) tables of the tail probabilities the arguments were the integer values of χ^2 and the degrees of freedom. In the tables of the *Statistical Methods* of R. A. Fisher and in the *Statistical Tables* of R. A. Fisher and F. Yates, the arguments are P and the degrees of freedom. Later tables have followed this practice. More detailed tables have been compiled by Thompson (1941), by Hartley and Pearson (1950) reprinted in the *Biometrika Tables* (1954), by Hald and Sinbaek (1950), Khamis (1963) and Harter (1964). Further references are available on pages 140–147 of Greenwood and Hartley (1962).

5. THE DISTRIBUTION OF CERTAIN QUADRATIC FORMS IN NORMAL VARIABLES

The definition of the χ^2 variable as the sum of the squares of independent standard normal variables suggests that we should consider the distribution of quadratic forms in normal variables. The following theorem was the first significant result in this direction. (Pearson, 1900*a*)

Theorem 5.1. *If an n-component random variable* $(X_1, \ldots, X_n)$ *has a distribution specified by*

$$(5.1) \quad f(x_1, \ldots, x_n) = C \exp\left[-\tfrac{1}{2}\mathbf{x}^T\mathbf{Q}\mathbf{x}\right], \quad \text{where } C \text{ is a constant,}$$

where $\mathbf{Q}$ *is a positive definite matrix, then the random variable* $\mathbf{X}^T\mathbf{Q}\mathbf{X}$ *is distributed as* χ^2 *with n degrees of freedom.*

    Proof. There exists a non-singular linear transformation

$$(5.2) \quad\quad\quad\quad \mathbf{x} = \mathbf{A}\mathbf{y}$$

under which the positive definite quadratic form $\mathbf{x}^T\mathbf{Q}\mathbf{x}$ is transformed to $\mathbf{y}^T\mathbf{y}$. Since the Jacobian of this transformation is a constant, the density function becomes proportional to $\exp -\tfrac{1}{2}\mathbf{y}^T\mathbf{y}$ and so the transformed variables $Y_1, Y_2, \ldots, Y_n$ form a set of mutually independent standardized normal variables. $\mathbf{x}^T\mathbf{Q}\mathbf{x} = \mathbf{y}^T\mathbf{y}$ is therefore the sum of squares of the type required by Theorem 1.1.

Theorem 5.2. (Cochran, 1934). *A necessary and sufficient condition that a quadratic form,* $\mathbf{X}^T\mathbf{Q}\mathbf{X}$, *in n independently distributed, standardized normal variables,* X_i, *should have the* χ^2-*distribution for* $p \leqslant n$ *degrees of freedom is that the matrix of the form be idempotent of rank p.*

    Proof. The moment generating function of a quadratic form, $\mathbf{X}^T\mathbf{Q}\mathbf{X}$, is $|1 - 2t\mathbf{Q}|^{-\frac{1}{2}}$ and if this is equal to $(1 - 2t)^{-\frac{1}{2}p}$, the characteristic function of χ^2 with p degrees of freedom, then $\mathbf{Q}$ has p latent roots equal to $+1$ and $(n - p)$ latent roots equal to zero. The sufficiency is proved by noting that if $\mathbf{Q}$ is an idempotent of rank, p, then the characteristic function takes the required form.

Theorem 5.3. *Let a set of mutually independent, standard normal variables be written as the elements of a vector,* $\mathbf{X}$. *Let* $\mathbf{A}$ *and* $\mathbf{B}$ *be real symmetric* $n \times n$ *matrices. Then necessary and sufficient conditions for the independence of*

$\mathbf{X}^T\mathbf{AX}$ *and* $\mathbf{X}^T\mathbf{BX}$, *are*

(5.3) (i) $|\mathbf{1} - 2\rho\mathbf{A} - 2\sigma\mathbf{B}| = |\mathbf{1} - 2\rho\mathbf{A}|\,|\mathbf{1} - 2\sigma\mathbf{B}|$, (Cochran, 1934);

(5.3′) (ii) $\mathbf{AB} = \mathbf{0}$ (Craig, 1943).

Proof. The joint moment generating function $G(\rho, \sigma)$ of $\mathbf{X}^T\mathbf{AX}$ and $\mathbf{X}^T\mathbf{BX}$ is the integral of $\exp(\rho\mathbf{X}^T\mathbf{AX} + \sigma\mathbf{X}^T\mathbf{BX})$ over the distribution of $\mathbf{X}$.

(5.4) $$G(\rho, \sigma) = |\mathbf{1} - 2\rho\mathbf{A} - 2\sigma\mathbf{B}|^{-\frac{1}{2}}.$$

The condition $G(\rho, \sigma) = G(\rho, 0)G(0, \sigma)$ is equivalent to (i), so that (i) is a necessary and sufficient condition for independence. Condition (ii) implies condition (i) for if (ii) holds, $\mathbf{1} - 2\rho\mathbf{A} - 2\sigma\mathbf{B} = (\mathbf{1} - 2\rho\mathbf{A})(\mathbf{1} - 2\sigma\mathbf{B})$ and (i) follows after taking determinants. (i) implies (ii), for we can suppose that a transformation with an orthogonal matrix, $\mathbf{H}$, has been carried out so that the transformed matrix, $\mathbf{H}^T\mathbf{AH} = \mathbf{A}^*$, is in canonical diagonal form, $\operatorname{diag}(\alpha_1, \alpha_2, \ldots, \alpha_p, 0^{n-p})$. The condition, (i) may be written in homogeneous form (Aitken, 1950)

(5.5) $$\lambda^n\,|\lambda\mathbf{1} - 2\rho\mathbf{A}^* - 2\sigma\mathbf{B}^*| = |\lambda\mathbf{1} - 2\rho\mathbf{A}^*|\,|\lambda\mathbf{1} - 2\sigma\mathbf{B}^*|,$$

where $\mathbf{B}^* = \mathbf{H}^T\mathbf{BH}$. Let $\mathbf{A}^*$ be partitioned so that $\mathbf{A}^*_{11}$ is a $p \times p$ matrix, and

(5.6) $$\mathbf{A}^* = \begin{bmatrix} \mathbf{A}^*_{11} & \mathbf{A}^*_{12} \\ \mathbf{A}^*_{21} & \mathbf{A}^*_{22} \end{bmatrix}.$$

Let $\mathbf{B}^*$ be partitioned in a similar manner. Then equating the coefficients of ρ^p on both sides yields

(5.7) $$\lambda^n\,|\lambda\mathbf{1}_{n-p} - 2\sigma\mathbf{B}^*_{22}| = \lambda^{n-p}\,|\lambda\mathbf{1} - \sigma\mathbf{B}^*|.$$

This relation implies that the latent roots of $\mathbf{B}^*$ are the same as those of $\mathbf{B}^*_{22}$ together with p zeroes. But the sum of squares of the elements of a real symmetric matrix are equal to the sum of squares of the latent roots for, if $\mathbf{K}^T\mathbf{BK} = \operatorname{diag}(\beta_i)$, then

$$\beta_i^2 = \sum_{i,j} b_{ij}^2, \quad \text{since} \quad \sum \beta_i^2 = \operatorname{tr}\operatorname{diag}\beta_i^2 = \operatorname{tr}(\mathbf{K}^T\mathbf{BK})(\mathbf{K}^T\mathbf{BK}) = \operatorname{tr}\mathbf{K}^T\mathbf{B}^2\mathbf{K}$$

$$= \operatorname{tr}\mathbf{B}^2\mathbf{KK}^T = \operatorname{tr}\mathbf{B}^2 = \operatorname{tr}\mathbf{BB}^T = \sum_{i,j} b_{ij}^2.$$

This implies that all the elements of $\mathbf{B}^*$ not in $\mathbf{B}^*_{22}$ are zero and so $\mathbf{B}^*_{11}$, $\mathbf{B}^*_{12}$ and $\mathbf{B}^*_{21}$ are matrices of zeros.

(5.8) $$\mathbf{A}^*\mathbf{B}^* = \begin{pmatrix} \mathbf{A}^*_{11} & 0 \\ 0 & 0 \end{pmatrix}\begin{pmatrix} 0 & 0 \\ 0 & \mathbf{B}^*_{22} \end{pmatrix} = 0$$

$$\mathbf{AB} = \mathbf{H}(\mathbf{H}^T\mathbf{AH})(\mathbf{H}^T\mathbf{BH})\mathbf{H}^T = \mathbf{HA}^*\mathbf{B}^*\mathbf{H}^T = 0.$$

Corollary. *If* **A** *and* **B** *are non-negative definite, a necessary and sufficient condition for independence is*

(5.9) $\operatorname{tr} \mathbf{AB} = 0.$ (Matérn, 1949)

Proof. $\mathbf{AB} = \mathbf{0}$ implies $\operatorname{tr} \mathbf{AB} = 0$. For the sufficiency, make a transformation as in the theorem. Then every diagonal element of both $\mathbf{A}^*$ and $\mathbf{B}^*$ is non-negative. $0 = \operatorname{tr} \mathbf{A}^*\mathbf{B}^* = \sum_{i=1}^{n} a_{ii}^* b_{ii}^*.$ b_{ii}^* must be zero for $a_{ii}^* > 0$. But since $\mathbf{B}$ and hence $\mathbf{B}^*$ is non-negative definite, $b_{ii}^* = 0$ implies $b_{ij}^* = 0$ for every j, so that every element of $\mathbf{B}_{11}^*$, $\mathbf{B}_{12}^*$ and $\mathbf{B}_{21}^*$ is a zero.

Note that this corollary applies in particular to those quadratic forms which are distributed as χ^2.

Another theorem of Cochran (1934) has been treated by James (1952) and Lancaster (1954). It is given here as a matrix result.

Theorem 5.4. (Cochran, 1934). *If* $\mathbf{A}_i$, $i = 1, 2, \ldots, k$ *are real symmetric matrices of ranks,* n_i, *such that the sum is the unit matrix,* $\sum_{i=1}^{k} \mathbf{A}_i = \mathbf{1}_n$, *then any of the three following conditions implies the other two.*

(5.10)

(i) $\sum_{1}^{k} n_i = n.$

(ii) $\mathbf{A}_i\mathbf{A}_j = \mathbf{0}$ *for* $i \neq j$.

(iii) *each* $\mathbf{A}_i$ *is idempotent.*

Proof. (i) implies (iii). For let $\mathbf{S} = \sum_{2}^{k} \mathbf{A}_i$. Then there exists an orthogonal matrix $\mathbf{H}$ such that $\mathbf{H}^T\mathbf{A}_1\mathbf{H} = \operatorname{diag} \{\alpha_j, 0^{n-n_1}\}$. Since $\mathbf{A}_1 + \mathbf{S} = \mathbf{1}_n$, $\mathbf{H}^T\mathbf{S}\mathbf{H}$ is also diagonal and has the form $\operatorname{diag} \{1 - \alpha_j, 1^{n-n_1}\}$. But each of the $1 - \alpha_j$ must be zero otherwise $\mathbf{S}$ would have rank greater than the sum of the ranks of its components. The latent roots of $\mathbf{A}_1$ are thus all units and zeroes. $\mathbf{A}_1$ is thus idempotent. Similarly each $\mathbf{A}_i$ is idempotent.

(iii) implies (ii) for $\operatorname{tr} (\mathbf{A}_1 + \mathbf{A}_2 + \ldots + \mathbf{A}_k)^2 = \operatorname{tr} \mathbf{1} = n$ or $\sum_i \operatorname{tr} \mathbf{A}_i^2 + \sum_{i \neq j} \operatorname{tr} \mathbf{A}_i\mathbf{A}_j = n$. But $\operatorname{tr} \mathbf{A}_i^2 = \operatorname{tr} \mathbf{A}_i$ and so this last relation implies $\sum_{i \neq j} \operatorname{tr} \mathbf{A}_i\mathbf{A}_j = 0$. Now $\mathbf{A}_i^2$ and $\mathbf{A}_j$ are non-negative definite, therefore for each pair of indices, $i \neq j$, $\operatorname{tr} \mathbf{A}_i\mathbf{A}_j$ is non-negative and so must be zero. But this is sufficient to give $\mathbf{A}_i\mathbf{A}_j = \mathbf{0}$.

(ii) implies (i). For (ii) implies (iii) because $\mathbf{A}_i = \mathbf{A}_i\mathbf{1} = \mathbf{A}_i(\mathbf{A}_1 + \mathbf{A}_2 + \ldots + \mathbf{A}_k) = \mathbf{A}_i^2$; and (iii) implies (i) because $n = \operatorname{tr} \mathbf{1} = \operatorname{tr} \sum \mathbf{A}_i = \sum \operatorname{tr} \mathbf{A}_i = \sum n_i$, since the trace of a symmetric idempotent is its rank.

Definition. *Let* **A**, **B** *and* **C** *be symmetric, non-negative definite matrices. If* $\mathbf{C} = \mathbf{A} + \mathbf{B}$ *and if* $\mathbf{X}^T\mathbf{AX}$, $\mathbf{X}^T\mathbf{BX}$ *and* $\mathbf{X}^T\mathbf{CX}$ *each has the* χ^2 *distribution, then*

$\mathbf{X}^T\mathbf{C}\mathbf{X}$ *is said to be partitioned into* $\mathbf{X}^T\mathbf{A}\mathbf{X}$ *and* $\mathbf{X}^T\mathbf{B}\mathbf{X}$. *By Theorem* 5.4 $\mathbf{X}^T\mathbf{A}\mathbf{X}$ *is independent of* $\mathbf{X}^T\mathbf{B}\mathbf{X}$.

Theorem 5.5. (Lancaster, 1954). *Let* $X_1, X_2, \ldots, X_n$ *be a set of n independently distributed standard normal variables. If* $\mathbf{X}^T\mathbf{A}\mathbf{X}$ *and* $\mathbf{X}^T\mathbf{C}\mathbf{X}$ *are* χ^2 *variables with a and c degrees of freedom a necessary and sufficient condition that* $\mathbf{X}^T\mathbf{A}\mathbf{X}$ *should be a component of* $\mathbf{X}^T\mathbf{C}\mathbf{X}$ *is that*

$$(5.11) \qquad \text{corr}\,(\mathbf{X}^T\mathbf{A}\mathbf{X}, \mathbf{X}^T\mathbf{C}\mathbf{X}) = \sqrt{(a/c)}.$$

Proof. *Necessity.* We are given that $\mathbf{X}^T\mathbf{C}\mathbf{X}$ can be exhibited as a sum

$$(5.12) \qquad \mathbf{X}^T\mathbf{C}\mathbf{X} = \sum_1^a y_i^2 + \sum_{a+1}^c y_i^2 = \mathbf{X}^T\mathbf{A}\mathbf{X} + \mathbf{X}^T\mathbf{B}\mathbf{X},$$

where the y's are derived from the x's by an orthogonal transformation. Straightforward calculation then yields the required result. Alternatively, we have

$$\text{cov}\,(\mathbf{X}^T\mathbf{A}\mathbf{X}, \mathbf{X}^T\mathbf{C}\mathbf{X}) = 2\,\text{tr}\,\mathbf{A}\mathbf{C} = 2a;$$
$$\text{var}\,\mathbf{X}^T\mathbf{C}\mathbf{X} = 2\,\text{tr}\,\mathbf{C}^2 = 2\,\text{tr}\,\mathbf{C} = 2c;$$
$$\text{var}\,\mathbf{X}^T\mathbf{A}\mathbf{X} = 2a.$$

Sufficiency. It is easily verified that neither a diagonal element nor the sum of squares in any row of an idempotent matrix can be greater than unity. Suppose that a transformation is carried out so that $\mathbf{A}^*$, the matrix of the quadratic form $\mathbf{X}^T\mathbf{A}\mathbf{X}$ in the new variables is in diagonal canonical form and suppose the same transformation is applied to $\mathbf{B}$ and $\mathbf{C}$ to yield $\mathbf{B}^*$ and $\mathbf{C}^*$. Then from the condition on the correlation, it follows that tr $\mathbf{C}^*\mathbf{A}^* = a$. But tr $\mathbf{A}^*\mathbf{C}^* = \sum_1^a c_{ii}$, each $c_{ii}^* = 1$ for $i = 1, 2, \ldots, a$. From the first remark above, $c_{ij}^* = c_{ji}^* = 0$ for $j \neq i$, when $i = 1, 2, \ldots, a$. $b_{ii}^* = 0$ for $i = 1, 2, \ldots, a$. The non-zero parts of $\mathbf{A}^*$ and $\mathbf{B}^*$ are disjunct. A further orthogonal transformation on the variables other than the first a of them can be found to bring $\mathbf{X}^T\mathbf{C}\mathbf{X}$ to diagonal form and in this final set of variables, the two quadratic forms are exhibited as distinct sums of squares as in (5.12).

EXERCISES AND COMPLEMENTS

1. Let the mode of a probability distribution be taken as the origin, then the Pearson system of random variables satisfies the differential equation,

$$\frac{\mathrm{d}}{\mathrm{d}u}\,[\log f_U(u)] = u/(B_0 + B_1 u + B_2 u^2).$$

Verify that if $B_2 = 0$, $B_1 = 1$, $B_0 = -(m - 1)$ and $X = U - (n - 1)$, then the frequency function of X is given by (1.1). The Γ-variable and χ^2 are said to belong to Pearson's Type III. (K. Pearson, 1895 and 1901a).

2. Suppose that the random variables $X_1, X_2, \ldots, X_n$ are mutually independent standard normal random variables. Write down the moment generating function of $X_i^2 \equiv Y_i$ and of $Y = \sum X_i^2$ and hence determine the moments and the cumulants of the sum, Y.

3. Prove the additive law for Γ-variables, U and V, and so for the χ^2 variables by means of the transformation,

$$U = RW$$
$$U + V = W$$

(Sawkins, 1949).

4. Verify that the mean and mode of the χ^2 distribution are n and $(n - 2)$ respectively. Assuming the rule,

$$2(\text{median-mean}) = \text{mode-median},$$

determine approximately the median (i.e. the 50% point) of χ^2. Verify that this is a good approximation from the tables of χ^2, in which it appears that the 50% point of χ^2 is close to $n - 1 + 0 \cdot 333 \ldots$ as n becomes large. (Doodson, 1917; Haldane, 1942b; Lancaster, 1968; Pearson, 1895).

5. Show that if (2.4) be written

$$\mathbf{u} = x^{\frac{1}{2}} \mathbf{h}_1$$

the Jacobian can be evaluated as follows: write

$$\frac{\partial \mathbf{h}_1}{\partial \theta_i} = \mathbf{h}_{i+1}, i = 1, 2, \ldots, (n - 1);$$

noting that $\mathbf{h}_1^T \mathbf{h}_1 = 1$, prove that $\mathbf{h}_1^T \mathbf{h}_{i+1} = 0$. Prove that for $i < k$

$$\frac{\partial}{\partial \theta_i} \frac{\partial \mathbf{h}_1}{\partial \theta_k} = c \frac{\partial \mathbf{h}_1}{\partial \theta_k},$$

where c is a multiplier. Hence, by partially differentiating $\mathbf{h}_1^T \mathbf{h}_{k+1} = 0$ with respect to θ_i, show that $\mathbf{h}_{i+1}^T \mathbf{h}_{k+1} = 0$ for $i < k$. The rows of the matrix of the Jacobian are thus mutually orthogonal for they are

$$\frac{\partial \mathbf{h}_1^T}{\partial x} = \frac{1}{2} x^{-\frac{1}{2}} \mathbf{h}_1^T$$

$$\frac{\partial \mathbf{h}_1^T}{\partial \theta_i} = x^{\frac{1}{2}} \mathbf{h}_{i+1}^T, \quad \text{for} \quad i > 1.$$

Now evaluate the determinant using the proposition that the determinant of a matrix with mutually orthogonal rows is equal to the product of the norms of the rows. (Lancaster, 1965b).

6. Evaluate the Jacobian of (2.4) by induction. (van der Waerden, 1957).

7. Find the frequency function of χ. Show that its kth moment is

$$\mu'_k = 2^{\frac{1}{2}k}\Gamma(\tfrac{1}{2}\{v + k\})/\Gamma(\tfrac{1}{2}v),$$

where v is the number of degrees of freedom. (Johnson and Welch, 1939).

8. Suppose n is large in (2.1) and consider the random variable,

$$V = n^{-\frac{1}{2}}(Z - n) = n^{-\frac{1}{2}}\sum (U_i^2 - 1).$$

Prove that for a fixed value k, $\mathscr{E}V^k$ is asymptotically equal to the kth moment of a normal variable, $N(0, 2)$. [*Hint.* Consider all the partitions of k and show that the factor $n^{-\frac{1}{2}k}$ renders asymptotically negligible all terms of the expansion of $\mathscr{E}V^k$ except those of the form,

$$\mathscr{E}X_{i_1}^2\mathscr{E}X_{i_2}^2 \ldots \mathscr{E}X_{i_m}^2,$$

where $m = \tfrac{1}{2}k$, k is even and $\{i_j\}$ is an m-combination of the set, $1, 2, 3, \ldots, n$.] (Ellis, 1844*b*).

9. Suppose that the probability of a sovereign dying is independent of the length of his reign and is equal to $\delta\mu$ in a small length of time, δ. Then the frequency function of the length of reign is given by

$$f(u) = \mu\exp(-\mu u), \qquad 0 \leqslant u < \infty.$$

Let the time unit be chosen by the substitution $\mu u = x$, then the frequency function is

$$g(x) = \exp(-x), \qquad 0 \leqslant x < \infty.$$

The distribution of Y, the sum of the lengths of n successive reigns is then given by

$$h(y) = y^{n-1}\exp(-y)/\Gamma(n)$$

(Ellis, 1844*a*).

10. In Theorem 5.1, show that $\mathbf{A}$ can be taken to be $\mathbf{Q}^{-\frac{1}{2}}$, where

$$\mathbf{Q}^k = \mathbf{H}\operatorname{diag}(q_i^k)\mathbf{H}^T,$$

and

$$\mathbf{H}^T\mathbf{Q}\mathbf{H} = \operatorname{diag}(q_i); \qquad q_i > 0.$$

(Page 175 of Turnbull and Aitken, 1932).

11. Prove that $-\log|1 - t\mathbf{C}|$ can be expanded in the form

$$t\operatorname{tr}\mathbf{C} + \tfrac{1}{2}t^2\operatorname{tr}\mathbf{C}^2 + \tfrac{1}{3}t^3\operatorname{tr}\mathbf{C}^3 + \ldots.$$

12. Let the elements of $\mathbf{X}$ be a set of n mutually independent standard normal variables. Derive the *c.g.f.* of a quadratic form, $\mathbf{X}^T\mathbf{A}\mathbf{X}$, by transforming the variables to a co-ordinate system in which the quadratic form is represented by $\alpha_1 Y_1^2 + \alpha_2 Y_2^2 + \ldots + \alpha_r Y_r^2$, where $r = \text{rank of } \mathbf{A}$ and $\{Y_i\}$ are mutually independent normal $(0, 1)$. (Lancaster, 1954).

13. Prove the following propositions on real symmetric idempotent matrices.

(i) the non-zero latent roots are equal to unity;
(ii) the rank of an idempotent matrix is equal to its trace;

(iii) if a diagonal element is unity, all other elements in the same row or in the same column are zeroes.

(iv) a diagonal element must lie in [0, 1].

(v) any idempotent matrix of rank greater than one can be decomposed into a sum of idempotent matrices of unit rank which have the orthogonal property $A_i A_j = 0$ for $i \neq j$.

(vi) every idempotent matrix of unit rank is the product, aa^T, of a unit vector postmultiplied by its transpose.

14. Interpret Theorem 5.4 as a theorem in the distribution of χ^2, namely that (ii) is the condition of independence and (iii) implies that each component has the χ^2 distribution.

15. Prove that if $aa^T B = 0$, then $a^T B = 0^T$. Hence or otherwise prove that if $\{X_i\}$ is a set of mutually independent standard normal variables, the necessary and sufficient condition that $a^T X$ should be independent of $X^T BX$ is that $a^T B = 0^T$.

16. Prove that if $\{X_i\}$ is a set of mutually independent standard normal variables, then independence in pairs of a set of linear forms $\{Y_i\}$ implies complete mutual independence of the Y_i. Prove that this still holds if the $\{X_i\}$ are jointly normal with covariance matrix, V. Prove a similar proposition for quadratic forms.

17. Let $\{X_1, X_2, \ldots, X_n, \ Y_1, Y_2, \ldots, \ Y_n\} \equiv \{Z_i\}$ be a set of $2n$ mutually independent standard normal variables. Consider the quadratic forms, $\sum X_i^2$, $\sum Y_i^2$ and $\sum (X_i \cos \theta + Y_i \sin \theta)^2$. Determine the correlation between the first and third of these forms. Calculate this correlation when $\theta = \pi/4$. Generalise the problem to k sets of n variables $\{X_i^{(j)}\}$ and to the quadratic forms

$$\frac{1}{k-1} \sum_{i=1}^{n} \left[\sum_{j=1}^{k-1} X_i^{(j)} \right]^2 \quad \text{and} \quad \frac{1}{k} \sum_{i=1}^{n} \left[\sum_{j=1}^{k} X_i^{(j)} \right]^2 .$$

18. Derive the Cochran and Craig conditions of Theorem 5.3 as follows.

Let μ_{ij} and κ_{ij} be the multivariate moments about the respective means and the cumulants of $X^T AX = Q_1$ and $X^T BX \equiv Q_2$, quadratic forms in an independent set of standard normal variables. Evaluate μ_{20}, μ_{11}, μ_{02} and μ_{22} by a consideration of the expectations of the products, $\mathscr{E}(Q_1^r Q_2^s)$. Then calculate κ_{11} and κ_{22} by the formulae, $\kappa_{11} = \mu_{11}$ and

$$\kappa_{22} = \mu_{22} - \mu_{20}\mu_{02} - 2\mu_{11}^2$$
$$= \mathscr{E}(Q_1 - \mu_{10})^2 (Q_2 - \mu_{01})^2 - \mathscr{E}(Q_1 - \mu_{10})^2 \mathscr{E}(Q_2 - \mu_{01})^2 - 2\mu_{11}^2$$

Prove that $AB = 0$ is thus a necessary condition for independence, since it follows from $\kappa_{22} = 0$. Prove also that it is sufficient since it can be given the interpretation that there is an orthogonal H such that $H^T AH$ and $H^T BH$ are both diagonal with no non-zero terms appearing at the same position in each matrix.

19. Let $X_1, X_2, \ldots, X_n$ be standard normal variables, jointly normal with covariance matrix, V. Prove that the distribution of $X^T V^{-1} X$ conditional on $a^T X = 0$ is that of χ^2 with $(n-1)$ d.f. If A is an $n \times s$ matrix of rank s, $s < n$, prove that the distribution of $X^T V^{-1} X$, conditional on $A^T X = 0$, is that of χ^2 with $(n-s)$ d.f. What modifications have to be made to the above if $a^T X = c$ or $A^T X = c$, where c and $c^T c$ are not zero. (Pearson, 1916b).

20. Show that it is not possible to replace the condition for the independence of two random variables

$$G(\rho, \sigma) = G(\rho, 0)G(0, \sigma)$$

by the statement that the multivariate cumulants are zeroes. Use as your counter-example marginal distributions of a form due to Stieltjes

$$p(x) = \{1 + a \sin (2\pi \log x)\}q(x), \qquad |a| < \cdot 1, \qquad 0 \leqslant x < \infty$$

$$q(x) = kx^{-\log x}, \qquad 0 \leqslant x < \infty.$$

$p(x)$ has the same moments as $q(x)$. Now form a bivariate distribution

$$f(x, y) = q(x)q(y)\{1 + a \sin (2\pi \log x) \sin (2\pi \log y)\}$$

$f(x, y)$ has the same bivariate moments as $q(x)q(y)$ and hence $\mathscr{E}(X^s Y^r) = \mathscr{E}X^s \mathscr{E}Y^r$ for $r, s = 0, 1, 2, \ldots$. The multivariate cumulants all vanish but X is not independent of Y since $f(x, y)$ does not factorize. (See Shohat and Tamarkin, 1950, p. 22).

Note that in this case the m.g.f. of $q(x)$ does not exist even though the distribution has moments of all orders. The Poincaré transform, $\mathscr{E}(Xz)$, $z \equiv u + it$, is not defined for $u > 0$, and so is not an analytic function, even though it can be differentiated at the origin an arbitrary number of times.

CHAPTER III

Discrete Distributions

1. CONDENSATION AND RANDOMIZED PARTITIONS

A random variable assumes a set of values, collectively termed the space. These values need not be numerical; they may form an abstract set, a set of qualities with special properties. In this chapter, a random variable is supposed to have a set of values on the real line; in other words, the space of a particular random variable will be a subset of the real line. The random variable, X say, in such cases has a distribution function,

$$(1.1) \qquad F(x) = \mathscr{P}\{X \leqslant x\}.$$

The following properties of distribution functions are set down without proof (see Loève, 1955).

$F(x)$ is a monotone increasing function of x, continuous to the right, such that $F(-\infty) = 0$, $F(+\infty) = 1$. Further, by the Lebesgue decomposition theorem, $F(x)$ can be represented as a convex sum of a continuous distribution function and a discrete distribution function. The continuous distribution function can in turn be decomposed into an absolutely continuous distribution function and a singular distribution function. $F(x)$ can be represented in the form

$$(1.2) \qquad F(x) = a_d F_d(x) + a_c F_{ac}(x) + a_s F_s(x),$$

where $a_d + a_c + a_s = 1$ and each a is non-negative. By taking $a_d = 1$, a discrete distribution function is obtained; $a_c = 1$ gives an absolutely continuous distribution function; $a_s = 1$ yields a singular distribution function. In all other cases, the distribution function will be of a mixed type. In the absolutely continuous case,

$$(1.3) \qquad F(x) = F_{ac}(x) = \int_{-\infty}^{x} f(x)\, dx; \qquad f(x) = F'(x), \quad a.e.$$

For a discrete distribution, the distribution function takes the form

$$(1.4) \qquad F(x) = \sum_{c_j \leqslant x} p_j, \quad \text{where} \quad p_j = \mathscr{P}(X = c_j).$$

There can be at most a countable number of such points c_j.

An important class of discrete distributions is defined on the set of non-negative integers, well-known examples being the binomial, hypergeometric and Poisson distributions. For such integer valued random variables, the $\{c_j\}$ of (1.4) becomes $\{0, 1, 2, \ldots\}$.

The distribution of a continuous random variable X may be *condensed* into a finite or enumerable discrete distribution. The reverse process, the derivation of a continuous distribution from a discrete distribution, is known as a *randomized partition* and is effected with the aid of an auxiliary sampling experiment to be explained below.

Condensation is a familiar process in elementary applied statistics when measurements of a continuous random variable are rounded off to the nearest unit or when they are arbitrarily "grouped" for convenience in computation. The reverse process, which is less common and may be used to obtain plausible values for the original measurements from the condensed measurements, is a feature of the Neyman-Pearson system of inference.

These procedures are now considered more formally, beginning with the condensation process.

Theorem 1.1. *Any distribution can be condensed into a finite discrete distribution.*

Proof. Let X be a random variable with distribution function $F(x)$. Choose $n - 1$ arbitrary real numbers, $-\infty = d_0 < d_1 < d_2 < \ldots < d_{n-1} < d_n = \infty$ and then set $Y = e_j$ when $d_{j-1} < x \leqslant d_j$, so that Y takes at most n distinct values with probabilities $F(d_j) - F(d_{j-1})$. If for some j this difference vanishes, then the corresponding value of Y has zero probability and so the distribution of Y has less than n distinct values.

Remark. It may also happen that some of the e_j coincide. If so, there will be less than n distinct points in the resulting distribution; for example some e_j might be set equal to zero, the remainder equal to unity.

Theorem 1.2. *Any discrete distribution can be randomly partitioned to give an absolutely continuous distribution. If the discrete random variable Y can take the values y_j, $y_{j'}$ with non-zero probabilities and $y_j > y_{j'}$, then a corresponding continuous random variable $X(Y)$ can be defined so that $x_j = X(y_j) > x_{j'} = X(y_{j'})$.*

Proof. Suppose that Y takes the values $y_1 < y_2 < \ldots < y_n$ with positive probabilities, $p_1, p_2, \ldots, p_n$. Choose $d_0 < y_1$, and d_i such that $y_{i-1} < d_i \leqslant y_i$ for $i = 2, 3, \ldots, n$. Now let absolutely continuous frequency functions $F_j(x)$ be chosen such that $F_j(d_{j-1}) = 0$ and $F_j(d_j) = 1$ and define the random variable X as follows. For an observation on Y, there will result a value y_j. Let Z be an independently distributed rectangular variable from which an observed z is obtained. Set x equal to the least solution of $F_j(x) = z$. Then the random variable, X, so defined has a continuous distribution function.

In fact

(1.5)

$$F(x) = \mathscr{P}(X \leqslant x) = p_1 + p_2 + \ldots + p_{j-1} + p_j F_j(x), \quad \text{for} \quad d_{j-1} < x \leqslant d_j$$

and the derivative is given by

(1.6) $$F'(x) = p_j F'_j(x), \quad \text{for} \quad x \neq d_j.$$

This procedure also ensures that if $y_i < y_j$, then $X(y_i) < X(y_j)$. It is evident that the transformation,

(1.7) $$Y \rightarrow X$$

can be represented also as

(1.8) $$X(Y_j) = d_{j-1} + (d_j - d_{j-1})W_j, \quad j = 1, 2, \ldots, n; \quad d_0 = 0$$

where W_j is a random variable on the unit interval possessing an absolutely continuous distribution function.

Theorem 1.3. *The discrete distribution of a random variable, Y, can be randomly partitioned so that a rectangular distribution defined on the interval $(0, 1)$ results, in which the ordering of the variables is preserved, so that*

(1.9) $$U(y) > U(y^*) \quad \text{if} \quad y > y^*.$$

Proof. Make the transformation $Y \rightarrow X$ as in Theorem 1.2 and set $U = F(X)$. U has a rectangular distribution on $(0, 1)$. The ordering is preserved since the transformations from Y to X and from X to U are both monotone.

2.   SIGNIFICANCE TESTS IN DISCRETE DISTRIBUTIONS

Consider a discrete distribution with probabilities, $\{p_j\}$, at a finite or countable number of distinct points. There is no loss of generality in assuming that the random variable takes only non-negative integer values. Define

(2.1) $$P(j) = \sum_{i \geqslant j} p_i, \quad \text{and set}$$

(2.2) $$P'(j) = P(j + 1) = \sum_{i > j} p_j.$$

Now consider a test of significance with significance level α. The critical region for the exact test consists of all values of j for which $P(j) \leqslant \alpha$. There is usually no value of j for which $P(j) = \alpha$ but there will always be a marginal value of j, say J, for which

(2.3) $$P(J) \geqslant \alpha > P'(J).$$

In the Neyman-Pearson theory of hypothesis testing, the situation in which the equality sign in (2.3) does not hold, is dealt with by considering the random variable

$$(2.4) \qquad P^* = P' + \theta(P - P'),$$

where θ is a random variable uniformly distributed over $(0, 1)$, whose value is determined by means of a sampling experiment using random numbers (Stevens (1950), Tocher (1950), Lehmann (1959)). If $P^* < \alpha$, the hypothesis is rejected. For a fixed distribution this is the most powerful test among those with significance level, α, but the increase in size and power is bought at a price. Given the same experimental data, the statistician will not always assign to identical experimental results the same decision.

To deal with this deficiency a *median probability* has been proposed (Lancaster, 1949b and 1961a), defined by

$$(2.5) \qquad P_m(j) = \tfrac{1}{2}\{P(j) + P(j + 1)\} = \tfrac{1}{2}\{P(j) + P'(j)\}$$

with the rule of rejection,

$$(2.6) \qquad P_m(j) \leqslant \alpha.$$

For a given hypothesis and experimental result this test will always lead to the same decision. It is easily seen that the rule of rejection, (2.6), applied to the event J is equivalent to a rule of rejection when $\theta \geqslant \tfrac{1}{2}$, where θ is defined by $\theta = \{\alpha - P'(j)\}/\{P(j) - P'(j)\}$.

Let a test, $T(\theta_0)$, be defined for the marginal event so that the null hypothesis is rejected when $\theta > \theta_0$ and let us determine how often it would agree with the auxiliary random sampling method in a mixture of populations such that θ is a rectangularly distributed variable on $[0, 1]$. Agreement will occur in a proportion of cases

$$(2.7) \qquad \int_0^{\theta_0}(1 - \theta)\,\mathrm{d}\theta + \int_{\theta_0}^1 \theta\,\mathrm{d}\theta = 0.75 - (0.5 - \theta_0)^2,$$

under these idealised conditions. For with $\theta \leqslant \theta_0$, $T(\theta_0)$ will accept the null hypothesis, and the auxiliary sampling method will accept the null hypothesis with relative frequency $(1 - \theta)$. Similarly for $\theta > \theta_0$, the test, $T(\theta_0)$, will reject the null hypothesis, and the auxiliary sampling method test will reject the null hypothesis with relative frequency, θ. Of all such tests with choice of θ_0, the median probability obtained by taking θ_0 to be 0.5 will agree most often with the auxiliary random sampling method; for the maximum amount of agreement is obtained by setting θ_0 equal to 0.5, that is, by using the median probability test.

3. THE NORMAL APPROXIMATION
TO THE BINOMIAL DISTRIBUTION

The approximation to the binomial distribution by means of the normal distribution is well known and has been treated by many authors. It is easy to apply Stirling's approximation to the individual terms of the binomial distribution, to equate this discrete value to an appropriate area under the normal curve. From the sums of such areas it is possible to obtain the tail probabilities, but other approaches are possible. K. Pearson (1895) set up a frequency polygon for the terms of the binomial distribution and considered the possibility of fitting a smooth curve which would have a slope parallel to the slope of the polygon at the mid-points of the intervals. He solved a differential equation and obtained a normal curve with variance $(n + 1)pq$ in place of the usual npq. A third method is to invoke the central limit theorem and note that a binomial variable can be represented as the sum of n independently distributed variables, each taking values of zero or unity with probabilities, q and p respectively. Romanovsky (1923) calculated the moments of the standardized binomial variable and showed that they approach those of a normal distribution.

Here Stirling's approximation,

$$(3.1) \qquad \log_e k! \simeq (k + \tfrac{1}{2}) \log_e k - k + \tfrac{1}{2} \log_e (2\pi)$$

will be applied to the binomial terms,

$$(3.2) \qquad b\{m \mid n, p\} = \binom{n}{m} p^m (1 - p)^{n-m}, \quad \text{to obtain}$$

(3.3)

$$b\{m \mid n, p\} \simeq (\sigma^2 2\pi)^{-\frac{1}{2}} \exp \left\{ -\tfrac{1}{2}(m - np)^2 / \sigma^2 \right\}, \quad \text{where} \quad \sigma^2 = np(1 - p)$$

$$\simeq (2\pi\sigma^2)^{-\frac{1}{2}} \int_{m-np-\frac{1}{2}}^{m-np+\frac{1}{2}} \exp \left\{ -\tfrac{1}{2} y^2 / \sigma^2 \right\} dy$$

$$= (2\pi)^{-\frac{1}{2}} \int_{(m-np-\frac{1}{2})/\sigma}^{(m-np+\frac{1}{2})/\sigma} \exp -\tfrac{1}{2} x^2 \, dx.$$

From (3.3), the tail probability $P(j)$ is given by

$$(3.4) \qquad P(j) \simeq (2\pi)^{-\frac{1}{2}} \int_{(j-np-\frac{1}{2})/\sigma}^{\infty} \exp -\tfrac{1}{2} x^2 \, dx,$$

where j is the observed number of successes. A formula essentially equivalent to (3.4) was given by Laplace. The discrepancy between the exact probability and the normal approximation using the uncorrected value, $(j - np)/\sigma$, in place of the corrected value, $(j - np - \tfrac{1}{2})/\sigma$, remains appreciable even when

n is $O(10^3)$. The corrected normal deviate gives a very good approximation to the exact probability.

We have thus the corrected and crude values,

$$(3.5) \qquad \begin{aligned} \chi_c &= \chi_c(j) = (j - np - \tfrac{1}{2})/\sigma, \\ \chi &= \chi(j) = (j - np)/\sigma \end{aligned}$$

of the standardized deviate. The integral of (3.4) may be thought of as a probability, which can be written

$$(3.6) \qquad \mathscr{P}(\chi_c(j)) \approx P(j), \qquad \mathscr{P}(\chi_c(j+1)) \simeq P'(j) = P(j+1)$$

where P' and P are defined in (2.1) and (2.2) of this chapter. Both the approximations of (3.6) are found to be excellent in any reasonable cases examined.

Write

$$(3.7) \qquad \chi(j) = \tfrac{1}{2}[\chi_c(j) + \chi_c(j+1)]$$

so that the uncorrected χ is midway between the corrected χ of the observation and the corrected χ of the observation next to it. Linear interpolation will be adequate so that

$$(3.8) \qquad \begin{aligned} \mathscr{P}(\chi(j)) &\simeq \tfrac{1}{2}[P(j) + P(j+1)] \\ &= P_m(j). \end{aligned}$$

$\mathscr{P}(\chi(j))$ may, therefore, be expected to approximate to the median probability. There is ample empirical evidence that this is so (Lancaster (1961a)).

4. THE NORMAL APPROXIMATION TO THE HYPERGEOMETRIC AND POISSON DISTRIBUTIONS

Let $\{a_{ij}\}$ be the entries in the contingency table, $a_{i.}$, $a_{.j}$ be the marginal totals and $a_{..}$ be the over-all total. Then the general term of the hypergeometric distribution associated with a fourfold contingency table may be written

$$(4.1) \qquad p(a_{11} \mid \{a_{i.}\}, \{a_{.j}\}) = \frac{a_{1.}! a_{2.}! a_{.1}! a_{.2}!}{a_{..}! a_{11}! a_{12}! a_{21}! a_{22}!}$$

$$\simeq (2\pi\sigma^2)^{-\frac{1}{2}} \exp\left\{-\tfrac{1}{2}\Delta^2/\sigma^2\right\}$$

where $\Delta = a_{11} - a_{1.}a_{.1}/a_{..}$

$$\sigma^2 = a_{1.}a_{2.}a_{.1}a_{.2}/\{a_{..}^2(a_{..} - 1)\}$$

and once again the discrete probability can be written in the form of an integral. Feller (1957) recommends that the normal approximation to the binomial be applied to the probability (4.1) in the form,

$$(4.2) \qquad p(a_{11} \mid \{a_{i.}\}, \{a_{.j}\}) = \binom{a_{1.}}{a_{11}}\binom{a_{2.}}{a_{21}} \Big/ \binom{a_{..}}{a_{.1}}.$$

Alternatively, the mean and variance of a_{11} may be computed to obtain

$$(4.3) \qquad \mathcal{E}a_{11} = a_{1.}a_{.1}/a_{..}; \qquad \text{var } a_{11} = (a_{1.}a_{2.}a_{.1}a_{.2})/\{a_{..}^2(a_{..} - 1)\},$$

by the consideration of the factorial moments,

$$(4.4) \qquad \mathcal{E}a_{11}^{(r)} = a_{1.}^{(r)}a_{.1}^{(r)}/a_{..}^{(r)},$$

where $a^{(r)} = a(a - 1) \ldots (a - r + 1)$. We may then derive a test function,

$$(4.5) \qquad \chi = (a_{11} - \mathcal{E}a_{11})/\sqrt{(\text{var } a_{11})},$$

and determine that its higher cumulants tend to zero when the marginal totals are all large. This justifies the use of χ as a standardized normal variable (Laplace, 1774; Romanovsky, 1925).

An approximation is also available for the Poisson distribution with mean λ

$$(4.6) \qquad p\{a \mid \lambda\} = e^{-\lambda}\lambda^a/a! \simeq (2\pi\lambda)^{-\frac{1}{2}} \exp\{-\tfrac{1}{2}(a - \lambda)^2/\lambda\}.$$

In conclusion, for the hypergeometric distribution, the crude and corrected χ's may be written

$$(4.7) \qquad \chi = a_{..}(a_{..} - 1)^{\frac{1}{2}}\{a_{11} - a_{1.}a_{.1}/a_{..}\}/(a_{1.}a_{2.}a_{.1}a_{.2})^{\frac{1}{2}}, \quad \text{and}$$

$$(4.8) \qquad \chi_c = a_{..}(a_{..} - 1)^{\frac{1}{2}}\{|a_{11} - a_{1.}a_{.1}/a_{..}| - \tfrac{1}{2}\}/(a_{1.}a_{2.}a_{.1}a_{.2})^{\frac{1}{2}};$$

for the Poisson distribution

$$(4.9) \qquad \chi = (a - \lambda)/\sqrt{\lambda}, \qquad \chi_c = (a - \lambda - \tfrac{1}{2})/\sqrt{\lambda},$$

where for simplicity, we suppose that a one way test and the upper tail probabilities are being considered, so that it is not necessary to introduce the absolute sign to modify (4.9) to

$$(4.9') \qquad \chi_c = (|a - \lambda| - \tfrac{1}{2})/\sqrt{\lambda}.$$

5.   THE NORMAL OR χ^2 APPROXIMATION TO THE MULTINOMIAL DISTRIBUTION

In Chapter 5, the approximation to discrete multi-dimensional probability distributions by the continuous χ^2 distributions will be justified in several different ways. Some requirements on such an approximation are now considered.

The approximate test at any arbitrary significance level, α, should reject about a proportion, α, of the observations. Now for a given discrete distribution it is usually impossible for any exact or approximate test that this should be so. Discrepancies between α and the size of the test (the proportion rejected under the null hypothesis) will arise up to at least the value of the relative frequency of the marginal observation. Further, whether the test function is χ^2 or whether the probabilities are exactly enumerated there will be coincidences between different sets of readings. For example, in the multinomial expansion $(0\cdot5 + 0\cdot3 + 0\cdot2)^{N}$* we shall have configurations of the form, $\{x_i\} = \{x_1, x_2, x_3\}$, with $x_1 + x_2 + x_3 = N$. The probabilities associated with such a configuration will be the multinomial

(5.1) $p(\{x_i\}) = p\{x_1, x_2, x_3\} = 0\cdot5^{x_1}0\cdot3^{x_2}0\cdot2^{x_3}N!/(x_1!\,x_2!\,x_3!).$

There can be different configurations for which $p(\{x_i\}) = p(\{x_i'\})$ although $\{x_i\} \neq \{x_i'\}$. Or when using χ^2, where

(5.2) $$\chi^2 = \sum (x_i - Np_i)^2/(Np_i)$$

takes the same values for two different sets of values of $\{x_i\}$. Such coincidences will cause the cumulative distributions to have larger jumps than would occur without them. The effect is particularly marked for symmetrical multinomial distributions. The distribution (5.1) was considered for special values, $N = 10$ and 20, by el Shanawany (1936) and for $N = 10$ by Neyman and Pearson (1931a). These authors used the exact value of $p(\{x_i\})$ to order the configurations and found that the χ^2 approximations were accurate. More extensive tables have since been given, in which the configurations are ranked by means of the χ^2 values. Table II of Lancaster (1961) gives the sizes of the test at the 1%, 5%, 10(20)90% levels and Table III of the same paper gives the number of configurations falling in each probability class, of which the following table for $N = 24$ is a sample.

Probability Class	Number of Configurations Observed	Number of Configurations Calculated	Size of the Test for Lower Limit of the Class
0·9 to 1·0	2	3	93·05
0·7 to 0·9	7	7	71·36
0·5 to 0·7	9	8	50·77
0·3 to 0·5	13	13	30·74
0·1 to 0·3	27	29	10·92
0·05 to 0·1	17	18	5·29
0·01 to 0·05	42	42	0·96
0·00 to 0·01	208	∞	—
Total	325	—	—

* We centre the decimal point following the English custom.

The values in the table were calculated from an approximate formula which showed that the number of configurations in any probability class, π_1 to π_2, should be

$$2\pi N(0{\cdot}03)^{\frac{1}{2}} \log_e (\pi_1/\pi_2), \quad \text{approximately.}$$

The borderline configurations at the level α should be approximately

$$(0{\cdot}03)^{-\frac{1}{2}}(2\pi N)^{-1} \exp -\tfrac{1}{2}\chi_\alpha^2 = 0{\cdot}9189\alpha/N,$$

or $0{\cdot}03929\alpha$ when $N = 24$. Now these satisfactory approximations have resulted in distributions where the value of the smallest expectation was $4{\cdot}8$. Tables II and III of the paper cited gave similar computations for $N = 9(1)31$. R. W. Bennett (1962) has extended these computations up to $N = 50$. Cochran (1936a and 1942) has carried out similar computations for some special multinomial distributions and for contingency tables.

An important application of χ^2 is to the test for the consistency of parallel counts, introduced by Fisher, Thornton and Mackenzie (1922). Sukhatme (1938a) tested with the aid of a random sampling experiment whether the χ^2 distribution was adequate in such cases with a small value of the mean (as low as unity). Lancaster (1952a) and Lancaster and Brown (1965) re-examined the problem carrying out a complete enumeration of all possible configurations. This work confirmed Sukhatme's conclusions as to the appropriateness of the χ^2-approximation even for low values of the mean and suggested that Sukhatme's favorable results were obtained because Sukhatme was in effect sampling from a mixture of symmetric multinomial distributions with index, N, and N was a Poisson variable with a certain parameter determined by the method of computation. In the symmetric multinomial distribution, with class probabilities each equal to $1/n$, the configurations become partitions repeated $n!(n_0!n_1!\ldots n_n!)^{-1}$ times, where n_j is the number of times j occurs in the configuration. The formula for χ^2 in this symmetrical case is

$$(5.3) \qquad \chi^2 = n \sum x_i^2/\sum x_i - \sum x_i,$$

summation being over the integers $1, 2, \ldots, n$. Now for a fixed $N = \sum x_i$, $\sum x_i^2$ can only take integer values between the integral part of $N^2 n^{-1}$ and N^2, which, moreover must have the same parity as N. There may thus be many different partitions with the same value of χ^2. This might be expected to disturb the distribution but the paper of Lancaster (1952a) shows that the approximating distribution comes out of the test quite well. Lancaster and Brown (1965) after calculating the sizes of the test for individual values of $\sum x_i = N$ equal to 10(5)100, gave N a Poisson probability distribution with values of the parameter equal to 5(1)25. The agreements between sizes and significance levels were very close.

EXERCISES AND COMPLEMENTS

Condensation

1. Suppose U is a Poisson variable with parameter, λ. After condensation, one can obtain a random variable, Z, taking values, 0 and 1, according as to whether U takes the value, zero, or not. $\mathscr{P}(Z = 0) = \exp(-\lambda)$. Show that this procedure can be used for estimation of the parameter in counting experiments. (Fisher (1921); Peierls (1935); Tippett (1932); Gordon (1939)).

2. Let X be standard normal. Define Y so that $Y = 1$ if $X \geqslant 0$ and $Y = -1$ if $X < 0$. The space of Y results from a condensation of the space of X. Prove that the correlation between X and Y is $(2/\pi)^{\frac{1}{2}}$.

3. Let X be standard normal. Define $Z = +1$ if $X^2 \geqslant 1$ and $Z = -1$ if $X^2 < 1$. Prove that Z is uncorrelated with X and with Y as defined in Exercise 2.

4. In practice, condensation is usually carried out so that there are a number of equal class intervals and one or two tail classes which may be of infinite width. Show that there is always a transformation from the rectangular distribution on the unit interval to an arbitrary continuous distribution. Hence show that for any discrete probability distribution, a random partition can be carried out so that the resulting variable has any arbitrary continuous distribution. As an example, show how to set up a correspondence between a binomial variable and the rectangular distribution on the unit interval and hence between the binomial distribution and that of χ^2 with any arbitrary number of degrees of freedom.

5. Compute the cumulative distribution function of the binomial distribution $(\frac{1}{2} + \frac{1}{2})^7$. With the aid of pseudo-random numbers, imitate sampling from this distribution. Note that this procedure is a condensation from the rectangular distribution to the binomial distribution.

6. Determine the power of the one sided, upper-tail test at the 5% level of significance in the binomial $(\frac{1}{2} + \frac{1}{2})^7$, when the alternative distribution is $(\frac{1}{4} + \frac{3}{4})^7$ using the exact test and the Neyman-Pearson auxiliary randomization procedure.

Stirling's Approximation

7. Given the Stirling formula

$$\log n! \simeq (n + \tfrac{1}{2}) \log n - n + \log c,$$

determine c by applying the formula to the terms of the binomial, $(\frac{1}{2} + \frac{1}{2})^n$, and equating the terms to appropriate areas under the normal curve. [*Note.* c is obtained after the application of the Wallis formula

$$\tfrac{1}{2}\pi = 2 . 2 . 4 . 4 . 6 . 6 \ldots / (1 . 3 . 3 . 5 . 5 . 7 . 7 \ldots).]$$

8. Show that a knowledge of the Wallis formula is not necessary by applying Stirling's approximation to the terms of the Poisson distribution with large integer

value of the parameter and equating the terms of the Poisson to areas under the normal curve. c is then found to be $(2\pi)^{\frac{1}{2}}$. Deduce the Wallis formula.

9. Equate the ratios, mean deviation to standard deviation, in the Poisson and normal distributions to obtain the Stirling formula.

The Poisson Distribution

10. Let Y be a Poisson variable with parameter, λ. Show that $(Y - \lambda)^2/\lambda$ is distributed approximately as χ^2 with 1 d.f. as $\lambda \to \infty$ by showing that the cumulants of the standardized variable $(Y - \lambda)/\sqrt{\lambda}$ approach those of the standardized normal variable.

11. Let $Y_1, Y_2, \ldots, Y_n$ be Poisson variables with parameter $\lambda p_1, \lambda p_2, \ldots, \lambda p_n$, $\sum p_i = 1$. Show that $\sum (Y_i - \lambda p_i)^2/(\lambda p_i)$ is distributed approximately as χ^2 with n degrees of freedom as $\lambda \to \infty$.

The Hypergeometric Distribution

12. Calculate the factorial moments of the hypergeometric distribution and hence show that the moments of the number of observations falling in one cell of the fourfold contingency table approximate to the normal. (Romanovsky, 1925.)

13. With fixed marginal totals, say 10, 10 and 11, 9 of the fourfold contingency table calculate the probabilities of obtaining 0, 1, . . . , 9 in the cell with the smallest expectation. Verify that Yates' correction gives a good approximation to the tail probabilities. (Yates, 1934.)

Miscellaneous

14. Plot the possible configurations of results from a multinomial expansion $(0.2 + 0.3 + 0.5)^N$, with $N = 12$ say, on a trilinear graph. Associate the probabilities, $p(\{x_i\})$, of (5.1) with each point. Draw the contour lines of χ^2 corresponding to the probability levels, 0.01, 0.05 and 0.10 (0.20)0.90. Observe that the total frequencies in the various contours approximate fairly well to the theoretical values. (Lancaster, 1961; Bennett, 1962; see also Neyman and Pearson, 1931 and el Shanawany, 1936.)

15. Consider the symmetrical multinomial expansion, $(\frac{1}{3} + \frac{1}{3} + \frac{1}{3})^{12}$, in the same way as in the previous case. Show that there are symmetries, which diminish the number of distinct values of χ^2. With a true null hypothesis, calculate the relative frequency of assignment to the probability classes. (Lancaster, 1952a; Lancaster and Brown, 1965.)

16. Show that if a sample of n parallel independent drawings from a Poisson population with mean, λ, is taken and the sampling repeated, the joint sampling distribution will tend towards a mixture of symmetrical multinomial distributions such that the parameter, N, of the multinomial distribution is a Poisson variable with mean $n\lambda$. (This is the interpretation that can be given to the sampling experiment of Sukhatme, 1938a.)

CHAPTER IV

Orthogonality

1. ORTHOGONAL MATRICES

A square real matrix of finite size is said to be *orthogonal* if

(1.1) $H^T H = 1$,   the unit matrix, or equivalently,

(1.2) $HH^T = 1$.

The weaker definition requiring only that HH^T or $H^T H$ be diagonal should be avoided.

An orthogonal matrix is obviously non-singular. Orthogonal matrices of a given size form a group under matrix multiplication. It is possible to obtain orthogonal matrices from any non-singular matrix by post-multiplication with an appropriate upper triangular matrix, the Schmidt or Gram-Schmidt orthogonalization process.

Theorem 1.1. *Let* **B** *be a real non-singular matrix of size, n. Then there is an upper triangular matrix* **U**, *such that*

(1.3) $M = BU,$

is orthogonal.

Proofs can be found in the usual texts on matrix algebra. This theorem has many important corollaries. Its geometric meaning is that if a set of n linearly independent vectors defining a space of n dimensions be given, orthogonal axes can be chosen so that the first k of them always lie in the space determined by the first k vectors.

Corollary 1. *Let the vector* **b** *be given, then an orthogonal matrix can always be found with the elements of its first column proportional to those of* **b**.

Proof. Let **b** be the first column of a matrix **B**; suppose that the i_1^{th} element of **b** is non-zero. Complete **B** by inserting units at the position i_j in the jth column for $j = 2, 3, \ldots, n$ such that $i_1, i_2, \ldots, i_n$ is a permutation of $1, 2, 3, \ldots, n$. Insert zeroes elsewhere. **B** is now of rank n and the Gram-Schmidt process may be applied, with $u_{11} = (b^T b)^{-\frac{1}{2}}$.

Corollary 2. *Let k $n \times 1$ linearly independent column vectors be given. Then an orthogonal matrix can be formed which has its last $(n - k)$ columns all orthogonal to the given k vectors.*

Proof. Write the k vectors as the first k columns of a matrix **B**. Since they are linearly independent, at least one $k \times k$ minor has non-zero determinant, say, the minor with rows $i_1, i_2, \ldots, i_k$. Complete **B** by writing units at the i_j place in the jth column $j = k + 1, k + 2, \ldots, n$ so that

$$i_1, i_2, \ldots, i_k, i_{k+1}, \ldots, i_n$$

is a permutation of $1, 2, \ldots, n$. **B** is non-singular, and the Gram-Schmidt process may be applied.

Corollary 3. *Given k mutually orthogonal and normalised $n \times 1$ vectors, an $n \times n$ orthogonal matrix can always be formed which has the k vectors for the first k columns.*

Proof. This is a special case of Corollary 2. The first k columns of the post-multiplying matrix, **U**, have units in the first k of the diagonal positions, and zeroes above the diagonal.

2. THE FORMATION OF ORTHOGONAL MATRICES FROM OTHER ORTHOGONAL MATRICES

The *direct sum* of two square matrices **A** and **B** of sizes, m and n, is defined by

$$A + B = \begin{bmatrix} A & 0 \\ 0^T & B \end{bmatrix}$$

where **0** is the $m \times n$ zero matrix.

The following properties, S_1 to S_6, follow almost directly from the definition; **A** and **C** are assumed to be size, m, and **B** and **D** of size, n, except in S_1.

S_1. *The taking of the direct sum is associative.*

$$A + (B + C) = (A + B) + C.$$

S_2. *Ordinary sums of direct sums of matrices are given by*

$$(A + B) + (C + D) = (A + C) + (B + D).$$

S_3. *The matrix product of direct sums is the direct sum of matrix products,*

$$(A + B)(C + D) = AC + BD.$$

S_4. *The transpose of a direct sum is the direct sum of the transposes,*

$$(A + B)^T = A^T + B^T.$$

S_5. *The direct sum of two unit matrices is a unit matrix*,

$$\mathbf{1}_m \dotplus \mathbf{1}_n = \mathbf{1}_{m+n}.$$

S_6. *The inverse of a direct sum is the direct sum of the inverses*

$$(\mathbf{A} \dotplus \mathbf{B})^{-1} = \mathbf{A}^{-1} \dotplus \mathbf{B}^{-1}.$$

S_6 follows from S_3, the product rule and S_5. Note that the transposition and inversion operations do not reverse the order of the matrices in S_4 and S_6.

Theorem 2.1. *The direct sum of orthogonal matrices is orthogonal.*

Proof. Let $\mathbf{A}$ and $\mathbf{B}$ be orthogonal and of sizes, m and n, respectively. Then

$$(\mathbf{A} \dotplus \mathbf{B})^T(\mathbf{A} \dotplus \mathbf{B}) = (\mathbf{A}^T \dotplus \mathbf{B}^T)(\mathbf{A} \dotplus \mathbf{B})$$
$$= \mathbf{A}^T\mathbf{A} \dotplus \mathbf{B}^T\mathbf{B} = \mathbf{1}_m \dotplus \mathbf{1}_n = \mathbf{1}_{m+n}.$$

Corollary 1. *If $\mathbf{H}$ is orthogonal, so is $(1 \dotplus \mathbf{H})$, where 1 is the scalar unit.*

Corollary 2. *If $\mathbf{H}_1$ and $\mathbf{H}_2$ are orthogonal and have the same elements in the first row (column) they are related by*

$$(2.1) \qquad \mathbf{H}_1 = (1 \dotplus \mathbf{H}_3)\mathbf{H}_2 \quad or \quad \mathbf{H}_1 = \mathbf{H}_2(1 \dotplus \mathbf{H}_3).$$

Proof. $\mathbf{H}_1 = \mathbf{H}_1\mathbf{H}_2^T\mathbf{H}_2$. Now $\mathbf{H}_1\mathbf{H}_2^T$ has a unit in the $(1, 1)$ position since the first rows of $\mathbf{H}_1$ and $\mathbf{H}_2$ are the same. $\mathbf{H}_1\mathbf{H}_2^T$ is therefore of the form $(1 \dotplus \mathbf{H}_3)$ where $\mathbf{H}_3$ is orthogonal.

Corollary 3. *If $\mathbf{H}_1$ and $\mathbf{H}_2$ are orthogonal and have the same elements in the first k rows then*

$$(2.2) \qquad \mathbf{H}_1 = (\mathbf{1}_k \dotplus \mathbf{H}_4)\mathbf{H}_2$$

where $\mathbf{H}_4$ is orthogonal.

Proof. $\mathbf{H}_1 = \mathbf{H}_1\mathbf{H}_2^T\mathbf{H}_2$. $\mathbf{H}_1\mathbf{H}_2^T$ has units along the diagonal in the first k places and is orthogonal. It has, therefore, the form, $(\mathbf{1}_k \dotplus \mathbf{H}_4)$, where $\mathbf{H}_4$ is orthogonal.

The *direct product* of square matrices, $\mathbf{A}$ and $\mathbf{B}$ of sizes, m and n, respectively is a square matrix of size mn defined by

$$(2.3) \qquad \mathbf{A} \times \mathbf{B} = \begin{bmatrix} a_{11}\mathbf{B} & a_{12}\mathbf{B} & \ldots & a_{1m}\mathbf{B} \\ a_{21}\mathbf{B} & a_{22}\mathbf{B} & \ldots & a_{2m}\mathbf{B} \\ \cdots & & \cdots & \\ a_{m1}\mathbf{B} & a_{m2}\mathbf{B} & \ldots & a_{mm}\mathbf{B} \end{bmatrix}$$

The element of $\mathbf{A} \times \mathbf{B}$ at the intersection of the row, $(i - 1)n + i'$, and of the column, $(j - 1)n + j'$, is $a_{ij}b_{i'j'}$, where i, j run over the indices

$1, 2, \ldots, m$ and i', j' run over the indices $1, 2, \ldots, n$. The direct product has the following properties.

P_1. *The operation of taking the direct product is associative*

$$\mathbf{A} \times (\mathbf{B} \times \mathbf{C}) = (\mathbf{A} \times \mathbf{B}) \times \mathbf{C}$$

and so either expression can be written as $\mathbf{A} \times \mathbf{B} \times \mathbf{C}$.

P_2. *The general term of* $\mathbf{A} \times \mathbf{B} + \mathbf{C} \times \mathbf{D}$ *will be of the form* $a_{ij}b_{i'j'} + c_{ij}d_{i'j'}$, *if* $\mathbf{A}$ *and* $\mathbf{C}$ *have the same size and* $\mathbf{B}$ *and* $\mathbf{D}$ *have the same size*, the ordinary matrix sum.

P_3. *The matrix product of direct products is given by*

$$(\mathbf{A} \times \mathbf{B})(\mathbf{C} \times \mathbf{D}) = \mathbf{AC} \times \mathbf{BD}.$$

Proof. Note that $\mathbf{A} \times \mathbf{B}$ is of the form, $(a_{ij}\mathbf{B})$, and $\mathbf{C} \times \mathbf{D}$ is of the form, $(c_{ij}\mathbf{D})$. The rule for multiplying partitioned matrices applies and we obtain a matrix of the form, $(f_{ij}\mathbf{BD})$, where $f_{ij} = \sum_k a_{ik}c_{kj}$ and the result follows.

P_4. *The transpose of a direct product is the direct product of the transposes*,

$$(\mathbf{A} \times \mathbf{B})^T = \mathbf{A}^T \times \mathbf{B}^T.$$

This follows from the definition.

P_5. *The direct product of unit matrices is again a unit matrix.*

P_6. *The inverse of a direct product is the direct product of the inverses*,

$$(\mathbf{A} \times \mathbf{B})^{-1} = \mathbf{A}^{-1} \times \mathbf{B}^{-1}.$$

Proof. $(\mathbf{A} \times \mathbf{B})(\mathbf{A}^{-1} \times \mathbf{B}^{-1}) = \mathbf{AA}^{-1} \times \mathbf{BB}^{-1} = \mathbf{1}_m \times \mathbf{1}_n = \mathbf{1}_{mn}$.

P_7. *The direct product of orthogonal matrices is orthogonal.*

Proof. Let $\mathbf{A}$ and $\mathbf{B}$ be orthogonal; then,

$$(\mathbf{A} \times \mathbf{B})(\mathbf{A} \times \mathbf{B})^T = \mathbf{AA}^T \times \mathbf{BB}^T = \mathbf{1}_m \times \mathbf{1}_n = \mathbf{1}_{mn}.$$

Note that neither direct sum nor direct product is a commutative operation.

It is sometimes convenient to define a *direct sum of vectors*,

$$(2.4) \qquad \mathbf{x} \dotplus \mathbf{y} = \{x_1, x_2, \ldots, x_n\} \dotplus \{y_1, y_2, \ldots, y_n\}$$
$$= \{x_1, x_2, \ldots, x_n, y_1, y_2, \ldots, y_n\},$$

and a *direct product of vectors*,

$$(2.5) \qquad \mathbf{x} \times \mathbf{y} = \{x_1 y_1, x_1 y_2, \ldots, x_1 y_n, x_2 y_1, \ldots, x_m y_n\}.$$

Sometimes this result is usefully written in the symbolic form, $\mathbf{x} \times \mathbf{y} = \mathbf{z}$, $\mathbf{z}$ having as elements z_{ij}, $i = 1, 2, \ldots, m$, $j = 1, 2, \ldots, n$, arranged in dictionary order. Two results are important in considering transformations of variables arranged in classes or in a hierarchy.

(2.6) $$|A + B| = |A| |B|, \quad \text{and}$$

(2.7) $$|A \times B| = |A|^n |B|^m.$$

The result (2.6) is readily proved from the definition. The proof of (2.7) is a little longer. Note that

(2.8) $$A \times B = (A \times 1_n)(1_m \times B).$$

The determinant of the second factor on the right is $|B|^m$. A re-arrangement of rows and columns can be found that will give $J(A \times 1_n)J^T = 1_n \times A$ where J is a permutation matrix. The required permutation is $[i - 1]n + i' \to [i' - 1]m + i$. The units are at the positions $([i' - 1]m + i, [i - 1]n + i')$ of J. But $JJ^T = 1$, so that

(2.9) $$|A \times B| = |A \times 1_n| |1_m \times B|$$
$$= |J| |A \times 1_n| |J^T| |B|^m$$
$$= |1_n \times A| |B|^m = |A|^n |B|^m.$$

3. SETS OF ORTHOGONAL FUNCTIONS ON A FINITE SET OF POINTS

Let the real functions, $g_i(x)$, have finite variance on a statistical distribution, so that

(3.1) $$\int g_i^2(x) \, dF(x) < \infty, \qquad i = 0, 1, 2, \ldots .$$

It is usual to write $g_i \in L_F^2$ in the language of the theory of Hilbert space. $\{g_i(x)\}$ will be said to form an *orthonormal set* with respect to $F(x)$ if

(3.2) $$\int g_i(x)g_j(x) \, dF(x) = \delta_{ij}, \qquad i, j = 0, 1, 2, \ldots .$$

The integral of (3.2) always exists by the Bunyakovsky-Schwarz inequality. It is convenient to define $g_0(x) = 1$ which ensures that every other function is standardized to have zero mean and unit variance. The functions are said to form an *orthogonal set* if δ_{ij} is replaced by $c_i\delta_{ij}$, $c_i > 0$. If X has a discrete distribution, (3.2) can be replaced by a summation

(3.3) $$\sum_x g_i(x)g_j(x)p(x) = \delta_{ij}$$

where the random variable, X, takes values, $x_1, x_2, \ldots, x_n$, with probabilities, $p_1, p_2, \ldots, p_n$; $p_j > 0$. In the partition of χ^2, orthogonal matrices with $\sqrt{p_i}$ as elements of the first column play an important role.

Theorem 3.1. *To every orthogonal* $\mathbf{H}$ *with* $h_{i1} = \sqrt{p_i}$ *there corresponds a set of* $(n-1)$ *non-constant orthonormal functions on the statistical distribution defined on the space of points,* x_i, *by*

$$\mathscr{P}(X = x_i) = p_i, \qquad p_i > 0, \qquad i = 1, 2, \ldots, n,$$

and conversely.

Proof. Let $g_{j-1}(x_i) = h_{ij}h_{i1}^{-1}$.

It is easily verified that

$$\sum_i g_j(x_i)g_{j'}(x_i)p_i = \delta_{jj'}, \quad \text{so that}$$

$g_0(x)$, $g_1(x)$, $g_2(x)$, $\ldots$, $g_{n-1}(x)$ form an orthonormal set. Conversely given the orthonormal set $\{g_i(x)\}$ including $g_0(x) = 1$, the matrix, defined by

(3.4) $$h_{ij} = g_{j-1}(x_i)\sqrt{p_i},$$

is orthogonal.

Corollary. *Any two sets of* n *orthonormal functions* $\{g_i(x)\}$ *and* $\{h_i(x)\}$ *defined on a distribution,* $p_j, j = 1, 2, \ldots, n$, *of* n *points are related by an orthogonal transformation, in such a way that* $g_0(x) = 1 = h_0(x)$ *and* $g_i(x) = \sum_j a_{ij}h_j(x)$, $i, j = 1, 2, \ldots, n-1$ *and* (a_{ij}) *is an orthogonal matrix of size,* $n-1$.

It is convenient for some purposes to write $\{x^{(j)}\}$ for the orthonormal set on the distribution of X rather than $g_j(x)$ as we have done in the theorem above. Orthogonal matrices of a simple form due to Helmert can be readily computed (Lancaster, 1965a). Let $x^{(1)}, x^{(2)}, \ldots, x^{(n-1)}$ be orthonormal functions defined recursively so that $x^{(j)}$ takes non-zero values only on the points, $x_1, x_2, \ldots, x_{j+1}$ and let $x^{(j)}$ take the value $x_i^{(j)}$ at the point x_i. $x^{(0)} = 1$.

Define $P_k = \sum_1^k p_k$. $x^{(1)}$ may be computed by

(3.5) $$x_1^{(1)}p_1 + x_2^{(1)}p_2 = 0, \qquad x_1^{(1)2}p_1 + x_2^{(1)2}p_2 = 1,$$

and so

(3.6)

$$x_1^{(1)} = -\sqrt{p_2}/\sqrt{(P_2 P_1)}, \ x_2^{(1)} = +\sqrt{P_1}/\sqrt{(p_2 P_2)}, \qquad x_i^{(1)} = 0 \ \text{if} \ i > 2.$$

Now suppose that we have obtained $(k-1)$ of these orthonormal functions, $x^{(k)}$ is then orthogonal to $x^{(1)}, x^{(2)}, \ldots, x^{(k-1)}$; however, these orthogonality conditions place restriction only on the values, $x_i^{(k)}$, $i = 1, 2, \ldots, k$, since all the functions $x^{(1)}, x^{(2)}, \ldots, x^{(k-1)}$ take zero values on $x_{k+1}, x_{k+2}, \ldots, x_n$. But the constant function is also orthogonal to the same functions so that $x^{(k)}$ must have a constant value at the points $x_1, x_2, \ldots, x_k$. $x^{(k)}$ has further to obey the conditions of being orthogonal to the constant term and the normalization condition, so that

(3.7) $\sum_{i=1}^{k+1} x_i^{(k)} x_i^{(0)} p_i = 0,$ or since $x_j^{(k)}$ is constant for $j = 1, 2, \ldots, k,$

$$x_1^{(k)} P_k + x_{k+1}^{(k)} p_{k+1} = 0,$$

and

(3.8) $$\sum_{i=1}^{k+1} x_i^{(k)^2} p_i = x_1^{(k)^2} P_k + x_{k+1}^{(k)^2} p_{k+1} = 1.$$

Hence,

(3.9) $\begin{aligned} x_i^{(k)} &= -\sqrt{p_{k+1}}/\sqrt{(P_{k+1}P_k)} \quad \text{for} \quad i = 1, 2, \ldots, k \\ x_i^{(k)} &= +\sqrt{P_k}/\sqrt{(p_{k+1}P_{k+1})}, \quad \text{for} \quad i = k+1 \\ x_i^{(k)} &= 0 \quad \text{for} \quad i > (k+1), \text{ is the general solution.} \end{aligned}$

At each stage in the proof above, we have used the fact that if a distribution consists of two points with probabilities, p and q, then $-\sqrt{(q/p)}$ and $+\sqrt{(p/q)}$ are the two values of the orthonormal function.

The elements of the Helmert matrix are, therefore,

(3.10) $\begin{aligned} h_{i1} &= \sqrt{p_i}, \quad i = 1, 2, \ldots, n, \\ h_{ij} &= -\sqrt{p_i}\sqrt{p_j}/\sqrt{(P_{j-1}P_j)} \quad \text{for} \quad i = 1, 2, \ldots, (j-1), \quad j > 1 \\ h_{jj} &= +\sqrt{P_{j-1}}/\sqrt{P_j} \quad \text{for} \quad j > 1 \\ h_{ij} &= 0 \quad \text{for} \quad i > j > 1. \end{aligned}$

Helmert (1876) introduced a transformation which has a matrix, $\mathbf{H}$, given by

(3.11) $\begin{aligned} h_{i1} &= n^{-\frac{1}{2}} \\ h_{ij} &= -\{j(j-1)\}^{-\frac{1}{2}} \quad \text{for} \quad i = 1, 2, \ldots, (j-1), \quad j > 1 \\ h_{jj} &= +\sqrt{(j-1)}/\sqrt{j}, \, j > 1 \\ h_{ij} &= 0 \quad \text{for} \quad i > j > 1. \end{aligned}$

$\mathbf{H}$ may be called Helmertian in the strict sense. It is still in common use in the analysis of variance. In fact, the first element of $\mathbf{H}^T\mathbf{a}$ is $n^{\frac{1}{2}}$ times the sample mean, and each successive row measures the difference between the kth variable and the mean of the first $(k-1)$ variables. Use was made of such a matrix by Burnside (1936) and by Irwin (1942) in his estimate of the variance of the weighted mean. Irwin (1949) first applied the matrix to obtain an exact partition of χ^2 in the multinomial.

4. ORTHONORMAL POLYNOMIALS AND FUNCTIONS ON STATISTICAL DISTRIBUTIONS

Let $\mathbf{x}$ be the vector, the ith element of which is x^{i-1} for $i = 1, 2, 3, \ldots,$ and $\mathbf{x}_k$ be the vector with elements x^{i-1} for $i = 1, 2, 3, \ldots, (k+1).$

Lemma 4.1. *If a statistical distribution has more than n points of increase, the polynomial,* $\mathbf{a}^T\mathbf{x}_n$, *where the elements of* $\mathbf{a}$ *are* $(n + 1)$ *arbitrary real elements, cannot vanish at each point of increase without being identically zero.*

Proof. This is a well-known theorem.

Square matrices, $\mathbf{D}_k$ and $\mathbf{D}$, are defined of sizes, $(k + 1)$ and infinity, with elements, $d_{ij} = \mu_{i+j-2}$, where the moments may be taken either about the mean or an arbitrary origin. $F(x)$ in the discussion following is implied by the notation to have an infinite number of points of increase but the theorems apply to finite discrete distributions if only polynomials of degree less than the number of points of increase are considered. It is assumed that all moments are finite.

Lemma 4.2. $\mathbf{D}_k$ *is positive definite.*

Proof. Let $\mathbf{a}$ be a vector with arbitrary real coefficients. Then

$$(4.1) \qquad \mathbf{a}^T\mathbf{D}_k\mathbf{a} = \mathbf{a}^T\mathscr{E}\mathbf{x}_k\mathbf{x}_k^T\mathbf{a} = \mathscr{E}(\mathbf{a}^T\mathbf{x}_k\mathbf{x}_k^T\mathbf{a})$$

$$= \mathscr{E}(\mathbf{a}^T\mathbf{x}_k)^2 = \int (\mathbf{a}^T\mathbf{x}_k)^2 \, dF(x),$$

and this last expression is positive by Lemma 4.1.

Let $p^{(k)}(x) \equiv p^{(k)}$ be a polynomial, in which the coefficient of x^k is positive but no higher power of x appears. $\{p^{(k)}\}$ will be said to form an orthogonal set if

$$(4.2) \qquad \int p^{(k)}p^{(k')} \, dF(x) = c_k\delta_{kk'}, \qquad c_k > 0.$$

If $c_k = 1$ for each k the set is said to be orthonormal. The orthonormal set may be written in the form of a vector, $\mathbf{p}$. Each element of $\mathbf{p}$ is a linear function of the form $\mathbf{a}^T\mathbf{x}$ and so

$$(4.3) \qquad \mathbf{p} = \mathbf{A}\mathbf{x},$$

where $\mathbf{A}$ is lower triangular and has positive diagonal elements. $\mathbf{p}_k$ and $\mathbf{A}_k$ are defined by a convention similar to that for $\mathbf{x}_k$. By definition

$$(4.4) \qquad \mathscr{E}\mathbf{p}_k\mathbf{p}_k^T = \mathbf{1}_{k+1}, \qquad k = 0, 1, 2, \ldots$$

where $\mathbf{1}_{k+1}$ is the unit matrix of size, $k + 1$.

$$(4.5) \qquad \mathbf{1}_{k+1} = \mathscr{E}\mathbf{p}_k\mathbf{p}_k^T = \mathscr{E}(\mathbf{A}_k\mathbf{x}_k\mathbf{x}_k^T\mathbf{A}_k^T) = \mathbf{A}_k\mathbf{D}_k\mathbf{A}_k^T.$$

Taking determinants on both sides,

$$(4.6) \qquad |\mathbf{A}_k| = |\mathbf{D}_k|^{-\frac{1}{2}} = D_k^{-\frac{1}{2}}, \quad \text{say.}$$

$|\mathbf{A}_k|$ is the product of the diagonal terms and moreover the first k diagonal terms of $\mathbf{A}_k$ are identical with those of $\mathbf{A}_{k-1}$, so that the coefficient of x^k in $p^{(k)}(x)$ is $D_k^{-\frac{1}{2}} D_{k-1}^{\frac{1}{2}}$. Further

(4.7) $$\mathbf{A}_k \mathbf{D}_k = (\mathbf{A}_k^T)^{-1}, \quad \text{from (4.5).}$$

And $\mathbf{A}_k \mathbf{D}_k$ is therefore upper triangular since $\mathbf{A}_k^T$ and its inverse are upper triangular. Considering the last row of (4.7) as a row vector, $\mathbf{a}^T$ say

(4.8) $$\mathbf{a}^T \mathbf{D}_k = [0, 0, \ldots, D_k^{-\frac{1}{2}} D_{k-1}^{-\frac{1}{2}}].$$

Thus $\mathbf{a}$ is orthogonal to every column of $\mathbf{D}_k$ except the last and so the elements of $\mathbf{a}$ are proportional to the cofactors of the last column of $\mathbf{D}_k$. The orthogonal polynomial can thus be written as the determinant of a matrix formed by replacing the elements of the last column of $\mathbf{D}_k$ by x^{i-1} or more usually by transposing this matrix. In this determinant, the coefficient of x^k is D_{k-1}, so that the determinant must be multiplied by a factor to obtain

(4.9) $$p^{(k)}(x) = D_k^{-\frac{1}{2}} D_{k-1}^{-\frac{1}{2}} \begin{vmatrix} \mu_0 & \mu_1 & \cdots & \mu_k \\ \mu_1 & \mu_2 & \cdots & \mu_{k+1} \\ \cdots & & & \cdots \\ \mu_{k-1} & \mu_k & \cdots & \mu_{2k-1} \\ 1 & x & \cdots & x^k \end{vmatrix}$$

It can be easily verified that (4.9) is the solution of equation (4.2) with $c_i = 1$.

If the distribution has only n points of increase, a set of $(n-1)$ orthonormal polynomials can be determined with the constant term adjoined. An orthogonal $\mathbf{H}$ can be constructed with

(4.9′) $h_{i1} = \sqrt{p_i}, \quad h_{i,k+1} = p^{(k)}(x_i)\sqrt{p_i}, \quad k = 1, 2, \ldots, (n-1),$

where the random variable X takes the values, x_i, with probability, p_i. Every set of orthonormal functions, defined on the n points, will contain linear combinations of the polynomial orthonormal set. A matrix corresponding to $\mathbf{H}$ formed from the new orthonormal set will be of the form $\mathbf{H}(1 + \mathbf{H}_1)$, with $\mathbf{H}_1$ orthogonal, if the constant term is the first function.

Theorem 4.1. *If a distribution has finite moments of all orders, then the set of orthonormal polynomials is given by (4.9) with $D_{-1} = 1$, $D_0 = 1$ and $D_k = |\mathbf{D}_k|$ for $k > 1$, where $\mathbf{D}_k = (\mu_{i+j-2})$, $i = 1, 2, \ldots, (k+1)$; $j = 1, 2, \ldots (k+1)$.*

For some distributions, the set of orthogonal polynomials may not exist, for example, in the Cauchy. In some important distributions such as the normal and the χ^2 distributions, orthonormal sets of polynomials, the

Hermite and the Laguerre respectively, exist and have the desirable property of completeness.

$\xi(x)$ is said to belong to L_F^2 if

(4.10)
$$\int \xi^2(x)\,dF = a^2 < \infty.$$

$\{x^{(i)}\}$ is said to be a *statistical orthonormal set* if $x^{(0)} = 1$ and

(4.11)
$$\int x^{(i)}x^{(j)}\,dF = \delta_{ij}, \qquad i, j = 0, 1, 2, \ldots.$$

Let

(4.12)
$$s_n = a_0 + a_1 x^{(1)} + a_2 x^{(2)} + \ldots + a_n x^{(n)},$$

and

(4.13)
$$R_n = \int [\xi(x) - s_n(x)]^2\,dF.$$

R_n is minimized by setting

(4.14)
$$a_i = \int \xi(x)x^{(i)}\,dF,$$

an integral that always exists by (4.10) and (4.11) and an application of the Bunyakovsky-Schwarz inequality.

The set $\{x^{(i)}\}$ is said to be *complete* if any one of the three equivalent conditions holds, for arbitrary $\xi(x) \in L_F^2$,

(A) $R_n \to 0$,   as $n \to \infty$

(B) $\underset{n \to \infty}{\mathrm{Lt}} \sum_0^n a_i^2 = \int \xi^2(x)\,dF(x) = a^2$,   (Parseval's equality)

(C) $\int \xi(x)x^{(i)}\,dF(x) = 0$, $i = 0, 1, 2, \ldots$,

implies that

$$\int \xi^2(x)\,dF(x) = 0.$$

For further details on completeness the reader is referred to the texts and to the exercises.

Theorem 4.2. *Complete sets orthonormal on the same distribution of infinitely many points of increase may be transformed into one another by means of an infinite orthogonal matrix.*

Proof. Let the non-constant elements, $y^{(i)}$ and $z^{(i)}$, $i = 1, 2, 3, \ldots$ respectively of complete sets on the same measure space be written as elements of vectors $\mathbf{y}$ and $\mathbf{z}$ and further write $y^{(0)} = z^{(0)} = 1$. Let $\mathbf{A}$ be defined by $a_{ij} = \mathscr{E}y^{(i)}z^{(j)}$, $ij \neq 0$. Each a_{ij} is bounded in absolute value by unity from the

Schwarz inequality. $y^{(i)}$ can be expanded as a Fourier series in the $z^{(j)}$.

(4.15) $$y^{(i)} = \sum a_{ij} z^{(j)}, \quad \text{and} \quad \sum_j a_{ij}^2 = 1$$

by the Parseval equation. Further

(4.16) $$\sum a_{ij} a_{i'j} = \mathscr{E}\left(\sum a_{ij} z^{(j)} \sum a_{i'k} z^{(k)}\right) = \mathscr{E}(y^{(i)} y^{(i')}) = \delta_{ii'}.$$

(4.16) is equivalent to $\mathbf{A}\mathbf{A}^T = \mathbf{1}$. A similar argument gives $\mathbf{A}^T\mathbf{A} = \mathbf{1}$. $\mathbf{A}$ is an orthogonal infinite matrix.

Theorem 4.3. *Let $x^{(0)} = y^{(0)} = 1$ and suppose that $\{x^{(i)}\}$ and $\{y^{(i)}\}$ are complete orthonormal sets on two probability measure spaces. Then the product set $\{x^{(i)} y^{(j)}\} \equiv \{x^{(i)}\} \times \{y^{(j)}\}$ is complete on the product measure.*

Proof. Suppose that the measures are μ_1 and μ_2. Then it is required to prove that if a square summable measurable function, $q(x, y)$, is such that

(4.17) $$\iint q(x, y) x^{(i)} y^{(j)} \, d\mu_1 \, d\mu_2 = 0,$$

$$\text{for} \quad i = 0, 1, 2, 3, \ldots; \quad j = 0, 1, 2, 3, \ldots,$$

then $q(x, y)$ is a null function. Let there be defined

(4.18) $$\int q(x, y) x^{(i)} \, d\mu_2 = q_i(y).$$

By hypothesis,

$$\int q_i(y) y^{(j)} \, d\mu_2 = 0 \quad \text{for} \quad j = 0, 1, 2, \ldots.$$

$q_i(y)$ is thus a null function. However, this implies that for almost all y, $q(x, y)$ is a null function, and hence is zero for almost all x for almost all y. [See p. 34 of Zygmund, 1959.] A more satisfying proof is as follows. Because the property of completeness is transitive, we can assume that the orthonormal sets have been obtained by orthonormalizing indicator variables, χ_A and χ_B, respectively; $A \in \mathscr{A}$ and $B \in \mathscr{B}$, where $\mathscr{A}$ and $\mathscr{B}$ are dense classes of sets. Let us suppose that the partitions of the marginal spaces into A and its complement and into B and its complement are proper in the sense that $\mathscr{E}\chi_A$ and $\mathscr{E}\chi_B$ are neither 1 nor 0. Let ξ and η be orthonormal functions, ξ taking one value in A and a second value in the complement of A and η taking one value in B and a second value in the complement of B. We are given that

$$\int q(x, y) \, d\mu_1 \, d\mu_2 = \int q\xi \, d\mu_1 \, d\mu_2 = \int q\eta \, d\mu_1 \, d\mu_2 = \int q\xi\eta \, d\mu_1 \, d\mu_2.$$

The averaged value of $q(x, y)$ over each of the four quadrants, $A \times B$, $A \times \bar{B}$, $\bar{A} \times B$ and $\bar{A} \times \bar{B}$ is thus zero. But since A and B can be chosen arbitrarily $q(x, y)$ must be zero a.e.

EXERCISES AND COMPLEMENTS

Orthogonal Matrices

1. Let H be an orthogonal matrix in the sense of the definition, $HH^T = 1$; let D be a non-singular real diagonal matrix of the same size as H and such that $D^2 \neq 1$. Prove that DH is orthogonal by rows and HD is orthogonal by columns but neither is an orthogonal matrix as defined. If $a^2 - a \neq 0$, then aH is orthogonal by rows and by columns but is not an orthogonal matrix by the definition used above.

2. Prove the equivalence of $H^T H = 1$ and $HH^T = 1$ for finite square matrices; show that this equivalence does not hold for infinite matrices. [*Hint*. Write $h_{ij} = 1$, if $j = i + 1$, $h_{n1} = 0$ and $h_{ij} = 0$ otherwise.] Give other examples obtained by deleting rows or columns of an infinite H, such that $H^T H = 1$ or $HH^T = 1$.

3. Let b_{ij} be the elements of a real non-singular matrix B of size n. Let $C = B^T B$ and C_k be the leading submatrix of C of size k, $k \leqslant n$, and $|C_k| = C_k$. Show that the orthogonalization process can be represented as giving the following set of vectors

$$\mathbf{m}_1 = C_1^{-\frac{1}{2}}\mathbf{b}_1$$

$$\mathbf{m}_2 = C_1^{-\frac{1}{2}}C_2^{-\frac{1}{2}} \begin{vmatrix} \mathbf{b}_1^T\mathbf{b}_1 & \mathbf{b}_1^T\mathbf{b}_2 \\ \mathbf{b}_1 & \mathbf{b}_2 \end{vmatrix}$$

$$\cdots$$

$$\mathbf{m}_k = C_{k-1}^{-\frac{1}{2}}C_k^{-\frac{1}{2}} \begin{vmatrix} \mathbf{b}_1^T\mathbf{b}_1 & \cdots & \mathbf{b}_1^T\mathbf{b}_k \\ \mathbf{b}_{k-1}^T\mathbf{b}_1 & \cdots & \mathbf{b}_{k-1}^T\mathbf{b}_k \\ \mathbf{b}_1 & \cdots & \mathbf{b}_k \end{vmatrix}$$

where the "determinants" are to be expanded by the terms of the last row.

4. Give a matrix proof of Theorem 1.1 using the notation of Exercise 3 as follows. It is required to find an upper triangular matrix U such that $U^T C U = 1$. It is easily verified that $U_k^T C_k U_k = 1_k$ since $u_{ij} = 0$ if $i > k$, $j \leqslant k$. These results $1 = |1_k| = |U_k^T C_k U_k| = |C_k| \, u_{11}^2 u_{22}^2 \ldots u_{kk}^2$ and consequently $u_{kk}^2 = C_k C_{k-1}^{-1}$, $C_k U_k = (U_k^T)^{-1}$, a lower triangular matrix $U_k = C_k^{-1}U_k^T$ and the kth column of U_k and consequently of U is given by $C_k^{\frac{1}{2}}C_{k-1}^{-\frac{1}{2}}$.

Direct Sum

5. Show how to justify the usual t-test in the comparison of two means (one-way analysis of variance) by arranging the variables x_{ij} in dictionary order as the

elements of a vector $\mathbf{x}$, $i = 1, 2$; $j = 1, 2, \ldots, n_i$. Let $\mathbf{A}_1$ and $\mathbf{A}_2$ be orthogonal matrices having $n_1^{-\frac{1}{2}}$ and $n_2^{-\frac{1}{2}}$ respectively as elements of the first row. Make the transformation $(\mathbf{A}_1 + \mathbf{A}_2)\mathbf{x} = \mathbf{y}$ and then make a second transformation

$$\sqrt{n_1}y_1 + \sqrt{n_2}y_{n_1+1} = \sqrt{(n_1 + n_2)}z_1$$

$$-\sqrt{n_2}y_1 + \sqrt{n_1}y_{n_1+1} = \sqrt{(n_1 + n_2)}z_2$$

$$z_3 = y_2, \ldots, z_{n_1+1} = y_{n_1}$$

$$z_{n_1+i} = y_{n_1+i}, \qquad i > 1.$$

If the X_{ij} are specified by the null hypothesis to have a mean equal to μ and a variance, σ^2 and to be normal and mutually independent, prove that the z_i are mutually independent and each has variance σ^2 but only z_1 has expectation not equal to zero.

Generalize to the one-way analysis of variance.

Direct Product

6. Let x_{ij} and y_{ij} be real numbers $i = 1, 2, \ldots, m$; $j = 1, 2, \ldots, n$ forming matrices, $\mathbf{X}$ and $\mathbf{Y}$. Let the numbers of each set be rearranged in dictionary order as the elements of a vector, $\mathbf{x}$ or $\mathbf{y}$. Prove that the transformations,

$$\mathbf{M}^T\mathbf{X}\mathbf{N} = \mathbf{Y} \quad \text{and} \quad \mathbf{M} \times \mathbf{N}\mathbf{x} = \mathbf{y}$$

are the same. Hence regarding the first of these as two successive transformations namely, $\mathbf{X} \to \mathbf{X}\mathbf{N}$ and $\mathbf{X}\mathbf{N} \to \mathbf{M}^T(\mathbf{X}\mathbf{N})$, use the rules for determining Jacobians to prove (2.7). (Hsu, 1943)

7. Use the transformation,

$$\mathbf{Y} = \mathbf{M}^T\mathbf{X}\mathbf{N},$$

$\mathbf{X}$ and $\mathbf{Y}$ being $m \times n$ matrices, to justify the distribution theory and to obtain the parameters of non-centrality in the two-way analysis of variance, specialising the orthogonal $\mathbf{M}$ and $\mathbf{N}$ to have $m_{i1} = m^{-\frac{1}{2}}$, $n_{i1} = n^{-\frac{1}{2}}$. The alternative hypothesis specifies that

$$X_{ij} = \mu + \alpha_i + \beta_j + \epsilon_{ij},$$

where $\{\epsilon_{ij}\}$ is a set of mutually independent normal $(0, \sigma^2)$ variables and α_i and β_j are certain unknown constants. The null hypothesis specifies $\alpha_i = \beta_j = 0$.

8. Let $\{X_{ij}\}$ be a set of m independent observations on a set of n variables distributed in a normal correlation with unknown means, μ_i, and known covariance matrix, $\mathbf{V}$. Show that a transformation can be made from this set of variables to a fresh set $\{Y_{ij}\}$ by an orthogonal transformation such that only $Y_{11}, Y_{12}, \ldots,$ Y_{1n} have a distribution involving the parameters, μ_i, and such that $\{Y_{ij}\}$ is a set of mutually independent normal variables with variances, σ_j^2. (Aitken, 1949.)

Helmert Matrices

9. Show that the matrix

$$
\mathbf{H_0} = \begin{bmatrix}
1 & 1 & 1 & & 1 \\
1 & -1 & 1 & & 1 \\
1 & 0 & -2 & \cdots & 1 \\
& & & \cdots & \\
1 & 0 & 0 & & 1 \\
1 & 0 & 0 & & -(n-1)
\end{bmatrix}
$$

is orthogonal by columns. If $\mathbf{D^2}$ is a diagonal matrix with elements n, $1 \cdot 2$, $2 \cdot 3$, $\ldots$ *, $(n-1) \cdot n$, show that $\mathbf{H} = \mathbf{H_0 D^{-1}}$ is orthogonal. This is the original matrix of Helmert (1876b). (Helmertian in the strict sense.) (Lancaster, 1965a.)

10. Verify that any Helmert matrix $\mathbf{H}_n$, of size n, can be factored into matrices of planar rotations of the form

$$
\mathbf{R}_{k+1} = \begin{bmatrix}
\cos\theta_k & \mathbf{0}^T & -\sin\theta_k \\
\mathbf{0} & \mathbf{1}_{k-2} & \mathbf{0} \\
\sin\theta_k & \mathbf{0}^T & \cos\theta_k
\end{bmatrix} + \mathbf{1}_{n-k}
$$

$$
\mathbf{H}_n = R_2 R_3 \ldots R_n.
$$

Use this rule to calculate the elements of a Helmert matrix with prescribed first column.

11. Prove that any orthogonal matrix can be exhibited as a product of Helmert matrices of the form,

$$
\mathbf{A} = \mathbf{H}_n(1 \dotplus \mathbf{H}_{n-2})(1_2 \dotplus \mathbf{H}_{n-2}) \ldots (1_{n-2} \dotplus \mathbf{H}_2)
$$

and hence prove that any general orthogonal matrix $\mathbf{A}$ of size n can be factored into the product of $\frac{1}{2}n(n-1)$ matrices, each of which is a rotation in the plane of a pair of coordinate axes. (Schur, 1924; Brauer, 1929.)

This procedure generalizes the Euler factorization of 3×3 matrices.

12. The frequencies in a statistical distribution on three points are proportional to 9030, 8921 and 9023. Calculate a Helmert matrix with elements of the first column proportional to the square roots of these frequencies. Calculate the values of the corresponding orthonormal set of functions on the three points. (Lancaster, 1949a.)

Sets of Orthogonal Functions

13. Let a random variable, X, take n distinct real values, a_j, with positive probabilities, p_j, $\sum_1^n p_j = 1$. Let μ_k be the kth moment of the distribution of X. Define the matrix, $\mathbf{U}$, by $u_{ij} = \mu_{i+j-2}$ and let $\mathbf{U}_k$ be the leading submatrix of size,

*We use the full point as a multiplication sign.

k. Probe that U_k is singular if $k > n$ and positive definite if $k \leqslant n$. Write

$$
A = \begin{bmatrix}
1 & 1 & \cdots & 1 \\
a_1 & a_2 & \cdots & a_n \\
& \cdots & & \cdots \\
a_1^{n-1} & a_2^{n-1} & \cdots & a_n^{n-1}
\end{bmatrix}
$$

Evaluate $|U_n|$ by showing that

$$
U_n = A \operatorname{diag}(p_i) A^T.
$$

[*Hint.* For an arbitrary real vector, t, consider the expectation of

$$
(t_1 + t_2 X + t_3 X^2 + \ldots + t_n X^{n-1})^2.]
$$

14. Show that the elements of a real vector a can be represented as the values of a function from a domain of n points to a range, the real axis; in symbols, $a_i = g(i)$. Show that the ordinary definition of orthogonality, $a^T b = 0$, can be generalized to a statistical distribution on n distinct points by writing, $a^T \operatorname{diag}(p_i) b = 0$ as the definition of orthogonality. Show that the appropriate extension to a general statistical distribution is to define $\int g_i(x) g_j(x)\, dF(x) = (g_i, g_j)$, provided that (g_i, g_i) and (g_j, g_j) are each finite; the orthogonality condition is then $(g_i, g_j) = 0$ if $i \neq j$.

15. Exercise 14 enables the process of orthogonalization to be carried out on any sequence of square summable linearly independent functions defined on a statistical distribution. It is convenient in statistics to take $g_0(x) \equiv 1$, if $\{g_i(x)\}$ is to be the set of orthogonal functions. Show that the generalization of Exercise 13 is then to define

$$
C = \begin{bmatrix}
1 & (1, g_1) & \cdots & (1, g_k) & \cdots \\
(g_1, 1) & (g_1, g_1) & \cdots & (g_1, g_k) & \cdots \\
& \cdots & & \cdots & \cdots \\
(g_k, 1) & (g_k, g_1) & \cdots & (g_k, g_k) & \cdots
\end{bmatrix}
$$

(Szegö, 1949).

Completeness of Orthonormal Sets

16. Let $x^{(0)} = 1$, and $\{x^{(i)}\}$, $i = 0, 1, 2, \ldots$ be an orthonormal set of functions defined with respect to a probability measure, μ. Let $f(x)$ be an arbitrary function in L^2, so that

$$
\int f^2\, d\mu
$$

is finite. Then prove that the following three conditions are equivalent:
(i) for every square summable f,

$$
\int f^2\, d\mu = \sum_0^\infty a_i^2, \quad \text{where } a_i = \int f x^{(i)}\, d\mu.
$$

(ii) f can be arbitrarily closely approximated in mean square by its Fourier series,
$\sum_0^n a_i x^{(i)}$.

(iii) $\int f x^{(i)} \, d\mu = 0$ for $i = 0, 1, 2, \ldots$ implies that f is a null function. (Alexits, 1961; Szegö, 1949; Tricomi, 1955.)

17. Prove that if an orthonormal set $\{\phi\}$ is complete with respect to two functions, $g(x)$ and $h(x)$, it is complete with respect to any linear combination of them. (Tricomi, 1955.)

18. Prove that if an orthonormal $\{\phi\}$ is complete with respect to every function of an orthonormal $\{\psi\}$ and if $\{\psi\}$ is complete with respect to an orthonormal set or class $\{\zeta\}$, then $\{\phi\}$ is complete with respect to $\{\zeta\}$. In other words the property of completeness is transitive. (Tricomi, 1955.)

19. A necessary and sufficient condition (NASC) for the completeness of an orthonormal set $\{\phi\}$ in L^2 is the completeness of $\{\phi\}$ with respect to any other complete orthonormal set. (Lauricella, 1912.)

20. A NASC for the completeness of an orthonormal set $\{\phi\}$ on the rectangular distribution on the unit interval is that

$$\sum_{i=0}^{\infty} \left(\int_0^{\xi} \phi_i(x) \, dx \right)^2 = \xi, \qquad 0 \leqslant \xi \leqslant 1.$$

(In other words, Parseval's equality holds for every step function equal to 1 on $[0, \xi]$ and zero on $(\xi, 1]$). (Vitali, 1921.)

21. Prove the equivalence of the condition in Exercise 20 with

$$\sum_{i=0}^{\infty} \int_0^1 \left[\int_0^{\xi} \phi_i(x) \, dx \right]^2 d\xi = \tfrac{1}{2}. \qquad \text{(Dalzell, 1945).}$$

Existence of Complete Orthonormal Sets

22. Let Z take n distinct values with non-zero probabilities, p_j. Show that an orthonormal set can be formed on the space of Z. Now let a new space be formed by randomizing the event, $Z = z_k$, for a certain k; the new r.v., Z^*, now takes $(n + 1)$ values with probabilities as did Z except that the two events, $Z = z_{k'}$ and $Z = z_{k''}$ with probabilities, θp_k and $(1 - \theta)p_k, 0 < \theta < 1$, replace the event $Z = z_k$. Show that an orthonormal set can be formed on the new distribution which consists of the previous set, of which each function takes the same value on $z_{k'}$ and $z_{k''}$ as it did on z_k and an additional function, which is zero if $Z^* \neq z_{k'}$ or $z_{k''}$. Hence prove that an orthonormal set can always be defined on any space of a denumerable number of points.

23. Use criterion (i) of Exercise 16 to prove that an orthonormal set $\{z^{(i)}\}$ is complete on a discrete probability measure if at every point, for which $p_j > 0$,
$$p_j^{-1} = \sum_0^{\infty} z_j^{(i)2},$$ where $z_j^{(i)}$ is the value of the ith function at the jth point.

24. Let a r.v., Z, take values $z_1, z_2, \ldots$ with probabilities, $p_1, p_2, \ldots, \sum p_j = 1$. Form a Helmert matrix corresponding to these probabilities and hence derive a set of orthogonal functions $\{z^{(i)}\}$ on the probability distribution. If $z_j^{(i)}$ is the value of the ith function at the point z_j, prove that $\sum z_j^{(i)2} = p_j^{-1}$, where the summation is over the index, i. Hence prove that the function, ζ_j, which is equal to 1 if $Z = z_j$ and zero otherwise, can be approximated arbitrarily closely by a certain linear form in the functions, $\{z^{(i)}\}$; $\{z^{(i)}\}$ is therefore a complete orthonormal set.

25. Prove that $\{z^{(i)}\}$ cannot be a complete orthonormal set on any distribution for which $p_j > 0, p_{j'} > 0$ unless $z_j^{(i)} z_{j'}^{(i)}$ is negative for at least one i. [*Hint.* Consider the functions χ_j and $\chi_{j'}$ which are equal to 1 at the points, j and j', respectively and zero elsewhere. $\mathscr{E}\chi_j\chi_{j'} = 0$.]

26. Let a space of a random variable Z be partitioned into sets, A_i, and suppose $\mu(A_i) > 0$ for $i = 1, 2, \ldots, n$. $\sum_1^n \mu(A_i) = 1$. Suppose $\{z^{(i)}\}$ is complete on this probability measure space or as is usually stated $\{z^{(i)}\}$ is an orthonormal basis. Now suppose an arbitrary set, A, is partitioned into sets, B and C, and measure is assigned so that $\mu(B) + \mu(C) = \mu(A)$. Define a new function

$$z^{(n)} = -\sqrt{\frac{\mu(C)}{\mu(B)\mu(A)}} \quad \text{when} \quad Z \in B$$

$$z^{(n)} = +\sqrt{\frac{\mu(B)}{\mu(C)\mu(A)}} \quad \text{when} \quad Z \in C$$

$$z^{(n)} = 0 \quad \text{otherwise.}$$

Prove that $z^{(0)}, z^{(1)}, \ldots, z^{(n)}$ is an orthonormal basis on the $(n + 1)$ points of the new partition. Show by induction that there is always an orthonormal basis or a complete orthonormal set on any statistical distribution. (This process is used in obtaining the Haar series and less evidently in obtaining the Walsh series.)

27. Show how a complete orthonormal set can be defined on a general statistical distribution using the results of the preceding exercises and the decomposition of a general distribution function into its discrete, singular and absolutely continuous components.

Some Special Complete Orthonormal Sets

28. The Haar orthonormal system is defined as follows

$$\chi_0^{(0)}(x) \equiv 1, \qquad \chi_0^{(1)}(x) = \begin{cases} 1, & x \in [0, \tfrac{1}{2}) \\ 0, & x = \tfrac{1}{2} \\ -1, & x \in (\tfrac{1}{2}, 1] \end{cases}$$

$$\chi_m^{(k)}(x) = \begin{cases} 2^{\frac{1}{2}m}, & x \in ((k-1)2^{-m}, (k-\tfrac{1}{2})2^{-m}) \\ -2^{\frac{1}{2}m}, & x \in ((k-\tfrac{1}{2})2^{-m}, k2^{-m}) \\ 0, & x \in ((l-1)2^{-m}, l2^{-m}), \quad l \neq k, \end{cases}$$

$$1 \leqslant k \leqslant 2^m; \qquad 1 \leqslant l \leqslant 2^m.$$

At points of discontinuity set $\chi_m^{(k)}(x)$ equal to the arithmetic mean of the values on the two sides of the discontinuity.

Prove that $\{\chi_m^{(k)}\}$ is an orthonormal set. Note the resemblance of the procedure of definition to that of Exercise 22 above. Prove that the Haar system is complete on the unit interval. (Haar, 1910; Alexits, 1961.)

29. Define the Rademacher system by $r_0(x) = 1$,

$$r_n(x) = \text{sign sin } (2^n \pi x).$$

Prove that the set is orthonormal, but not complete, on the rectangular (0, 1) distribution. (Rademacher, 1922; Alexits, 1961.)

30. Define the Walsh functions as follows $w_0(x) \equiv 1$. If

$$2n = 2^{n_1} + 2^{n_2} + 2^{n_3} + \ldots + 2^{n_p},$$

$$n_1 < n_2 < n_3 < \ldots < n_p, \quad \text{set}$$

$$w_n(x) = r_{n_1}(x) r_{n_2}(x) \ldots r_{n_p}(x)$$

where the $r_k(x)$ are the Rademacher functions. Prove that $\{w_n(x)\}$ is orthonormal. Prove that any step function equal to unity in the interval $[0, k2^{-m}]$ and zero in the interval $(k2^{-m}, 1]$ is identically equal to a certain sum in the Walsh functions of order 2^m or less. Hence prove that every step function, taking the values 1 in $[0, \xi)$ and zero elsewhere, can be approximated arbitrarily closely by the Walsh functions. Hence prove that the Walsh system is complete on $[0, 1]$ with respect to the rectangular distribution. (Walsh, 1923; Kaczmarz, 1929; Kac, 1959; Alexits, 1961.)

Trigonometric Series

31. Show that the trigonometric series,

$$1, \quad \sqrt{2} \sin (2\pi m x), \quad \sqrt{2} \cos (2\pi m x), \quad m = 1, 2, \ldots,$$

is complete with respect to the functions of L^2 by the use of Dalzell's condition, the functions being defined on the unit interval. See Exercise 21. (Dalzell, 1945.)

Orthogonal Polynomials

32. Use the density function

$$B(\tfrac{1}{2}, m - \tfrac{1}{2})(1 + x^2)^{-m}, \quad -\infty < x < \infty,$$

to show that it may be possible to define an orthogonal polynomial of precise degree, k, but not to define one of degree, $k + 1$. In particular, the Cauchy distribution has no orthogonal polynomials of any order.

33. If a distribution possesses finite moments of all orders, prove that it is possible to define orthogonal polynomials of arbitrarily high order. The matrix, C, of Exercise 15 then becomes (μ_{i+j-2}), where the moments are measured either about the mean or the origin.

In the following exercises, the existence of all moments and hence of all orthogonal polynomials is assumed. Unless otherwise obvious it is also assumed that the distribution is not wholly concentrated on a finite set of points. Many of the following results are classical.

34. If the convention is adopted that the leading coefficient of the orthonormal polynomials is positive, then the set of orthonormal polynomials, $\{\phi_n(x)\}$, on a given distribution is unique. [If $\phi_n(x) = \alpha_n x^n + \alpha_{n-1} x^{n-1} \ldots$; we have assumed $\alpha_n > 0$.]

35. Among all the polynomials, $\pi_n(x)$ of degree not greater than n, the integral,

$$\int (f(x) - \pi_n(x))^2 \, dF(x),$$

attains its minimum value for

$$\pi_n(x) = \sum_0^n a_i \phi_i(x), \qquad \text{where } a_i = \int f(x)\phi_i(x) \, dF(x).$$

[Orthogonal polynomials owe their importance in statistics to such applications in the theory of least squares.]

36. Among all the polynomials of degree n with leading coefficient equal to unity, the integral,

$$\int \pi_n^2(x) \, dF(x),$$

attains its minimum for $\pi_n(x) = \alpha_n^{-1}\phi_n(x)$.

37. The exact upper limit of the values assumed at a point, x_0, by the square $\pi_n^2(x)$ of a polynomial of degree not greater than n and satisfying

$$\int \pi_n^2(x) \, dF(x) \leqslant 1$$

is $\sum_{k=0}^n \phi_k^2(x_0)$.

38. Prove the recurrence relation,

$$\frac{\alpha_k}{\alpha_{k+1}} \phi_{k+1}(x) = (x - \gamma_k)\phi_k(x) - \frac{\alpha_{k-1}}{\alpha_k} \phi_{k-1}(x),$$

where α_j is the coefficient of the leading term of $\phi_j(x)$.

39. If the distribution is symmetric, then $\phi_n(-x) = (-1)^n \phi_n(x)$, so that $\phi_n(x)$ contains only powers of x, which are congruent to n (mod 2).

40. Assume the formula

$$\int_0^\infty y^{c-1} e^{-by} \, dy = b^{-c}\Gamma(c), \qquad c > 0, \qquad b = k + il, \qquad k > 0.$$

Put $c = (n+1)/\lambda$, $n = 0, 1, 2, \ldots$; $l/k = \tan \frac{1}{2}\alpha\pi$, $0 < \alpha < \frac{1}{2}$, $y = x^\alpha$.

Hence obtain

$$\int_0^\infty x^n \exp{(-kx^\alpha)} \cos{(kx^\alpha \tan{\tfrac{1}{2}\alpha\pi})}\, dx = 0. \qquad n = 0, 1, 2, \ldots$$

However, $\cos{(kx^\alpha \tan{\tfrac{1}{2}\alpha\pi})}$ is square summable and not a null function, so that the powers of x and hence the orthogonal polynomials are not complete on the distribution with frequency function proportional to $\exp{(-kx^\alpha)}$. (Stieltjes, 1894–5; Shohat and Tamarkin, 1943.)

41. Let X be normal $N(0, 1)$. Then $\mathscr{E}[\exp{(Xr)}\sin{(2\pi X)}] = 0$, for $r = 0, 1, 2, \ldots$. However, $\sin{(2\pi X)}$ is not a null function so the system of functions $\exp{(Xr)}$ is not complete on the normal distribution. Suppose now that $Y = \exp X$, so that $X = \log Y$. Then on the distribution of Y, $\mathscr{E}[Y^r \sin{(2\pi \log Y)}] = 0$, $r = 0, 1, 2, \ldots$. So that the set of powers of Y and hence the set of orthogonal polynomials on the distribution of Y is not complete. The frequency function of Y is given by $g(y)\, dy = (2\pi)^{-\frac{1}{2}} y^{-\frac{1}{2}\log y - 1}\, dy$, $0 < y < \infty$. (Stieltjes, 1894–5; Shohat and Tamarkin, 1943.)

Legendre Polynomials

42. Verify that the expansion of $(1 - 2zh + h^2)^{-\frac{1}{2}}$ generates a set of polynomials in z, the Legendre polynomials, $P_n(z)$ and that

$$P_0(z) = 1; \qquad P_1(z) = z; \qquad P_2(z) = \tfrac{1}{2}(3z^2 - 1);$$
$$P_3(z) = \tfrac{1}{2}(5z^3 - 3z); \qquad P_4(z) = \tfrac{1}{8}(35z^4 - 30z^2 + 3).$$

43. By setting $z = \pm 1$ in the generating function verify that $P_n(1) = 1$ and $P_n(-1) = (-1)^n$. By setting $z = 0$, verify that $P_{2n+1}(0) = 0$,

$$P_{2n}(0) = (-1)^n \binom{2n}{n} 2^{-2n}.$$

44. Assuming Rodrigues' formula,

$$P_n(z) = \frac{1}{2^n n!} \frac{d^n}{dz^n} (z^2 - 1)^n,$$

prove that $\int_{-1}^1 P_m(x) P_n(x)\, dx = 2\delta_{mn}(2n + 1)^{-1}$.

45. Prove that, if Y is a rectangular variable on the unit interval, the set $\{y^{(i)}\}$, $y^{(i)} = \sqrt{(2n + 1)} P_n(2Y - 1)$ is an orthonormal set. Prove that the set is complete on the distribution with respect to the functions of L^2 by the use of Dalzell's condition. (Dalzell, 1945.)

46. Prove the completeness of the set $\{y^{(i)}\}$ of Exercise 45 with respect to continuous functions on the unit interval by the use of the Weierstrass theorem, or alternatively by the use of the Bernstein polynomials. (Alexits, 1961; Achieser, 1956; Meixner, 1938.)

47. Verify from the definition of Exercise 42 that

$$P_n(z) = \sum_{r=0}^{m} (-1)^r \frac{(2n - 2r)!}{2^n r! \, (n - r)! \, (n - 2r)!} z^{n-2r},$$

where m is the integral part of $\frac{1}{2}n$. (Whittaker and Watson, 1944.)

48. Suppose Y has a singular distribution function $G(y)$. Define a new random variable $U(Y) = G(Y)$. Then U has the rectangular distribution on the unit interval. Hence deduce that since the Legendre polynomials form a complete set of orthogonal functions on the unit interval, there is also at least one complete set on the distribution of Y.

49. Let a series of polynomials be defined by

$$t_n(x) = n! \, \Delta^n \binom{x}{n} \binom{x - N}{n}.$$

Prove that $\{t_n(x)\}$ is orthogonal on the points, $0, 1, 2, \ldots, (N - 1)$, to each of which is assigned measure N^{-1}. Use the mean value theorem

$$\Delta^n f(x) = f^{(n)}(x + \theta_n), \qquad |\theta| < 1,$$

where $f^{(n)}$ is the derivative of nth order, to prove that

$$\lim_{N \to \infty} N^{-n} t_n(Nx) = P_n(2x - 1).$$

(Chebyshef, 1864; Szegö, 1949.)

50. Derive the result of Exercise 49 by showing that for a fixed n every moment of the discrete variable $N^{-1}X$ up to order $2n$, can be approximated arbitrarily closely by those of the continuous variable, Y, on the unit interval.

51. Verify that the orthogonal polynomials of Exercise 49 are those tabulated by Fisher and Yates (1938 etc.).

Laguerre Polynomials

52. The Laguerre polynomials may be defined (i) by the generating function,

$$(1 - w)^{-\alpha-1} \exp \left(-xw/[1 - w]\right),$$

(ii) by being the unique set of orthogonal polynomials on the gamma distribution with parameter $(\alpha + 1)$, so that

$$\frac{1}{\Gamma(\alpha + 1)} \int_0^\infty e^{-x} x^\alpha L_n^{(\alpha)}(x) L_m^{(\alpha)}(x) \, \mathrm{d}x = \delta_{nm} \binom{n + \alpha}{n}$$

and (iii) by the Rodrigues formula,

$$e^{-x} x^\alpha L_n^{(\alpha)}(x) = \frac{1}{n!} \left(\frac{\mathrm{d}}{\mathrm{d}x}\right)^n (e^{-x} x^{n+\alpha}).$$

Prove that these definitions are equivalent. (Szegö, 1959.)

53. Verify that the first few Laguerre polynomials are given by $L_0^{(\alpha)}(x) = 1$, $L_1^{(\alpha)} = -x + \alpha + 1$, $L_2^{(\alpha)}(x) = \frac{1}{2}x^2 - (\alpha + 2)x + \frac{1}{2}(\alpha + 1)(\alpha + 2)$.

54. Prove that

$$\int_0^\infty e^{-x} x^\alpha L_m^{(\alpha)}(x) L_n^{(\alpha)}(x) \, dx = \begin{cases} 0 & \text{if } m \neq n \\ \Gamma(\alpha + n + 1)/\Gamma(n + 1) & \text{if } m = n \end{cases}$$

by a consideration of the generating function in (i) of Exercise 52.

55. Prove that the Laguerre polynomials $\{L_n^{(\alpha)}(x)\}$ are complete on the gamma distribution with parameter, $\alpha + 1$. (Dalzell, 1945; Szegö, 1959.)

Hermite or Hermite-Chebyshev Polynomials

56. Prove that $\exp(tx - \frac{1}{2}t^2)$ generates a series of orthogonal polynomials $\{H_n(x)\}$ on the normal $N(0, 1)$ distribution by considering the expectation of $\exp(tx - \frac{1}{2}t^2) \exp(ux - \frac{1}{2}u^2)$. By integrating $\exp(tx - \frac{1}{2}t^2) \exp(wy - \frac{1}{2}w^2)$ over the bivariate normal distribution determine the expectation of $H_m(X)H_n(Y)$. [$\exp(tx - \frac{1}{2}t^2) = 1 + tH_1(x)/1! + t^2H_2(x)/2!\ldots$, where the $H_m(x)$ are the Hermite polynomials.] Verify that $H_0(x) = 1$, $H_1(x) = 1$, $H_2(x) = x^2 - 1$, $H_3(x) = x^3 - 3x$, $H_4(x) = x^4 - 6x^2 + 3$.

57. Prove Runge's identity for the Hermite series, namely $2^{\frac{1}{2}n} H_n(2^{-\frac{1}{2}}x + 2^{-\frac{1}{2}}y) = \sum_0^n \binom{n}{i} H_i(x) H_{n-i}(y)$ by a consideration of the generating function given in Exercise 56. (Runge, 1914.)

58. Generalize Runge's identity—If W_1 and W_2 are independent and additive variables whose orthogonal polynomials are generated by a function of the form $f(t) \exp[w_i u(t)]$ and if $X = W_1 + W_2$, then

$$P_n(x) = \sum_{i=0}^n \binom{n}{i} P_i(w_1) P_{n-i}(w_2.)$$

("Additive" means that $u(t)$ is the same for W_1, W_2 and X.) (Eagleson, 1964.)

59. Prove that the Hermite polynomials are complete on the normal distribution with respect to the functions of L^2 by the following methods.

(i) By the completeness of the Laguerre polynomial series. (Szegö, 1959.)
(ii) By noting that the moment problem is determined for the normal distribution. (Shohat and Tamarkin, 1943.)
(iii) By an application of a theorem of Picone, which yields a proof for all ortho-normal sets generated by a function of the form, $f(t) \exp[xu(t)]$. (Eagleson, 1964.)

60. Prove that the Hermite polynomials may be defined by

$$\exp\left(-\frac{1}{2}x^2\right) H_n(x) = (-1)^n \left(\frac{d}{dx}\right)^n \exp\left(-\frac{1}{2}x^2\right).$$

(Szegö, 1949.)

61. Prove that the Hermite polynomials can be expressed in terms of the Laguerre polynomials,

$$H_{2m}(2^{-\frac{1}{2}}x) = (-1)^m 2^{2m} m! \, L_m^{(-\frac{1}{2})}(\tfrac{1}{2}x^2)$$
$$H_{2m+1}(2^{-\frac{1}{2}}x) = (-1)^m 2^{2m-1} m! \, x L_m^{(\frac{1}{2})}(\tfrac{1}{2}x^2).$$

(Szegö, 1949.)

The Kravčuk Polynomials

62. Show that $(1 + qw)^x (1 - pw)^{n-x}$ generates a set of orthogonal polynomials (Krawtchouk) on the binomial distribution with parameters, p, q and n. (Krawtchouk, 1929.)

63. Let $g = -p^{\frac{1}{2}}q^{-\frac{1}{2}}$, $h = p^{-\frac{1}{2}}q^{\frac{1}{2}}$. Prove that the function taking the values g when $X = 0$ and h when $X = 1$ is orthonormal on the distribution of X, where $\mathscr{P}(X = 0) = q$, $\mathscr{P}(X = 1) = p$. Define

$$\mathbf{G} = \begin{bmatrix} 1 & g \\ 1 & h \end{bmatrix}, \qquad \mathbf{D} = \operatorname{diag}(q^{\frac{1}{2}}, p^{\frac{1}{2}}).$$

Show that $\mathbf{GD}$ is orthogonal. Define $\mathbf{H} = \mathbf{G} \times \mathbf{G} \times \ldots \times \mathbf{G}$, ($n$ terms). Then the columns of $\mathbf{H}$ define an orthonormal set on the product space of n independent random variables. Suppose now that the rows and columns of $\mathbf{G}$ are numbered $(0, 1)$. Let A_k be the set of all integers $0, 1, 2, \ldots, 2^n - 1$ which possess precisely k units in their binary expansion. Prove that $P_s(k) = \binom{n}{s}^{-\frac{1}{2}} \sum\limits_{j \in A_s} h_{ij}$ is the same for every $i \in A_k$. Hence $P_0(k), P_1(k), \ldots, P_n(k)$ forms an orthonormal set on the sets A_k, i.e. the points of a binomial variable, $Y = \sum X_i$, where each X_i has the distribution of X above. Hence obtain

$$P_s(k) = \binom{n}{s}^{-\frac{1}{2}} \sum \binom{k}{t} \binom{n-k}{s-t} h^t g^{s-t},$$

with summation over the permissible values of t, namely $\max(0, s + k - n) \leqslant t \leqslant \min(k, s)$. Verify that the coefficient of k^s in $P_s(k)$ is not identically zero and so the set $\{P_s(k)\}$ are the normalized Kravčuk polynomials. (Bahadur, 1961; Lancaster, 1965b.)

The Poisson–Charlier Polynomials

64. Show that $e^{-w}(1 + \lambda^{-1}w)^x = G(w)$ generates a system of orthogonal polynomials (Poisson–Charlier) on the Poisson distribution with parameter, λ.

65. In Exercise 63 let $n \to \infty$ in such a way that $np = \lambda$. Show that the orthonormal polynomial becomes

$$P_s(k) = \lambda^{\frac{1}{2}s}(s!)^{-\frac{1}{2}} \sum (-1)^{s-t} \binom{k}{t} t! \, \lambda^{-t} \binom{s}{t},$$

the Poisson–Charlier polynomial or order, s.

66. Write $\psi_s(k) = (-1)^s \lambda^{-\frac{1}{2}s}(s!)^{\frac{1}{2}}P_s(k)$. Verify that $\psi_s(k) = \psi_k(s)$ for $k = 0, 1, 2, \ldots$. (This is the dual property) (Meixner, 1938; Eagleson, 1967).

67. Use the identity of Exercise 66 to prove the completeness of the Poisson–Charlier polynomials on the Poisson distribution.

Miscellaneous

68. Let $S = \sum\limits_{0}^{\infty} c_n \phi_n(x)$, where $\{\phi_n(x)\}$ is a general orthonormal series on a statistical distribution, taken without loss of generality to be defined on $(0, 1)$. Then, if $\sum\limits_{0}^{\infty} |c_n| < \infty$, S converges absolutely almost everywhere on $(0, 1)$.

[*Hint*.

$$\sum_{n=0}^{\infty} \int_0^1 |c_n \phi_n(x)| \, d\mu(x) \leqslant \sum |c_n| \int_0^1 |\phi_n(x)| \, d\mu(x) \leqslant \sum |c_n| \left[\int \phi_n^2(x) \, dF \right]^{\frac{1}{2}} = \sum |c_n|.]$$

(See Alexits 1961, page 63.)

69. Derive the generating function of Exercise 64 from that in Exercise 62 by a limiting process.

70. A NASC for the completeness of a set $\{\phi\}$, orthonormal w.r.t. the measure $\mu(x)$, is that for every real x

$$\sum_{n=1}^{\infty} \left[\int e^{itx} \phi_n(t) \, d\mu(t) \right]^2 = 1.$$

(Picone, 1934.)

71. Let $\{x^{(i)}\}$ be complete on the distribution of X and $\{y^{(j)}\}$ be complete on the distribution of Y. Let A and B be measurable sets on the distributions of X and Y, respectively. Let ψ_A and ψ_B be the indicator variables of A and B, respectively. Show that the Parseval equation holds for the product $\psi_A \psi_B$ with respect to the set $\{x^{(i)} y^{(j)}\}$, $i = 0, 1, 2, \ldots$, and $j = 0, 1, 2, \ldots$. Hence prove that this product set is complete with respect to the indicator variables of rectangles in the product space. Hence prove that the set $\{x^{(i)} y^{(j)}\}$ is complete with respect to the measurable functions with finite variance on the product space.

CHAPTER V

The Multinomial Distribution

1. INTRODUCTORY

The normal approximation to the multinomial distribution can be justified in a number of ways, each of which can be shown to be a generalization of the more familiar theory of the normal approximation to the binomial distribution. For example, the expression,

$$(1.1) \qquad \chi^2 = \sum (\text{observed-expected})^2/(\text{expected}),$$

of Karl Pearson (1900a) is a generalization of

$$(1.2) \quad X^2 = (m - Np)^2/(Npq) = (m - Np)^2/(Np) + (n - m - Nq)^2/(Nq),$$

where X is the standardized variable of the binomial distribution, $(p + q)^N$. X can also be represented as the standardized sum of N independently distributed variables assuming two values with probabilities, p and q; $p + q = 1$. These variables obey the conditions necessary for the validity of the central limit theorem and so $X = (m - Np)/(Npq)^{\frac{1}{2}}$ is asymptotically standard normal. This form of argument was generalized by Pearson (1900a), to obtain an asymptotically normal multivariate distribution.

Let Z be a random variable taking distinct values $z_1, z_2, \ldots, z_n$ with probabilities $p_1, p_2, \ldots, p_n$ such that $p_j > 0$ and $\sum p_j = 1$. Then if N stochastically independent observations are made on Z, there is a probability that Z assumes the value z_i on a_i occasions, $i = 1, 2, \ldots, n$, equal to

$$(1.3) \qquad \mathscr{P}(\{a_i\} \mid N, \{p_i\}) = N! \prod_i (p_i^{a_i}/a_i!).$$

In other words, a_i is the frequency in the ith class of the multinomial distribution and the expression (1.3) is the general term of the multinomial expansion

$$(1.4) \qquad (p_1 + p_2 + \ldots + p_n)^N,$$

and $\sum a_i = a_{.} = N$. The result (1.3) or (1.4) is well known. Note that since Z can take one of n distinct values at any observation, the sample space for the N observations contains n^N points. However, only the number of times of occurrence of $Z = z_i$ is noted, so that the multinomial sample space consists of the compositions of N into n parts, zero parts being permitted. There are then $\binom{N + n - 1}{N}$ sample points. This is a generalization from the binomial distribution where there are 2^N points in the sample space but, if the order in which successes and failures occur is neglected and only the number of successes (i.e. $Z = z_1$) or failures (i.e. $Z = z_2$) counted, there are $\binom{N + 2 - 1}{N}$ or $(N + 1)$ sample points.

The following conventions are used. N independent observations on Z generate a set of nN variables, $Y_j^{(i)}$, consisting of N independently distributed n-sets or vectors of observations, $\{Y_j^{(1)}, Y_j^{(2)}, \ldots, Y_j^{(n)}\} \equiv \mathbf{Y}_j$, arising from a common distribution, that of the set $\{Y^{(1)}, Y^{(2)}, \ldots, Y^{(n)}\} \equiv \mathbf{Y}$. The subscript, j, denotes at the jth observation. Summation over the subscript index will be denoted by the operator, $\mathscr{S}$, and written in the form,

$$(1.5) \qquad \mathscr{S}Y^{(i)} = \mathscr{S}Y_j^{(i)} = Y_1^{(i)} + Y_2^{(i)} + \ldots + Y_N^{(i)}.$$

Summation over the superscript index will be denoted by the operator, $\sum$,

$$(1.6) \qquad \sum b_i Y_j^{(i)} = b_1 Y_j^{(1)} + b_2 Y_j^{(2)} + \ldots + b_n Y_j^{(n)} = \mathbf{b}^T \mathbf{Y}_j.$$

In the rest of this chapter, various proofs of the asymptotic distribution of the Pearson χ^2 are given. The most satisfactory proofs are based on the multivariate form of the central limit theorem due to Bernstein (1926) and such a proof is given in Section 2. This approach has the advantage that the distribution of some analogues of the Pearson χ^2 can be derived in the same manner; further, the assumption latent in Pearson's proof, that variables individually normal are jointly normal, is thereby avoided. In Section 3, a simplified version of Pearson's proof is given. The succeeding sections are devoted to other methods of proof, which have their interest in some applied and historical contexts. In Section 4, Stirling's approximation is used; in Section 5, a supernumerary random variable is introduced which leads to the consideration of certain Poisson variables; in Section 6, the distribution is obtained by factorization of the expression (1.3); in Section 7, the moments of the Pearson χ^2 distribution are shown to be asymptotically equal to the moments of the theoretical χ^2 distribution. After these various proofs, the methods of Section 2 are applied in Section 8 to determine the asymptotic distribution of some analogues of the Pearson χ^2, of which the best known examples arise in Neyman's smooth test of goodness of fit. In Section 9, empirical verifications of the asymptotic theory are examined to see whether the theory or tests need adjustment for moderate values of N. In Section 10 are given examples of the use of χ^2 in the multinomial distribution with parameters specified by hypothesis. This section supplements material given later in Chapter VIII. It is hoped that this early introduction will help to explain the use of χ^2 without introducing problems of estimation.

2. THE MULTIVARIATE CENTRAL LIMIT THEOREM

Lemma 2.1. *Let* $\{Y^{(1)}, Y^{(2)}, \ldots, Y^{(n)}\}$ *be a set of random variables, functions of a random variable, Z, having zero expectation and a covariance matrix, $\mathbf{V}$. Suppose that N independent observations are made on Z, then the standardized sums,*

$$(2.1) \quad X^{(i)} = N^{-\frac{1}{2}} \mathscr{S}_j Y_j^{(i)} \equiv N^{-\frac{1}{2}} \mathscr{S} Y^{(i)}, \quad i = 1, 2, \ldots, n; \ j = 1, 2, \ldots, N,$$

have the same covariance matrix as the variables, $Y^{(1)}, Y^{(2)}, \ldots, Y^{(n)}$.

Proof. An arbitrary linear form, $L = \sum b_i Y^{(i)}$ in the variables, has a finite variance, $\mathbf{b}^T \mathbf{V} \mathbf{b}$ and zero expectation. Now the standardized sum, $N^{-\frac{1}{2}} \mathscr{S} L$, has zero expectation and the same variance as its components for

$$\mathscr{E}(N^{-\frac{1}{2}} \mathscr{S} L)^2 = N^{-1} \mathscr{E}(L_1 + L_2 + \ldots + L_N)^2,$$

$$N^{-1} \mathscr{S}(\mathscr{E} L_j^2 + \mathscr{E}(L_j L_{j'})) = \mathscr{E} L^2 = \mathbf{b}^T \mathbf{V} \mathbf{b},$$

since $\mathscr{E}(L_j L_{j'}) = \mathscr{E} L_j \mathscr{E} L_{j'} = 0$, the L's being mutually independent. We may now give arbitrary values to the elements of $\mathbf{b}$. Setting $b_i = 1$ and all other $b_k = 0$, var $X^{(i)} = v_{ii}$. Setting $b_i = b_m = 1$ and other $b_k = 0$, we have that $\mathscr{E}(X^{(i)} + X^{(m)})^2 = v_{ii} + v_{mm} + 2v_{im}$ and so $\mathscr{E}(X^{(i)} X^{(m)}) = v_{im}$. The result of this lemma can be written concisely as

$$(2.2) \qquad\qquad \mathscr{E} \mathbf{X} \mathbf{X}^T = \mathscr{E} \mathbf{Y} \mathbf{Y}^T = \mathbf{V}.$$

Lemma 2.2. *Under the conditions of Lemma 2.1, the covariance matrix of variables,* $\mathbf{MX}$*, is* $\mathbf{M} \mathbf{V} \mathbf{M}^T$.

Proof. $\mathscr{E}(\mathbf{MX})(\mathbf{MX})^T = \mathscr{E} \mathbf{M}(\mathbf{X} \mathbf{X}^T) \mathbf{M}^T = \mathbf{M} \mathscr{E}(\mathbf{X} \mathbf{X}^T) \mathbf{M}^T = \mathbf{M} \mathbf{V} \mathbf{M}^T$.

Lemma 2.3. *Let* $W_1, W_2, \ldots, W_n$ *be a set of random variables such that every linear form in them is normally distributed. Then the distribution is jointly normal, in the sense that if there is an n'-subset of them, which has a positive definite covariance matrix, $\mathbf{V}$, then the joint frequency function of the variables of the n'-subset can be written in the form,*

$$(2.3) \qquad\qquad f(\mathbf{w}) = C \exp\left(-\tfrac{1}{2} \mathbf{w}^T \mathbf{V}^{-1} \mathbf{w}\right),$$

where C is a constant.

Proof. $\mathbf{b}^T \mathbf{W}$ is normal for arbitrary real $\mathbf{b}$. Put $b_i = 1$ and $b_k = 0$ if $i \neq k$; W_i is therefore normal. The means are thus all finite and each may be taken without loss of generality to be zero. Further the variances are finite. Suppose that the covariance matrix has rank $n' < n$. It is now possible to take new variables $U_1, U_2, \ldots, U_{n'}$, defined by

(2.4) $$U = AW, \quad \text{where} \quad A \text{ is } n' \times n,$$

such that

(2.5) $$\mathscr{E} UU^T = A\mathscr{E} WW^T A^T = 1_{n'},$$

by a suitable choice of A. Every form linear in the W_i is linear in the U_i and conversely. For an arbitrary vector c, $c^T U$ is normal by hypothesis.

(2.6) $$\mathscr{E} \exp(tc^T U) = \exp(\tfrac{1}{2}t^2 B(c)),$$

where $B(c) > 0$ if c is not a null vector. But $B(c)$ is the variance of the linear form and hence $B(c) = c^T c$ by an elementary computation. Setting $t = 1$ in (2.6), we have for arbitrary real c

(2.7) $$\mathscr{E}(\exp c^T U) = \exp \tfrac{1}{2} c^T c.$$

But the expression on the right of (2.7) is just the moment generating function of a set of n mutually independent standardized normal variables. Assuming that the moment generating function uniquely determines the distribution, the set, $U_1, U_2, \ldots, U_{n'}$ are thus mutually independent standardized normal variables and possess the joint density,

(2.8) $$g(u) = (2\pi)^{-\frac{1}{2}n'} \exp(-\tfrac{1}{2}u^T u).$$

Any linearly independent n'-subset of $\{W_i\}$ is related to $\{U_i\}$ by a non-singular matrix transformation and hence possesses a joint density of the required form.

Remark. In Chapter X an alternative definition of joint normality will be introduced which does not require that $\mathscr{E} WW^T$ should be positive definite.

Theorem 2.1. *Let* $Y^{(1)}, Y^{(2)}, \ldots, Y^{(n)}$ *be a set of n random variables, each having zero expectation, and suppose that* $\mathscr{E} YY^T = V$ *is positive definite. Let N independent observations, $\{Y_j^{(1)}, Y_j^{(2)}, \ldots, Y_j^{(n)}\}, j = 1, 2, \ldots, N$, be made on the set. If $X^{(k)}$ is the standardized sum*

(2.9) $$X^{(k)} = N^{-\frac{1}{2}} \mathscr{S} Y_j^k,$$

the set $\{X^{(i)}\}$ has the same covariance matrix as $\{Y^{(i)}\}$. Further as $N \to \infty$, the joint distribution of the set $\{X^{(i)}\}$ is normal.

Proof. The first statement has been proved as Lemma 1. Let $L = b^T Y$ be an arbitrary linear form in the variables, Y_i. Then L has finite variance and zero expectation and obeys the conditions necessary for the validity of the central limit theorem. The standardized sum, $N^{-\frac{1}{2}} \mathscr{S}(L) = b^T X$ is therefore asymptotically normal if N is large. An arbitrary linear form in the $X^{(i)}$ is thus normal and hence by Lemma 2.3, the set $\{X^{(i)}\}$ is jointly normal.

Corollary. *The standardized sums of uncorrelated variables are asymptotically normal and independent.*

Proof. The covariance matrix is diagonal and hence with an appropriate change in scale $V = 1$ and in a normal correlation this ensures that the variables are independent.

Lemma 2.4. *Let Z be a random variable taking values $1, 2, \ldots, n$ with positive probabilities, $p_1, p_2, \ldots, p_n$, $\sum p_i = 1$. Define $\xi^{(i)}, i = 1, 2, \ldots, n$, functions of Z by*

$$(2.10) \qquad \xi^{(i)} = (1 - p_i)p_i^{-\frac{1}{2}}, \quad \text{when } Z = i,$$
$$= -p_i^{\frac{1}{2}}, \quad \text{when } Z \neq i.$$

Then

$$(2.11) \qquad \mathscr{E}\xi^{(i)} = 0,$$

$$(2.12) \qquad \mathscr{E}\xi^{(i)^2} = (1 - p_i), \qquad \mathscr{E}\xi^{(i)}\xi^{(j)} = -p_i^{\frac{1}{2}}p_j^{\frac{1}{2}},$$

and

$$(2.13) \qquad \mathscr{E}\boldsymbol{\xi}\boldsymbol{\xi}^T = 1 - (p_i^{\frac{1}{2}}p_j^{\frac{1}{2}}) \quad \text{where } \boldsymbol{\xi} = (\xi^{(1)}, \ldots, \xi^{(n)})$$

Proof. $\mathscr{E}\xi^{(i)^2} = (1 - p_i)^2 p_i^{-1} p_i + p_i(1 - p_i) = (1 - p_i)$. For $i \neq j$ and $n > 2$,

$$\mathscr{E}\xi^{(i)}\xi^{(j)} = [(1 - p_i)p_i^{-\frac{1}{2}}(-p_j^{\frac{1}{2}})]p_i + [(1 - p_j)p_j^{-\frac{1}{2}}(-p_i^{\frac{1}{2}})]p_j$$
$$+ [p_i^{\frac{1}{2}}p_j^{\frac{1}{2}}](1 - p_i - p_j) = -p_i^{\frac{1}{2}}p_j^{\frac{1}{2}}.$$

Then (2.13) follows.

Lemma 2.5. *Let $\mathbf{Y} \equiv \{Y^{(1)}, Y^{(2)}, \ldots, Y^{(n)}\}$ be defined by*

$$(2.14) \qquad \mathbf{Y} = \mathbf{H}\boldsymbol{\xi},$$

where $\mathbf{H}$ is orthogonal and $h_{1j} = p_j^{\frac{1}{2}}$. Then

$$(2.15) \qquad \mathbf{Y}^T\mathbf{Y} = \boldsymbol{\xi}^T\boldsymbol{\xi};$$
$$(2.16) \qquad Y^{(1)} = 0;$$
$$(2.17) \qquad \mathscr{E}Y^{(j)^2} = 1 \qquad \text{if } j \neq 1;$$
$$(2.18) \qquad \mathscr{E}Y^{(i)}Y^{(j)} = 0 \qquad \text{if } i \neq j,$$

and consequently

$$(2.19) \qquad \mathscr{E}\mathbf{Y}\mathbf{Y}^T = \text{diag}(0, 1^{n-1}).$$

Proof. (2.15) follows immediately from (2.14). (2.16) follows from the definitions, (2.10) and (2.14), since at a given point $Z = j$,

$$Y_1 = \sum_{i=1}^{n} p_i^{\frac{1}{2}}\xi^{(i)} = p_j^{\frac{1}{2}}(1 - p_j)p_j^{-\frac{1}{2}} + \sum_{i \neq j} p_i(-p_i^{\frac{1}{2}}) = (1 - p_j) - \sum_{i \neq j} p_i = 0;$$

Y_1 is identically zero.

$$\mathscr{E}(YY^T) = \mathscr{E}H\xi\xi^T H^T = H\mathscr{E}\xi\xi^T H^T = H[1 - (p_i^{\frac{1}{2}}p_j^{\frac{1}{2}})]H^T$$
$$= 1 - \text{diag}(1, 0^{n-1}) = \text{diag}(0, 1^{n-1}).$$

Theorem 2.2. *Suppose that a. independent observations are made on a random variable, Z, which can take values, $1, 2, 3, \ldots, n$ with probabilities, $p_i > 0$,*
$\sum_{i=1}^{n} p_i = 1$ *and suppose that Z takes the value, i, A_i times, then if a. is large, the quadratic form,*

$$(2.20) \qquad Q^T Q = a_.^{-1} \sum (A_i - a_. p_i)^2 / p_i,$$

is distributed asymptotically as χ^2 with $(n - 1)$ degrees of freedom. [Note. This important theorem states that the expression, (observed-expected)^2/ (expected), summed over the classes $Z = j$, $j = 1, 2, \ldots, n$ is distributed asymptotically as χ^2 with $(n - 1)$ degrees of freedom. The classes are mutually exclusive and exhaustive by the manner of definition. In the proof $Q^T Q$ is exhibited as the sum of squares of $(n - 1)$ asymptotically normal standardized variables, mutually independent.]
    Proof.

$$Q_i \equiv (A_i - a_. p_i)(a_. p_i)^{-\frac{1}{2}} \equiv a_.^{-\frac{1}{2}}[A_i(1 - p_i)p_i^{-\frac{1}{2}} - (a_. - A_i)p_i^{\frac{1}{2}}] = \mathscr{S}\xi^{(i)}$$

so that Q_i is the standardized sum of the variable $\xi^{(i)}$. Each Q_i is asymptotically normal as $a_. \to \infty$. By Lemma 2.2, HQ has the covariance matrix of $H\xi$ namely $\text{diag}(0, 1^{n-1})$ and by the corollary to Theorem 2.1, the elements of $H\xi$ are mutually independent and normally distributed. $Q^T Q = (H\xi)^T H\xi$ is thus asymptotically χ^2 with $(n - 1)$ degrees of freedom.

3. THE PROOFS OF K. PEARSON

We now briefly review the famous article in which K. Pearson (1900a) introduced χ^2 as a test of goodness of fit. In his first section, Pearson (1900a) shows by a geometrical argument that if a density function of n variables is proportional to $\exp -\frac{1}{2}x^T R^{-1}x$ at all points x of an n-dimensional infinite space, then $x^T R^{-1}x$ has the distribution of χ^2 with n degrees of freedom. R is the correlation matrix of the variables. Later in Section 3, Pearson ensures that R has full rank and so R is positive definite. Although he does not mention it explicitly it seems clear from the argument that he knew that χ^2 could be exhibited as the sum of the squares of n mutually independent standardised normal variables. In his second section, the probability integral of χ^2 is obtained by integration by parts. In the third section, the expected numbers in the $(n + 1)$ classes are computed with the differences, (observed-expected) being written e_i. The identity $\sum e_i = 0$ is noted and so Pearson takes only n of the e_i as "variables". He then writes down the variances and covariances of the first n of the e_i and inverts the correlation matrix by a trigonometrical substitution. This is the most difficult section of the paper to read and one that he could have omitted, for once he has obtained the inverse, he would only need to verify that the presumed inverse obeyed the relation, $\mathbf{R}\mathbf{R}^{-1} = \mathbf{1}$. In the fourth section, Pearson outlines rather tersely the use of the test. In the fifth section, he concludes (erroneously) that estimating the parameters makes little difference to the test. In the sixth and seventh sections, he applies χ^2 to distributions with theoretically given and estimated parameters respectively. Finally he gives a table of probabilities for integral values of χ^2 and $n = 2(1)19$. Much of the necessary preliminary work for Pearson's proof has already been carried out in the preceding section.

The covariance matrix of the first n variables

$$(3.1) \qquad A^{(i)} = N^{-\frac{1}{2}}e_i,$$

where e_i is the difference, (observed-expected), is

$$(3.2) \qquad \mathbf{R} = (p_i\delta_{ij} - p_ip_j).$$

The inverse matrix is

$$(3.3) \qquad \mathbf{W} = (p_i^{-1}\delta_{ij}) + p_{n+1}^{-1}\mathbf{U}, \qquad \text{where } u_{ij} = 1.$$

To verify this write $\mathbf{D}_n = \text{diag}(p_i^{\frac{1}{2}})$ and $\mathbf{p}$, the vector with elements

$$p_j^{\frac{1}{2}}(1 - p_{n+1})^{-\frac{1}{2}}, \, j = 1, 2, \ldots, n.$$

$\mathbf{p}\mathbf{p}^T$ is thus idempotent.

$$(3.4) \qquad \begin{aligned} \mathbf{R} &= \mathbf{D}_n(\mathbf{1} - (1 - p_{n+1})\mathbf{p}\mathbf{p}^T)\mathbf{D}_n \\ \mathbf{W} &= \mathbf{D}_n^{-1}(\mathbf{1} + p_{n+1}^{-1}(1 - p_{n+1})\mathbf{p}\mathbf{p}^T)\mathbf{D}_n^{-1} \end{aligned}$$

and it is easily verified that $\mathbf{R}\mathbf{W} = \mathbf{1}$. It now follows that

$$(3.5) \qquad \begin{aligned} \sum_{i=1}^{n+1} &\{(a_i - a.p_i)^2/(a.p_i)\} \\ &= \sum_1^{n+1} A^{(i)^2}/p_i \\ &= \sum_1^n A^{(i)^2}/p_i + \left[\sum_1^n A^{(i)}\right]^2 \Big/ p_{n+1}, \qquad \text{since } \sum_1^n A^{(i)} = -A^{(n+1)}, \\ &= \mathbf{A}^T\mathbf{W}\mathbf{A}, \end{aligned}$$

which has the distribution of χ^2 with n degrees of freedom by Pearson's Lemma of his first section.

W can be determined by the following device. Let **J** be any idempotent matrix and let a and b be constants

Then

(3.6) $$(1 + a\mathbf{J})(1 + b\mathbf{J}) = 1 + (a + b + ab)\mathbf{J}.$$

If $b = -a(1 + a)^{-1}$, $a \neq -1$, $1 + a\mathbf{J}$ has the inverse, $1 + b\mathbf{J}$. Pearson (1916a) later felt that this method of finding the inverse could be simplified by an application of the theory of partial correlation. In the same year, Pearson (1916b) incorporated a suggestion from H. E. Soper, which we give in Section 6 below, and which has been used by Fisher (1922a).

4. STIRLING'S APPROXIMATION

Stirling's approximation may be applied directly to the general term of the multinomial distribution but it is more neatly applied as follows. For notational convenience Δ, p and a are written in place of Δ_i, p_i and a_i.

Lemma 4.1. *If* $\Delta = a - Np$ *is of order* $(Np)^{\frac{1}{2}}$

(4.1) $\log \{(Np)^a \exp(-Np)/a!\} \simeq \frac{1}{2} \log(2\pi Np) - \frac{1}{2}(a - Np)^2/(Np).$

Proof. By Stirling's approximation

(4.2) $\log a! = \log(Np + \Delta)!$

$$\simeq (Np + \Delta + \tfrac{1}{2}) \log(Np + \Delta) - (Np + \Delta) + \tfrac{1}{2} \log(2\pi).$$

So that

(4.3) $\log \{(Np)^a \exp(-Np)/a!\} \simeq -\tfrac{1}{2} \log(2\pi) - \tfrac{1}{2} \log(Np)$

$$- (Np + \Delta + \tfrac{1}{2}) \log(1 + \Delta/[Np]) + \Delta.$$

After expanding $\log(1 + \Delta/[Np])$ as a logarithmic series and neglecting terms $\Delta^k N^{-k}$, where $k > 2$, we obtain the expression on the right of (4.1). Subscripts may be introduced so that Δ, a and p become Δ_i, a_i and p_i.

Theorem 4.1. *The general term of the multinomial expansion,* $(p_1 + p_2 + \cdots + p_n)^N$, *can be approximated by*

(4.4) $\log p(\{a_i\} \mid N, \{p_i\}) = -\tfrac{1}{2}(n - 1) \log(2\pi N) -$

$$- \tfrac{1}{2} \sum \log p_i - \tfrac{1}{2} \sum (a_i - Np_i)^2/(Np_i)$$

for terms such that $(a_i - Np_i) = 0(N^{\frac{1}{2}})$.

Proof.

$$p(\{a_i\} \mid N, \{p_i\}) = \prod_{i=1}^{n} \{(Np_i)^{a_i} \exp(-Np_i)/a_i!\}/\{N^N \exp(-N)/N!\}.$$

Taking logarithms of both sides and using (4.1) to obtain the logarithm of the factors of the numerator on the right and Stirling's approximation to the logarithm of the denominator, the result (4.4) follows.

Using the notation of the previous section, we have thus obtained an expression for those general terms of the multinomial distribution for which $(a_i - a.p_i)/\sqrt{(a.p_i)}$ remains finite as $a.$ is taken to be very large; but this means, in view of the law of large numbers, for every term with non-negligible probability, (4.4) can be expressed alternatively as

$$(4.5) \quad p(\{a_i\} \mid a., \{p_i\}) = (2\pi a.)^{-\frac{1}{2}(n-1)}(p_1 p_2 \ldots p_n)^{-\frac{1}{2}} \exp\left(-\tfrac{1}{2}\mathbf{Q}^T\mathbf{Q}\right),$$

where

$$\mathbf{Q}^T\mathbf{Q} = \sum_1^n (a_i - a.p_i)^2/(a.p_i).$$

But we have seen in Section 3 that $\mathbf{Q}^T\mathbf{Q}$ can also be written

$$(4.6) \qquad \mathbf{Q}^T\mathbf{Q} = \mathbf{A}^T\mathbf{W}\mathbf{A}, \qquad \text{where } \mathbf{A}^{(i)} = (a_i - a.p_i)/a.^{\frac{1}{2}}$$

is the ith component of the vector, $\mathbf{A}$, of $(n-1)$ components

$$(4.5') \quad p(\{a_i\} \mid a., \{p_i\}) \simeq |\mathbf{W}|^{\frac{1}{2}} (2\pi)^{-n/2} \exp\left(-\tfrac{1}{2}\mathbf{x}^T\mathbf{W}\mathbf{x}\right)$$

$$\simeq \int \ldots \int_D (2\pi)^{-\frac{1}{2}(n-1)} \exp\left(-\tfrac{1}{2}\mathbf{x}^T\mathbf{W}\mathbf{x}\right) \mathrm{d}x_1 \ldots \mathrm{d}x_{n-1}$$

where D is a region of volume $|\mathbf{W}|^{\frac{1}{2}}$ surrounding the point,

$$(4.7) \qquad\qquad\qquad\qquad \mathbf{x} = \mathbf{A}.$$

The choice of D is to some extent arbitrary. Now make the transformation

$$(4.8) \qquad\qquad\qquad\qquad \mathbf{x} = \mathbf{W}^{-\frac{1}{2}}\mathbf{y} = \mathbf{V}^{\frac{1}{2}}\mathbf{y}.$$

Then

$$(4.9) \quad p(\{a_i\} \mid a., \{p_i\}) \simeq \int \ldots \int_{D'} (2\pi)^{-\frac{1}{2}(n-1)} \exp\left(-\tfrac{1}{2}\mathbf{y}^T\mathbf{y}\right) \mathrm{d}y_1 \ldots \mathrm{d}y_{n-1}$$

where D' corresponds to D in the previous space. The discrete probability is thus identified with a region of the multivariate normal distribution and all the probability associated with the points such that $\sum (a_i - a.p_i)^2/(a.p_i) > \chi_\alpha^2$ are identified with the points of the theoretical distribution such that $\chi^2 > \chi_\alpha^2$, where χ_α^2 is the value of the theoretical χ^2 with n d.f. corresponding to the α level of significance.

The region D, is not unique because we can always choose the orthogonal linear forms in different ways if $n > 1$. This point is more readily made when the factorization proof is considered in Section 6.

5. THE PROOF OF H. E. SOPER

K. Pearson (1916b) noted that H. E. Soper had suggested to him an alternative method of proof, the essential feature of which was the introduction of a supernumerary variable so that the multinomial expansion $(p_1 + p_2 + \ldots + p_r)^{a.}$, had $(r + 1)$ classes with probabilities, $sp_1, sp_2, \ldots, sp_r, 1 - s$, where $0 < s < 1$. The new multinomial expansion

$$(5.1) \qquad \{sp_1 + sp_2 + \ldots + sp_r + (1 - s)\}^\lambda,$$

was then to be considered allowing $\lambda \to \infty$ and $\lambda s \to a.$. The aim was to obtain r variables with negligible correlations. This procedure gives r Poisson variables with parameters, $a.p_i$, and zero correlation. The general term of the multinomial distribution is evaluated then as the probability of r independent Poisson variables, subject to the condition that their sum is $a.$

$$(5.2) \quad p(\{a_i\} \mid \{p_i\}, a., \text{multinomial}) = a.! \prod_i (p_i^{a_i}/a_i!)$$

$$= \prod \{e^{-\lambda p_i}(\lambda p_i)^{a_i}/a_i!\}/\{e^{-\lambda \lambda^{a.}}/a.!\}$$

Now the left of (5.2) is independent of λ, which can therefore be chosen arbitrarily. λ is set equal to $a.$. The expression (5.2) shows that the multinomial distribution can be interpreted as the conditional probability of n Poisson variables given that their sum is $a.$. $\{a_i\}$ can be considered as a set of Poisson variables with parameters, $\{a.p_i\}$, subject to the constraint $\sum a_i = a.$. As $a.p_i \to \infty$, $(a_i - a.p_i)/\sqrt{(a.p_i)}$ is asymptotically a standardised normal variable. But then $\sum (a_i - a.p_i)^2/(a.p_i) = \sum Q_i^2$, say, is the sum of n such squares subject only to the constraint $\sum Q_i\sqrt{p_i} = 0$. Let $\mathbf{B}$ be an orthogonal matrix with elements of the first column, $\sqrt{p_i}$. The variables defined by $\mathbf{U} = \mathbf{B}^T\mathbf{Q}$ consist of one identically zero and $(r - 1)$ variables asymptotically normal, standardised and subject to no restraint. Moreover $\sum Q_i^2 = \sum U_i^2$ and the expression $\sum Q_i^2$ is thus distributed as χ^2 with $(r - 1)$ degrees of freedom. This method of proof has been used to great effect by Fisher (1922a) and Fisher, Thornton and MacKenzie (1922).

If the variables are asymptotically independent, the conditional distribution of the $(r - 1)$ linear forms is the same as the unconditional distribution of the same $(r - 1)$ linear forms. The parameter, λ, does not appear in the numerator of the U_i. Since its "observed" value, $a.$, will only differ from λ by terms $0(\lambda^{\frac{1}{2}})$ or $0(a^{\frac{1}{2}})$, the value of $\chi^2 = \sum_2^r U_i^2$, is not altered greatly by substituting $a.$ for λ in the denominator.

6. THE FACTORIZATION PROOF

It may be readily proved by induction that the general term of the multinomial distribution can be exhibited as the product of binomial probabilities.

$$(6.1) \quad p(\{a_i\} \mid \{p_i\}, a.)$$
$$= b\{a_1 \mid (a_1 + a_2), p_1(p_1 + p_2)^{-1}\} b\{(a_1 + a_2) \mid (a_1 + a_2 + a_3),$$
$$(p_1 + p_2)(p_1 + p_2 + p_3)^{-1}\} \dots b\{(a_1 + a_2 \dots a_{n-1}) \mid a., (p_1 + p_2 + \dots p_{n-1})\}.$$

So that the joint probability of $\{a_i\}$ is equal to the product of binomial probabilities of $(a_1 + a_2 + \dots a_i)$ given $(a_1 + a_2 + \dots a_{i+1})$ and the ratio, $(p_1 + p_2 + \dots + p_i)(p_1 + p_2 + \dots + p_{i+1})^{-1}$.

Let Stirling's approximation be applied to the multinomial expression on the left and to each binomial expression on the right. Since with each p_i fixed the ratio $a^{-1}(a_1 + a_2 \dots a_i)/(p_1 + p_2 \dots p_i) \to 1$ as $a. \to \infty$ for each i, the exponents on the two sides may be equated. The usual Pearson expression, $\sum (a_i - a.p_i)^2/(a.p_i)$, is thus exhibited as the sum of $(r - 1)$ standardised normal variables, asymptotically independent. The component χ's here correspond to an inexact partition of χ^2 due to Lancaster (1949a), but the corresponding exact partition has been given by Irwin (1949) and the most convenient form for the matrix of the transformation is the Helmertian. The partition shows that we might be better off with suitably normalised variables and suggests the proof given above in Section 2, which is based on the Bernstein theorem. The factorization method can be shown to be closely related to the theory of partial correlation. In fact, the component χ^2's of the inexact partition may be written as follows setting $A_i = a_1 + a_2 \dots a_i$, $P_i = p_1 + p_2 \dots p_i$.

$$(6.2) \quad U_j^2 = (A_j - A_{j+1}P_{j+1}^{-1})/\sqrt{\{A_{j+1}P_j p_{j+1}/P_{j+1}^2\}}.$$

U_{n-1} is then the unconditional standardised binomial variable, asymptotically normal. U_{n-2} is then the value of the standardised variable, given the value of U_{n-1} or of a_n. U_{n-3} is the standardised variable given $a_n + a_{n-1} \dots U_1$ is the standardised variable given $a_n + a_{n-1} \dots + a_3$. χ^2 is thus exhibited as the sum of $(n - 1)$ squares of asymptotically independent standardised normal variables. It is easy in this proof to mark out the region to which the integral in Section 4 would correspond; if we mark off the midpoints the appropriate boundaries are the hyperplanes corresponding to values $A_j \pm \frac{1}{2}$ in (6.2). This parallelopiped is not unique, however, as the variables may have been permuted to obtain another set of $(r - 1)$ mutually independent asymptotically normal variables. The volume of the parallelopiped would, however, be the same.

7. PROOF BY CURVE FITTING (LEXIS THEORY)

Let

(7.1) $Q = \sum$ (observed-expected)^2/(expected)

for the n classes of a multinomial distribution, and compute the moments
(Pearson, 1932a) or cumulants (Haldane, 1937a). The mean and cumulants
of Q are

$$\kappa_1 = \mathscr{E}(Q) = n - 1$$

(7.2)
$$\kappa_2 = 2(n - 1) + (\sum p_i^{-1} - n^2 - 2n + 2)a^{-1}$$
$$\kappa_3 = 8(n - 1) + 2[11 \sum p_i^{-1} - (9n^2 + 18n - 16)]a^{-1} + O(a^{-2})$$
$$\kappa_4 = 48(n - 1) + 96(4 \sum p_i^{-1} - 3n^2 - 6n + 5)a^{-1} + O(a^{-2}).$$

As $a \to \infty$, the first four cumulants tend to those of χ^2 with $(n - 1)$ degrees
of freedom. We can, therefore, use χ^2 with $(n - 1)$ degrees of freedom as an
approximation to Q. This is in the Lexis tradition of the derivation of the
moments of χ^2 in contingency tables by Tschuprow (1918/19).

8. ANALOGUES OF THE PEARSON χ^2

Let $x^{(0)} = 1$, and $x^{(1)}, x^{(2)}, \ldots$ be a set of orthonormal functions on the
distribution of a random variable, X. The methods of Section 2 show that
we can regard the Pearson χ^2 as a sum of squares of standardized asymptotically normal sums. The standardized sums are of the form

(8.1) $X^{(i)} = N^{-\frac{1}{2}}\mathscr{S}_j x_j^{(i)}, \qquad j = 1, 2, \ldots, N,$

and

(8.2) $$\chi^2 = \sum_{i=1}^{n} X^{(i)^2},$$

where $x_j^{(i)}$ is the value of the ith function at the jth observation. The $x^{(i)}$ take
the special form in the Pearson χ^2 of orthonormalized indicator variables.
But other systems are possible. Neyman (1937) made a preliminary transformation on the random variable so as to obtain a rectangular distribution.
He then set $x^{(i)}$ equal to the standardized Legendre polynomial of the ith
degree. This is the basis of the "smooth test" of goodness of fit. It is remarkable that no fresh tests of this kind had been developed before the paper of
Lancaster (1953a) who suggested that any orthogonal system might be used,
in particular the Hermite-Chebyshev system in the normal case and other
sets of orthogonal polynomials on finite discrete distributions. Barton (1955)
has since shown that the absolute continuity of the distribution is not a
necessary condition for such tests to be devised. Hamdan (1962, 1964) has
shown that other orthonormal systems, such as the Walsh, may also be used

and that these have the advantage of simplicity. We shall return to such χ^2 analogues after a consideration of parameters of non-centrality.

9. EMPIRICAL VERIFICATIONS OF THE DISTRIBUTION OF THE DISCRETE χ^2

Does the distribution of the continuous χ^2 give a good approximation to that of the discrete χ^2 (or X^2 as Cochran (1952) writes it)? The opinion of many writers has been that since χ^2 has been obtained by the use of Stirling's approximation, great care should be taken not to use the χ^2 test if the expected number falls below 5 or even 10 in any one cell. It is usually acknowledged now (Vessereau, 1958; Cochran, 1954) that this advice is unduly conservative.

In Section III.3, we have already shown that the χ^2 or χ-approximations are very effective in the binomial distribution; χ_c gives a close approximation to the exact probability and the uncorrected or crude χ gives a close approximation to the medium probability.

In the multinomial distribution, it is not easy to devise a satisfactory analogue of the Yates correction although some authors such as Wise (1963 and 1964) claim to have done so. There has been special interest in the multinomials with $\{p_i\} = \{0\cdot2, 0\cdot3, 0\cdot5\}$ because if the smallest p_i is not to be $0\cdot1$, this particular set is the only partition of unity into three distinct one digit decimals. Thus without taking N too large, the smallest expectation is not too small and at the same time the computations are relatively easy on a desk machine. Neyman and Pearson (1931a) ordered the configurations from the above multinomials with index 10 by means of the relative frequency, their P_c, by the likelihood ratio, their P_λ, and by χ^2. El Shanawany (1936) ordered the configuration by P_c for $N = 10$ and $N = 20$. Both papers expressed surprise at the good agreement between P_{χ^2}, the cumulative (discrete) probability of χ^2, and P_c. It seems more logical to compare the discrete χ^2 distribution with the theoretical, and the computations of Lancaster (1961) and Bennett (1962) are based on this principle. Certain significance levels, $0\cdot01$, $0\cdot05$, $0\cdot1(0\cdot2)0\cdot9$ were chosen and the rejection rate or size of the discrete χ^2 test was compared at each significance level with the continuous or theoretical. Lancaster (1961) gave comparisons for $N = 9(1)31$ and Bennett (1962) for $N = 31(1)50$ at these chosen significance levels. Both sets of tables show a reasonable approximation of the discrete to the continuous distribution. Stirling's approximation was also used in both these papers to give an estimate of the number of configurations to be found in each probability class.

Of greater interest from the practical point of view are the symmetrical multinomial distributions which arise naturally in counting experiments. Sukhatme (1938a and b) imitated, by a random sampling experiment, the taking of n parallel counts from a Poisson population with parameter values as low as unity. N was the sample total and the index of the multinomial distribution and the probabilities were n^{-1}. The population was essentially a mixture of multinomials, the index of which had the Poisson distribution with parameter $n\lambda$, where λ was the Poisson parameter of a single count and n was the number of counts. It is easily verified that with $n = 15$ and $N = 15$ there is not a great number of distinct values of $\chi^2 = n \sum x_i^2/\sum x_i - \sum x_i$, $N = \sum x_i$. Indeed the 5% point of χ^2 with 14 degrees of freedom is 23·685. The largest value of $\sum x_i^2$ permissible if $\chi^2 < 23·685$ is 37 and the smallest possible value is 15. All permissible values of $\sum x_i^2$ are odd, so that there can be at most 12 permissible values of $\sum x_i^2$. Similarly with $\sum x_i = N = 30$, $n = 15$, there are at most 24 permissible values with $\chi^2 < 23·685$. Evidently, for a fixed N, here 30, the distribution function of the discrete χ^2 will have some large saltuses. If, however, mixtures are taken so that N is given the Poisson distribution the sizes of the saltuses will be less and the degree of approximation improved. It would be of interest to see further computations done on the multinomials with other values of the probabilities.

It was formerly recommended that pooling should be carried out in the tails of a distribution, for example, when the goodness of fit by a Poisson or normal distribution to empirical data was being considered. It was recommended that no cell should have an expectation less than 5. This recommendation can be ignored. It is probably always possible to add several cells with expectations little above unity without disturbing the theoretical distribution. Whether it is desirable to do so, that is, whether it gives a more powerful test, is dependent on the forms of the null and alternative hypotheses. Formally, if it be supposed that Z has the distribution of χ^2 with k degrees of freedom and another variable W has the distribution of the sum of squares of m standardized Poisson variables with parameters, $\lambda_1, \lambda_2, \ldots, \lambda_m$, the sum $Z + W$ will have approximately the distribution of χ^2 with $k + m$ degrees of freedom provided that each $\lambda_i > 1$ and that $k > 2m$. This rule would usually be fulfilled in practice as there might be two or three classes in each tail of the distribution with an expectation of less than 5, and perhaps, 8 or 10 other classes. In such a case k might be 8 and m would be 4, depending on the hypothesis and the number of parameters estimated. This recommendation is consistent with that of Cochran (1952).

10. APPLICATIONS OF χ^2 TO THE MULTINOMIAL DISTRIBUTION

We give now a short description of the application of χ^2 to a physical problem, in which a knowledge of the unknown parameters is irrelevant. The symmetrical multinomial distribution appears in a natural manner in the analysis of counting experiments such as occur in haematology, in bacteriology, in physics (as scintillator counts) and in many other fields of applied science. Although Abbe (1889) counting red cells in his new haemocytometer chamber and Student (1908) counting yeast cells, both gave the appropriate formula, neither mentioned Poisson and it seems that the distribution has been independently rediscovered many times. The large sample test used by Student (1908) is too laborious to be used as a routine. A small sample test is as follows.

Suppose that $Y_1, Y_2, \ldots, Y_n$ is a set of n independent observations on a Poisson distribution with parameter, λ, and that $N = Y_1 + Y_2 + \ldots + Y_n$. Then

$$(10.1) \quad \mathscr{P}\{Y_1 = y_1; Y_2 = y_2; \ldots Y_n = y_n\} = \prod_{i=1}^{n} \{e^{-\lambda}\lambda^{y_i}/y_i!\}$$

$$= e^{-n\lambda}\lambda^{\Sigma y_i} \Big/ \prod_{i=1}^{n} y_i!$$

N is also a Poisson variable with parameter, $n\lambda$, so that

$$(10.2) \quad \mathscr{P}\{N = y\} = e^{-n\lambda}(n\lambda)^y/y!.$$

The probability of the event $\{Y_1 = y_1, Y_2 = y_2 \ldots\}$ conditional on $N = y = \sum y_i$ can now be obtained as a quotient of the absolute probabilities in the usual way.

$$(10.3) \quad \mathscr{P}\{\{y_i\} \mid y\} = \mathscr{P}\{\{y_i\} \mid \lambda, \text{Poisson}\}/\mathscr{P}\{N = y \mid n\lambda, \text{Poisson}\}$$

$$= n^{-y}y! \Big/ \prod_{i=1}^{n} y_i!$$

But the expression on the right of (10.3) is simply the multinomial probability, the general term of $(n^{-1} + n^{-1} + \ldots + n^{-1})^y$. This expression does not contain λ; the possibility of this being so characterizes the Poisson distribution (Moran, 1952).

The small sample test of the consistency of counts is thus to examine whether $\{y_i\}$ falls into a certain critical region. As a test function, we may choose

$$(10.4) \quad \chi^2 = \sum (\text{observed expected})^2/(\text{expected})$$

$$= \sum (y_i - \bar{y})^2/\bar{y}, \quad \text{where } \bar{y} = \sum y_i/n;$$

$$= n \sum y_i^2/(\sum y_i) - \sum y_i,$$

the last expression being the most convenient for computation.

This test has been found very useful in bacteriology (Fisher, Thornton and McKenzie, 1922; Eisenhart and Wilson, 1943) and in haematology (Lancaster, 1950b), from which articles further references can be obtained.

The null hypothesis is given in the discussion of the equations (10.1) to (10.4). As a rule there is no specific alternative hypothesis. It is usually thought that the alternatives will be associated with high values of χ^2. Thus in parallel counts of bacterial dilutions, there may be clumping of bacteria in some sample aliquot volumes which will lead to a high value of χ^2. A full discussion is available in the references cited above. Similarly in white cell counting the stroma of red cells may be inadequately broken down by the acid solution and white cells may be clumped together. It is easily verified that white blood cell counting is technically quite feasible in which agreement between parallel counts of an order suggested by the Poisson hypothesis can be obtained. The same is true of red blood cell counting if the unit of area taken for comparison is 0·2 mm square, that is a square area consisting of 16 small haemocytometer squares each of 0·05 mm side. If a smaller area, the small haemocytometer square, is taken as the unit of area and if the cell count is relatively high, say 6 red cells per small haemocytometer square, then there is correlation between neighbouring squares and the theoretical conditions no longer hold.

EXERCISES AND COMPLEMENTS

1. Let Z be the variable taking values i with probabilities, $p_i > 0$, $i = 1, 2, \ldots, n$. Write ψ_i for the indicator variable for the point, i; $\psi_i = 1$ when $Z = i$, $\psi_i = 0$ otherwise. Find the mean and variance of ψ_i. Find also the variance of $(\psi_i + \psi_j)$ and hence the covariance of ψ_i and ψ_j, when $i \neq j$.

2. (Continuation). Show that the set, $\{\psi_i - \mathscr{E}\psi_i\}$, is linearly dependent. Show that there are $(n - 1)$ non-constant linearly independent functions on the space of n points.

3. Show that the correlation between $\xi^{(i)}$ and $\xi^{(j)}$ of (2.8) is

$$-p_i^{\frac{1}{2}}p_j^{\frac{1}{2}}(1 - p_i)^{-\frac{1}{2}}(1 - p_j)^{-\frac{1}{2}} \quad \text{if} \quad i \neq j.$$

The value is -1 if there are only two classes. (Pearson, 1900a, 1903b).

4. Prove that if a correlation or covariance matrix is of full rank, it is positive definite.

5. Obtain the central limit theorem by the use of characteristic functions (see Loève (1961) and Lukacs (1959)).

6. Let $Y_1, Y_2, \ldots, Y_N$ be mutually independent and each have the same distribution. If there is a limiting form for the distribution of the standardized variable

$$X = N^{-\frac{1}{2}} \sum_1^N Y_i, \quad \text{as } N \to \infty,$$

which possesses finite moments of all orders, then all the cumulants of all orders greater than 2 are zero. [*Hint.* Let $X_1 = N^{-\frac{1}{2}} \sum_1^N Y_i$, $X_2 = N^{-\frac{1}{2}} \sum_1^N Y_{N+i}$, then as

$N \rightarrow \infty$, X_1, X_2 and $X = 2^{-\frac{1}{2}}[X_1 + X_2]$ have the same distribution. Prove that the assumption that $K_s \neq 0$ for $s > 2$ leads to a contradiction. This is a weak form of a theorem by Pólya (1922).]

7. Prove that if a random variable U has the moments of a standard normal distribution, its Poincaré transform $\mathscr{E}[\exp zU]$, where z is a complex variable $x + iy$, is defined for z in any finite part of the plane. [*Hint.* $\mathscr{E}(1 + zU/1! + z^2U^2/2! + \ldots + z^nU^n/n!)$ is equal to the first $n + 1$ terms of the expansion of $\exp \frac{1}{2}z^2$.] The Poincaré transform is therefore an analytic function which coincides with the characteristic function of the normal distribution when $x = 0$. Assuming that the characteristic function uniquely determines a distribution function, prove that the distribution is normal.

8. With the notation of Section 2, determine the moments of order k of the standardized sum X in terms of the moments of the components, Y_j. For k fixed and N indefinitely large, evaluate the moments of X neglecting terms of order, $N^{-\frac{1}{2}}$. Hence show that the kth moment of X is asymptotically equal to the kth moment of the standardized normal distribution. (Laplace, 1819; Bienaymé, 1852; Ellis, 1844b; Sheppard, 1898a; Hansen, Hurwitz and Madow, 1953).

9. With the conditions on $\{Y_i\}$ as in Exercise 6, prove that, for a fixed r, the rth cumulant of the standardized sum tends to zero as $N \rightarrow \infty$.

10. Let the variables, Y_k, of Exercise 6 take the values ± 1 with probabilities $\frac{1}{2}$. Write down the m.g.f. and c.g.f. of Y_k. Verify that there is a pole of the logarithm of the Poincaré transform, namely $\log \mathscr{E}(\exp zY_k)$ at the point $z = \pi\mu$. Hence verify that the cumulants of Y_k increase without limit. So the use of Exercise 9 and Exercise 7 does not yield a proof of the central limit theorem even when all moments of the component Y_k exist. [The odd moments of Y_k are zero and the even moments are all unity.]

11. Prove Pearson's Lemma by considering the m.g.f. of $\mathbf{x}^T\mathbf{R}^{-1}\mathbf{x}$ when the joint density function of the elements of $\mathbf{X}$ is proportional to $\exp -\frac{1}{2}\mathbf{x}^T\mathbf{R}^{-1}\mathbf{x}$. Make a transformation $\mathbf{X} = \mathbf{R}^{\frac{1}{2}}\mathbf{Y}$ and identify the individual degrees of freedom.

12. Let $f(x, y; \rho)$ be the frequency function of a bivariate normal distribution with correlation, ρ, in which the marginal variables are standardized to have zero mean and unit variance. Now define $g(x, y) = \theta f(x, y; \rho) + (1 - \theta)f(x, y; -\rho)$, $0 < \theta < 1$. Show that $g(x, y)$ is a bivariate frequency function and that both marginal distributions are standardized normal. However, $g(x, y)$ is not the frequency function of a normal correlation. Verify this statement by determining the m.g.f. of $a_1x + a_2y$. If $\theta = \frac{1}{2}$, show that the marginal variables can be normal and the correlation zero without the marginal variables being independent. (Lancaster, 1959; Sarmanov, 1960.)

13. Pearson (1900a) gave a table of χ^2 for d.f. 2(1)19. Pearson thought that χ^2 might not be applicable if the d.f. were too numerous. Show that this need not be so by a consideration of (7.2) where with a fixed n, the number of classes, a, can be chosen sufficiently large. Consider also the implications of the sampling experiments.

14. Carry through the details of Soper's proof in Section 5. Set

$$Q_j = (a_j - \lambda p_j)/(\lambda p_j)^{\frac{1}{2}},$$

and consider $Q_1, Q_2, \ldots, Q_n$ to be standardized Poisson variables, asymptotically normal. Now make the transformation, $\mathbf{U} = \mathbf{BQ}$. Verify that U_1 is $(N - \lambda)/\lambda^{\frac{1}{2}}$ and that λ does not appear in the numerator of any element of $\mathbf{U}$. Since the Q's are asymptotically normal, so are the U's. The distribution of the last $(n - 1)$ U's is independent of U_1. Set $\lambda = N$, then U_1 is zero identically and $(U_2^2 + U_3^2 + \ldots + U_n^2)$ is χ^2 with $(n - 1)$ degrees of freedom. Justify this last step by either (i) $\lambda = N + o(N^{\frac{1}{2}})$ and so the change from λ to N in the denominators makes little change to the magnitudes of the elements, or (ii) the multinomial probability expressed as a conditional probability is independent of the value of the unknown λ.

15. Give practical computing routines for a desk computer to evaluate the possible values of χ^2 and their relative frequencies in the multinomials, $(\frac{1}{3} + \frac{1}{3} + \frac{1}{3})^N$ and $(0 \cdot 2 + 0 \cdot 3 + 0 \cdot 5)^N$, where $N = 15(1)20$.

16. An important source of discrepancy between the size of the test and the significance level is the magnitude of the individual probabilities concentrated at one point. The discrepancies may become greater if such points are combined. For example, if two or more of p_i are equal or if there are equalities of the form $\sum a_i^2 p_i^{-1} = \sum b_i^2 p_i^{-1}$ possible for $\{a_i\} \neq \{b_i\}$ where $\sum a_i = \sum b_i = N$, a_i and b_i being positive integers. Such condensations are less likely to occur if the p_i are rationals when the likelihood ratio is the test function. Of course, the converse would be true if the p_i were chosen so that $\log p_i$ would not all be incommensurable but the p_i would be.

17. Typical haemocytometer counts are as follows:

Counts $\{y_i\}$

39	28	43	36	33
126	133	144	121	140
35	44	56	58	48
99	101	105	103	103
117	119	114	135	122

Calculate $\sum y_i$, $\sum y_i^2$ and χ^2 for each set of counts.

18. The hour of birth has been recorded for all the births in a year in a maternity hospital. Commencing at the hour 12 midnight to 1 a.m. the results are given in the following table, which is to be read by columns.

97	117	97	92
136	80	93	102
151	125	100	100
110	87	93	101
144	101	131	127
136	107	105	118

(D. H. Sutton, "Gestation period", *Med. J. Aust.*, 1, 1945, 611–613.) Test the null hypothesis that births occur uniformly throughout the 24 hours.

19. The results of observers counting red cells on five areas of side, 0·2 mm, using the methods outlined above are summarized in the following table.

χ^2 in Red Cell Counting

Probability Class	Value of χ^2 (4 d.f.) at upper end of interval	Counter				
		A	B	C	D	E
1·0–0·9	0	6	15	15	24	13
0·9–0·7	1·064	15	16	23	25	31
0·7–0·5	2·195	14	15	24	24	22
0·5–0·3	3·357	13	16	11	25	20
0·3–0·1	4·878	14	22	18	2	7
0·1–0	7·779	2	16	9	0	7
Total	—	64	100	100	100	100

A is the author. B and C were well trained technician and physician respectively. D was a technician who avoided counts far from the mean. E was the same technician after being instructed to count on certain pre-determined areas. Comment on the consistency of the counts. (Lancaster, 1950b.)

20. In the multinomial of (5.2), compute the correlation of a_i and a_j for fixed a_k and hence for fixed totals of any other set of the $a_1, a_2, \ldots, a_n$. Hence determine the coefficients and partial coefficients of correlation of the forms,

$$\rho_{12} = -\{p_1 p_2 / (1 - p_1)(1 - p_2)\}^{\frac{1}{2}}$$

$$\rho_{12\cdot34\ldots j} = -\{p_1 p_2 / (1 - p_1 - p_3 - \ldots - p_j)(1 - p_2 - p_3 - \ldots - p_j)\}^{\frac{1}{2}}.$$

(Pearson, 1916a; Cramér, 1946, p. 318; Lancaster, 1949a.)

21. Give an alternative proof of Theorem 2.2, in which the central limit theorem is used with orthogonalised variables consisting of linear forms of the first two, the first three ... the first $(n - 1)$ variables and finally the whole set of n variables, $\xi^{(i)}$. Relate the values of these linear forms to the elements of a Helmert matrix, of which the first column has elements equal to $\sqrt{p_i}$.

22. Let A be the number of events out of N falling into a certain class with probability, p, in the multinomial. Prove that $\mathscr{E}A^{(r)} = N^{(r)}p^r$, where $A^{(r)}$ and $N^{(r)}$ are the factorial products of the form $N(N - 1) \ldots (N - r + 1)$. Hence find the mean and variance of Q in equation (7.1). (Pearson, 1932a; Haldane, 1937a.)

23. The partition of χ^2 in the multinomial distribution may be of interest as in studying the effect of bacteriostatic (i.e. "disinfectant") agents. If it is ineffective in dilutions $1, 2, \ldots, k$ then the comparison of the k counts can be carried out

with the aid of the χ^2 test. It it is effective at the $(k + 1)^{\text{th}}$ level, then the χ^2 comparing the counts at the first k levels with that at the $(k + 1)^{\text{th}}$ level may be unduly large. Prove that this χ^2 can be obtained by computing the difference,

$$(k + 1) \sum_1^{k+1} a_i^2 \bigg/ \sum_1^{k+1} a_i - \sum_1^{k+1} a_i - k \sum_1^{k} a_i^2 \bigg/ \sum_1^{k} a_i - \sum_1^{k} a_i.$$

This is an inexact partition. (Irwin, 1949; Cochran, 1952.)

24. The digits 0(1)9 in the decimal expansion of $\pi - 3$ have the frequency 968, 1026, 1021, 974, 1014, 1046, 1021, 970, 948, 1014.   Test the null hypothesis that in the expansion each digit should be represented approximately 1000 times.   [$\chi^2 = 11\cdot370$ for 9 d.f.] (Pathria, 1962.)

CHAPTER VI

Canonical or Standard Forms for Probability Distributions

1. PROBABILITY MEASURES

Often in statistical applications, the sample space is the real line or a subset of it. However, the sample space may be of more general form; for example, there may be a qualitative classification such as by hair colour or sex. Even in such cases, a correspondence can be set up between the qualitatively defined classes and the points on the real line; so that, it is possible to assume that a statistical distribution possesses a distribution function, although this is not assumed to be the case in the following discussion. In the general theory, there is a class, $\mathscr{B}$, of sets $\{B_i\}$, such that the whole space B is the union of all sets, B_i. Although the events of probability theory are equated with the sets of the measure theoretical language, it would be more precise to say that there is a random variable, Z, and that the events are $Z \in B_i$. A set of rules is given for the formation of countable unions, complements and intersections; this is the algebra or σ-algebra, $\mathscr{S}$. A *measurable space* $(B, \mathscr{S})$ is obtained and by assigning a probability measure to certain classes of the sets, a *measure space* $(B, \mathscr{S}, \mu)$.

Any number of *probability measure spaces* can be defined on the same measurable space. We suppose that t is an index belonging to a set, T, and that for each $t \in T$, there is a rule for assigning the measure to each $B_i \in B$. Then we have

$$(1.1) \qquad \mu_t(B_i) = \mathscr{P}\{Z \in B_i\}, \; B_i \subset B, \; B_i \in \mathscr{B}, \; t \in T.$$

T, for example, may be the space of parameters. In the Neyman-Pearson, theory, $T = T_0 \cup T_1$, where values of μ_t are specified by the null hypothesis when $t \in T_0$ and by the alternative hypothesis when $t \in T_1$. It is convenient to write μ in place of μ_t when $t \in T_0$ is specified by the null hypothesis, H_0, and ν for μ_t when $t \in T_1$ is specified by the alternative hypothesis, H_1. $T = T_0 \cup T_1$ is the set of possible values of t under the null and alternative hypotheses. If T_i contains only one member, the hypothesis is *simple*. An important test of hypothesis occurs when T_0 contains only one member.

Sometimes it is convenient to consider not the original space $(B, \mathscr{S}, \mu)$ defined by the σ-algebra, $\mathscr{S}$, but a "condensed" space, $(B, \mathscr{S}^*, \mu)$ induced by a sub-algebra, $\mathscr{S}^*$ say, of $\mathscr{S}$. For example, $\mathscr{S}$ may be the algebra of the Borel sets on the real line and $\mathscr{S}^*$ the algebra on the set of a finite number of classes, of the form, $d_{i-1} < x \leqslant d_i$; the infinite space has been condensed into a discrete space having a finite number of points.

It is desirable to have some measure of the closeness of one distribution to another. In the theory of this monograph, the ϕ^2 of Pearson (1904) is of outstanding importance. Here a rather broader definition of ϕ^2 is given for pairs of measures, μ and ν, on the same measurable space. Let us consider first the whole space and define

$$(1.2) \qquad \phi^2(0) + 1 = [\nu(B)]^2/\mu(B) = 1,$$

and generally

$$\phi^2(i) + 1 = \sum_j [\nu(M_{ij})]^2/\mu(M_{ij})$$

$$= \sum_j [\nu(M_{ij})/\mu(M_{ij})]^2\mu(M_{ij}), \qquad M_{ij} \in \mathscr{B},$$

with the convention that $\phi^2(i)$ is infinite if $\nu(M_{ij})$ is positive when $\mu(M_{ij})$ is zero for any j. In (1.2), i is the index of the partition of B into n_i parts, so that

$$(1.3) \quad B = \bigcup_j M_{ij}; \quad M_{ij} \cap M_{ik} = \phi, \qquad \text{if } j \neq k, \quad j = 1, 2, \ldots, n_i.$$

In other words, the space, B, is partitioned into classes of mutually exclusive and exhaustive sets. A sequence of partitions is chosen so that the space is progressively refined, the sets of any partition being *nested* within those of any earlier partition, that is, if $i' > i$, then for any given j', $M_{i'j'}$ is included wholly within one set, M_{ij}, of the earlier partition. The refinement at each step can be supposed not to be trivial, so that for at least one j, the set M_{ij} with $\mu(M_{ij}) > 0$ is partitioned with at least two of its subsets, $M_{i+1,j'}$, associated with a positive measure by either μ or ν; in other words, it is supposed that at the $(i + 1)^{\text{th}}$ partition there are two sets, M_1 and M_2, each subsets of an M of the ith partition and such that $\mu(M_1) + \nu(M_1) > 0$ and $\mu(M_2) + \nu(M_2) > 0$. It is further required of the sequence of partitions, that for every set M_{ij} of the ith partition $\mu(M_{ij}) < \epsilon$ for sufficiently large, i, or that the set M_{ij} cannot be further partitioned in a non-trivial fashion; in this last case, M_{ij} is an "atom" or indecomposable set. For an "atom", M, $\mu(A) = 0 = \nu(A)$, or $\mu(A) = \mu(M)$, $\nu(A) = \nu(M)$ for every subset A of M.

Lemma 1.1. *The functionals,* $\phi^2(1)$, $\phi^2(2), \ldots,$ *of a nontrivial nested sequence of partitions form a non-decreasing sequence.*

Proof. At the kth partition, let measures be assigned, $\mu(M) = f$ and $\nu(M) = g$; $f \neq 0, g \neq 0$, to some set, M. The contribution of this set to the sum (1.2) of the kth partition is then g^2/f. In the $(k + 1)^{\text{th}}$ partition, let M be partitioned into M_1 and M_2 with μ-measures, f_1 and f_2, and ν-measures, g_1 and g_2. If either f_1 or f_2 is zero and the corresponding g_i is not zero the contribution will be infinite by convention. If either f_1 and g_1 or f_2 and g_2 are zero the partition is not a refinement of the previous partition and there is nothing to prove; so that we can assume that $f_1 \neq 0$ and $f_2 \neq 0$. The contribution to the $(k + 1)^{\text{th}}$ partition is thus $g_1^2/f_1 + g_2^2/f_2$. The contribution has been increased by $g_1^2/f_1 + g_2^2/f_2 - g^2/f = (f_1 g_2 - f_2 g_1)^2/(f_1 f_2 f)$, which is non-negative. $\phi^2(k)$ is thus a never decreasing function of k.

Theorem 1.1. *Let a non-trivial sequence of nested partitions,* $\{M_{kj}\}$, $k = 1$, $2, \ldots$; $j = 1, 2, \ldots, n_k$, *be applied to the measurable space,* B, *with which are associated two measures,* μ *and* ν. *Then, if the sets cannot be further partitioned or if the measures assigned to each set capable of subpartition tend to zero, the functional,* $\phi^2(k)$, *is either infinite or tends to a unique limit as* $k \to \infty$.

Proof. First, any sequence of values of the sums (1.2) is non-decreasing by Lemma 1.1 and so as $k \to \infty$, $\phi^2(k)$ is infinite or tends to a definite limit. We suppose that it is possible to form a sequence of partitions such that the sum (1.2) tends to a limit η_1 for every refinement of the partition. If now in a second sequence of partitions, the sum (1.2) is $\eta_2 > \eta_1$ for some finite partition, a contradiction would result; for, consider any partition with sets, $E_1, E_2, \ldots, E_m$ in the first sequence. Then no refinement of this partition can give a value to the sum (1.2) greater than η_1. On the other hand a partition into a finite number of sets $B_1, B_2, \ldots, B_n$ in the second sequence can be found such that the sum (1.2) is greater than $\eta_2 - \epsilon$, where ϵ can be chosen arbitrarily and so less than $\eta_2 - \eta_1$. But the partition into the sets $\{E_i \cap B_j\}$, $i = 1, 2, \ldots, m$, $j = 1, 2, \ldots, n$, is a refinement of the first sequence of partitions and so the sum is less than η_1 but it is also a refinement of the second sequence and so is greater than $\eta_2 - \epsilon$ and so greater than η_1; a contradiction has been reached. The limiting value of $\phi^2(k)$, if finite, is unique and may be written as ϕ^2.

It is thus permissible to define ϕ^2, the *Pearson functional* for measures, μ and ν, as follows. *If* μ *and* ν *are two probability measures on a given measurable space,*

$$(1.4) \qquad \phi^2 + 1 = \int (d\nu)^2/d\mu = \int (d\nu/d\mu)^2 \, d\mu = \int \Omega^2 \, d\mu$$

where the first two integrals are to be interpreted in the sense of the limiting values of the two sums of (1.2) and $\Omega = d\nu/d\mu$ *is the Radon-Nikodym derivative of* ν *with respect to* μ. ϕ^2 can thus be interpreted as the variance of the Radon-Nikodym derivative or likelihood ratio since $\int \Omega \, d\mu = 1$ and so

$\phi^2 = \int \Omega^2 \, d\mu - [\int \Omega \, d\mu]^2$. Integrals of the form of (1.4) are sometimes (e.g. Cramér, 1946) named after Hellinger (1909).

Theorem 1.2. *The measure ν coincides with μ if and only if $\phi^2 = 0$.*

Proof. If $\nu \equiv \mu$, then $\Omega \equiv 1$ and the conclusion $\phi^2 = 0$ follows. If $\phi^2 = 0$, an arbitrary countable partition of the measurable space may be taken. Suppose that μ and ν assign measures, a_j and b_j, respectively to the sets. Then

$$0 \leqslant \sum_j (b_j - a_j)^2/a_j = \sum_j b_j^2/a_j - 1 \leqslant \phi^2 = 0$$

so that $a_j = b_j$ for every j.

If ϕ^2 is bounded, ν is said to be ϕ^2-bounded with respect to μ. ϕ^2-boundedness is reflexive but not symmetrical; it may or may not be transitive. By convention, ϕ^2 is infinite if the ν-measure is positive for any set for which the μ-measure is zero. It follows from this convention that ϕ^2-bounded ν-measures are absolutely continuous with respect to μ but the converse need not be true. It is not required in the present theory that μ should be absolutely continuous with respect to ν, although it is convenient in information theory to deal with such *homogeneous* measures (Kullback, 1959).

Theorem 1.3. *If ν is ϕ^2-bounded with respect to μ, then the likelihood ratio or Radon-Nikodym derivative can be expanded as a series (1.6) convergent in the (quadratic) mean, in a complete orthonormal set, defined on μ, and there exists a canonical form for any such measure, ν, namely*

(1.5) $$d\nu \equiv \Omega(x) \, d\mu = (1 + \phi\xi^{(1)}) \, d\mu,$$

where $\xi^{(1)}$ is the first member of a complete orthonormal set.

Proof. There exists a complete orthonormal set, $\{x^{(i)}\}$, say on μ. $\Omega(x)$ is square summable by the hypothesis of the theorem and so can be expanded as a series,

(1.6) $$\Omega(x) = 1 + \sum_1^\infty a_i x^{(i)}, \qquad \sum_{i=1}^\infty a_i^2 = \phi^2.$$

Now let a second orthonormal set $\{\xi^{(i)}\}$ be derived from $\{x^{(i)}\}$ by an orthogonal transformation such that

(1.7) $$\xi^{(1)} = \phi^{-1} \sum_{i=1}^\infty a_i x^{(i)}.$$

Then (1.5) follows by substituting (1.7) into (1.6) and, by the mode of formation, $\{\xi^{(i)}\}$ is complete since completeness is preserved under orthogonal transformations by Theorem IV.4.2. An alternative proof is as follows. $(\Omega(x) - 1)^2$ has an expectation equal to ϕ^2. $\xi^{(1)} \equiv (\Omega(x) - 1)\phi^{-1}$ then has zero mean and unit variance and may be taken as the first member of an orthonormal set which can be completed. It will often be convenient to write $\mu(-\infty, x) = F(x)$ and $\nu(-\infty, x) = G(x)$, with the usual statistical convention that, when $F(x)$ and $G(x)$ are absolutely continuous, $F'(x) = f(x)$ and $G'(x) = g(x)$. The canonical form is especially simple if the hypothesis is made that $g(x)$ or $G(x)$ depends on a parameter, α, in such a way that the Taylor expansion can be written

$$(1.8) \qquad g(x) \equiv f(x, \alpha) = f(x, \alpha_0) + (\alpha - \alpha_0)f'(x, \alpha_0) + O(\alpha - \alpha_0)^2$$

$$\simeq f(x, \alpha_0)\left(1 + (\alpha - \alpha_0)\frac{\partial}{\partial\alpha_0}\log f(x, \alpha_0)\right).$$

If $(\alpha - \alpha_0)^2$ and terms of higher order can be neglected, $\xi^{(1)}$ can be approximated by

$$(1.9) \qquad \xi^{(1)} = \frac{\partial}{\partial\alpha}\log f(x) \Big/ \left[\int\left[\frac{\partial}{\partial\alpha}\log f(x)\right]^2 f(x)\,dx\right]^{\frac{1}{2}}.$$

Theorem 1.4. *The ϕ^2-bounded property is preserved under all transformations of the variable. ϕ^2 is invariant under measure preserving transformations.*

Proof. A transformation of the variable cannot refine the partition into sets and so cannot increase ϕ^2. The second statement follows directly from the definition.

2. FINITE DISCRETE DISTRIBUTIONS IN TWO DIMENSIONS

Let X and Y, respectively row and column variables, take precisely m and n distinct values with positive probabilities; $m \leqslant n$. Without loss of generality the spaces of X and Y can be taken to be the set of row and column numbers. A bivariate distribution on the mn points can be defined by

$$(2.1) \quad \mathscr{P}\{X = i, Y = j\} = f_{ij}, \qquad \sum_{i,j} f_{ij} = 1, \qquad f_{ij} \geqslant 0,$$

$$i = 1, 2, \ldots, m;$$

$$j = 1, 2, \ldots, n.$$

$$(2.2) \qquad \sum_j f_{ij} = f_{i\cdot}, \qquad \sum_i f_{ij} = f_{\cdot j}, \qquad \sum_i f_{i\cdot} = \sum_j f_{\cdot j} = 1.$$

To avoid trivialities, it is assumed that $f_{i\cdot} > 0, f_{\cdot j} > 0$.

The f_{ij} can be written as the elements of a matrix, $\mathbf{F}$.

Lemma 2.1. *Given an arbitrary real $m \times n$ matrix, $\mathbf{B}$, there exists a pair of orthogonal matrices, $\mathbf{M}$ and $\mathbf{N}$, of sizes m and n, such that*

$$(2.3) \qquad\qquad \mathbf{M}^T\mathbf{B}\mathbf{N} = [\text{diag}\,(c_i), 0_{m,n-m}].$$

Further $\mathbf{M}$ and $\mathbf{N}$ are such that $\mathbf{M}^T\mathbf{B}\mathbf{B}^T\mathbf{M}$ and $\mathbf{N}^T\mathbf{B}^T\mathbf{B}\mathbf{N}$ are in diagonal form.

Proof. Select an orthogonal $\mathbf{N}$ such that $\mathbf{N}^T\mathbf{B}^T\mathbf{B}\mathbf{N}$ is in diagonal form with the positive latent roots arranged in descending order of magnitude. At most m latent roots of $\mathbf{B}^T\mathbf{B}$ can be positive since the rank of $\mathbf{B}^T\mathbf{B}$ is at most m. Since $(\mathbf{BN})^T\mathbf{BN} = \mathbf{N}^T\mathbf{B}^T\mathbf{BN}$ is a diagonal matrix, the k non-zero columns of $\mathbf{BN}$ are mutually orthogonal. They can be normalized and then taken to be the first k columns of an orthogonal $\mathbf{M}$. Let $\mathbf{M}$ be completed with due regard for the orthogonal conditions. $\mathbf{M}^T\mathbf{BN}$ has the required canonical form since the elements of its first k rows are simply those of $\mathbf{N}^T\mathbf{B}^T\mathbf{BN}$ multiplied by normalizing factors, and all remaining elements are zeroes.

Theorem 2.1. *Given a discrete distribution on $m \times n$ points with proportion f_{ij} in the cell of the ith row and j^{th} column, let an $m \times n$ matrix $\mathbf{B}$ be defined by*

$$(2.4) \qquad b_{ij} = f_{ij} f_{i.}^{-\frac{1}{2}} f_{.j}^{-\frac{1}{2}}.$$

Then orthogonal matrices $\mathbf{M}$ and $\mathbf{N}$ exist with elements of the first column $\sqrt{f_{i.}}$ and $\sqrt{f_{.j}}$ respectively such that $\mathbf{M}^T\mathbf{BN}$ is in canonical form, namely

$$(2.5) \qquad \mathbf{M}^T\mathbf{BN} = \mathbf{C} = [\operatorname{diag}(1, \rho_1, \rho_2, \ldots, \rho_{m-1}), \mathbf{0}_{m,n-m}]$$

with $1 \geqslant \rho_1 \geqslant \rho_2 \geqslant \cdots \geqslant \rho_{m-1} \geqslant 0$.

Proof. Let $\mathbf{M}_1$ and $\mathbf{N}_1$ be any orthogonal matrices of sizes, m and n, respectively with the required first columns. Then it is easy to verify that

$$(2.6) \qquad \mathbf{M}_1^T\mathbf{BN}_1 = \begin{bmatrix} 1 & 0 \\ 0 & \mathbf{B}_2 \end{bmatrix},$$

where $\mathbf{B}_2$ is $(m-1) \times (n-1)$, since the first row of $\mathbf{M}_1^T\mathbf{B}$ is $\sqrt{f_{.j}}$ and the first column of $\mathbf{BN}_1$ is $\sqrt{f_{i.}}$. By the Lemma, there are two orthogonal matrices, $\mathbf{M}_2$ and $\mathbf{N}_2$, of sizes $(m-1)$ and $(n-1)$ respectively, such that $\mathbf{M}_2^T\mathbf{B}_2\mathbf{N}_2$ has the required diagonal form (2.3). We write

$$(2.7) \qquad \mathbf{M} = \mathbf{M}_1(1 \dotplus \mathbf{M}_2), \qquad \mathbf{N} = \mathbf{N}_1(1 \dotplus \mathbf{N}_2)$$

and the result (2.5) follows. To prove that no ρ is greater than unity, we note that $\rho_i = c_{i+1,i+1} = \sum_k \sum_l m_{k,i+1} b_{kl} n_{l,i+1} = \sum_{k,l} x_k^{(i)} f_{kl} y_l^{(i)}$, where $x^{(i)}$ and $y^{(i)}$ are orthonormal functions by the discussion of Section IV.4. But then $\rho_i^2 \leqslant 1$ by the Bunyakovsky-Schwarz inequality. In particular, $\rho_1 \leqslant 1$ and so the ρ_i can be arranged as in the theorem.

Corollary 1. 1 *and* ρ_i^2 *are the non-zero latent roots of* $\mathbf{BB}^T$ *and* $\mathbf{B}^T\mathbf{B}$.

Proof. $\mathbf{CC}^T$ and $\mathbf{C}^T\mathbf{C}$ are diagonal matrices and from (2.5) it follows that they are the diagonal canonical forms of $\mathbf{BB}^T$ and $\mathbf{B}^T\mathbf{B}$, respectively.

Corollary 2. (Fisher's identity; Maung, 1942).

$$(2.8) \qquad f_{ij} = f_{i.} f_{.j} \left(1 + \sum_1^{m-1} \rho_k x_i^{(k)} y_j^{(k)} \right).$$

Proof. Let $\mathbf{D}_r = \text{diag}\,(f_{i.}^{\frac{1}{2}})$, $\mathbf{D}_c = \text{diag}\,(f_{.j}^{\frac{1}{2}})$

$$\mathbf{F} = \mathbf{D}_r\mathbf{B}\mathbf{D}_c = \mathbf{D}_r\mathbf{M}\mathbf{C}\mathbf{N}^T\mathbf{D}_c = \mathbf{D}_r^2\mathbf{X}\mathbf{C}\mathbf{Y}^T\mathbf{D}_c^2$$

where $\mathbf{X}$ is the matrix of values x_{ij} of the $(j-1)^{\text{th}}$ orthogonal function $j = 1, 2, \ldots, m$ on the ith row and $\mathbf{Y}$ is similarly defined. (2.8) follows.

Corollary 3. *The marginal variables, X and Y, are mutually independent if and only if $\rho_i = 0$ for $i = 1, 2, \ldots, m - 1$.*

Proof. If $\rho_i = 0$ in (2.8), then $f_{ij} = f_{i.}f_{.j}$. If this last expression holds then the sum on the right of (2.8) is identically unity and hence $\rho_i = 0$ for the set $\{x_i^{(k)}x_j^{(k')}\}$ is linearly independent on the mn points, since it is orthonormal on the mn points.

Definition. *ϕ^2 is defined in a contingency table by*

$$(2.9) \qquad \phi^2 + 1 = \sum_{i,j} f_{ij}^2/(f_{i.}f_{.j}).$$

Theorem 2.2.

$$\phi^2 = \sum_1^{m-1} \rho_i^2.$$

Proof.

$$\phi^2 = \sum_{i,j} b_{ij}^2 - 1 = \sum_{i,j} c_{ij}^2 - 1 = \sum_1^{m-1} \rho_i^2;$$

alternatively

$$1 + \phi^2 = \text{tr}\,\mathbf{B}\mathbf{B}^T = \text{tr}\,(\mathbf{M}^T\mathbf{B}\mathbf{N})(\mathbf{M}^T\mathbf{B}\mathbf{N})^T = \text{tr}\,\mathbf{C}\mathbf{C}^T = 1 + \sum_1^{m-1} \rho_i^2$$

Theorem 2.3. *A necessary and sufficient condition for independence is $\phi^2 = 0$.*

Proof. In Theorem 1.2 set $\mu(i,j) = f_{i.}f_{.j}$ and $\nu(i,j) = f_{ij}$. The theorem can also be deduced from Theorems 2.1 and 2.2. Theorem 2.2 is the finite analogue of Parseval's equality.

3. ϕ^2-BOUNDED BIVARIATE DISTRIBUTIONS

In Section 2, the Pearson functional, ϕ^2, has been defined for a finite discrete distribution. Pearson (1904) extended the definition of ϕ^2 to a general bivariate distribution by refining the partitions of the marginal distributions and passing to the limit. It is convenient to modify Pearson's definition by using the integral sign in the sense of Lebesgue-Stieltjes and adopting the notation of Hellinger (1909). We therefore define ϕ^2 by

$$(3.1) \qquad \phi^2 + 1 = \iint\limits_{-\infty}^{+\infty} [dF(x, y)]^2/[dG(x)\,dH(y)],$$

or

$$(3.2) \qquad \phi^2 + 1 = \int\!\!\int\limits_{-\infty}^{+\infty} \Omega^2(x, y) \, dG(x) \, dH(y),$$

where

$$(3.3) \qquad \Omega(x, y) = dF(x, y)/[dG(x) \, dH(y)].$$

$\Omega(x, y)$ is to be taken as zero, if the point (x, y) does not correspond to points of increase of both $G(x)$ and $H(y)$, the marginal distribution functions of a general bivariate distribution function, $F(x, y)$. ϕ^2 can be regarded as the limit of sums, $\sum\limits_{i,j} f_{ij}^2/(f_{i.}f_{.j}) - 1$ where f_{ij} is the measure of the bivariate distribution of the intersection of sets, A_i and B_j, and $f_{i.}$ and $f_{.j}$ are the measures of the marginal distributions for the same sets.

It follows from the definition by an analysis similar to that used in Section 1, that ϕ^2 is uniquely determined by the passage to the limit if it is bounded. The bivariate distribution defined by the distribution function, $G(x)H(y)$, is sometimes referred to as the conjugate distribution, an unfortunate term; preferable terms are *product distribution* or the *Cartesian product* of the marginal distributions, often written $G(x) \times H(y)$. It is assumed in this section that the bivariate distributions are ϕ^2-bounded with respect to the product distribution. The steps taken can be justified by the theory of integral equations and orthogonal functions. Many authors have contributed to the development of the ideas of a canonical form in a statistical context, of whom we may cite Fisher (1940), Friede and Münzner (1948), Gebelein (1941 and 1952), Hirschfeld (1935), Hotelling (1936), Lancaster (1957, 1958, 1963a and b), Maung (1941), Pearson (1900b and c, 1904), Rényi (1959), Richter (1949), Sarmanov (1941–1966) and Sarmanov and Zaharov (1960a and b).

Let $\{x^{(i)}\}$ and $\{y^{(i)}\}$ be complete sets of orthonormal functions defined on the marginal distributions, $G(x)$ and $H(y)$, respectively obeying the orthonormalizing conditions,

$$(3.4) \qquad x^{(0)} = y^{(0)} = 1, \qquad \int x^{(i)} x^{(j)} \, dG(x) = \int y^{(i)} y^{(j)} \, dH(y) = \delta_{ij}.$$

Then the coefficients of correlation, ρ_{ij},

$$(3.5) \qquad \rho_{ij} = \int\!\!\int x^{(i)} y^{(j)} \, dF(x, y),$$

are bounded in absolute value by unity by the Bunyakovsky-Schwarz inequality since G and H can be replaced by F in the integrals of (3.4). Further

$$(3.6) \qquad \rho_{00} = 1, \qquad \rho_{0k} = \rho_{k0} = 0, \qquad k \neq 0.$$

Theorem 3.1. *If $F(x, y)$ is a ϕ^2-bounded distribution and if*

$$(3.7) \qquad S_{mn} = S_{mn}(x, y) = \sum_{i=0}^{m} \sum_{j=0}^{n} \lambda_{ij} x^{(i)} y^{(j)},$$

then

(3.8)
$$Q_{mn} = \iint (\Omega - S_{mn})^2 \, dG(x) \, dH(y)$$

is minimized by taking

(3.9) $\lambda_{ij} = \rho_{ij}, \qquad i = 0, 1, 2, \ldots, m; \quad j = 0, 1, 2, \ldots, n.$

Writing S for S_{mn} as $m \to \infty$ and $n \to \infty$,

(3.10)
$$\Omega(x, y) = S(x, y),$$

in the mean square, and

(3.11)
$$\sum_{i=1}^{\infty} \sum_{j=1}^{\infty} \rho_{ij}^2 = \phi^2,$$

Proof. The set $\{x^{(i)}\} \times \{y^{(i)}\}$ is complete over the distribution $G(x) \times H(y)$, and $\Omega(x, y)$, as defined in (3.3), is square summable by (3.2) and the hypothesis of the theorem. The result (3.9) follows by differentiating (3.8) with regard to λ_{ij} for $i = 0, 1, 2, \ldots, m; \; j = 0, 1, 2, \ldots, n$ or using the device of completing the square. For any finite m and n, the sum $\sum\limits_{i,j} \rho_{ij}^2 \leqslant \phi^2$, so that $\sum\limits_{i,j} \rho_{ij}^2$ converges. The completeness ensures the truth of (3.11), which is the Parseval equality. The correlations may be written in the form of a (possibly) infinite matrix,

(3.12) $\mathbf{R} = (\rho_{ij}), \qquad i = 1, 2, \ldots, \qquad j = 1, 2, \ldots.$

It is our aim now to obtain by orthogonal transformations, complete sets of orthonormal functions in the two marginal variables such that $\mathbf{R}$ is diagonal. This problem has already been solved in Section 2 for the finite case. The definitions of Fisher (1940) and Maung (1942) may be extended as follows. The canonical variables (or functions) are two sets of orthonormal functions defined on the marginal distributions in a recursive manner such that the correlation between corresponding members of the two sets is a maximum. Unity may be considered as a member of zero order of each set of variables. Symbolically, the orthogonal and normalising conditions are

$$\xi^{(i)} \equiv \xi^{(i)}(x), \qquad \eta^{(i)} \equiv \eta^{(i)}(y),$$

$$\int \xi^{(i)} \, dG(x) = \int \eta^{(i)} \, dH(y) = 0, \qquad i = 1, 2, \ldots,$$

(3.13) $\int \xi^{(i)^2} \, dG(x) = \int \eta^{(i)^2} \, dH(y) = 1, \qquad i = 1, 2, \ldots,$

$$\int \xi^{(i)} \xi^{(j)} \, dG(x) = \int \eta^{(i)} \eta^{(j)} \, dH(y) = 0 \quad \text{for} \quad i \neq j,$$

and the maximisation conditions are that

(3.14) $$\rho_i = \operatorname{corr}(\xi^{(i)}, \eta^{(i)}) = \iint \xi^{(i)} \eta^{(i)} \, dF(x, y)$$

should be maximal for each i, given the preceding canonical variables. The ρ_i are the canonical correlations and can by convention be taken always to be positive, multiplying $\eta^{(i)}$ by -1 if necessary.

Theorem 3.2. *The canonical variables obey a second set of orthogonal conditions*

(3.15) $$\mathscr{E}(\xi^{(i)}\eta^{(j)}) \equiv \iint \xi^{(i)} \eta^{(j)} \, dF(x, y) = 0, \qquad \text{if } i \neq j.$$

Conversely suppose that $\{\xi^{(i)}\}$ and $\{\eta^{(i)}\}$ are complete sets of orthonormal functions on the respective marginal distributions and that $\rho_i = \int \xi^{(i)} \eta^{(i)} \, dF \geq 0$ for $i = 1, 2, \ldots$, where $\rho_1 \geq \rho_2 \geq \rho_3 \ldots \geq 0$, $\sum \rho_i^2 < \infty$. Then if (3.15) holds, the ρ_i are the canonical variables and $\{(\xi^{(i)}, \eta^{(i)})\}$ is the set of pairs of canonical variables.

Proof. For definiteness, let $j > i$. By hypothesis $\mathscr{E}(\xi^{(i)}\eta^{(i)})$ is maximal in the sense of the definition above and is equal to ρ_i, say. Suppose that $\mathscr{E}(\xi^{(i)}\eta^{(j)})$ is not zero but equal to $\rho_i \tan \theta$ for some $\tan \theta \neq 0$. Now $\eta^{(j)}$ has been defined according to (3.13) and so the function, $\cos \theta \eta^{(i)} + \sin \theta \eta^{(j)}$, obeys all the necessary orthogonal and normalizing conditions, and its correlation with $\xi^{(i)}$ is easily found to be $\rho_i \sec \theta$ and this is greater than ρ_i in absolute value, a contradiction results and so the theorem is proved.

Conversely, suppose that the first canonical variables are ξ and η. Then $\xi = \sum a_i \xi^{(i)}$ and $\eta = \sum b_i \eta^{(j)}$, $\sum a_i^2 = \sum b_i^2 = 1$ from the assumed completeness. $\int \xi \eta \, dF = \sum a_i b_i \rho_i \leq \frac{1}{2} \sum (a_i^2 + b_i^2) \rho_i \leq \rho_1$; the first canonical variables are obtained by setting $a_1 = b_1 = 1$, $a_j = 0$, $b_j = 0$ for $j \neq 1$. The second pair can be found similarly; $a_1 = b_1 = 0$ from (3.13), $a_2 = b_2 = 1$, $a_j = b_j = 0$ if $j \neq 2$. The canonical variables can thus be obtained recursively.

By (3.13) the canonical functions are necessarily square summable and so can be written as linear forms in any complete set of orthonormal functions, defined on the marginal distributions. Thus we can write

(3.16) $$\xi^{(i)} = \sum_{k=1}^{\infty} a_{ik} x^{(k)}, \qquad \sum_k a_{ik}^2 = 1$$

$$\eta^{(i)} = \sum_{k=1}^{\infty} b_{ik} y^{(k)}, \qquad \sum_k b_{ik}^2 = 1.$$

If the sets $\{x^{(i)}\}$ and $\{y^{(i)}\}$ were canonical in Theorem 3.1, the expansion would take a particularly simple form. But it may happen that the canonical sets are not complete. In this case, we can adjoin sets $\{\xi^{*(i)}\}$ and $\{\eta^{*(i)}\}$ to

$\{\xi^{(i)}\}$ and $\{\eta^{(i)}\}$ to form complete orthonormal sets on the marginal distributions and proceed as in Theorem 3.1.

Theorem 3.3. *If $F(x, y)$ is a ϕ^2-bounded bivariate distribution with marginal distributions, $G(x)$ and $H(y)$, then complete sets of orthonormal functions can be defined on the marginal distributions such that each canonical variable appears as a member of the complete set of orthonormal functions. The element of frequency can be expressed in terms of the marginal distributions,*

$$(3.17) \qquad dF(x, y) = \left\{1 + \sum_1^\infty \rho_i x^{(i)} y^{(i)}\right\} dG(x)\, dH(y),$$

in mean square, and

$$(3.18) \qquad \phi^2 = \sum_{i=1}^\infty \rho_i^2.$$

Proof. Let us take as the orthonormal sets, the direct sums $\{\xi^{(i)}\} + \{\xi^{*(i)}\}$ and $\{\eta^{(i)}\} + \{\eta^{*(i)}\}$, respectively, and call them $\{x^{(i)}\}$ and $\{y^{(i)}\}$. Then taking any finite sums with $m = m_1 + m_2$, and $n = n_1 + n_2$,

$$(3.19) \qquad S_{mn} = S_{mn}(x, y) = \sum_{i=0}^m \sum_{j=0}^n \lambda_{ij} x^{(i)} y^{(j)}$$

$$= \sum_{i=0}^{m_1} \sum_{j=0}^{n_1} \lambda_{ij} \xi^{(i)} \eta^{(j)} + \sum_{i=0}^{m_1} \sum_{j=1}^{n_2} \alpha_{ij} \xi^{(i)} \eta^{*(j)}$$

$$+ \sum_{i=1}^{m_2} \sum_{j=0}^{n_1} \beta_{ij} \xi^{*(i)} \eta^{(j)} + \sum_{i=1}^{m_2} \sum_{j=1}^{n_2} \gamma_{ij} \xi^{*(i)} \eta^{*(j)}$$

and minimizing the sum of squares Q_{mn} of (3.8), we obtain in the first expression $\lambda_{00} = 1$, $\lambda_{ij} = \delta_{ij} \rho_i$. All the terms, α_{ij} and β_{ij}, in the second and third expressions will vanish by Theorem 3.2 and we now show that all the γ_{ij} are zero. It follows as in Theorem 3.1 that

$$(3.20) \qquad \sum_i \rho_i^2 + \sum_{i,j} \gamma_{ij}^2 \leqslant \phi^2.$$

Either the number of functions in the set $\{\xi^{(i)}\}$ of canonical variables is finite or $\rho_i \to 0$ as $i \to \infty$ and the number in the set $\{\xi^{(i)}\}$ is infinite. In the first case, a contradiction results if any $|\gamma_{ij}| > 0$ because there is some linear function in the $\{\xi^{*(i)}\}$ which has correlation with some linear function of the $\{\eta^{*(i)}\}$ not less than $|\gamma_{ij}|$. These linear functions, contrary to hypothesis, would thus be canonical functions. In the second case, some pairs of linear functions would again have correlation greater than the ρ_i for $i > i_0$ where i_0 is sufficiently large; again a contradiction has been reached. The γ_{ij} are thus all zero. When the integral (3.8) is minimized therefore, only terms involving the canonical variables appear and (3.17) holds; (3.18) follows from the general theory and is the Parseval equation.

It may be proved that the choice of orthonormal funetions is unique except for a convention as to sign if the ρ_i form a pair-wise different set; once $x^{(i)}$ is chosen, $y^{(i)}$ is defined so as to give the expectation of $x^{(i)}y^{(i)}$ a positive value. If, however, $\rho_{j+1}, \rho_{j+2}, \ldots, \rho_{j+k}$ are of equal magnitude and $x^{(j+1)}, x^{(j+2)}, \ldots, x^{(j+k)}$ is one solution for the corresponding canonical variables, then any other solution is given by a particular orthogonal transformation of these $x^{(j+1)} \ldots x^{(j+k)}$ and the same transformation on the $y^{(j+1)} \ldots y^{(j+k)}$. A converse of Theorem 3.3 holds.

Theorem 3.4. *If a bivariate distribution can be written in the form* (3.17) *with* $\{x^{(i)}\}$ *and* $\{y^{(i)}\}$ *forming complete sets on the marginal distributions and if* $\sum_i \rho_i^2$ *is finite, then the* ρ_i *are the canonical correlations,* $x^{(i)}$ *and* $y^{(i)}$ *are the canonical variables and* $\sum_i \rho_i^2 = \phi^2$.

Proof. We suppose first that the ρ_i are pairwise different. Then if ξ and η are the first pair of canonical variables

$$(3.21) \qquad \text{corr}\,(\xi, \eta) = \text{corr}\left(\sum_i a_i x^{(i)}, \sum_j b_j y^{(j)}\right) = \sum_i a_i b_i \rho_i,$$

$$\leqslant \tfrac{1}{2} \sum (a_i^2 + b_i^2)\rho_i.$$

The sum on the right of (3.21) is clearly maximized by taking $a_1 = b_1 = 1$. If $\rho_1 = \rho_2 \ldots = \rho_k > \rho_{k+1}$, the solution is not unique because the first k canonical correlations have the same value ρ_i. The first statement of the theorem is then proved by induction.

The integral of the square of $\Omega(x)$ with respect to the distribution function $G(x) \times H(y)$ is equal to the sum of squares of the coefficients, ρ_i, the Parseval equality.

4. THE GENERAL BIVARIATE DISTRIBUTION

In Section 3, it has been assumed that the bivariate distributions are ϕ^2-bounded with respect to the product distribution. However, it is desirable to have theorems which hold without that restriction.

Definition. *Let* ξ *be a two valued function on the space of X. Then the two values of* ξ *define a partition of the space of X. Further* var $\xi \neq 0$. *Similarly let* η *be a two valued function on the space of Y. Then* ξ *and* η *possess a joint distribution on a* 2×2 *space. Corr* (ξ, η) *may be referred to as a Boas or Boas-Yule coefficient of correlation.*

Example   (i) ξ is the indicator variable of a set A. $\mathscr{P}(X \in A)\mathscr{P}(X \notin A) \neq 0$.

Example   (ii) ξ is a two valued function with zero mean and unit variance. If η also has zero mean and unit variance, then corr $(\xi, \eta) = \mathscr{E}(\xi \eta) = \iint \xi\eta \, dF$. See also the discussion in Section XI.3.

Theorem 4.1. *Let $\{x^{(i)}\}$ and $\{y^{(j)}\}$ be orthonormal sets, complete on given marginal distributions, $G(x)$ and $H(y)$. Then the joint bivariate distribution is determined, almost everywhere with respect to the product measure, by the measures of the marginal distribution and the matrix of correlations (ρ_{ij}), where ρ_{ij} is the expectation of $x^{(i)}y^{(j)}$ with respect to the joint distribution.* (Lancaster, 1963a)

Proof. Suppose that there are two distinct distributions, $F_1(x, y)$ and $F_2(x, y)$, with marginal distributions $G(x)$ and $H(y)$ and possessing coefficients of correlation given by

$$(4.1) \quad \rho_{ij} = \int x^{(i)}y^{(j)} \, dF_k(x, y),$$

$$i = 1, 2, \ldots ; \qquad j = 1, 2, \ldots ; \qquad k = 1, 2.$$

If F_1 and F_2 are not identical then there is a product set $A \times B$ such that $\iint dF_1 \neq \iint dF_2$ for integration over the set $A \times B = \{x, y : x \in A, y \in B\}$. For the marginal distribution of X we can define ξ to be a standardized two valued function taking one of its values in A and the other in the complement of A. Similarly, η can be defined taking values in B and its complement. Corresponding to F_1 and F_2 there are thus two distinct 2×2 probability distributions with the same marginal distributions by hypothesis. Let the Boas coefficient of correlation in the 2×2 distribution corresponding to F_k be ρ_k. Then $\rho_1 \neq \rho_2$. A contradiction will now be demonstrated with the aid of (4.1).

Let us now approximate to ξ and η by series.

$$(4.2) \qquad S_m = \sum_1^m a_i x^{(i)}, \qquad S_n' = \sum_1^n b_j y^{(j)}$$

so that $\int (\xi - S_m)^2 \, dG$ and $\int (\eta - S_n')^2 \, dH$ are each less than ϵ^2. By the Schwarz inequality, for $k = 1, 2$,

$$(4.3) \qquad \left| \int \xi(\eta - S_n') \, dF_k \right| \leqslant \left[\int \xi^2 \, dF_k \int (\eta - S_n')^2 \, dF_k \right]^{\frac{1}{2}}$$

$$= \left[\int \xi^2 \, dG \int (\eta - S_n')^2 \, dH \right]^{\frac{1}{2}} < \epsilon.$$

Further $\left| \int S_n'(\xi - S_m) \, dF_k \right| < \epsilon$, since $\int S_n'^2 \, dH \leqslant \int \eta^2 \, dH = 1$. We now have

(4.4)

$$\rho_1 - \rho_2 = \int \xi\eta \, dF_1 - \int \xi\eta \, dF_2,$$

$$= \int (\xi\eta - S'_n S_m) \, dF_1 - \int (\xi\eta - S'_n S_m) \, dF_2, \qquad \text{since } \rho_{ij}^{(1)} = \rho_{ij}^{(2)}$$

$$= \int \xi(\eta - S'_n) \, dF_1 + \int S'_n(\xi - S_m) \, dF_1$$

$$- \int \xi(\eta - S'_n) \, dF_2 - \int S'_n(\xi - S_m) \, dF_2.$$

Taking absolute values and using (4.3)

(4.5) $$\qquad\qquad |\rho_1 - \rho_2| < 4\epsilon.$$

Since ϵ is arbitrary a contradiction has been reached if $\rho_1 \neq \rho_2$. F_1 and F_2 are therefore functions identical a.e.

Corollary 1. *A necessary and sufficient condition for independence of the marginal variables of a bivariate statistical distribution is that* $\rho_{ij} = 0$, *for* $i > 0$ *and* $j > 0$, *where the* ρ_{ij} *are defined by* (4.1).

Proof. Let $F_1(x, y)$ be the bivariate distribution and let $G(x)$ and $H(y)$ be the marginal distributions. Let the product distribution, F_2, be defined so that $F_2(x, y) = G(x)H(y)$. Then F_1 and F_2 have the same marginal distributions and further the same correlation matrix; for the matrix of correlations of F_1 consists of zeroes by hypothesis and the matrix of correlations of F_2 is easily verified to be the null matrix. F_1 and F_2 are therefore identical a.e. and so $F_1(x, y) = F_2(x, y) = G(x)H(y)$ which shows that the two random variables are mutually independent.

Remark. Corollary 1 can be deduced from the main theorem of Sarmanov (1960a) since if the maximal correlation is zero every ρ_{ij} must be zero. Sarmanov assumed ϕ^2-boundedness but this restriction was removed by Lancaster (1961b), who used the equivalence of statistical independence to the vanishing of the Boas coefficients for every choice of two-valued step functions. The method of proof of Theorem 4.1 can yield more general results, namely, that if the matrix of coefficients of correlation is determined for one pair of complete sets of orthonormal functions, it is determined for any other pair of complete sets.

It follows from Theorem 4.1 that, although the Radon-Nikodym derivative may not be expressible as a series, the measure assignable to any product set, of the form $A \times B$ in the bivariate space can be obtained from the hypotheses of the theorem, even though the ϕ^2 is not bounded.

The assumption of ϕ^2-boundedness can be removed from Theorem 3.3 to give a stronger theorem.

Theorem 4.2. *Let $F(x, y)$ be a bivariate distribution function and $G(x)$ and $H(y)$ the corresponding marginal distributions. Suppose that the canonical correlations are $\rho_1, \rho_2, \ldots$, and the canonical variables, $\{\xi^{(i)}\}$ and $\{\eta^{(i)}\}$, and that $\sum \rho_i^2 < \infty$. Then complete sets of orthonormal functions can be defined on the marginal distributions such that each canonical variable appears as a member of the complete set of orthonormal functions. The element of frequency can be expressed in terms of the marginal distributions,*

$$(4.6) \qquad \mathrm{d}F(x, y) = \left\{ 1 + \sum_1^\infty \rho_i x^{(i)} y^{(i)} \right\} \mathrm{d}G(x)\, \mathrm{d}H(y),$$

in mean square, and

$$(4.7) \qquad \phi^2 = \sum_{i=1}^\infty \rho_i^2,$$

where ϕ^2 is defined by (3.1).

Proof. The proof goes through after the style of Theorem 3.3. First, $\rho_i = 0$ or $\rho_i \to 0$ as $i \to \infty$ since the sum $\sum \rho_i^2$ is finite. Once again the coefficients α_{ij}, β_{ij}, and γ_{ij} can be shown to be zero or a contradiction would result. Now the proof of Theorem 4.1 shows that if the correlations are determined for any pair of complete sets on the marginal distributions they are also determined for any other and so are determined for the complete sets formed from the indicator functions on the margins. The sum of squares of the correlations is invariant and so every finite partition of the marginal spaces yields a ϕ^2 which is not greater than $\sum \rho_i^2$. ϕ^2 is thus bounded.

In a bivariate distribution on a finite number of points, Theorem 2.1 shows that the canonical correlations and variables can always be calculated. In the general case, approximate solutions can be obtained by considering increasingly fine partitions of the marginal spaces and calculating the canonical variables of these finite discrete spaces. The theorem that the kth latent root of a positive definite square symmetric matrix does not exceed in absolute value the kth latent root of the same matrix "bordered" by additional rows and columns enables us to state that in ϕ^2-bounded distributions the sequence of values of the kth canonical correlation converges from below to the theoretical value of the kth canonical correlation. For we can consider that we are looking for the latent roots of a sequence of matrices $\mathbf{BB}^T$ with $\mathbf{B}$ defined as in (2.4). A finite partition can be chosen so that $\sum b_{ij}^2 - 1 > \phi^2 - \epsilon^2$, where ϵ is chosen arbitrarily. Then no latent root, $\rho_k'^2$ say, of $\mathbf{BB}^T$ in the finite partition can differ by more than ϵ from the true value, ρ_k^2, since $\sum (\rho_j^2 - \rho_j'^2) < \epsilon$ and all the differences are non-negative so that $\rho_k^2 - \rho_k'^2 < \epsilon$.

If ϕ^2 is not bounded, this reasoning does not apply and the only conclusion available is that $\rho_j \geqslant \rho_j'$ for $j = 1, 2, \ldots$.

There is no general method for determining the canonical variables in the distributions on infinitely many points in both variables. However, it may be possible to determine them in special cases.

Example   (iii) In the bivariate normal distribution, $\mathscr{E}(x^{(i)}y^{(j)}) = \delta_{ij}\rho^i$, and so Theorem 3.2 could be used to prove that $|\rho|^i$ are the canonical correlations and $x^{(i)}$ and $y^{(i)}$ the orthonormal (Hermite) polynomials on the marginal distributions.

5. MULTIVARIATE DISTRIBUTIONS

Let the distribution function, $F(x, y, z)$, of three random variables, X, Y and Z, have marginal distribution functions, $G(x)$, $H(y)$ and $K(z)$; these functions are written as F, G, H and K whenever there is no possibility of ambiguity. The discussion is limited to ϕ^2-bounded distributions, that is, those obeying the condition,

$$(5.1) \quad \phi^2 + 1 = \int (\mathrm{d}F)^2/(\mathrm{d}G\,\mathrm{d}H\,\mathrm{d}K) = \int (\mathrm{d}F/\mathrm{d}G\,\mathrm{d}H\,\mathrm{d}K)^2\,\mathrm{d}G\,\mathrm{d}H\,\mathrm{d}K < \infty$$

with the integral used in the sense of Hellinger and justified by a passage to the limit or alternatively thought of as the integral of the square of a Radon-Nikodym derivative. The marginal distributions may have a finite or infinite number of points of increase. Complete sets of orthonormal functions can always be defined on the marginal distribution and the set of products of one term from each marginal set may be written as a product set $\{x^{(i)}\} \times \{y^{(i)}\} \times \{z^{(i)}\}$ which is complete on the product distribution, written symbolically as $G \times H \times K$ or GHK. To bring the discussion more into relation with the analysis of variance we define

$$(5.2) \qquad \rho_{ijk} = \int x^{(i)}y^{(j)}z^{(k)}\,\mathrm{d}F$$

as a generalized coefficient of correlation and note that analogues can be defined for higher dimensions also. ρ_{000} is a unity. ρ_{00k} and two similar expressions are zero and may be thought of as zero-order interactions. ρ_{ij0} and two similar expressions are ordinary coefficients of correlation. ρ_{ijk} for $i \neq 0$, $j \neq 0$ and $k \neq 0$ is a generalized coefficient of correlation. In the most general case a new feature emerges, for ρ_{ijk} may not be bounded in absolute value by unity.

Theorem 5.1. *A necessary and sufficient condition for independence is that ϕ^2 of* (5.1) *should be zero.*

Proof. This is proved in the same way as the two-dimensional case.

Theorem 5.2. *If F is ϕ^2-bounded with respect to its product distribution, $G \times H \times K$, then dF can be expanded in the form*

$$(5.3) \quad dF = \left\{ 1 + \sum_{j,k} \rho_{0jk} y^{(j)} z^{(k)} + \sum_{i,k} \rho_{i0k} x^{(i)} z^{(k)} + \sum_{i,j} \rho_{ij0} x^{(i)} y^{(j)} \right.$$
$$\left. + \sum_{i,j,k} \rho_{ijk} x^{(i)} y^{(j)} z^{(k)} \right\} dG \, dH \, dK,$$

in mean square, and the sum of the squares of the coefficients is ϕ^2.
Proof. The proof follows as in the two-dimensional case.

Theorem 5.3. *A necessary and sufficient condition for independence in the class of multivariate distributions, ϕ^2-bounded with respect to the product distribution, is that the sum of squares of the coefficients in (5.3) should be zero.*
Proof. The proof is obvious.

Theorem 5.4. *For the class of three dimensional ϕ^2-bounded measures, the two dimensional marginal measures are ϕ^2-bounded and the bivariate element of distribution can be expanded in the form,*

$$(5.4) \quad dL(x, y) = \left\{ 1 + \sum_{i,j} \rho_{ij0} x^{(i)} y^{(j)} \right\} dG \, dH, \qquad L(x, y) \equiv F(x, y, \infty),$$

almost everywhere.
Proof. ϕ^2 is not increased by pooling sets (i.e. condensation). In particular, all the sets of the Z-space can be pooled. The marginal measure is thus ϕ^2-bounded with respect to the product measure $G \times H$. Alternatively, integrating both sides of (5.3) with respect to the variable, z, the left side becomes $dL(x, y)$ and the right side has the required form since all the terms involving $z^{(k)}$, for $k > 0$, become zero on integration.

Theorem 5.5. *For any choice of complete sets of orthonormal functions on the marginal distributions,*

$$\sum \rho_{ij0}^2 = (XY0), \qquad \sum \rho_{i0k}^2 = (X0Z), \qquad \sum \rho_{0jk}^2 = (0YZ), \qquad \sum \rho_{ijk}^2 = (XYZ),$$

and consequently their sum, are invariants.
Proof. This depends on applications of Parseval's equality. $(XY0)$ is obviously equal to the two dimensional ϕ^2, ϕ^2_{XY} say of (3.1), and ϕ^2_{XY} does not depend on the particular complete orthonormal sets chosen. Similar considerations apply to $(X0Z)$ and $(0YZ)$. Similarly the total sum of squares of (5.3) is also independent of the complete orthonormal sets chosen because of the mode of definition in (5.2).

The Parseval equality permits us to write ϕ^2_{XY} or $(XY0)$. Probability measures in three dimensions can now be classified into $2^4 = 16$ classes since each of the sums $(XY0)$, $(X0Z)$, $(0YZ)$, (XYZ) can be either zero or positive. Neglecting cases which are obtained by permutation of the variables, there

Table 5.1

Independence Table for a Three Dimensional Statistical Distribution

	(i)	(ii)	(iii)	(iv)	(v)	(vi)	(vii)	(viii)
Interactions			Values	of the	Interactions			
(XYZ)	0*	0	0	0	+†	+	+	+
$(XY0)$	0	+	+	+	0	+	+	+
$(0YZ)$	0	0	+	+	0	0	+	+
$(X0Z)$	0	0	0	+	0	0	0	+
Variables			Dependence	Values				
X, Y	*I*	*D*	*D*	*D*	*I*	*D*	*D*	*D*
Y, Z	*I*	*I*	*D*	*D*	*I*	*I*	*D*	*D*
X, Z	*I*	*I*	*I*	*D*	*I*	*I*	*I*	*D*
$X, (Y, Z)$‡	*I*	*D*	*D*	*D*	*D*	*D*	*D*	*D*
$Y, (X, Z)$	*I*	*D*	*D*	*D*	*D*	*D*	*D*	*D*
$Z, (X, Y)$	*I*	*I*	*D*	*D*	*D*	*D*	*D*	*D*

* 0 signifies $(XYZ) = 0$ *et cetera.*
† + signifies $(XYZ) \neq 0$ *et cetera.*
‡ *I* in this part of the table is to be read, X is independent of the joint distribution of Y and Z. *D* signifies that this is not so.

are eight types, for four of which (XYZ) is zero. Corresponding to each of the possibilities $(XYZ) = 0$ and $(XYZ) > 0$, none, one, two or three of the first order ϕ^2's may be non-zero. The eight possibilities are listed in Table 5.1. Note that permutation of the variables X, Y and Z yields two other cases in (ii), (iii), (vi) and (vii) which brings the total cases up to 16.

Example. There exist three dimensional distributions of all eight types. For if we define each marginal variable to assume only the two values, ± 1, with equal probabilities, then

$$(5.5) \qquad p(x, y, z) = (1 + \rho_{110}xy + \rho_{011}yz + \rho_{101}xz + \rho_{111}xyz)/8$$

is a probability distribution on the eight points if $|\rho_{110}| + |\rho_{011}| + |\rho_{101}| + |\rho_{111}| \leqslant 1$. Under these conditions $p(x, y, z) \geqslant 0$ and the sum of the frequencies is unity.

The different cases can now be interpreted with the aid of Table 5.1 in which is written *I* for independence and *D* for dependence. A dependence exists if the independence conditions are not fulfilled. The various cases given above may be considered in turn: (It should be noted that in (v) to (viii) no variable is independent of the joint distribution of the other two.)

(i) Complete independence.

(ii) Z is independent separately of X and of Y and of them jointly. All other possible relations show dependencies.

(iii) X is independent of Z. All other relations are dependencies.

(iv) No pair of marginal variables is mutually independent. No variable is independent of the joint distribution of the other two.

(v) The marginal variables are independent in pairs but no variable is independent of the other two jointly.

(vi) Two pairs of marginal variables are mutually independent but no variable is independent of the other two jointly.

(vii) One pair of marginal variables is mutually independent.

(viii) All relations are dependencies.

The relations may be gathered together as a theorem.

Theorem 5.6. ϕ^2-*bounded three dimensional distributions can be classed into 8 types according to whether the* ϕ^2_{XYZ} *is zero or not and according to the number of the* ϕ^2_{XY}, ϕ^2_{YZ} *and* ϕ^2_{XZ} *which are zero.*

The dependence or independence can be read off from Table 5.1.

Theorem 5.7. *A necessary and sufficient condition for the mutual independence of a set of n random variables is the vanishing of all the ordinary and generalized coefficients of correlation of the form,*

$$(5.6) \qquad \rho^{\{i_j\}} = \mathscr{E}[x_1^{(i_1)} x_2^{(i_2)} \ldots x_n^{(i_n)}], \qquad i_j = 0, 1, 2, \ldots, \quad j = 1, 2, \ldots, n.$$

Note. A related theorem was given by Sarmanov (1958) for ϕ^2-bounded distributions. It was proved in the general case by Lancaster (1961b). We introduce and prove two lemmas and give as a Lemma a result already proved.

Lemma 5.1. *Let* $X_1, X_2, \ldots, X_n$ *be a set of n random variables with joint probability function,* $P(A)$, *and product probability function,* $P^*(A)$, *defined by*

$$(5.7) \qquad P(A) \equiv P(A_{\{j\}}^{\{i_j\}}) = P(X_1 \in A_1^{(i_1)}, X_2 \in A_2^{(i_2)}, \ldots, X_n \in A_n^{(i_n)})$$

$$(5.8) \qquad P^*(A) \equiv P^*(A_{\{j\}}^{\{i_j\}}) = \prod_j P(X_j \in A_j^{(i_j)}).$$

The affixes $\{j\}$ and $\{i_j\}$ are to be interpreted as $1, 2, \ldots, n$ and $i_1, i_2, \ldots, i_n$. We shall write sums and integrals using the integral sign and dP or dP^* as the case may be. Real orthonormalised functions can be defined on the distribution of each X_j and will be denoted by $x_j^{(i_j)}$, $i_j \neq 0$; $x_j^{(0)} = 1$.

Lemma 5.2. *If the random variables,* $X_1, X_2, \ldots, X_n$, *do not form a mutually independent set, then step functions* $\zeta_1, \zeta_2, \ldots, \zeta_n$, *everywhere finite and normalised to have zero mean and unit variance on the appropriate marginal distributions, can be found such that at least one of the products of the* X_i *taken two or more at a time has non-zero expectation.*

Proof. Let normalised step functions having only two values be defined as follows on each of the marginal distributions of the form,

$$(5.9) \qquad \zeta(x) = -\sqrt{\{P(X \in A^{(2)})/P(X \in A^{(1)})\}} \qquad \text{for } X \in A^{(1)}$$
$$= +\sqrt{\{P(X \in A^{(1)})/P(X \in A^{(2)})\}} \qquad \text{for } X \in A^{(2)}$$

where $P(A^{(1)}) + P(A^{(2)}) = 1$, $A^{(1)}$ and $A^{(2)}$ are mutually exclusive sets, neither $P(A^{(i)})$ being null. ζ is normalised on the distribution of X as well as on its own two-point distribution.

Let it now be supposed that the X_j do not form a mutually independent set. Then there exists at least one division of each of the marginal distributions into two proper mutually exclusive sets so that the $P(A) \equiv P(A^{\{i_j\}}_{\{j\}})$ of (5.7) is not equal to $P^*(A)$ of (5.8) for some set $\{i_j\}$. The superscripts i_j take values, 1 and 2. Now let us define a set of variables $\zeta_j \equiv \zeta_j(X_j)$ corresponding to this partition and consider the joint distribution of the $\{\zeta_i\}$. These new variables do not form a mutually independent set and the probability distribution is defined on 2^n points. The product set of functions $\{1, \zeta_1\} \times \{1, \zeta_2\} \times \ldots \times \{1, \zeta_n\}$ are all mutually orthogonal and indeed orthonormal on P^* and form an orthonormal basis for finite functions on the partition of P^* into 2^n sets of positive probability. In particular, the ratio, $Q \equiv Q(\{\zeta_j\})$, of the measures assigned to the 2^n "quadrants" by P and P^* by making the transformation, $\{X_j\} \to \{\zeta_j\}$, is finite and can be expressed as a linear form in the elements of the orthonormal basis. The coefficients of the elements in this linear form will be equal to $\int Q\zeta_1^{i_1}\zeta_2^{i_2} \ldots \zeta_n^{i_n} dP^*$, which is equal to $\int \zeta_1^{i_1}\zeta_2^{i_2} \ldots \zeta_n^{i_n} dP$, since on P and P^*, Q is a step function constant over any given "quadrant" and so is the product, $\zeta_1^{i_1}\zeta_2^{i_2} \ldots \zeta_n^{i_n}$; $i_j = 0$ or 1. The coefficient of the constant term is unity and all the other coefficients cannot vanish since if they did the P-measure assigned to each quadrant would be P^*-measure. This would constitute a contradiction since the hypothesis given assumes that the probability function $P(A)$ does not factorise for this particular partition. The coefficients of $\zeta_1, \zeta_2, \ldots, \zeta_n$ taken singly in the linear form are all zero.

Lemma 5.3. *If $X_1, X_2, \ldots, X_k$ form a mutually independent set, then a product of orthonormal functions, $\zeta_1\zeta_2 \ldots \zeta_k$, can be approximated in quadratic mean arbitrarily closely by expressions of the form, $S_1S_2 \ldots S_k$, where*

$$(5.10) \qquad S_j = \sum_{i=1}^{N_j} a_j^{(i)} x^{(i)}, \qquad j = 1, 2, \ldots, k$$

The proof is by induction. Given that the lemma is true for $(k-1)$, $S_1, S_2, \ldots, S_k$ can be chosen such that, for arbitrary ϵ, $\mathscr{E}(X_1 \ldots X_{k-1} - S_1 \ldots S_{k-1})^2 \leqslant \epsilon^2/4$ and also $\mathscr{E}(X_k - S_k)^2 \leqslant \epsilon^2/4$. Then since for real numbers, $(a + b)^2 \leqslant 2(a^2 + b^2)$,

$$(5.11) \quad (\zeta_1\zeta_2 \ldots \zeta_k - S_1S_2 \ldots S_k)^2 \leqslant 2[\{\zeta_1 \ldots \zeta_{k-1}(\zeta_k - S_k)\}^2$$
$$+ \{S_k(\zeta_1\zeta_2 \ldots \zeta_{k-1} - S_1S_2 \ldots S_{k-1})\}^2]$$

Taking expectations, we find that the expectation on the left of (5.11) is not greater than ϵ^2, for the independence of the k variables enables the expectations of the two expressions on the right to be evaluated as products of expectations. Further $\mathscr{E}(S_k^2) \leqslant \mathscr{E}(\zeta_k^2) = 1$.

Lemma 5.4. *Let $\{x_1^{(i_1)}\}$ and $\{x_2^{(i_2)}\}$ be complete orthonormal sets on the marginal distributions of two random variables, which have a joint probability measure, $P \equiv P(x_1, x_2)$, and let $x_1^{(0)} = x_2^{(0)} = 1$. Then a necessary and sufficient condition for the independence of X_1 and X_2 is that every coefficient of correlation, $\rho_{i_1 i_2} = \int x_1^{(i_1)} x_2^{(i_2)} \, dP$, should be zero for every $i_1 > 0$, $i_2 > 0$.*
    Proof. Lemma 5.4 is a version of Corollary 1 to Theorem 4.1.
    Proof *of Theorem* 5.7. The necessity is evident as in Theorem 4.1. For the sufficiency, it is impossible that any product of the form, $\zeta_i\zeta_j$, $i \neq j$ should have a non-zero expectation by Lemma 5.4. Let us now suppose the theorem true for sets of k variables and prove it true for a set of $k + 1$ variables, $2 \leqslant k < n$. Without loss of generality, we take the variables to be $X_1, X_2, \ldots, X_k$. Let us suppose that $X_1, X_2, \ldots, X_{k+1}$ do not form a mutually independent set. By Lemma 5.2 at least one product of the form $\zeta_{j_1}\zeta_{j_2} \ldots$ can be found which has a non-zero expectation. This can only be $\zeta_1\zeta_2 \ldots \zeta_{k+1}$ since any such product with less than $k + 1$ factors would have been zero. But now as before we can consider the identity

$$(5.12) \quad (\zeta_1\zeta_2 \ldots \zeta_k)\zeta_{k+1}$$
$$= \zeta_{k+1}(\zeta_1\zeta_2 \ldots \zeta_k - S_1S_2 \ldots S_k) + \zeta_1\zeta_2 \ldots \zeta_k(\zeta_{k+1} - S_{k+1})$$
$$- (\zeta_{k+1} - S_{k+1})(\zeta_1\zeta_2 \ldots \zeta_k - S_1S_2 \ldots S_k) - S_1S_2 \ldots S_{k+1}$$

and there follows

$$(5.13) \qquad 0 \neq \mathscr{E}(\zeta_1\zeta_2 \ldots \zeta_{k+1}) < 2\epsilon + \epsilon^2$$

and a contradiction has been reached. The vanishing of the generalised coefficients of correlation ensures the independence of the marginal variables taken in pairs, then in threes, fours, $\ldots$, and finally in a set of n. They form a completely independent set of random variables.

Corollary. *A necessary and sufficient condition for independence of two variables, X and Y, is that the maximum correlation between any two functions $\xi(X)$ and $\eta(Y)$ should be zero.* (Sarmanov 1958.)
    Proof. The necessity is obvious. The sufficiency follows as in Lemma 5.3 by noting that $\mathscr{E}\{\xi(X)\eta(Y)\} = 0$ forces all the $\rho_{ij} = \mathscr{E}(x_1^{(i)}x_2^{(j)})$ to be zero since otherwise $\xi(X)$ and $\eta(Y)$ would not have maximal correlation.

6. INDEPENDENCE AND ASSOCIATION

Yule (1932) in his well-known text devoted some five chapters to the theory of attributes, giving an account of the work of de Morgan (1847), Boole (1854), Jevons (1870) and Yule (1900, 1901 and 1903). He uses the Jevons notation, an observation being classed into one of two exhaustive and mutually exclusive sets, denoted by the corresponding pairs of Latin and Greek letters, such as A and α. Intersections of sets are denoted by notations such as AB, $A\beta$. The total numbers are denoted by N or (U), U being the universal class. In other words, there are random variables, $X, Y, Z, \ldots,$ $X \in A \cup \alpha, Y \in B \cup \beta$ and so on. Although Yule writes in terms of frequencies, much of his discussion is applicable to probability distributions; indeed, to any set of observed frequencies there corresponds a theoretical distribution defined by assigning a measure to any class equal to the observed relative frequencies and the measures so defined have the required sum of unity and are non-negative. The number of classes in a three-fold classification is 2^3 or 8 (in the general case, 2^n) and the numbers (ABC), $(AB\gamma)$, $\ldots$ are referred to as the ultimate class frequencies. They are necessarily consistent in the sense that no negative frequencies are implied. But the table may be given in alternative form; for example when only Latin letters appear as in (U), (A), (AB), $\ldots$; these are termed the positive-class frequencies. In the general case, there would be 2^n positive-class frequencies. The numbers in the ultimate classes can always be computed from the positive-class frequencies. A set of numbers describing a theoretical or observed distribution will be said to be consistent, if it does not imply that any ultimate class frequency is negative.

Some inequalities are obtained

$$(6.1) \qquad (A) \leqslant (U), \qquad (A) \geqslant 0.$$

With a twofold classification, we get further sets of inequalities; for example, since $(\alpha\beta) \geqslant 0$

$$(6.2) \qquad (U) - (A) - (B) + (AB) \geqslant 0.$$

These conditions have their generalization in the consistency conditions of Kolmogorov (1933). In the older literature, two variables are said to be associated or disassociated according as

$$(6.3) \qquad (AB) > (A)(B)/N \quad \text{or} \quad (AB) < (A)(B)/N,$$

independence holding when equality occurs. Boas (1909) appears to have been among the first to introduce the notion of correlation to such a 2×2 table. If X takes the value 1 when A occurs (or $X \in A$) and zero when α occurs and Y is similarly defined, then X and Y have positive, zero or negative correlations when they are associated, independent, or disassociated. Alternatively, it may be said that X and Y are the indicator variables of the sets, A and B, respectively. The Boas coefficients of correlation, as we may conveniently name them, can be written with theoretical frequencies,

$$(6.4) \qquad r = (p_{11} - p_{1.}p_{.1})/\sqrt{(p_{1.}p_{2.}p_{.1}p_{.2})} \, .$$

or with the observed frequencies,

$$(6.5) \qquad r = \{(AB)N - (A)(B)\}/\{(A)(\alpha)(B)(\beta)\}^{\frac{1}{2}}.$$

Boas coefficients of correlation have been used above, as in Theorem 4.1 and its corollaries, to prove propositions about dependence and the relations between independence and zero correlation in the more general cases.

Measures of association have been defined as follows.

$$(6.6) \qquad Q = \frac{ad - bc}{ad + bc} \, , \qquad \text{Yule (1900),}$$

$$(6.7) \qquad Y = \frac{\sqrt{(ad)} - \sqrt{(bc)}}{\sqrt{(ad)} + \sqrt{(bc)}} \, , \qquad \text{Yule (1900),}$$

in which $a = (AB)$, $b = (A\beta)$, $c = (\alpha B)$ and $d = (\alpha\beta)$.

Similar coefficients can be defined in the universe of (C) observations, and are known as partial coefficients. It must be admitted that these coefficients have had only a limited value. Q and Y have an important difference from r. All three coefficients can vary from -1 to $+1$. However, r can only take the extreme values if two diagonally opposite cells of the table are empty, whereas the coefficients, Q and Y, can take the extreme values when only one of the cells is empty. One must agree with Pearson and Heron (1913) that this is a serious defect in Q and Y. The multidimensional aspects of this section are amplified in Chapter XII.

EXERCISES AND COMPLEMENTS

Univariate Distributions

1. Suppose a general distribution (or measure space) has been condensed to a distribution on n points. Demonstrate the correspondence between the points of the "condensed space" and a simple function taking n distinct values.

2. Let B be the unit interval $[0, 1]$. Let μ be Lebesgue measure. Suppose B is partitioned into n intervals each of positive length, $A_1, A_2, \ldots, A_n$. Define indicator variables of the sets, A_i, and form an orthonormal basis on the condensed distribution. Describe how to complete this to a set of functions on B, which is a complete orthonormal set. [cf. definition of Haar functions.]

3. Prove that if a condensation of the finite line be made, the Boolean algebra, $\mathscr{S}^*$, on the sets so determined is an algebra for

(i) if $A, B \in \mathscr{S}^*$, then $A \cup B \in \mathscr{S}^*$

(ii) if $A, B \in \mathscr{S}^*$, then $AB \in \mathscr{S}^*$

(iii) if $A \in \mathscr{S}^*$, $-A \in \mathscr{S}^*$

[(iii) and either (i) or (ii) imply the remaining condition.] $\mathscr{S}^*$ is a subalgebra of $\mathscr{S}$, the Boolean algebra on the Borel sets of the real line, since if $A \in \mathscr{S}^*$, $A \in \mathscr{S}$.

4. Let t take values $t \in T$, T being some subset of the real numbers. Consider the normal $(t, 1)$ distributions. If $T_0 = \{0\}$, write down μ prescribed by H_0. Write down B, $\mathscr{B}$ and $\mathscr{S}$ of Section 1. Let B now be partitioned at the points, $d_1, d_2, \ldots,$ d_{n-1}, $-\infty < d_i > d_{i-1} < \infty$, $i = 1, 2, \ldots, n$. What is the subalgebra, $\mathscr{S}^*$, and the induced measures, μ_i^*? Calculate the ϕ^2 of $\nu = \mu_t$ with respect to μ.

5. Let the measurable space be the interval $[0, 1]$. Suppose that the rationals in this interval are numbered and that measure, p_i, is assigned to the ith rational; $p_i \geqslant 0$, $\sum p_i = 1$ and suppose that $p_i > 0$ for an infinity of values of i. Let a sequence of partitions be chosen so that the sets at the ith partition are the points $k2^{-i}$, $k = 0, 1, \ldots, 2^i$ and the open sets, whose endpoints are these points. What are the atoms? Is the sequence of partitions non-trivial in the sense of Section 1?

6. Let μ be Lebesgue measure on $[0, 1]$ and suppose that $g(x) \geqslant 0$, $\int g(x)\, d\mu = 1$. Then $g(x)$ is a probability frequency function. If further, $\int g^2(x)\, d\mu < \infty$, and $\int g(x) x^{(i)}\, d\mu = a_i$, where $\{x^{(i)}\}$ is a complete orthonormal set on the measure space, then $\sum a_i^2 < \infty$ is by the Bessel inequality. Is it possible to define a measure if the condition $\sum a_i^2 < \infty$ is given instead of $\int g^2(x)\, d\mu < \infty$? Is this density given by $g(x)$ ϕ^2-bounded with respect to the uniform (or rectangular) density on $[0, 1]$.

7. Let the measurable space be the unit interval $(0, 1]$. Let μ be Lebesgue measure and $\nu(0, x] = x^2$, $0 < x \leqslant 1$. Let the space be partitioned into n_i half-open intervals of equal length, such that n_{i+1} is a multiple of n_i. Show that $\phi^2(i) = (4n_i^2 - 1)/(3n_i^2) - 1$; $\phi^2 = \frac{1}{3}$. Note that the ratio $d\nu/d\mu$ is $2x$ and ϕ^2 can be obtained by a simple integration.

8. Let three measures ν_i be defined on the interval $(0, 1]$ by

$$\nu_i = k_i x^{k_i - 1}, \qquad k_i > 0, \qquad i = 1, 2, 3,$$

where $k_1 = 1$, $k_2 = 0.6$ and $k_3 = 0.4$. Verify that ν_2 is ϕ^2-bounded with respect to ν_1 and also ν_3 is ϕ^2-bounded with respect to ν_2, but ν_3 is not ϕ^2-bounded with respect to ν_1. The property is not transitive.

9. Show that ν_3 of Exercise 8 is absolutely continuous with respect to ν_2 and to ν_1.

10. If two probability distributions possess frequency functions, say $f(x)$ and $g(x)$, then the Radon–Nikodym derivative is the ratio, $g(x)/f(x)$.

11. Let the measurable space, B, consist of n points, b_i, with n finite. If $\mu(b_i)$ is positive for each i, show that any other probability measure defined on B is ^{2}-bounded with respect to μ.

12. Show that if $F(x)$ and $G(x)$ are distribution functions of two different pure

types (discrete, singular or absolutely continuous), then $G(x)$ is not ϕ^2-bounded with respect to $F(x)$.

13. Prove that if μ_1 and μ_2 are orthogonal probability measures on subsets A and B of a measurable space, possessing complete orthonormal sets, $\{u^{(i)}\}$ and $\{v^{(i)}\}$, respectively, and if $\beta_1 > 0$, $\beta_2 > 0$ and $\beta_1 + \beta_2 = 1$, a complete set can be defined on the "mixture", with measure $\beta_1\mu_1 + \beta_2\mu_2$ by means of the following

$$
\begin{aligned}
k^{(0)}(x) &= 1 \\
k^{(1)}(x) &= -\beta_1^{-\frac{1}{2}}\beta_2^{\frac{1}{2}} && \text{if } x \in A \\
&= +\beta_1^{\frac{1}{2}}\beta_2^{-\frac{1}{2}} && \text{if } x \in B \\
&= 0 \text{ elsewhere} \\
k^{(2i)}(x) &= \beta_1^{-\frac{1}{2}}u^{(i)} && \text{if } x \in A, && i = 1, 2, \ldots \\
&= 0 && \text{if } x \notin A \\
k^{(2i+1)}(x) &= \beta_2^{-\frac{1}{2}}v^{(i)} && \text{if } x \in B, && i = 1, 2, \ldots \\
&= 0 && \text{if } x \notin B
\end{aligned}
$$

If μ_1 or μ_2 is an atomic measure on a finite number of points, suggest the appropriate changes of notation.

14. Prove that a complete orthonormal basis can be defined on an arbitrary statistical distribution.

15. Prove that if $(X, \mathscr{A}, \mu)$ is a probability measure space and if $(X, \mathscr{A}, \nu)$ is another probability measure space, which is absolutely continuous with respect to μ, then there exists a finite valued measurable function f on X such that

$$
\nu(A) = \int_A f \, d\mu, \quad \text{for every measurable set } A.
$$

The function f is unique in the sense that if also $\nu(A) = \int_A g \, d\mu \ A \in \mathscr{A}$, then $f = g$, a.e. with respect to μ-measure. [This is a weakened form of Theorem 31B of Halmos (1950).] f is the Radon–Nikodym derivative of ν with respect to μ. (Halmos and Savage, 1949.)

16. Let $g(x)$ be a frequency function and suppose that $\Omega(x) = g(x)/f(x)$ where $f(x)$ is the frequency function of X, a normal $(0, 1)$ variable. Suppose that $g(x)$ is approximated by the series, $\sum_{j=0}^{\infty} c_j H_j(x)f(x)$. Then we might require that

$$
\mathscr{E}\left(\Omega(x) - \sum_0^n c_j H_j(x)\right)^2
$$

should be minimized. Differentiation of $\int \left[\Omega(x) - \sum_0^n c_j H_j(x)\right]^2 f(x) \, dx$ with respect to c_k yields $\int \left[\Omega(x) - \sum_0^n c_j H_j(x)\right] H_k(x)f(x) \, dx = 0$, or

$$
c_k = \int \Omega(x) H_k(x) f(x) \, dx = \int H_k(x) g(x) \, dx.
$$

This yields

$$c_0 = 1$$
$$c_1 = \mu_1'$$

and generally

$$c_k = (k!)^{-1}(b_k\mu_k' + b_{k-2}\mu_{k-2}' + b_{k-4}\mu_{k-4}' + \ldots)$$

where $b_k, b_{k-2}, \ldots$, are the coefficients of $x^k, x^{k-2}, \ldots$ in $H_k(x)$. This is the Gram–Charlier series of Type A.

17. In Exercise 16, no attempt has been made to minimize the expression $\sum (g_i - f_i)^2/f_i$, where g_i and f_i are the measures assigned by $g(x)$ and $f(x)$ to the ith class of a collection of classes resulting from a partition of the space. Discuss whether it is reasonable to require the Gram–Charlier series of Type A to minimize $\sum (g_i - f_i)^2/f_i$, or equivalently to test the goodness of fit by χ^2.

18. If X is a random variable it is possible to have functions of X, $\alpha(X)$ and $\beta(X)$, mutually independent. Verify that if X is rectangular, there is an infinity of mutually independent functions of X, namely the Rademacher functions. (Kac, 1959.)

19. Prove that if $X_1, X_2, X_3, \ldots$ form an orthonormal basis or a complete orthonormal set on a measure space and if X_i is stochastically independent of X_j for $i \neq j$, then each X_i can take only two distinct values with non-zero probability. (Jamnik, 1962, 1964; Lancaster, 1965c.)

Bivariate Distributions

20. Prove that the measure of any joint distribution of two random variables taking a finite number of distinct values is ϕ^2-bounded with respect to the product measures of the marginal distributions.

21. Suppose a joint bivariate distribution is given as in (2.2). Let a random variable $W = g(X)$ be defined. Determine the function, $Z = h(Y)$ which has maximum correlation with W. (Yates, 1948; Lancaster, 1958.) [*Hint.* Expand $g(x)$ as a linear form in an orthonormal system, which includes the canonical variables.]

22. For $|\rho| < 1$, the bivariate normal distribution is ϕ^2-bounded with respect to its marginal distributions. An integration gives

$$\phi^2 = \iint f^2(x, y)/[g(x)h(y)] \, dx \, dy - 1 = \rho^2/(1 - \rho^2).$$

(Pearson, 1904; Lancaster, 1957.)

23. For $|\rho| = 1$, the ϕ^2 of Exercise 22 is unbounded.

24. ϕ^2 is unbounded for any bivariate distribution, which consists of an infinite number of points of increase along a straight line, not parallel to either axis.

25. Use the canonical form (3.17) to prove that

$$\mathscr{E}(x^{(i)} \mid Y = y) = \rho_i y^{(i)} \equiv \rho_i y^{(i)}(y).$$

26. Let $\xi = \xi(X)$ and $\eta = \eta(Y)$. Then $\mathscr{E}(\xi \mid Y = y)$ is said to be the regression function of ξ on Y. Similarly $\mathscr{E}(\eta \mid X = x)$ is the regression function of η on X.

Suppose now that $\{x^{(i)}\}$ is a complete orthonormal set on the distribution of X and $\{y^{(i)}\}$ is a complete orthonormal set on the distribution of Y. Then a NASC that the regression function of the $x^{(i)}$ should be a multiple of $y^{(i)}$ is that $\{x^{(i)}\}$ and $\{y^{(i)}\}$ should be the sets of canonical variables. (Hirschfeld, 1935.)

27. Prove that a NASC for the mutual independence of X and Y is that every function of X with finite variance should be uncorrelated with every function of Y with finite variance, and prove that this is equivalent to the condition, $\mathscr{E}x^{(i)}y^{(j)} = \rho_{ij} = 0$ for any complete orthonormal sets $\{x^{(i)}\}$ and $\{y^{(j)}\}$ and also equivalent to the condition of Exercise 14 in the finite case. (Lancaster, 1959, 1961b.)

28. Prove that if X has unit correlation with Y then $X = Y$ a.e. (Gebelein, 1942b and 1952.)

29. Prove that in a 2×2 distribution if $\alpha(X)$ and $\beta(Y)$ are non-constant functions of X and Y respectively, corr $(\alpha, \beta) = 1$ implies complete mutual dependence so that $\alpha(X) = \gamma(\beta(Y))$, $\beta(Y) = \gamma^{-1}(\alpha(X))$ and corr $(\alpha, \beta) = 0$ implies independence of X and Y.

30. Prove that in an $r \times s$ distribution, it is possible for corr $(\alpha, \beta) = 1$, without there being complete dependence, where $\alpha(X)$ and $\beta(Y)$ are some functions of X and Y other than linear forms in X or Y. As an example consider a 4×4 table with $f_{i.} = f_{.j} = \frac{1}{4}$. Set up a bivariate measure such that the second orthogonal polynomials have unit correlation but X is not determined by Y nor Y by X.

31. In a bivariate distribution prove that the existence of a correlation of unity between $\xi(X)$ and $\eta(Y)$ implies that the bivariate distribution consists of at least two disjunct pieces. If there are k orthogonal functions in X which have unit correlation with k orthogonal functions in X then the distribution consists of at least $k + 1$ disjunct pieces. (Richter, 1949; Lancaster, 1958 and 1963b.)

32. Let p_{ij} and π_{ij} be two bivariate product measures on four points $i = 1, 2$; $j = 1, 2$. Prove that if $p_{ij} = p_{i.}p_{.j}$, $\pi_{ij} = \pi_{i.}\pi_{.j}$ then in general it is not true that $\theta p_{ij} + (1 - \theta)\pi_{ij} = (\theta p_{i.} + (1 - \theta)\pi_{i.})(\theta p_{.j} + (1 - \theta)\pi_{.j})$. Use this to prove that if $F(x, y)$ and $K(x, y)$ are distribution functions of independent random variables then $\theta F(x, y) + (1 - \theta)K(x, y)$ is not in general the distribution function of a pair of mutually independent random variables.

33. Prove that mixtures of distributions of independent variables may be again distributions of independent variables. Use, as your example, a 4×4 distribution, compounded of a mixture of four 2×2 distributions.

34. Let $\mathbf{A}$ be a positive definite matrix and suppose that $\mathbf{A}_{11}$ is a square symmetric submatrix of $\mathbf{A}$. Prove that the kth latent root of $\mathbf{A}_{11}$ is not greater than the kth latent root of $\mathbf{A}$, $k = 1, 2, \ldots, r$; $r < n$, where r and n are the sizes of $\mathbf{A}_{11}$ and $\mathbf{A}$ respectively. (Cauchy, 1829; Sarmanov and Zaharov, 1959.)

35. Prove that if $\{\rho_i\}$ is the set of canonical correlations of a bivariate distribution, then the kth canonical correlation in any bivariate distribution, obtained from it by condensation, is not greater than ρ_k. (Sarmanov and Zaharov, 1959.)

36. Form a bivariate distribution from a mixture of a uniform distribution along the straight line from $(-1, -1)$ to the origin with a second uniform distribution over the interior of a square with corners at $(0, 0)$ $(0, 1)$, $(1, 1)$ and $(1, 0)$, assigning

measures of β and $(1 - \beta)$ to them. Define complete orthonormal sets on the margins by means of Exercise 13, using the standardized Legendre polynomials as the sets $\{u^{(i)}\}$ and $\{v^{(i)}\}$ and setting the interval $A = [-1, 0)$, $B = (0, 1]$. Show that the pair of functions corresponding to $k^{(1)}(x)$ have unit correlation, as also do the functions $k^{(2i)}(x)$ but the pairs corresponding to $k^{(2i+1)}(x)$ have zero correlation. Prove also that $\mathscr{E}k^{(i)}(x)k^{(j)}(y) = 0$ if $i \neq j$. X and Y have thus an infinity of unit correlations but they are not mutually completely dependent. (Lancaster, 1963b.)

37. Modify the marginal distributions of Exercise 36 by condensing A and B into r_1 and r_2 points each possessing positive probability measure. Show that the joint distribution obtained by this condensation has r_1 canonical correlations of unity and all other correlations are zeroes.

38. Use Exercise 36 to prove that the criterion of dependence $\phi^2\{(r - 1)(c - 1)\}^{-\frac{1}{2}}$ suggested by Tschuprow, Steffensen and Weiler can be made to assume any value in $(0, 1)$ by appropriate choice of r_1 and r_2 and setting $r = c = r_1 + r_2$. (Tschuprow, 1925; Steffensen, 1934; Lancaster, 1963b and Weiler, 1966.)

39. Let measure β be uniformly distributed along the line from the origin to the point $(1, 1)$ and let measure $(1 - \beta)$ be uniformly distributed over the square which has this line as diagonal. Show that the marginal distributions are rectangular. Show that any square summable function on one marginal distribution, $\xi(X)$ say, has correlation, β, with a similar function in Y, $\xi(Y)$. Show that ϕ^2 is unbounded. Prove that no function of X has correlation greater than β with any function of Y. Prove that there is an infinity of pairs with maximal correlation. (Lancaster, 1963b.)

40. Prove that the square of the maximal coefficient of correlation is bounded by ϕ^2. Prove that ϕ^2 may be infinite although the maximal coefficient of correlation is less than unity. Is this possible in the jointly normal distribution?

41. The expansion (3.17) may be called the "canonical expansion". Prove that if X and Y have infinite range and if $\{x^{(i)}\}$ and $\{y^{(i)}\}$ are sets of orthogonal polynomials, the series in the canonical expansion cannot be finite. (Sarmanov, 1960a.)

Bivariate Gamma Distribution

42. $(X_1, Y_1) \ldots (X_n, Y_n)$ is a sample of n mutually independent observations from a bivariate normal with correlation coefficient, ρ. By considering the moment generating function and making the transformation $U = \frac{1}{2} \sum X_i^2$, $V = \frac{1}{2} \sum Y_i^2$, find a bivariate gamma distribution whose canonical variables are the Laguerre polynomials and whose n^{th} canonical correlation is ρ^{2n}. (Kibble, 1941.)

43. Suppose the elements of a population may be classified as being A or $\bar{A}$ and B or $\bar{B}$. Let a random sample be taken from the population. Prove that the numbers of occurrences of A and B are jointly distributed as a bivariate binomial distribution, if replacement be permitted, or as a bivariate hypergeometric distribution, if replacement be not permitted. By considering joint factorial moment generating functions, derive a canonical expansion in either case. (Aitken and Gonin, 1935.)

44. Obtain a canonical form for a bivariate Poisson by taking limits in the above example of a bivariate binomial. (Campbell, 1932b.)

Random Elements in Common

45. Generate a bivariate distribution by considering

$$X = U + V$$
$$Y = V + W,$$

where U, V and W are mutually independent and belong to the same type of distribution. By using either moment generating functions or the generalisation of Runge's identity of Exercise IV.57 show that the following distributions have polynomial canonical variables:

(a) Gamma   (Cherian, 1941, Eagleson, 1964.)
(b) Poisson   (Campbell, 1932b, Eagleson, 1964.)
(c) Binomial   (Eagleson, 1964.)
(d) Negative binomial   (Eagleson, 1964.)

Changes of Variable

46. Show that a transformation of the marginal variables of a bivariate distribution with a canonical expansion in polynomials will always give an expansion of the new bivariate distribution. Give examples where the canonical variables in the new variables are also polynomials.

Assume the truth of the Mehler identity and note the relation connecting the Hermite and Laguerre polynomials of (5.6.1) of Szegö (1959). Form a mixture of two normal distributions of the form, $\frac{1}{2}f(x, y; \rho) + \frac{1}{2}f(x, y; -\rho)$, where $f(x, y; \rho)$ is the joint bivariate normal density. Now consider the joint distribution of the squares of the marginal variables. Verify that the marginal variables are Γ-variables and that the joint distribution has a canonical expansion in Laguerre polynomials. (Sarmanov, 1960a.)

47. Transform only one marginal of the bivariate normal to a gamma variate. Use moment generating functions to obtain a canonical expansion for this "mixed" distribution in terms of the Hermite and Laguerre polynomials. (Kibble, 1941.)

48. Let a bivariate distribution in X and Y be given with orthonormal polynomials as the canonical variables. Let $\alpha(X)$ and $\beta(Y)$ be arbitrary polynomials in X and Y respectively. Determine the regression line for $\alpha(X)$ in terms of the orthonormal polynomials in Y and the canonical correlations. Determine similarly the regression line for $\beta(Y)$ in terms of polynomials in X and the canonical correlations. Determine the function in Y which has maximum correlation with $\alpha(X)$.

49. Let X and Y be random variables possessing a joint density,

$$h(x, y) = p_1a(x)a(y) + p_2[a(x)b(y) + b(x)a(y)] + p_3b(x)b(y),$$

where $p_1 + 2p_2 + p_3 = 1$, $p_1 > 0$, $p_2 > 0$ and $p_3 > 0$ and $a(x)$ and $b(y)$ are also probability density functions. Let $p = p_1 + p_2$, $q = p_2 + p_3$. Then the marginal densities are of the form,

$$h_1(x) = pa(x) + qb(x).$$

Verify that

$$h(x, y) = h_1(x)h_1(y) + (p_1p_3 - p_2^2)(a(x) - b(x))(a(y) - b(y)).$$

The canonical variables are thus $[a(x) - b(x)]/\{\mathscr{E}[a(x) - b(x)]^2\}^{\frac{1}{2}}$ and a similar expression in y. (Csaki and Fischer, 1960.)

50. The first canonical correlation may not be "attainable" in the sense that it can only be approached in the limit. Let the density of the joint distribution of X and Y be uniformly distributed over a region of the unit square, G. Let G be symmetric about the line $x = y$ and bounded by two curves meeting each other at the origin and having there a common tangent. The bounding curves of G can be taken to be convex. Choose step functions of the form,

$$\xi_\epsilon = 1 \quad \text{for} \quad 0 \leqslant x \leqslant \epsilon$$
$$\xi_\epsilon = 0 \quad \text{for} \quad x > \epsilon.$$

Show that corr $(\xi_\epsilon, \eta_\epsilon) \to$ q as $\epsilon \to 0$. (J. Czipszer cited by Rényi, 1959b.)

Multidimensional Distributions

51. Let X, Y and Z be three random variables such that $X = Y = Z$ with probability, 1. Assume further that X has an expectation of zero and unit variance but infinite third moment. Then $\mathscr{E}(XYZ)$ is a generalized coefficient of correlation, which is infinite. The Schwarz inequality cannot be used as in the two-dimensional case except in special cases.

52. Show that if X is independent of Y, then the generalized coefficients of correlation of X, Y and Z are bounded by unity. Prove that in such a case $|\mathscr{E}x^{(i)}y^{(i)}z^{(k)}| \leqslant 1$, since the Schwarz inequality may be used as follows,

$$|\mathscr{E}(x^{(i)}y^{(j)})z^{(k)}|^2 \leqslant [\mathscr{E}x^{(i)2}y^{(j)2}]\mathscr{E}z^{(k)2}$$
$$= \mathscr{E}x^{(i)2}\mathscr{E}y^{(j)2}\mathscr{E}z^{(k)2} = 1.$$

53. Prove that if the random variables, $X_1, X_2, \ldots, X_n$, do not form a mutually independent set, then step functions $\gamma_1(X_1), \gamma_2(X_2), \ldots, \gamma_n(X_n)$ everywhere finite and standardized to have zero mean and unit variance, can be found such that at least one of the products of $\gamma_1, \gamma_2, \ldots, \gamma_3$, taken two or more at a time has non-zero expectation.

54. Consider the three variables defined on 4 points, to each of which is assigned equal measure, $\frac{1}{4}$ and taking values as given in the table,

	X_1	X_2	X_3	$X_4 \equiv 1$
1	-1	-1	$+1$	$+1$
2	-1	$+1$	-1	$+1$
3	$+1$	-1	-1	$+1$
4	$+1$	$+1$	$+1$	$+1$

Show that the variables are independent in pairs but not in a set of three. In fact $X_1 = X_2X_3$. (Bernstein, 1934.)

55. Let a latin square be described by the n^2 triples $\{x, y, z\}$ where x, y, z take values $0, 1, 2, \ldots, n - 1$. Let a distribution in three dimensions be defined by setting $\mathscr{P}\{X = x, Y = y, Z = z\} = n^{-2}$ when $\{x, y, z\}$ occurs in the description

of the latin square and be zero otherwise. Prove that X, Y, Z are independent in pairs but not in a set of three. Relate this to the Bernstein example of Exercise 21. (Lancaster, 1965c).

56. Let $\mathbf{X} = (x_{ij})$ be a Hadamard matrix of size, $n \times n$, the last column of which consists of $+1$'s the remaining elements being ± 1. Define a random variable Z taking values, $1, 2, \ldots, n$ with equal probabilities and random variables, $X_j = x_{ij}$ if $Z = i, j = 1, 2, \ldots, n - 1$. Prove that the X_j are pairwise independent but not necessarily independent in sets of three, four, $\ldots$.

57. Let a random variable Z take values $1, 2, \ldots, n - 1$, each with probability $(n - 3)(n - 2)^{-2}$ and n with probability $(n - 2)^{-2}$. Let random variables X_1, $X_2, \ldots, X_{n-1}$ be defined by

$$X_j = (n - 3)^{\frac{1}{2}} \qquad \text{if } Z = j \text{ or } n,$$
$$X_j = -(n - 3)^{-\frac{1}{2}} \qquad \text{if } Z \neq j, Z \neq n.$$

Prove that the variables X_j are independent in pairs but not in sets of three, four, $\ldots$. (Lancaster, 1965.)

Correlation and Regression

58. Show that the regression of a function $\alpha(X)$ on $\beta(Y)$ can often be conveniently obtained from the canonical expansion. Show that independence of random variables, X and Y, is equivalent to the statement that every square summable function $\alpha(X)$ has zero regression on every square summable function $\beta(Y)$. [In particular, for $\alpha(X)$ and $\beta(Y)$, we can take step functions taking precisely two distinct values.]

59. Show by means of one of Stieltjes' examples given in Exercise II.14 that the existence of zero regression of every polynomial in X on every polynomial in Y is not sufficient for mutual independence of X and Y.

60. Show that a NASC for the existence of non-zero coefficients of correlation of the second order in a 3-dimensional distribution is that X should have non-zero regression coefficients on some function of the form, $\beta(Y)\gamma(Z)$.

61. It is usually not possible to obtain functions of X, Y and Z respectively so that the canonical variables in X are the same for the (X, Y)-distribution as for the (X, Z)-distribution. Suppose that X, Y, Z have the same distribution, weights of $\frac{1}{4}$ on each of 4 points. Define an orthonormal set $w_1(X)$, $w_2(X)$ and $w_3(X)$ on the distribution of X, with w_1, w_2 and w_3 taking the values of the columns in Exercise 54. Choose a set a_i such that $\sum |a_i| \leqslant 1$, then

$$f(x, y, z) = [1 + a_3 w_2(x)w_1(y) + a_2 w_3(x)w_1(z) + a_1 w_3(y)w_2(z)$$
$$+ aw_1(x)w_2(y)w_3(z)]/64$$

defines a probability distribution. Verify that the first canonical variable in X on the X, Y marginal distribution is not the first canonical variable in X on the X, Z marginal distribution.

62. Let $\{A_i\}, \{B_i\}, \ldots, \{C_i\}$ be partitions of a space, A. $A_i \cap A_j = \phi$, $A_1 \cup A_2 \ldots \cup A_a = A$. In set theory, the classifications are said to be independent if

no class $A_i \cap B_j \ldots \cap C_k$ is empty. Prove that this requirement is also necessary but not sufficient for statistical independence of random variables $X, Y, \ldots, Z$, where X is constant over the set $A_i, i = 1, 2, \ldots, a;$ B is constant over the set A_j, $j = 1, 2, \ldots, b; \ldots$.

63. Suppose that each of the spaces A, B, C of the random variables X, Y, Z is partitioned into two classes, A_1 and A_2, B_1 and B_2, C_1 and C_2. Prove that

$$\mathscr{P}(X \in A_1, Y \in B_1, Z \in C_1) = \mathscr{P}(X \in A_1)\mathscr{P}(Y \in B_1)\mathscr{P}(Z \in C_1)$$

is not sufficient for stochastic independence of X, Y and Z.

64. Suppose that a set, Z, of N observations is made and that the observations can be classified into complementary sets by a number of criteria so that $Z = A^+ \cup A^- = B^+ \cup B^- \ldots = C^+ \cup C^-$. Let A, B, C be the positive class frequencies of the first order so that A is the number of elements of Z which possess the first property, B is the number possessing the second property, $\ldots AB$ is the number possessing both the first and second properties and so on. Let $A^+B^- \ldots C^+$ be the number of elements possessing the first property but not the second $\ldots$ and possessing the last property. Let the number of properties be k then show that the number of positive classes is $2^k - 1$ which together with the universal class gives 2^k classes. Write

$$\mathbf{M} = \begin{bmatrix} 1 & 1 \\ 0 & 1 \end{bmatrix}, \qquad \mathbf{M}^{-1} = \begin{bmatrix} 1 & -1 \\ 0 & 1 \end{bmatrix},$$

and verify that

$$\mathbf{M} \begin{bmatrix} A^- \\ A^+ \end{bmatrix} = \begin{bmatrix} N \\ A \end{bmatrix}, \qquad \mathbf{M}^{-1} \begin{bmatrix} N \\ A \end{bmatrix} = \begin{bmatrix} A^- \\ A^+ \end{bmatrix};$$

$$\mathbf{M} \times \mathbf{M} \begin{bmatrix} A^-B^- \\ A^-B^+ \\ A^+B^- \\ A^+B^+ \end{bmatrix} = \begin{bmatrix} N \\ B \\ A \\ AB \end{bmatrix}; \quad \text{and} \quad \mathbf{M}^{-1} \times \mathbf{M}^{-1} \begin{bmatrix} N \\ B \\ A \\ AB \end{bmatrix} = \begin{bmatrix} A^-B^- \\ A^-B^+ \\ A^+B^- \\ A^+B^+ \end{bmatrix}.$$

Generalize this to an arbitrary number of dimensions by replacing A^+ and A^- by A_1^+ and A_1^-, B^+ and B^- by A_2^+ and $A_2^- \ldots$.

Non-Central χ^2

1. DISTRIBUTION THEORY

The notion of the displacement of the mean of a normal variable is familiar. Thus Z may be normal $(0, 1)$ under H_0 and normal $(c, 1)$ under H_1. Z^2 is χ^2 with 1 d.f. under H_0 and Z^2 is said to be distributed as *non-central χ^2 with 1 d.f. and parameter*, c^2, under H_1. This definition is readily extended. Let $Z_1, Z_2, \ldots, Z_n$ be a set of mutually independent standard normal random variables. Under H_0, each is normal $(0, 1)$; under H_1, each is normal $(c_i, 1)$. Under H_1, $\sum Z_i^2$ is said to be a *non-central χ^2 with parameter*, $\lambda \equiv \sum c_i^2$, and n d.f. This can be written

(1.1)
$$\chi'^2 = Z_1^2 + Z_2^2 + \ldots + Z_n^2$$

and it is readily verified that the expectation of χ'^2 is $n + \sum c_i^2$ or $n + \lambda$ under H_1. It is sometimes convenient when every $c_i = 0$, to refer to the variable as a *central χ^2*.

Theorem 1.1. *A non-central χ^2 with n degrees of freedom can be represented as a sum of a non-central χ^2 with one degree of freedom and the same parameter and a central χ^2 with $(n - 1)$ degrees of freedom, where the two variables are mutually independent.*

Proof. It is assumed that χ'^2 has been defined by (1.1) and that $\lambda \neq 0$. Let an orthogonal matrix $(b_{ij}) \equiv \mathbf{B}$ be chosen with

(1.2)
$$b_{1j} = c_j \lambda^{-\frac{1}{2}},$$

and define a new set of variables by the orthogonal transformation

(1.3)
$$\mathbf{Y} = \mathbf{BZ}, \text{ where the elements of } \mathbf{Z} \text{ are the } Z_i.$$

Taking expectations of (1.3) we find

(1.4)
$$\mathscr{E} Y_1 = \lambda^{\frac{1}{2}}, \qquad \mathscr{E} Y_j = 0 \quad \text{if } j > 1.$$

Further the variables Y_i form a mutually independent and normally distributed set. Writing $Y_1^2 = V, \sum_2^n Y_i^2 = U$, U is distributed independently of V

as χ^2 with $(n - 1)$ degrees of freedom, and Y_1 is normal $(\lambda^{\frac{1}{2}}, 1)$. So that $V = Y_1^2$, is non-central χ^2 with 1 d.f.

The moments and cumulants of χ'^2 can now be computed with ease. In particular,

(1.5) $$\mathscr{E}\chi'^2 = n + \lambda$$

(1.6) $$\text{var } \chi'^2 = 2n + 4\lambda.$$

We follow Patnaik (1949) in deriving the distribution of non-central χ^2. Using his notation, we have

Theorem 1.2. *Non-central χ^2 with n degrees of freedom and parameter, λ, has the density function of the form*

(1.7) $$g(w) = \frac{e^{-\frac{1}{2}w}e^{-\frac{1}{2}\lambda}w^{\frac{1}{2}n-1}}{2^{\frac{1}{2}n}\Gamma(\frac{1}{2}n)}\left(1 + \frac{1}{n}\frac{w\lambda}{2} + \frac{1}{n(n+2)}\frac{1}{2!}\left(\frac{w\lambda}{2}\right)^2 + \ldots\right),$$

$$0 \leqslant w < \infty.$$

Proof. Write $Y_1^2 = V$, $\sum_{2}^{n} Y_i^2 = U$; non-central χ^2 may be written, after Theorem 1.1, as $\chi'^2 = W = U + V$ where the Y_i form a set of mutually independent normal variables each with unit variance and of which only Y_1 has a non-zero expectation; $\mathscr{E}Y_1 = \lambda^{\frac{1}{2}}$. U has the distribution of χ^2 with $(n - 1)$ degrees of freedom and V has the distribution of non-central χ^2 with one degree of freedom and parameter, λ. U is independent of V. The density functions and the joint density functions can therefore be written

(1.8) $$f_1(u) \equiv f_U(u) = \frac{e^{-\frac{1}{2}u}u^{\frac{1}{2}(n-3)}}{2^{\frac{1}{2}(n-1)}\Gamma[\frac{1}{2}(n-1)]}, \qquad 0 \leqslant u < \infty$$

(1.9) $$f_2(v) \equiv f_V(v) = \frac{v^{-\frac{1}{2}}}{2\sqrt{(2\pi)}}\{e^{-\frac{1}{2}(v^{\frac{1}{2}}-\lambda^{\frac{1}{2}})^2} + e^{-\frac{1}{2}(v^{\frac{1}{2}}+\lambda^{\frac{1}{2}})^2}\}$$

(1.10) $$f_{U,V}(u, v) \equiv f(u, v) = f_1(u)f_2(v).$$

Make the transformation,

(1.11) $$W = U + V, \qquad U = U.$$

The Jacobian is unity and so the joint density function of U and W is

(1.12) $$g(u, w) = f(u, v) = \frac{e^{-\frac{1}{2}w}e^{-\frac{1}{2}\lambda}w^{\frac{1}{2}(n-4)}}{2^{\frac{1}{2}n}\Gamma(\frac{1}{2})\Gamma[\frac{1}{2}(n-1)]}\left(\frac{u}{w}\right)^{\frac{1}{2}(n-3)}$$

$$\times \left\{\left(1 - \frac{u}{w}\right)^{-\frac{1}{2}} + \frac{w\lambda}{2!}\left(1 - \frac{u}{w}\right)^{\frac{1}{2}} + \ldots\right\}.$$

Integrating with respect to u from 0 to w, we obtain

$$(1.13) \quad g(w) = \frac{e^{-\frac{1}{2}w}e^{-\frac{1}{2}\lambda}w^{\frac{1}{2}(n-2)}}{2^{\frac{1}{2}n}\Gamma(\frac{1}{2})\Gamma[\frac{1}{2}(n-1)]}\left\{B\left(\frac{n-1}{2},\frac{1}{2}\right) + \frac{w\lambda}{2!}B\left(\frac{n-1}{2},\frac{3}{2}\right) + \cdots\right\},$$

or

$$(1.14) \quad g(w) = \frac{e^{-\frac{1}{2}w}e^{-\frac{1}{2}\lambda}w^{\frac{1}{2}(n-2)}}{2^{\frac{1}{2}n}\Gamma(\frac{1}{2}n)}\left\{1 + \frac{1}{n}\left(\frac{w\lambda}{2}\right) + \frac{1}{n(n+2)}\frac{1}{2!}\left(\frac{w\lambda}{2}\right)^2 + \cdots\right\}.$$

The distribution of non-central χ^2 has been tabulated by Fix (1949), by Patnaik (1949) and by Owen (1962).

Various authors have introduced transformations of non-central χ^2 so as to give easy methods of tabulating the distribution.

2. THE COMPARISON OF TWO NORMAL POPULATIONS

A simplified version of this problem is considered first. Let $Z_1, Z_2, \ldots, Z_n$ be a set of mutually independent normal variables with unit variance under both H_0 and H_1. Suppose that H_0 specifies that the expections $\mathcal{E}_0 Z_i = 0$ and H_1 specifies that $\mathcal{E}_1 Z_i = c_i$. The problem is to distinguish between the hypotheses, given the sample. A wider class of problems can be reduced to this canonical form. We consider mutually independent normal variables, $Z_1^*, Z_2^*, \ldots, Z_n^*$, and let

$$(2.1) \quad \begin{aligned} \mathcal{E}_0 Z_i^* &= a_i, \quad \operatorname{var} Z_i^* = \sigma_i^2 \quad \text{under} \quad H_0 \\ \mathcal{E}_1 Z_i^* &= b_i, \quad \operatorname{var} Z_i^* = \sigma_i^2 \quad \text{under} \quad H_1, \quad i = 1, 2, \ldots, n. \end{aligned}$$

After the transformations

$$(2.2) \quad Z_i = (Z_i^* - a_i)/\sigma_i, \quad c_i = (b_i - a_i)/\sigma_i,$$

the problem takes a simpler form,

$$(2.3) \quad \mathcal{E}_0 Z_i = 0, \quad \mathcal{E}_1 Z_i = c_i; \quad \operatorname{var} Z_i = 1 \text{ under } H_0 \text{ and } H_1.$$

A further simplification is made possible by the transformation (1.3) and the hypotheses are then reduced to a canonical form, in which

$$(2.4) \quad \begin{aligned} &\mathcal{E}_0 Y_1 = 0, \quad \mathcal{E}_1 Y_1 = \lambda^{\frac{1}{2}}, \\ &\mathcal{E}_0 Y_j = \mathcal{E}_1 Y_j = 0 \quad \text{for } j = 2, 3, \ldots, n, \\ &\operatorname{var} Y_j = 1 \text{ under both } H_0 \text{ and } H_1. \end{aligned}$$

The canonical set of variables in (2.4) is evidently not unique if $n > 2$, for an arbitrary orthogonal transformation can be carried out on $Y_2, Y_3, \ldots, Y_n$ without altering the expectations or variances. Y_1 is, however, unique by its mode of definition and

$$(2.5) \quad Y_1 = \lambda^{-\frac{1}{2}} \sum c_i Z_i, \quad \sum c_i^2 = \lambda.$$

Theorem 2.1. *Of all quadratic forms in* $\{Y_i\}$ *of* (2.4) *with the* χ^2 *distribution under* H_0 *with* $f \leqslant n$ *degrees of freedom, the one with maximum parameter of non-centrality under* H_1 *is of the form*

$$(2.6) \qquad\qquad \chi^2 = Y_1^2 + Q,$$

where Q *is a quadratic form having the distribution* χ^2 *with* $(f - 1)$ *degrees of freedom in the variables* $\{Y_i\}$ *not including* Y_1.

Proof. A, the matrix of the required quadratic form, $\mathbf{Y}^T \mathbf{A} \mathbf{Y}$, is an idempotent and $\operatorname{tr} \mathbf{A} = f$. Under H_1, the expectation of $\mathbf{Y}^T \mathbf{A} \mathbf{Y}$ is $\sum_{i=1}^{n} a_{ii} \mathscr{E} Y_i^2$, since all the expectations of cross products, $Y_i Y_j$, vanish, and hence is equal to $f + a_{11} \lambda$. The maximum value of the non-centrality parameter is therefore attained when $a_{11} = 1$. A is thus the direct sum of 1 and an idempotent matrix of rank $(f - 1)$ and of size $(n - 1)$.

Remark. A special case of this theorem occurs when $f = 1$. The χ^2 with one degree of freedom with maximum parameter of non-centrality is Y_1^2. Similarly, among those linear forms in the variables, Y_i, which are normal $(0, 1)$ under H_0, Y_1 has maximum (absolute) expectation under H_1.

Corollary 1. *The parameter of non-centrality of a quadratic form in the set* $\{Y_i\}$ *is* $a_{11} \lambda$.

Theorem 2.2. *The expectation of any standardized linear form in the* $\{Z_i\}$ *or* $\{Y_i\}$ *is equal under* H_1 *to its correlation with* Y_1 *multiplied by* $\lambda^{\frac{1}{2}}$.
Proof. Let the linear form be $Y = \sum b_i Y_i$; $\sum b_i^2 = 1$ because of the standardization to give a variance of unity. The correlation of Y with Y_1 is simply b_1. Taking expectations the result follows.

From the general Neyman-Pearson theory, one would expect Y_1 to be the best function to differentiate H_1 from H_0 if the values of c_i are known for the set $\{Y_i\}$. χ^2 is, however, often used in cases where the c_i are not known and we deal with the general problem in the next section.

It is clear from Theorems 2.1 and 2.2 that χ^2 variables with the same number of degrees of freedom can vary in their parameter of non-centrality. The quadratic forms with the largest parameter of non-centrality will be those with Y_1^2 as a component of the quadratic form. These may be thought of as *fully efficient* test functions and we may say that such a function has unit efficiency. We define as the *efficiency* of a χ^2 variable, the coefficient of Y_1^2 in the expansion of χ^2 as $\mathbf{Y}^T \mathbf{A} \mathbf{Y}$ namely a_{11}. Further the correlation between $\mathbf{Y}^T \mathbf{A} \mathbf{Y}$ and Y_1^2 is $a_{11}/\sqrt{f}$, where f is the number of degrees of freedom (or rank) of χ^2.

Different quadratic forms may be compared by correlation as in Neyman (1940*a* and *b*). However, the analysis given above shows that there are only two criteria of value—the number of degrees of freedom and the parameter of non-centrality. As an example to show that quadratic forms, highly correlated under the null hypothesis, may have quite different efficiencies, consider the quadratic forms Q and $Q + Y_1^2$ of Theorem 2.1. The correlation will be $\sqrt{\{(f-1)/f\}}$ but Q will be quite ineffective as a test function since it has the same distribution under H_1 as under H_0.

Suppose now that there are N independent observations on the n variables and that H_0 and H_1 are fixed throughout the N observations. Theorem 2.1 is readily generalized.

Theorem 2.3. *Let N sets of n observations Y_{ij} (or Z_{ij}) be made on the variables $Y_{i1}, Y_{i2}, \ldots, Y_{in}$, $i = 1, 2, \ldots, N$. Of all quadratic forms having the χ^2 distribution under H_0 with $f \leqslant nN$ degrees of freedom, the one with maximum parameter of non-centrality under H_1 is of the form,*

$$(2.7) \qquad \chi^2 = N^{-1}(\mathscr{S} Y_{i1})^2 + Q$$

where Q is a quadratic form having the distribution of χ^2 with $(f-1)$ degrees of freedom in linear forms, each of which is orthogonal to $\mathscr{S} Y_{i1}$.

Proof. Suppose the nN variables $\{Y_{ij}\}$ are written out in an $N \times n$ array, **Y**. Then the transformation $\mathbf{KY} = \mathbf{Y}^*$, say with $k_{1j} = N^{-\frac{1}{2}}$ and **K** orthogonal, ensures that only Y_{11}^*, of all the variables Y_{ij}^*, has non-zero expectation under H_1. Theorem 2.1 therefore can be applied. But $Y_{11}^* = N^{-\frac{1}{2}}\mathscr{S} Y_{i1}$, where $\mathscr{S}$ is a summation operator over the row index, and the result of the theorem follows.

Theorem 2.4. *The expectation of any standardized linear form in the $\{Z_{ij}\}$ or $\{Y_{ij}\}$ is equal under H_1 to its correlation with Z_{11}^*, multiplied by $(N\lambda)^{\frac{1}{2}}$.*
Proof. The result goes through as in Theorem 2.3.

Theorem 2.5. *The efficiency of a χ^2 variable, which is a quadratic form in the $\{Z_{ij}\}$ and so in the $\{Y_{ij}\}$ is equal to the coefficient of Y_{11}^{*2} when χ^2 is expressed as a quadratic form in the set $\{Y_{ij}^*\}$. Further the correlation between χ^2 and Y_{11}^{*2} is $a_{11}/\sqrt{f}$, where f is the number of degrees of freedom of χ^2.*
Proof. As in Theorem 2.4.

3. ANALOGUES OF THE PEARSON χ^2, THE COMBINATION OF PROBABILITIES

The theory of Sections VI.1 and VII.2 is now applied to hypothesis testing. A sample of N independent observations is supposed to be available on a certain variable, Z. H_0 specifies a probability measure μ, on which there is a complete orthonormal set, $\{z^{(i)}\}$, and H_1 specifies a probability measure ν on the same measurable space. To avoid subscripts, summation over the N members of the sample is denoted by $\mathscr{S}$ without the introduction of a subscript. $\mathscr{E}_0$ and $\mathscr{E}_1$ are the operators yielding the expectations under H_0 and H_1 respectively. Any standardized function, U, can be represented in the form,

$$(3.1) \qquad U = \sum_1^\infty a_i z^{(i)}, \qquad \sum a_i^2 = 1$$

$$= \sum_1^\infty b_i \xi^{(i)}, \qquad \sum b_i^2 = 1,$$

where $\{\xi^{(i)}\}$ is the canonical set of (VI.1.5). Further

$$(3.2) \qquad \mathscr{E}_0 U = 0, \qquad \mathscr{E}_1 U = \phi b_1,$$

where ϕ^2 is as defined in (VI.1.4).

Standardized sums of such variables, U, can be written

$$(3.3) \qquad U^* = N^{-\frac{1}{2}} \mathscr{S} U.$$

It is readily verified that

$$(3.4) \qquad \mathscr{E}_0 U^* = 0, \qquad \mathscr{E}_0 U^{*2} = 1; \qquad \mathscr{E}_1 U^* = N^{\frac{1}{2}} \phi b_1.$$

$\mathscr{E}_1 U^{*2}$ can usually be evaluated for a given H_1. For many H_1 not too different from H_0, $\mathscr{E}_1 (U^* - \mathscr{E}_1 U^*)^2$, the variance of U^* under H_1 will not differ greatly from unity. Under these conditions U^* will be a displaced normal variable and U^{*2} will be non-central χ^2 with 1 d.f. and parameter, $N\phi^2 b_1^2$. The parameter of non-centrality is largest if U is taken equal to $\xi^{(1)}$ of (VI.1.5) so that $b_1 = 1$ and

$$(3.5) \qquad U = \phi^{-1}\left(\frac{d\nu}{d\mu} - 1\right) \equiv \phi^{-1}(\Omega - 1).$$

However, H_1 may not specify any particular measure ν or even class of measures. It is possible then to use the test function

$$(3.6) \qquad \chi^2 = \sum_1^n N^{-1}[\mathscr{S} z^{(i)}]^2,$$

which will have a parameter of non-centrality proportional to N and hence will almost always reject the null hypothesis if N is large enough. The parameter of non-centrality will depend on H_1 and will usually be non-zero unless it happens that all the $z^{(i)}$ chosen are orthogonal to $\xi^{(1)}$. The parameter of non-centrality of the χ^2 of (3.6) is at most equal to $N\phi^2$. Two special cases of the theory above are classical. The Pearson χ^2 is the test function based on the orthonormalized indicator variables, as has already been made clear in Section V.2. The test function, $\Psi_n^{\prime 2}$, of Neyman (1937) is based on the standardized Legendre functions. The Chebyshev series of orthonormal polynomials of the uniform distribution on n points also appears in a natural way if regression analysis is applied to the frequencies. Similarly there are the Hermite-Chebyshev, the Poisson-Charlier and the Laguerre sets of orthogonal polynomials on the normal, Poisson and gamma distributions respectively.

A problem that has received little attention is the relation between the different sets of orthonormal functions, for example, between the Hermite-Chebyshev series and the orthonormalized indicator variable series. Further a transformation is available from the normal distribution to the rectangular distribution. There is therefore a relation between the Hermite-Chebyshev series and the Legendre series. A discussion of some of these relations is given in the exercises. It should be remarked that the appropriate set is often determined by the form of H_1. Thus Neyman (1937) expressed his H_1 in terms of the Legendre polynomials and so the appropriate analysis is best expressed in the same terms. He experienced a difficulty that is common to many series expansions, namely that if the Radon-Nikodym derivative is expressed as a finite series (the infinite series truncated) in the orthonormal functions, it may take negative values. Neyman (1937) overcame this difficulty by expressing the Radon-Nikodym derivative as an exponential of a finite series but in doing so lost some of the advantages of the ortho-normality.

For a sample size of N, it is required to determine the best criterion for distinguishing between H_0 and H_1. If $N = 1$, the Neyman-Pearson theory requires that the test function should be $\Omega(x) \geqslant c$ or alternatively after (3.5), $U \geqslant \phi^{-1}(\Omega(x) - 1)$. Little has been gained for $N = 1$ by the notation but it will be found that ϕ^2 and standardized sums of $\Omega(x)$ play important roles in determining the best criteria among those which have the form of an analogue of χ or χ^2.

Theorem 3.1. ($\phi^2 + 1$) *has the multiplicative property, in the sense that if H_0 and H_1 specify pairs of density functions, $f_i(x)$ and $g_i(x)$, $i = 1, 2, \ldots, N$, or distribution functions, $F_i(x)$ and $G_i(x)$, and if*

$$(3.7) \qquad \phi_i^2 + 1 = \int (dG_i)^2/dF_i \equiv \int g_i^2(x) f_i^{-1}(x)\, dx,$$

then the $\phi^2 + 1$ of the joint distribution $G_1 \times G_2 \times \ldots \times G_n$ with respect to $F_1 \times F_2 \times \ldots \times F_n$ is given by

$$(3.8) \qquad \phi^2 + 1 = \prod_{i=1}^{N} (\phi_i^2 + 1).$$

Proof.

$$\phi^2 + 1 = \int \left\{ \prod_{i=1}^{N} dG_i \right\}^2 \Big/ \left\{ \prod_{i=1}^{N} dF_i \right\}$$

$$= \prod_{i=1}^{N} \int (dG_i)^2/dF_i$$

$$= \prod_{i=1}^{N} (\phi_i^2 + 1).$$

Corollary 1. *If the mutually independent variables are identically distributed*

(3.9) $\phi^2 + 1 = (\phi_i^2 + 1)^N.$

Corollary 2. *A necessary and sufficient condition that ϕ^2 of the product distribution should be finite is that each ϕ_i^2 should be finite.*

Corollary 3. $\log (\phi_i^2 + 1)$ *regarded as a functional of an H_1-distribution with respect to an H_0-distribution, is additive for independent observations.*

Example (i) Let H_0 and H_1 specify that Y is normal $(0, 1)$ and $(\theta, 1)$, respectively. Then

$$\phi^2 + 1 = (2\pi)^{-\frac{1}{2}} \int_{-\infty}^{\infty} \exp\left[-\tfrac{1}{2}(x^2 - 4\theta x + 2\theta^2)\right] dx = \exp \theta^2.$$

Example (ii) Suppose now that there is a sample of N independent observations in Example (i), then

$$\phi^2 + 1 = \exp N\theta^2.$$

In Example (ii), there is a single variable, $\sum Y_i$ say, for which the $\phi^2 + 1$ is the same as that for the whole set of variables, $\{Y_i\}$, that is for the whole product distribution. This is a property of *sufficient estimators*, which can be defined as follows. A statistic or estimator, V, is said to be *sufficient for the family*, $\{\mu_\theta, \theta \in T\}$ or *sufficient for* θ, if the conditional distribution of $\{Y_i\}$, the observations, given $V = v$, is independent of θ. It may be noted that the likelihood ratio is always a sufficient statistic. The next theorems indicate the close relation between the theory of sufficient statistics and ϕ^2. It may happen that there is a condensation of the space, B^*, on which the value of the Pearson functional is the same as on the space, B. Let μ and ν assign measures, μ^* and ν^*, to the sets of the condensed partition. If

(3.10) $$\int (d\nu)^2/d\mu = \int (d\nu^*)^2/d\mu^*,$$

the condensed partition will be said to be *sufficient*.

Theorem 3.2. *A partition of the space into sets, M_i, is sufficient if and only if*

(3.11) $\nu(M_i)/\mu(M_i) = \nu(M)/\mu(M), \qquad i = 1, 2, \ldots, k,$

where M is any set in the partition and M_i are its subsets in any arbitrary refinement of the partition.

Proof. *Necessity.* If this were not so the functional, ϕ^2, would increase in value after the refinement. The *sufficiency* is proved as follows. Let $\mu(M) = f$, $\mu(M_i) = f_i$, $\nu(M) = g$, $\nu(M_i) = g_i$. We are given that $g^2/f = \sum_i g_i^2/f_i$. This implies that $\sum (g_i - gf_i/f)^2/f_i$ is zero, which completes the proof.

Theorem 3.3. *Let $T_j(\mathbf{x})$ be a sufficient statistic for the parameter, θ_j, $j = 1, 2, \ldots, s$, where $\mathbf{x}$ is a vector of N independent observations on a random variable, X. Let H_0 specify zero values for the θ's and H_1 specify non-zero values, then the two following statements are equivalent.*

(i) *The ϕ^2 of the H_1-distribution of the $T_j(\mathbf{x})$ with respect to the H_0-distribution of $T_j(\mathbf{x})$ is equal to the ϕ^2 of the N-product H_1-distribution with respect to the N-product H_0-distribution.*

(ii) *$T_j(\mathbf{x})$ is a sufficient estimator of θ_j, $j = 1, 2, \ldots, s$. A proof is given for $s = 1$, which can readily extend to cases for which $s > 1$.*

Proof. (ii) implies (i). Consider the ratio of the product distributions, when the density is exhibited as a product, $\mathscr{P}(T(\mathbf{x}) = t) \; \mathscr{P}(\mathbf{x} \mid T(\mathbf{x}) = t)$. Suppose that the space of $T(\mathbf{x})$ is partitioned into sets, $A_i \equiv \{\mathbf{x} : T(\mathbf{x}) = t_i\}$, and the space of the conditional variable, $\mathbf{x}$ given t, into sets, C_j. The sufficiency definition then enables us to write for arbitrary measurable sets, A and C,

(3.12)
$$\frac{\mathscr{P}\{\mathbf{x} \in A \times C \mid H_1\}}{\mathscr{P}\{\mathbf{x} \in A \mid H_1\}} = \frac{\mathscr{P}\{\mathbf{x} \in A \times C \mid H_0\}}{\mathscr{P}\{\mathbf{x} \in A \mid H_0\}}$$

or after rearrangement,

(3.13)
$$\frac{\mathscr{P}\{\mathbf{x} \in A \times C \mid H_1\}}{\mathscr{P}\{\mathbf{x} \in A \times C \mid H_0\}} = \Omega(\mathbf{x})$$

constant for variations in C. But this is the sufficient condition that the sum (VI.1.3) should not be increased by the partition, $C_1 \cup C_2 \cup \ldots \cup C_n$.

(i) implies (ii). This follows by a reversal of the argument above. If the sum (VI.1.3) is not increased by the partition into sets, C_j, then (3.13) holds and hence (3.12) which is the sufficiency condition.

EXERCISES AND COMPLEMENTS

The Theoretical Non-central χ^2

1. Let Y_1 be normal $(\lambda^{\frac{1}{2}}, 1)$. Then by definition $V \equiv Y_1^2$ has the distribution of non-central χ^2 with 1 d.f. Verify that the moment generating function of V is

given by

$$\mathscr{E} \exp Vt = (1 - 2t)^{-\frac{1}{2}} \exp \{\lambda t/(1 - 2t)\} \quad \text{for} \quad t < \tfrac{1}{2}.$$

Hence show that the m.g.f. of W, a non-central χ^2 with ν d.f. is

$$\mathscr{E} \exp Wt = (1 - 2t)^{-\frac{1}{2}\nu} \exp \{\lambda t/(1 - 2t)\}.$$

Verify that the cumulants are

$$\kappa_1 = n + \lambda \qquad\qquad \kappa_3 = 8(n + 3\lambda)$$
$$\kappa_2 = 2(n + 2\lambda) \qquad \kappa_4 = 48(n + 4\lambda),$$

and generally $\kappa_s = (n + s\lambda)2^{s-1}(s - 1)!$ (Wishart, 1932; Tang, 1938.)

2. Prove that the cumulants of Exercise 1 obey the recurrence,

$$\kappa_r = 2(r + n\lambda) \frac{\mathrm{d}\kappa_{r-1}}{\mathrm{d}\lambda}.$$

(Chakrabarti, 1949.)

3. Prove that non-central χ^2 has the additive property, namely, if W_j is non-central χ^2 with parameter, λ_j, and ν_j d.f. and if the W_j form a set of mutually independent random variables, then $W = \sum W_j$ is non-central χ^2 with parameter, $\lambda = \sum \lambda_j$, and degrees of freedom, $\nu = \sum \nu_j$.

(Characteristic functions are often used to prove this result, but orthogonal transformations combined with the canonical form yield the result readily.)

4. Let Z_i be mutually independent and normal $(c_i, 1)$, $i = 1, 2, \ldots, n$; let the n variables further be mutually independent. Determine the distribution of $\mathbf{Z}^T\mathbf{Z}$ subject to the orthogonal set of restraints, $\mathbf{A}_1\mathbf{Z} = \mathbf{b}$, where $\mathbf{b}$ is a $p \times 1$ vector of constants and $\mathbf{A}_1$ consists of the first p rows of an orthogonal matrix, $\mathbf{A}$, $p < n$. [$\mathbf{Z}^T\mathbf{Z} - \mathbf{b}^T\mathbf{b}$ has the distribution of non-central χ^2 with $(n - p)$ d.f. and parameter, $\mathbf{c}^T\mathbf{c} - \mathbf{b}^T\mathbf{b}$. To prove this make the transformation $\mathbf{A}\mathbf{Z} = \mathbf{U}$. Then $\mathbf{U}^T\mathbf{U} = \mathbf{Z}^T\mathbf{Z}$. The distribution of the last $(n - p)$ variables, U_i, is independent of the distribution of the first p of them. Further the expectations of these last $(n - p)$ variables is given by $\mathbf{A}_2\mathbf{c}$ and so the parameter of non-centrality is $\mathbf{c}^T\mathbf{A}_2^T\mathbf{A}_2\mathbf{c} = \mathbf{c}^T(\mathbf{1} - \mathbf{A}_1^T\mathbf{A}_1)\mathbf{c} = \mathbf{c}^T\mathbf{c} - \mathbf{b}^T\mathbf{b}$.] (Patnaik, 1949.)

5. Prove that if the restraints are non-orthogonal so that $\mathbf{B}_1\mathbf{Z} = \mathbf{b}$ where $\mathbf{B}_1$ is $p \times n$ and of rank p, we can derive a similar result by forming an orthogonal matrix, $\mathbf{B}$, the first p rows of which are the p rows of $\mathbf{B}_1$ orthonormalized. The last $n - p$ rows of $\mathbf{B}$ form an $(n - p) \times n$ matrix $\mathbf{B}_2$. The parameter of non-centrality is then $\mathbf{c}^T\mathbf{B}_2^T\mathbf{B}_2\mathbf{c}$.

This can also be done by writing $\mathbf{B}_1\mathbf{B}^T = \mathbf{M}$ and noting that $\mathbf{M}$ is positive definite. The restraints can now be written $\mathbf{M}^{-\frac{1}{2}}\mathbf{B}_1\mathbf{Z} = \mathbf{M}^{-\frac{1}{2}}\mathbf{b}$ and are orthogonal for $(\mathbf{M}^{-\frac{1}{2}}\mathbf{B}_1)(\mathbf{M}^{-\frac{1}{2}}\mathbf{B}_1)^T = \mathbf{1}_p$. The parameter of non-centrality can now be written $\mathbf{c}^T\mathbf{c} - \mathbf{b}^T\mathbf{M}^{-1}\mathbf{b}$.

[Bateman (1949) obtains a solution to this problem by a different method.]

6. Prove that the "circular coverage" distribution is equivalent to that of non-central χ^2 with two degrees of freedom in the following form. Let X_1 and X_2 be normal $(0, 1)$ and mutually independent; express as an integral the probability

that (X_1, X_2) lies in a circle with centre (a, b) and radius, R. (See Owen, 1962 for references and tables.)

7. Let Z be normal $(0, 1)$ under H_0 and normal $(\theta, 1)$ under H_1. Let $U = Z$, $V = |Z|$ and $T = 2^{-\frac{1}{2}}(Z^2 - 1)$. Show that if θ is small, ϕ^2 is approximately θ^2 and the canonical function is Z approximately. Hence show that since V and T have zero correlation with Z, the test functions $V^* = N^{-\frac{1}{2}}\mathscr{S}V$ and $T^* = N^{-\frac{1}{2}}\mathscr{S}T$ have low power for discriminating between H_0 and H_1.

8. Suppose that ν is not small and λ is not large. Then the cumulants of Exercise 1 are not greatly different from those possessed by a normal $(\nu + \lambda, 2\nu + 4\lambda)$ variable in the sense that the higher cumulants of $(\chi'^2 - \nu - \lambda)/(2\nu + 4\lambda)^{\frac{1}{2}}$ are not greatly different from zero. The ratio, $(\nu + \lambda)/(2\nu + 4\lambda)^{\frac{1}{2}}$, can be used as approximate measure of the values of parameters of non-centrality reduced to a common standard.

9. Derive the non-central χ^2 distribution for 1 d.f. and parameter, μ^2, as follows. Let Z be normal $(\mu, 1)$.

$$\mathscr{P}(Z^2 < t^2) = \mathscr{P}(|Z| < t) = (2\pi)^{-\frac{1}{2}}\int_{-t^{\frac{1}{2}}}^{t^{\frac{1}{2}}} \exp\{-\tfrac{1}{2}(y - \mu)^2\}\, dy.$$

Expand the integrand as a series

(*) $$\mathscr{P}(|Z| < t) = \sum_{r=0}^{\infty} \exp(-\tfrac{1}{2}\mu^2)\frac{\mu^r}{r!}(2\pi)^{-\frac{1}{2}}\int_{-t}^{t} y^r \exp(-\tfrac{1}{2}y^2)\, dy.$$

Now use the duplication formula of the Γ-function, $(2s)! = 2^{2s}s!(s - \tfrac{1}{2})!/(-\tfrac{1}{2})!$; substitute u for y^2 in (*) and note that the integrals vanish for odd r.

(**) $$\mathscr{P}(Z^2 < t^2) = \sum_{s=0}^{\infty} \frac{\mathscr{P}(s \mid \tfrac{1}{2}\mu_i^2)}{2^{s+\frac{1}{2}}(s - \tfrac{1}{2})!}\int_0^{t^2} u^{s-\frac{1}{2}}\exp(-\tfrac{1}{2}u)\, du,$$

where $\mathscr{P}(s \mid \tfrac{1}{2}\mu_i^2) = \exp(-\tfrac{1}{2}\mu_i^2)(\tfrac{1}{2}\mu_i^2)^s/s!$, the probability of obtaining a value s from a Poisson distribution with parameter, $\tfrac{1}{2}\mu_i^2$. Therefore

(***) $$\mathscr{P}(Z^2 < t^2) = \sum_{s=0}^{\infty} \mathscr{P}(s \mid \tfrac{1}{2}\mu^{\frac{1}{2}})\mathscr{P}(\chi^2_{2s+1} < t^2)$$

since the second factor in (**) is simply the probability that χ^2 with $(2s + 1)$ degrees of freedom should be less than t^2. The non-central χ^2 with 1 d.f. is thus distributed as a central χ^2 with $(2s + 1)$ degrees of freedom where s has the Poisson distribution with parameter, $\tfrac{1}{2}\mu^2$. Alternatively, the distribution of the non-central χ^2 can be exhibited as a mixture of distributions, the weights being provided by the Poisson distribution with parameter, $\tfrac{1}{2}\lambda$. (Johnson, 1959; Kerridge, 1965; Guenther, 1964; Ruben, 1963.)

10. Prove by induction that the non-central χ^2 with ν d.f. and parameter, λ, has the distribution function given by

$$\mathscr{P}(W < t) = \sum_{q=0}^{\infty} \mathscr{P}(q \mid \tfrac{1}{2}\lambda)\mathscr{P}(\chi^2_{2q+\nu} < t).$$

[*Hint*. It is true for $\nu = 1$ and parameter $\lambda = 2\mu^2$ by Exercise 9. Suppose it true for ν_1 and λ_1, ν_2 and λ_2, the d.f. and parameters of non-central χ^2 variables, W_1 and W_2. Write $W = W_1 + W_2$.

$$\mathscr{P}(W < t) = \int_0^t \mathscr{P}(W_2 < t - t_1)\, d\mathscr{P}(W_1 < t_1), \qquad t_1 \leqslant t$$

$$= \sum_{s=0}^\infty \sum_{r=0}^\infty \mathscr{P}(s \mid \tfrac{1}{2}\lambda_2)\mathscr{P}(r \mid \tfrac{1}{2}\lambda_1) \int_0^t \mathscr{P}(\chi_{2s+\nu_2}^2 < t - t_1)\, d\mathscr{P}(\chi_{2r+\nu_1}^2 < t_1)$$

$$= \sum_{s=0}^\infty \sum_{r=0}^\infty \mathscr{P}(s \mid \tfrac{1}{2}\lambda_2)\mathscr{P}(r \mid \tfrac{1}{2}\lambda_1)\mathscr{P}(\chi_{2(r+s)+\nu_1+\nu_2}^2 < t),$$

by the additive property of χ^2 distribution,

$$= \sum_{q=0}^\infty \mathscr{P}(q \mid \tfrac{1}{2}\lambda)\mathscr{P}(\chi_{2q+\nu}^2 < t),$$

by the additive property of the Poisson distribution.] (Johnson, 1959; Kerridge, 1965.)

11. Let W be the non-central χ^2 of Exercise 1. Prove that $\sqrt{(2W)}$ has mean $\sqrt{\{2(\nu + \lambda) - (\nu + 2\lambda)/(n + \lambda)\}}$ to order $(n + \lambda)^{-\frac{3}{2}}$ and variance to order $(n + \lambda)^{-1}$. $\{2W(n + \lambda)/(n + 2\lambda)\}^{\frac{1}{2}}$ is approximately distributed normally with unit variance and mean $\{2(n + \lambda)^2/(n + 2\lambda) - 1\}^{\frac{1}{2}}$. (Patnaik, 1949.)

12. Write $r = n + \lambda$, $b = \lambda/(n + \lambda) \equiv \lambda/r$. Make the transformation

$$y = (w/r)^h \quad \text{in (1.7).}$$

The cumulants of $Y = (W/r)^h$ may now be calculated. Set $h = \frac{1}{3}$. Verify that measures of skewness and kurtosis of Y are given by

$$\beta_1(Y) = \frac{8b^4}{r(1 + b)^3} - \frac{64}{3^3}\frac{b^2(1 + 4b + 24b^2 - 41b^3 - 35b^4)}{r^2(1 + b)^4} + \frac{128}{3^6(1 + b)^5} O\left(\frac{1}{r^3}\right)$$

$$\beta_2(Y) = 3 - \frac{4}{3^2}\frac{(1 + 3b + 12b^2 - 44b^3)}{r(1 + b)^2} -$$

$$- \frac{64}{3^4}\frac{(1 + 5b + 7b^2 + 271b^3 - 220b^4 - 326b^5)}{r^2(1 + b)^3}$$

Compare these measures with $\beta_1(W) = 8(1 + 2b^2)/\{r(1 + b)^3\}$ and $\beta_2(W) = 3 + 12(1 + 3b)/\{r(1 + b)^2\}$. [The transformation given is an approximate normalization.] (Abdel-Aty, 1954.)

13. Set $h = 1 - 2(\nu + \lambda)(\nu + 3\lambda)/\{3(\nu + 2\lambda)^2\}$ in the transformation $Y = (W/r)^h$ of Exercise 12. Verify that the cumulants are then given by (5) of Sankaran (1959.)

14. Define a new variable

$$W^* = \frac{\chi_\xi^2 - \xi}{\sqrt{(2\xi)}}\, \sigma(W) + \mathscr{E}(W)$$

$$= \frac{\nu + 3\lambda}{\nu + 2\lambda}\chi_\xi^2 - \frac{\lambda^2}{\nu + 3\lambda}$$

where χ^2_ξ is distributed as a central χ^2 with fractional degrees of freedom

$$\xi = \frac{8}{\beta_1(W)} = \frac{(\nu + 2\lambda)^2}{(\nu + 3\lambda)^3} \, .$$

The distribution of χ^2_ξ can be obtained by interpolation. (Pearson, 1959.)

15. Write the non-central χ^2 distribution as in Exercise 10. Replace the incomplete integral $\mathscr{P}(\chi^2_{2j+\nu} < 2y)$ for the central χ^2 by an infinite series in y, set $\mu = \frac{1}{2}\lambda$ and obtain

$$G(2y; \nu, \lambda) = \sum_{j=0}^{\infty} e^{-\mu} \frac{\mu^j}{j!} \sum_{k=0}^{\infty} e^{-y} \frac{y^{\frac{1}{2}\nu+j+k}}{\Gamma(\frac{1}{2}\nu + j + k + 1)} = \sum_{k=0}^{\infty} e^{-(\mu+y)} y^{\frac{1}{2}\nu+k} \frac{I_{\frac{1}{2}\nu+k}[2\sqrt{(\mu y)}]}{(\sqrt{(\mu y)})^{\frac{1}{2}\nu+k}}$$

where

$$I_\alpha[x] = \sum_{m=0}^{\infty} \frac{(\frac{1}{2}x)^{\nu+2m}}{\Gamma(\nu + m + 1)m!} \, .$$

($2y$ has been used as the argument in place of y to avoid powers of 2 on the right; the properties of the Bessel functions can now be used to compute the required values for the tabulation of the distribution of non-central χ'^2.) (Seber, 1963.)

16. Use the previous exercises to determine the m.g.f. and c.g.f. of a general non-central χ^2 variable of the form $\mathbf{Z}^T \mathbf{A} \mathbf{Z}$, with $\mathbf{A}$ idempotent, in the variable $\mathbf{Z}$ of Equation (1.1). Suppose now that there are two idempotents, $\mathbf{A}$ and $\mathbf{B}$, with p and q degrees of freedom respectively, having moreover a component χ^2 of s degrees in common. (There is thus an orthogonal transformation which reduces $\mathbf{A}$ and $\mathbf{B}$ simultaneously to canonical form.) Derive the joint c.g.f. of $\mathbf{Z}^T \mathbf{A} \mathbf{Z}$ and $\mathbf{Z}^T \mathbf{B} \mathbf{Z}$.

17. Suppose now that the $\{c_i\}$ of (1.1) are realisations of random variables $\{C_i\}$. Discuss what meanings could be given to the phrase "the test functions, $\mathbf{Z}^T \mathbf{A} \mathbf{Z}$ and $\mathbf{Z}^T \mathbf{B} \mathbf{Z}$ are correlated". Give examples to show that there may be high correlation under H_0 between such test functions and yet $\mathbf{Z}^T \mathbf{A} \mathbf{Z}$ may be an effective test function for a given set $\{c_i\}$ whereas $\mathbf{Z}^T \mathbf{B} \mathbf{Z}$ may not be. Their effectiveness may depend on the $\{c_i\}$.

18. Suppose that $\{Z_{ij}\}$ is a set of N independent observations made on the set of normal variables $\{Z_1, Z_2, \ldots, Z_n\}$. Reduce the problem to a consideration of N independent observations on the variables, $\{Y_1, Y_2, \ldots, Y_n\}$, defined as in (2.4). Find the parameters of non-centrality of the variables,

$$W_i = N^{-\frac{1}{2}} \mathop{\mathscr{S}}_{j=1}^{N} Y_{ij}, \quad \text{for } i = 1, 2, \ldots, n.$$

Restate Theorem 2.1 for the variables, W_i. Show that the parameter of non-centrality of a quadratic form, $\mathbf{W}^T \mathbf{A} \mathbf{W}$, with $\mathbf{A}$ symmetric and idempotent, is $Na_{11}\lambda$ and the maximum is attained for $a_{11} = 1$.

19. For the standardized sums of Exercise 18, state and prove an analogue of Theorem 2.2, namely that if $\mathbf{a}^T \mathbf{a} = 1$ the expectation of $\mathbf{a}^T \mathbf{W}$ is equal to the correlation of $\mathbf{a}^T \mathbf{W}$ with W_1 multiplied by $(N\lambda)^{\frac{1}{2}}$ and so $\mathscr{E} \mathbf{a}^T \mathbf{W} = a_1 (N\lambda)^{\frac{1}{2}}$.

Sufficiency and Efficiency

20. Let $\{f_i\}$ and $\{g_i\}$ be probability measures assigned to n distinct points under H_0 and H_1 respectively. Suppose further that each f_i is positive and $\sum f_i = \sum g_i = 1$. Define $\Omega(i) = g_i/f_i$. Then $\sum \Omega(i)f_i = 1$ and $\sum (\Omega(i) - 1)f_i = 0$. Neither sum can measure the discrepancy between the measures under H_0 and H_1. [A similar phenomenon is met with in a consideration of the mean of the differences from zero.] If $\{\Omega(i) - 1\}$ is squared, its expectation is $\phi^2 = \sum g_i^2/f_i - 1$. [Pearson had considered measuring the discrepancy by its largest relative error by such a phrase as $\Omega(i)$ was 10% above its true value.] (Pearson, 1895.)

21. Write down the Radon–Nikodym derivates of the following pairs of measures:

	Distribution	H_0	H_1
(i)	Normal	$N(0, 1)$	$N(\theta, 1)$
(ii)	Normal	$N(0, 1)$	$N(0, \sigma^2)$
(iii)	Normal	$N(0, 1)$	$N(\theta, \sigma^2)$
(iv)	Poisson	Parameter, α	Parameter, γ
(v)	Binomial	$b(n, p)$	$b(n, \varpi)$
(vi)	Rectangular	$[0, 1]$	$[0, \theta]$

Compute ϕ^2 or the variance of the Radon–Nikodym derivative. Mention the conditions under which ϕ^2 is infinite.

22. Let $f(x) = 1$ for $0 \leqslant x \leqslant 1$ and $g(x) = 1 + \alpha$ for $0 \leqslant x \leqslant \frac{1}{2}$ and $g(x) = 1 - \alpha$ for $\frac{1}{2} < x \leqslant 1, |\alpha| \leqslant 1$. Then $\phi^2 + 1 = 1 + \alpha^2$. Prove that if the first partition is made at the point $x = \frac{1}{2}$, $\phi^2(1) = \alpha^2 + 1$, so that this is a sufficient partition. Show that $g(x) = \{1 + \alpha w_1(x)\}f(x)$ and that $\int w_s(x)g(x)\,dx = 0$ if $s > 1$, where $w_s(x)$ is the Walsh function of order, s.

23. With the notation of Exercise 21, suppose that a sample of N independent observations is made on the variables. Calculate the ϕ^2 of the product distributions. In cases (i), (iv), (v) and (vi) compute the ϕ^2 of the distribution of the sum of the N observations. Determine which of the distributions has a sufficient estimator of the form $N^{-1}\sum X_i$ for the parameter, θ.

24. Consider the ϕ^2 of the $N(\theta, 1)$ distribution with respect to the $N(0, 1)$ distribution. Show that there is no sufficient partition. Similarly with pairs of Poisson distributions with different parameters, show that there is no sufficient partition.

25. Let the probability densities of X under H_0 and H_1 be p_0 and p_1. Then $T(x) = p_1(x)/p_0(x)$ is sufficient for distinguishing the distributions. Prove that if $\Omega_i(x) = p_1^{(i)}(x)/p_0^{(i)}(x)$, where $p_0^{(i)}(x)$ and $p_1^{(i)}(x)$ are the densities of the ith member of a set of independently distributed variables, $\{X_i\}$, then $p_1(\mathbf{x})/p_0(\mathbf{x}) \equiv \Omega(\mathbf{x}) \equiv \prod_{i=1}^{N} [p_1^{(i)}(x)/p_0^{(i)}(x)]$ is also sufficient for distinguishing the product measures. Prove equivalently that $\log \Omega(\mathbf{x})$ and $\sum \log \Omega_i(x)$ are sufficient statistics.

26. Let $\zeta(X)$ be a measurable function of X such that $\mathscr{E}_0\zeta = 0$, $\mathscr{E}_0\zeta^2 = 1$ under the hypothesis that X is standard normal. Calculate $\mathscr{E}_1\zeta$ and $\mathscr{E}_1\zeta^2$ when H_1 specifies that X is normal $(\theta, 1)$.

27. Specialise $\zeta(X)$ of Exercise 26 by considering a finite partition of the space of X into sets, on each of which $\zeta(X)$ is constant. What function $\zeta(X)$ defined on the condensed space has maximum correlation with X under H_0? What is the efficiency of a suitably standardized sum of ζ, if N observations be made, if H_1 specifies that X is normal $(\theta, 1)$.

28. "The correlation between any estimate which satisfies the criterion of efficiency and any other consistent estimate of the same parameter tends for increasingly large values of N to $\sqrt{E}$." Prove this by supposing that $N^{-1}\mathscr{S}\zeta_1(X)$ is the most efficient estimator with variance $N^{-1}\sigma^2$ and $N^{-1}\mathscr{S}\zeta(X)$ is another estimator with variance, $(NE)^{-1}\sigma^2$. Show that ζ_1 is uncorrelated with any function which has zero expectation under H_1. [Take a complete orthonormal set which has ζ_1 as its first element. Express ζ as a linear form in the set. Assume that the class of estimators being considered consists of sums of the form, $N^{-1}\mathscr{S}\xi_j$.] (Fisher, 1924a and 1925a.)

29. Using the results (VI.1.7) and (VI.1.9) show that $\dfrac{\partial \log f(x; \theta)}{\partial \theta}$ may be expected to be an efficient test function, if θ is not large.

The Probability Integral Transformation

30. Prove that, if ϖ is distributed rectangularly $(0, 1)$, $-2 \log_e \varpi$ is distributed as χ^2 with 2 d.f. by a direct analytic method and also by the use of the m.g.f.

31. Three tests of significance have yielded probabilities $\{P\}$ of $0\cdot145$, $0\cdot263$ and $0\cdot087$. Compute $\chi^2 = -2 \log_e P$ for each P and test the significance of the sum. Alternatively compute $-2 \log_e (P_1 P_2 P_3)$, the product being $0\cdot003318$ approximately. (R. A. Fisher, *Statistical Methods.*)

32. Determine the mean and variance of $-2 \log_e P$ for a one sided test in the binomials, $(\tfrac{1}{2} + \tfrac{1}{2})^N$, giving N various values taking as an example the following computation:

Successes (m)	Relative Frequency of m	Cumulative Probability (P)	$-2 \log_e P$
4	0·0625	1·0000	0
3	0·2500	0·9375	0·12908
2	0·3750	0·6875	0·74939
1	0·2500	0·3125	2·32630
0	0·0625	0·0625	5·54518

$\mathscr{E}(-2 \log_e P)$ is $1\cdot241$ under a true null hypothesis. (Lancaster, 1949b.)

33. Show that if the space of ϖ in Exercise 30 is condensed, the function on the condensed distribution which has an expectation of 2 and maximum correlation with $\chi^2 = -2 \log_e \varpi$ is the mean value χ^2, namely

$$\zeta(A_i) = \int_{A_i} -2 \log_e \varpi \, d\varpi,$$

where A_i is a sub-interval of $[0, 1]$. (Lancaster, 1949b.)

34. With the partition into intervals as in Exercise 33, show that the median χ^2, defined by

$$\zeta(A_i) = -2\log_e\{\tfrac{1}{2}(P + P')\},$$

where P' and P are the end-points of the interval, A_i, if $P' \neq 0$.

$$\zeta(A_i) = -2\log_e P + 2, \quad \text{if } P' = 0.$$

is a good approximation to the mean value χ^2. (Lancaster, 1949b; Pearson, 1950.)

35. Let $K(z)$ be a continuous distribution function and suppose Z has zero mean and unit variance. Define a mean value function $\zeta(z)$ as follows. Suppose the space of Z is partitioned into sets $A_1, A_2, \ldots, A_n$ and that the value of $\zeta(z)$ on each A_i is constant

$$\zeta_{A_i}(z) = \int_{A_i} \zeta(z)\, dK(z) \Big/ \int_{A_i} dK(z).$$

Prove that $\mathscr{E}\zeta(z) = 0$, and that $\mathscr{E}\zeta^2(Z) \leqslant \mathscr{E}Z^2 = 1$. [This follows from the analysis of variance of $\zeta(z)$ into comparisons between mean values in the sets, A_i, and the variation in A_i about the mean value. Use such a mean value variable to correct the expectation of $-2\log_e P$ in a case such as that of Exercise 34.]

36. Let H_0 specify the rectangular $[0, 1]$ distribution. Let H_1 specify various density functions on $[0, 1]$, $g(w) \equiv g(w; \theta)$ as given below.

Let $\xi = (2w - 1)\sqrt{3}$; $\eta = -(1 + \log_e w)$; $\psi = -1$ for $0{\cdot}25 < w \leqslant 0{\cdot}75$ and $\psi = +1$ for other w. Verify and comment on the following:

(i) $g(x) = 1 + \alpha\xi, \quad |\alpha| \leqslant 3^{-\frac{1}{2}}$

$\phi^2 = \alpha^2$. The most efficient χ-analogue is $N^{-\frac{1}{2}}\mathscr{S}\xi$. η has a correlation of $\tfrac{1}{2}\sqrt{3}$ with ξ and so the χ-analogue $-N^{-\frac{1}{2}}\mathscr{S}\eta$ has a parameter of non-centrality of $0{\cdot}75\alpha^2$.

(ii) $g(x) = 1 - \beta(1 + \log_e x), \quad 0 \leqslant \beta \leqslant 1.$

The most efficient χ-analogue is $N^{-\frac{1}{2}}\mathscr{S}\eta$ with parameter of non-centrality $N\beta^2$.

(iii) $g(x) = 1 + \gamma\psi.$

The most efficient χ-analogue is $N^{-\frac{1}{2}}\mathscr{S}\psi$. (Lancaster, 1961$b$.)

37. Verify that except for a scale factor the distributions of Exercise 36 are (i) the sum of rectangular variables, for which see page 245 of Cramér (1946), (ii) χ^2 with $2N$ degrees of freedom, (iii) the binomial, $b(N, \tfrac{1}{2})$.

38. The observations in such a case as Exercise 36 may be of different "weights". For example, the α, β and γ may vary at different observations. To fix ideas consider case (i) in which α takes the value α_i at the ith observation. What χ-analogue, $\displaystyle\sum_{i=1}^{N} a_i\xi_i$, has maximum expectation under H_1, where ξ_i is the value of ξ at the ith observation? Verify that if the a_i are grouped closely about the mean value this most efficient χ-analogue will be highly correlated with the simple sum, $N^{-\frac{1}{2}}\sum \xi_i$.

39. The probability integral transformation can be generalized by taking as the χ-analogue, $\sum \lambda_i\eta_i$ in Exercise 36. (Good, 1958.)

40. Weighting can be achieved by transferring to χ^2 with n_i degrees of freedom at the ith observation and writing

$$\varpi_i = \int_{\chi^2}^{\infty} h_i(x)\,\mathrm{d}x,$$

where $h_i(x)$ is the density function of χ^2 with n_i degrees of freedom. (Lancaster, 1961b.)

χ-analogues based on Other Systems of Functions

41. Prove that in the binomial the expected value of the χ^2, corrected for continuity, is given by

$$\mathscr{E}(\chi_c^2) = 1 + (4npq)^{-1} - (\text{mean deviation})/(\text{variance}).$$

Compute the expected value of χ_c^2 in the binomial $(\frac{1}{2} + \frac{1}{2})^N$, for $N = 4$ and $N = 9$. Deduce that χ_c^2 is not a satisfactory test function for the combination of probabilities. (Lancaster, 1949b.)

42. Suppose the H_0 and H_1 specify p and P throughout all experiments as the binomial probabilities and the standardized variables,

$$X_i = (m_i - n_i p)/\sqrt{[np(1 - p)]},$$

and n_i are known. Show that an application of the result of Exercise 38 leads to the "natural result" that the most efficient χ-analogue is

$$X = (\mathscr{S} m_i - p\mathscr{S} n_i)/\sqrt{(p(1 - p)\mathscr{S} n_i.)}$$

43. Prove that for one way tests in the binomial, $b(n, p)$, the χ corrected for continuity

$$\chi = (m - np - \tfrac{1}{2})/\sqrt{(npq)}$$

has an expectation of $-\tfrac{1}{2}/\sqrt{(npq)}$ under H_0 and hence is unsuitable for use as a χ-analogue.

44. Let ϖ be rectangular $[0, 1]$. Define χ^2 with s d.f. by means of the equation

$$\varpi = \int_{\chi^2}^{\infty} h(x)\,\mathrm{d}x.$$

Give a heuristic proof that $\chi_{s_1}^2$ and $\chi_{s_2}^2$ so defined have a high correlation if $s_1 > 2$, $s_2 > 2$, by plotting the graph $(\chi_{s_1}^2(\varpi),\ \chi_{s_2}^2(\varpi))$ as ϖ varies. (Lancaster, 1961b.)

45. Suppose $X_1, X_2, \ldots, X_N$ are independently distributed Poisson variables, each with parameter, v, occurring for example at times, $1, 2, \ldots, N$. The joint distribution of the X_i conditional on a fixed value of their sum is a multinomial. Prove that $\sum (X_i - \bar{X})^2/\bar{X}$ is approximately χ^2 with $(N - 1)$ degrees of freedom. If H_1 specifies that v changes with time, the regression analysis may be used, conveniently with the aid of the Chebyshev polynomials as tabulated in the tables of *Fisher and Yates*. Identify the test of a regression coefficient of a given order as a test of the departure of the moments of the time variable from their expected value. Show that the sum of squares of the $(N - 1)$ standardized orthonormal polynomial functions is equal to the ordinary Pearson χ^2.

46. In Exercise 45, $N = 6$, and the observations are 41, 34, 54, 39, 49 and 45. Verify that the Pearson χ^2 is 5·9389. With the aid of tables calculate the values of the standardized polynomial sums and verify the values {0·90437, −0·41297, 0·28199, 0·65777, −2·10677} and that the sum of squares of these values is equal to the Pearson χ^2. (Lancaster, 1953.)

47. In an experiment where the results could be specified by an H_0 of the form, $p_j = N(\tfrac{1}{2})^n \binom{n}{j}, j = 0, 1, \ldots, n$, the numbers in the classes were 17, 81, 152, 180, 104, 17; $n = 5, N = 551$. Compute the values of the Kravčuk polynomials by the formula

$$P_s(k) = \binom{n}{s}^{-\frac{1}{2}} \sum \binom{k}{t}\binom{n-k}{s-t}(-1)^{s-t},$$

where summation is with respect to t, max $(0, s + k - n) \leqslant t \leqslant$ min (k, s) and hence partition the Pearson χ^2 calculated under H_0. The χ^2's corresponding to the standardized polynomial sums are {3·415245, 0·384029, 1·888203, 1·019601, 0·045372} and the total sum of these squares is equal to the Pearson χ^2 of 6·752450. (Lancaster, 1965b.)

48. The sum of squares of the Fourier coefficients of a function in L_2 with respect to an orthonormal basis or a complete orthonormal set on a measure space is an invariant for all choices of the basis or set. [This sum can be conveniently termed the *representation*.]

CHAPTER VIII

Tests of Goodness of Fit in The Multinomial Distribution

1. INTRODUCTORY

Tests of goodness of fit can often be reduced to a test of goodness of fit of the multinomial distribution. In other words, there are N observations, each falling into one of n classes with probabilities, p_i, $i = 1, 2, \ldots, n$. In the simplest cases, the p_i may be given by hypothesis as in the examples in Section V.10, where each $p_i = n^{-1}$. Usually, however, the p_i are specified as functions of some parameter which may or may not be given by hypothesis. For example, in a genetic experiment where the F_1 generation are all heterozygotes, the probabilities that a member of F_2 should have no, one or two dominants are given by

$$(1.1) \qquad p_0 = \tfrac{1}{4}, \qquad p_1 = \tfrac{1}{2}, \qquad p_2 = \tfrac{1}{4},$$

since these are the terms of the binomial, $(\tfrac{1}{2} + \tfrac{1}{2})^2$, which is relevant when an assumption is made that the probability that the offspring obtains a dominant from the first parent, and similarly from the second parent, is $\tfrac{1}{2}$, and a further hypothesis is made that these two events are independent.

A somewhat similar case arises in binovular twinning. Let p be the probability that a fertilized ovum becomes a male. If two ova are shed and fertilized, it is assumed that the sex of the first is independent of the sex of the second when the two ova are fertilized. The probabilities of no males, one male or two males occurring in a twin pair resulting from two ova fertilized at the same time are then, respectively,

$$q^2, 2pq \quad \text{and} \quad p^2, p + q = 1$$

For example, it may not be desirable to make the hypothesis, that the probability that a twin should be a male is the same as the probability that a singleton birth should be a male, for it may not be desirable to assume that male and female twins have the same viability as singletons. In such a case, p may be appropriately estimated from the data.

K. Pearson (1900a), as we have pointed out in the historical notes in Chapter I, did not recognize that the estimation of parameters from the data made a difference to the test, by inducing restraints and reducing the number of degrees of freedom. Pearson (1913b) was not quite consistent on this point, as he gave a correction for use in the estimation of the correlation coefficient by the formula $\rho^2 = \phi^2/(1 + \phi^2)$, which amounts to subtracting the expected value of χ^2 in the null case of zero correlation, which would, of course, be equal to the number of degrees of freedom. Pearson (1913b) gave $(r - 1)(c - 1)$ as the quantity to be subtracted and not $rc - 1$, as he might have been anticipated to suggest. Pearson (1916b) stated that estimation of the parameters from the data should reduce the number of "variables" (i.e. degrees of freedom) but he did not admit that writing the observed for the expected row or column totals in a contingency table reduces the number of "variables". Pearson (1916b) gave a discussion of the distribution of χ^2 in multidimensional contingency tables, which showed that he believed that some restrictions reduced the number of "variables" (d.f.). In some cases, the restrictions were non-homogeneous and Pearson believed that this would yield an increased expected value for the χ^2. However, Pearson and his school did not solve the problem of fitting estimates of the parameters. R. A. Fisher in a series of papers beginning in 1922 paid special attention to the estimation of parameters. For example, Fisher (1922a) shows that a fourfold table can be treated in several ways, depending on the model chosen to describe the experiment. In two of these models, which are described later in Section 11.2, certain parameters are introduced but do not appear in the expression for the probability of the distribution in the fourfold conditional on the marginal totals. This is another way of saying that with unrestricted sampling, the row and column parameters are estimated by *sufficient statistics*. Fisher (1922b) elaborates the notions of sufficiency and efficiency of estimators and applies his methods to the estimation of the parameters of certain populations with Pearson type distributions. Further papers showed that the method of estimation must be efficient if the χ^2 approximation is to be valid. A later section gives in detail Fisher's (1924) treatment of the estimation of parameters from the data.

2.   LEAST SQUARES AND MINIMUM χ^2

The distribution of the continuous χ^2 was discovered during their studies in least squares by Bienaymé (1852) and Helmert (1875b); the central limit theorem shows that problems in discrete variables may have an approximate solution in continuous variables. Consequently, Karl Pearson's derivation of χ^2 in discrete distributions can be seen as a development out of the least squares theory. A brief outline of some aspects of the theory are now given.

Misunderstandings have often arisen because the principle of least squares can be used without the introduction of a probability model. Thus there may be N points in the plane and the straight line of best fit to them may be required. The line may be chosen to minimize $\sum (y_i - y_i^*)^2$, the sum of the squares of the differences between the observed y_i and the fitted y_i^* values.

N independent observations are made on a random variable Y and on s other variables $X_1, X_2, \ldots, X_s$. What hyperplane of the form,

$$(2.1) \qquad y^* = b_1 x_1 + b_2 x_2 + \ldots + b_s x_s,$$

gives the best fit to these observations in the sense of least squares? Let $\mathbf{X}$ be the $N \times s$ matrix of the N observations on the s variables. The fitted values will be of the form

$$(2.2) \qquad y_i^* = x_{i1} b_1 + x_{i2} b_2 + \ldots + x_{is} b_s,$$

that is

$$(2.3) \qquad \mathbf{y}^* = \mathbf{Xb}$$

where the b_j are to be chosen to minimize the residuals,

$$(2.4) \qquad U = \mathbf{u}^T \mathbf{u} = (\mathbf{y} - \mathbf{y}^*)^T (\mathbf{y} - \mathbf{y}^*).$$

It is supposed that $N \geqslant s$ and that $\mathbf{X}$ is of rank s. By a process equivalent to completing the square, from (2.3) and (2.4), we obtain

$$
\begin{aligned}
(2.5) \quad U &= (\mathbf{y} - \mathbf{Xb})^T (\mathbf{y} - \mathbf{Xb}) \\
&= \mathbf{y}^T \mathbf{y} - \mathbf{y}^T \mathbf{X S^{-1} X^T y} + \mathbf{y}^T \mathbf{X S^{-1} X^T y} - 2\mathbf{b}^T \mathbf{X}^T \mathbf{y} + \mathbf{b}^T \mathbf{Sb} \\
&= \mathbf{y}^T \mathbf{y} - \mathbf{y}^T \mathbf{X S^{-1} X^T y} + (\mathbf{b} - \mathbf{S^{-1} X^T y})^T \mathbf{S} (\mathbf{b} - \mathbf{S^{-1} X^T y})
\end{aligned}
$$

where $\mathbf{S} = \mathbf{X}^T \mathbf{X}$; $\mathbf{S}$ is positive definite from the hypothesis that $\mathbf{X}$ is of rank s. The minimum value of U is evidently obtained by equating the final expression of (2.5) to zero. It follows that

$$(2.6) \qquad \mathbf{b} = \mathbf{S^{-1} X^T y},$$

is the best set of coefficients, in the sense that the sum of squares takes the least possible value

$$(2.7) \qquad U = \mathbf{y}^T \mathbf{y} - \mathbf{y}^T \mathbf{X S^{-1} X^T y}.$$

This result is usually obtained by differentiation and equating the vector whose elements are $\partial U / \partial b_i$ to the zero vector. Thus

$$(2.8) \qquad \mathbf{0} = \frac{\partial}{\partial \mathbf{b}} (\mathbf{y} - \mathbf{Xb})^T (\mathbf{y} - \mathbf{Xb}) = -2\mathbf{X}^T \mathbf{y} + 2\mathbf{X}^T \mathbf{Xb},$$

and

$$\mathbf{b} = \mathbf{S^{-1} X^T y} \qquad \text{as before.}$$

In the form,

$$(2.9) \qquad \mathbf{Sb} = \mathbf{X}^T\mathbf{y},$$

these k equations are usually referred to as the normal equations. The elements of $\mathbf{S}$ are given by

$$(2.10) \qquad s_{ii} = \sum_{j=1}^{N} x_{ji}^2; \qquad s_{ii'} = \sum_{j=1}^{N} x_{ji}x_{ji'}.$$

It is more common, however, to have a probability model, whereby each observation, Y, is a random variable, so that

$$(2.11) \qquad Y_i = \sum_{j=1}^{s} x_{ij}\beta_j + \epsilon_i,$$

or in matrix notation,

$$(2.12) \qquad \mathbf{Y} = \mathbf{X}\boldsymbol{\beta} + \boldsymbol{\epsilon},$$

where the elements of $\boldsymbol{\beta}$ are unknown parameters. The elements of $\boldsymbol{\epsilon}$ are taken to be N mutually independent normal variables. The problem is then to estimate $\boldsymbol{\beta}$ by a vector of estimates, $\mathbf{b}$, so as to minimize the sum of squares of the residuals, $\mathbf{e}^T\mathbf{e}$, where

$$(2.13) \qquad \mathbf{Y} = \mathbf{Xb} + \mathbf{e}.$$

The solution of this problem has already been obtained as (2.6). The fact that these estimates are unbiased and of minimum variance is left to be proved as an exercise. The joint distribution of the estimates, b_j, can now be obtained using (2.13) and (2.6)

$$(2.14) \qquad \begin{aligned} \mathbf{b} - \boldsymbol{\beta} &= \mathbf{S}^{-1}\mathbf{X}^T\mathbf{Y} - \boldsymbol{\beta} \\ &= \mathbf{S}^{-1}\mathbf{X}^T(\mathbf{X}\boldsymbol{\beta} + \boldsymbol{\epsilon}) - \boldsymbol{\beta} \\ &= \mathbf{S}^{-1}\mathbf{X}^T\boldsymbol{\epsilon}. \end{aligned}$$

The matrix $\mathbf{S}^{-1}\mathbf{X}^T$ is $s \times N$ and of rank, s. $\mathbf{b} - \boldsymbol{\beta}$ is therefore a vector, whose elements are linearly independent normal variables with joint covariance matrix

$$(2.15) \qquad \begin{aligned} \mathscr{E}(\mathbf{b} - \boldsymbol{\beta})(\mathbf{b} - \boldsymbol{\beta})^T &= \mathscr{E}(\mathbf{S}^{-1}\mathbf{X}^T\boldsymbol{\epsilon}\boldsymbol{\epsilon}^T\mathbf{X}\mathbf{S}^{-1}) \\ &= \sigma^2\mathbf{S}^{-1}. \end{aligned}$$

The inverse of this matrix is $\sigma^{-2}\mathbf{S}$. $\sigma^{-2}(\mathbf{b} - \boldsymbol{\beta})^T\mathbf{S}(\mathbf{b} - \boldsymbol{\beta})$ is thus distributed as χ^2 with k degrees of freedom by Theorem II.5.1.

It is sometimes convenient to consider the problem of least squares in a canonical form. Let $\mathbf{H}$ be an orthogonal matrix, whose first k rows are the k rows of $\mathbf{S}^{-\frac{1}{2}}\mathbf{X}^T$; then define $\mathbf{Z} = \mathbf{HY}$. Note that the remaining rows must be orthogonal to the rows of $\mathbf{X}^T$, i.e. to the columns of $\mathbf{X}$. The model then becomes

$$(2.16) \qquad \begin{aligned} \mathbf{Z} = \mathbf{HY} &= \mathbf{HX\beta} + \mathbf{H\epsilon} \\ &= \mathbf{\delta} + \mathbf{\gamma}, \end{aligned}$$

where only the first s elements of $\mathbf{\delta}$ are non-zero and because $\mathbf{H}$ is orthogonal the elements of $\mathbf{\gamma}$ are normal $(0, \sigma^2)$ and mutually independent. The vector $\mathbf{\delta}$ has to be estimated so as to minimize $\mathbf{g}^T\mathbf{g}$, where

$$(2.17) \qquad \mathbf{Z} = \mathbf{d} + \mathbf{g}.$$

But

$$(2.18) \qquad \mathbf{g}^T\mathbf{g} = (\mathbf{Z} - \mathbf{d})^T(\mathbf{Z} - \mathbf{d}) = \sum_{1}^{s}(Z_i - d_i)^2 + \sum_{s+1}^{N} Z_i^2.$$

This sum is minimized by setting $d_i = Z_i$ for $i = 1, 2, \ldots, s$, and the minimum is $\sum_{s+1}^{N} Z_i^2$. It is easily shown that the two solutions obtained are equivalent.

The results above can be generalized. Suppose that the mathematical model is specified as

$$(2.19) \qquad \mathbf{U} = \mathbf{M\beta} + \mathbf{\eta}, \quad \text{and} \quad \mathscr{E}\mathbf{\eta}\mathbf{\eta}^T = \mathbf{V}.$$

The transformation, $\mathbf{\epsilon} = \mathbf{V}^{-\frac{1}{2}}\mathbf{\eta}$, yields uncorrelated variables and reduces (2.19) to the form of (2.12) with $\mathbf{X} = \mathbf{V}^{-\frac{1}{2}}\mathbf{M}$, $\mathbf{Y} = \mathbf{V}^{-\frac{1}{2}}\mathbf{U}$, so that (2.8) becomes

$$(2.20) \qquad \mathbf{b} = \mathbf{S}^{-1}\mathbf{X}^T\mathbf{y} = (\mathbf{M}^T\mathbf{V}^{-1}\mathbf{M})^{-1}\mathbf{M}^T\mathbf{V}^{-\frac{1}{2}}\mathbf{U}.$$

3. THE FITTING OF SUFFICIENT STATISTICS

Statistics, which summarize the whole of the relevant information supplied by the sample, are termed *sufficient* (Fisher, 1922c). Fisher's definition is given in various forms. For example, Lehmann (1959) calls a statistic, T, sufficient for the family, $\{P_\theta, \theta \in \Omega\}$, if the conditional distribution of the sample, X, given $T = t$, is independent of the value of θ. If $P_\theta(\mathbf{x})$ is the probability of the sample value, $\mathbf{x}$, the factorization criterion is

$$(3.1) \qquad P_\theta(\mathbf{x}) = g_\theta[T(\mathbf{x})]h(\mathbf{x}),$$

where the first factor may depend on θ but depends on the sample values, $\mathbf{x}$, only through $T(\mathbf{x})$ and where $h(\mathbf{x})$ is independent of θ. It follows from the definition of sufficiency, that the distribution of the sample, $\mathbf{x}$, for given values of the sufficient estimator is independent of their true values, so that all class frequencies or distributions of functions defined in the conditional space under H_1 are identical with those holding in the conditional distribution under H_0. In the Poisson distribution, the conditional distribution of the sample counts given the sample total is obtained by elementary algebra and is found to be the multinomial, in which there is a linear restriction, whereby the class frequencies must add to the sum of the sample frequencies. Other slightly more complicated relations may hold if the observations are conditional on a number of restraints on the frequencies, for example, in contingency tables, which are treated in more detail in Chapter XI.

Example (*i*) *The Poisson distribution.* The sample mean is sufficient for the parameter of the Poisson distribution and the property of sufficiency has already been used in Section V.10 where the χ^2 test was used as a criterion of homogeneity among parallel counts. It is worth remarking that λ, the unknown parameter, did not appear in the computation of χ^2; further, the distribution of the counts for a given value of their mean, being independent of the distance of the mean from the theoretical value, can give no indication of such nearness under theoretical sampling conditions. However, under practical laboratory conditions, the existence of gross departures of the homogeneity χ^2 from its expected value may give a warning that the laboratory conditions have not been satisfactory. The probability of obtaining observations $a_1, a_2, \ldots, a_n$ from a Poisson distribution with parameter, λ, is given by

$$(3.2) \qquad \mathscr{P}\{\{a_i\} \mid \lambda, \text{Poisson}\} = \prod_{i=1}^{n} [e^{-\lambda} \lambda^{a_i}/a_i!]$$

$$= \frac{e^{-n\lambda}(n\lambda)^N}{N!} N! n^{-N} \Big/ \prod_{i=1}^{n} a_i!,$$

where $N = \sum a_i$. The first expression is of the form $g_\theta[T(x)]$ with $T(x) = N$. The second expression is the general term of the multinomial,

$$\left(\frac{1}{n} + \frac{1}{n} + \ldots + \frac{1}{n}\right)^N,$$

which corresponds to the $h(x)$ of (3.1).

The property of sufficiency can be used to give rather simp e proofs of χ^2, as was noted by H. E. Soper (1916). The method was very sk.ilfully used by R. A. Fisher in a number of his papers in the years just after 1921.

Example (*ii*) *The multinomial.* Suppose that there are n cells in the general multinomial distribution and that N independent observations have been made on a certain random variable which falls into the ith cell with probability, p_i. Then the multinomial probability is

$$(3.3) \qquad \mathscr{P}(\{a_i\} \mid \{p_i\}, N) = N! \prod_{i=1}^{n} [p_i^{a_i}/a_i!]$$

$$= \prod_{i=1}^{n} [e^{-\lambda p_i}(\lambda p_i)^{a_i}/a_i!]/[e^{-\lambda}\lambda^N/N!]$$

Now each term in the first expression can be approximated by

$$(3.4) \qquad \mathscr{P}(a_i \mid \lambda p_i, \text{Poisson}) = (2\pi\lambda p_i)^{-\frac{1}{2}} \exp\left[-\tfrac{1}{2}(a_i - \lambda p_i)^2/(\lambda p_i)\right]$$

$$\equiv (2\pi\lambda p_i)^{-\frac{1}{2}} \exp\left[-\tfrac{1}{2}q_i^2\right]$$

and if λp_i is large, $q_i^2 = (a_i - \lambda p_i)^2/(\lambda p_i)$ is approximately distributed as the square of a standardized normal variable and so as χ^2 with 1 d.f. Applying the same approximation to the denominator on the right of (3.3), we have as an approximation to the multinomial probability

$$(3.5)$$

$$\mathscr{P}(\{a_i\} \mid \{p_i\}, N) = \prod_{i=1}^{n} (\lambda p_i)^{-\frac{1}{2}} \exp\left[-\tfrac{1}{2}\sum_{1}^{n} q_i^2\right] \Big/ \left\{\lambda^{-\frac{1}{2}} \exp\left[-\tfrac{1}{2}(N-\lambda)^2/\lambda\right]\right\}$$

and the result follows from Section V.4.

In fact, define an orthogonal matrix, $\mathbf{H}$, with elements of the first row, $h_{1j} = p_j^{\frac{1}{2}}$, then

$$(3.6) \qquad\qquad\qquad \mathbf{H}q = \mathbf{y}, \quad \text{say.}$$

However, $\sum q_i^2 = \sum y_i^2$, because the transformation is orthogonal. y_1 is equal to $\sum q_i p_i^{\frac{1}{2}} = (\sum a_i - N)/N^{\frac{1}{2}} = 0$. A similar argument carries through if the q_i or a_i are submitted to further linear restriction. These linear restrictions can be orthogonalized and written as the first s rows of an orthogonal row of $\mathbf{H}$. The first s transformed variables, $y_1, y_2, \ldots, y_s$ are identically zero.

Example (iii) *The binomial distribution.* Let us take $\{P_\theta, \theta \in \Omega\}$ to be the family of binomial distributions with index n and with θ, the probability of a success in an individual trial taking values in a parameter space, $\Omega = (0, 1)$. Let X be the number of successes. X can take values $x = 0, 1, 2, \ldots, n$ with probabilities, $p_x = \theta^x(1 - \theta)^{n-x}\binom{n}{x}$. Suppose that N independent observations are made on X and that $X = j$ on a_j occasions. Then the probability of the set of values, $\{a_j\}, j = 0, 1, \ldots, n$ is given by

$$(3.7) \qquad\qquad \mathscr{P}(\{a_j\} \mid \theta, n, N) = \prod_{j=0}^{n} \left[\theta^j(1 - \theta)^{n-j}\binom{n}{j}\right]^{a_j}$$

Let $T = \sum_{j=0}^{n} j a_j$. Then $T/(Nn)$ is a sufficient statistic for θ, for (3.7) can be written in the form (3.1) with $g_\theta(T(x)) = \theta^T(1 - \theta)^{nN-T}$ and $h(x)$ equal to $N! \prod_{j=0}^{n} \binom{n}{j}^{a_j} / a_j!$. (3.7) can now be rewritten as the product

(3.8) $\mathscr{P}(\{a_j\} \mid \theta, n, N)$

$$= \frac{(nN)!}{T!(nN - T)!} \theta^T(1 - \theta)^{nN-T} T!(nN - T)! \prod_{j=0}^{n} \binom{n}{j}^{a_j} / nN!$$

$$= \mathscr{P}(T \mid \theta, n, N)\mathscr{P}(\{a_j\} \mid T).$$

$\mathscr{P}(\{a_j\} \mid T)$ does not contain the unknown parameter, θ, and so we could equally well write instead of (3.8),

(3.9) $\mathscr{P}(\{a_j\} \mid T) = \mathscr{P}(\{a_j\} \mid \hat{\theta}, n, N)/\mathscr{P}(T \mid \hat{\theta}, n, N),$

and set $\hat{\theta} = T/(nN)$. Let

(3.10) $q_j = (a_j - \hat{p}_j)/(N\hat{p}_j)^{\frac{1}{2}}$

and take as the first two rows of $\mathbf{H}$,

(3.11)
$$h_{1j} = \hat{p}_j^{\frac{1}{2}}$$
$$h_{2j} = j\hat{p}_j^{\frac{1}{2}}/(\sum j^2 \hat{p}_j)^{\frac{1}{2}},$$

where $\hat{p}_j \equiv \hat{p}_j(\hat{\theta})$ is the estimate of the binomial probabilities, p_j.

Let $\mathbf{H}$ then be completed with due regard for the orthogonal conditions. The transformation $\mathbf{Hq}$ now yields a vector $\mathbf{y}$ in which y_1 is identically zero. Since $\hat{\theta} = \sum j a_j/(Nn)$ and y_2 is a multiple of $\sum j(a_j - Np_j)$, it vanishes also.

4. χ^2 IN THE MULTINOMIAL DISTRIBUTION WITH ESTIMATED PARAMETERS (FISHER THEORY)

In this section Fisher's (1924a) discussion of the fitting of parameters is given. His notation is retained. Suppose that there are n independent observations of a random variable, which can fall into any given class with a probability, m/n, so that the expected number in a class is m; m is a function of some unknown parameter, θ. The maximum likelihood method requires the minimization, for variations in θ, of the likelihood function,

(4.1) $L = \mathscr{S}(x \log m),$

leading to an equation

(4.2) $\mathscr{S}\left(\frac{x}{m}\frac{\partial m}{\partial \theta}\right) = 0.$

The variance in random samples of the estimate so obtained is given by

(4.3)
$$-\frac{1}{\sigma^2} = \mathscr{S}\left(m\frac{\partial^2}{\partial\theta^2}\log m\right).$$

This expression simplifies since $\mathscr{S}m = n$ and so

(4.4)
$$\mathscr{S}\frac{\partial^2 m}{\partial\theta^2} = \mathscr{S}\frac{\partial m}{\partial\theta} = 0,$$

for a given sample size, n.

(4.5)
$$\frac{1}{\sigma^2} = \mathscr{S}\left(\frac{1}{m}\left(\frac{\partial m}{\partial\theta}\right)^2\right).$$

Let m' stand for the frequency calculated from an efficient estimate and define

(4.6)
$$\chi^2 = \mathscr{S}\left\{\frac{(x-m)^2}{m}\right\}, \qquad \chi'^2 = \mathscr{S}\left\{\frac{(x-m')^2}{m'}\right\}.$$

It is required to prove that the difference, $\chi^2 - \chi'^2$, is the square of a certain random variable divided by its standard deviation. If asymptotic normality is assumed an equivalent statement is that $\chi^2 - \chi'^2$ is approximately distributed as χ^2 with 1 d.f.

The estimate, $\hat{\theta}$, of θ is not exact but can be written

(4.7)
$$\hat{\theta} = \theta + \delta\theta,$$

where $\delta\theta$ is $0(n^{-\frac{1}{2}})$ if an efficient method of estimation is used. The minimum χ^2 method of estimation is efficient since the equation is

(4.8)
$$\mathscr{S}\left(\frac{x^2 - m^2}{m^2}\frac{\partial m}{\partial\theta}\right) = 0,$$

and this is equivalent to (4.2), the maximum likelihood equation above, since $(x + m)/(m - 2)$ is $0(n^{\frac{1}{2}})$, because of the law of large numbers.

χ^2 is now minimized by solving

(4.9)
$$0 = \frac{\partial\chi^2}{\partial\theta} = \frac{\partial}{\partial\theta}\mathscr{S}\{(x - m')^2/m'\} = \frac{\partial}{\partial\theta}\mathscr{S}\{x^2/m'\} - \frac{\partial}{\partial\theta}\mathscr{S}m'$$

$$= -\mathscr{S}\frac{x^2}{m'^2}\frac{\partial m'}{\partial\theta}.$$

An application of Taylor's theorem yields

(4.10)
$$\frac{1}{m} - \frac{1}{m'} = \delta\theta\frac{\partial}{\partial\theta}\frac{1}{m} + \frac{(\delta\theta)^2}{2!}\frac{\partial^2}{\partial\theta^2}\frac{1}{m'} + \cdots$$

$$= -\delta\theta\frac{1}{m'^2}\frac{\partial m'}{\partial\theta} + \frac{(\delta\theta)^2}{2!}\left(\frac{2}{m'^3}\left(\frac{\partial m'}{\partial\theta}\right)^2 - \frac{1}{m'^2}\frac{\partial^2 m'}{\partial\theta^2}\right).$$

Multiplication of both sides by x^2 and summation yields

(4.11) $$\chi^2 - \chi'^2 = \mathscr{S}\left(\frac{x^2}{m'^2}\frac{\partial m'}{\partial \theta}\right)\delta\theta + \mathscr{S}\left\{\frac{2x^2}{m'^3}\left(\frac{\partial m'}{\partial \theta}\right)^2 - \frac{x^2}{m'^2}\frac{\partial^2 m'}{\partial \theta^2}\right\}(\delta\theta)^2.$$

The first term on the right vanishes since the minimum χ^2 method of estimation is being used. Moreover x/m' and x^2/m' differ from unity by a factor of $O(n^{-\frac{1}{2}})$ so that

(4.12) $$\chi^2 - \chi'^2 = \tfrac{1}{2}(\delta\theta)^2\mathscr{S}\left\{\frac{2}{m'}\left(\frac{\partial m'}{\partial \theta}\right)^2 - \frac{\partial^2 m'}{\partial \theta^2}\right\}$$

$$= (\delta\theta)^2\mathscr{S}\left\{\frac{1}{m'}\left(\frac{\partial m'}{\partial \theta}\right)^2\right\},$$

the second term vanishing since $\mathscr{S}(m')$ is the sample size. However, (4.5) shows that the expression on the right of (4.12) is simply the square of deviation of $\hat{\theta}$ from its true value divided by the variance of $\hat{\theta}$. Asymptotically, $\chi^2 - \chi'^2$ has the distribution of the square of a standardized normal value and hence of χ^2 with 1 d.f.

In this proof, it has been assumed that the difference is a component χ^2 variable so that the remainder is also a χ^2 variable. G. S. James (1952) has shown that the χ^2 variables must be exhibited as quadratic forms in some basic set of variables having the joint normal distribution. To meet this objection, one could construct a vector,

(4.13) $$Q = \left(\frac{x - m'}{\sqrt{m'}}\right),$$

where m' is a function of $\hat{\theta}$, and then make a transformation $\mathbf{MQ}$, where $\mathbf{M}$ is orthogonal with elements of the first row equal to $\sqrt{m'}$ and with elements of the second row proportional to $(\partial m/\partial \theta)m^{\frac{1}{2}}$ at the point $\theta = \hat{\theta}$. The first two elements of $\mathbf{MQ}$ would then be identically zero.

5. ESTIMATED PARAMETERS (CRAMÉR THEORY)

The theory of Section VIII.4 illustrates clearly how χ^2 is diminished by the estimation of a parameter. A more general approach has been achieved by Cramér (1946). We retain the notation of Watson (1959) as far as possible but give a proof with emphasis on linear forms rather than quadratic forms.

Let N observations be made on a multinomial variable, Z, with class probabilities, $p_1, p_2, \ldots, p_k$; and let the observed class frequencies be $n_1, n_2, \ldots, n_k$, with $\sum n_i = N$. Suppose further that the class probabilities are functions of a single unknown parameter, θ, or of several parameters, the elements of a vector, $\boldsymbol{\theta}$. To simplify the notation we write q_i in place of

$p_i(\theta)$ or $p_i(\hat{\theta})$, where θ or $\hat{\theta}$ is a point in the parameter space. We write also

(5.1)
$$q_i^{(j)} = \left[\left(\frac{\partial}{\partial\theta}\right)^j p_i(\theta)\right]_{\theta=\hat{\theta}}$$

and later, when several parameters are being considered,

(5.2)
$$q_i^{(\alpha\beta)} = \left[\frac{\partial}{\partial\theta_\alpha}\frac{\partial}{\partial\theta_\beta} p_i(\boldsymbol{\theta})\right]_{\boldsymbol{\theta}=\hat{\boldsymbol{\theta}}}$$

$q_i, q_i^{(1)}, q_i^{(2)}$ are the values of p_i and its first and second derivatives at the point, $\hat{\theta}$. $q_i, q_i^{(\alpha 0)}, q_i^{(\alpha\alpha)}$ and $q_i^{(\alpha\beta)}$ are respectively the values of $p_i(\boldsymbol{\theta})$, the first and second partial derivatives with respect to θ_α and the mixed second partial derivative of p_i at the point, $\hat{\boldsymbol{\theta}}$. At present, we do not give any special interpretation to $\hat{\theta}$ and $\hat{\boldsymbol{\theta}}$. It is assumed now that $q_i^{(\alpha 0)}, q_i^{(\alpha\alpha)}$ and $q_i^{(\alpha\beta)}$ are all finite in some open region including the unknown parameter point, θ or $\boldsymbol{\theta}$.

The single parameter case is considered first. We write $\theta - \hat{\theta} = \delta$. (It will appear later that $\delta = O(N^{-\frac{1}{2}})$ in the cases of interest to us.) It is permissible to write

(5.3)
$$p_i = q_i + \delta q_i^{(1)} + O(\delta^2).$$

It is supposed that each q_i is bounded away from zero. An orthogonal matrix, $\mathbf{H}$, can be formed such that for $i = 1, 2, \ldots, k$,

(5.4)
$$h_{i1} = q_i^{\frac{1}{2}}$$
$$h_{i2} = cq_i^{-\frac{1}{2}}q_i^{(1)},$$

where c is an appropriate normalizing constant. Let the uth column of $\mathbf{H}$ be written as a vector $\mathbf{h}_u$. $\mathbf{h}_1^T\mathbf{h}_2 = 0$, so that (5.4) defines the first two columns of an orthogonal $\mathbf{H}$, which can be completed with due regard to the normalizing conditions. To simplify notation we suppose that $\mathbf{h}$ is a unit vector orthogonal to $\mathbf{h}_1$ and $\mathbf{h}_2$ and that $\mathbf{g}$ is a vector with elements such that $h_i = g_i q_i^{\frac{1}{2}}$. It is required now to determine the asymptotic joint distribution of linear forms, $\mathbf{h}^T\mathbf{X}$, where

(5.5)
$$X_i = (n_i - Nq_i)/(Nq_i)^{\frac{1}{2}}, \qquad i = 1, 2, \ldots, k.$$

Lemma 5.1. *Subject to the above conditions,*

(5.6)
$$Y = \mathbf{h}^T\mathbf{X}$$

has mean and variance differing from 0 and 1 by terms of $O(\delta)$ if $\delta = O(N^{-\frac{1}{2}})$. Further $\mathrm{cov}\,(Y_u, Y_v) = 0 + O(\delta)$ *where $u > 2, v > 2$ and $Y_u = \mathbf{h}_u^T\mathbf{X}, Y_v = \mathbf{h}_u^T\mathbf{X}$.*

Proof. The Nq_i in the numerator can be neglected since $\mathbf{h}^T\mathbf{q}^{\frac{1}{2}} = \mathbf{h}^T\mathbf{h}_1 = 0$.

$$\mathscr{E}Y = \mathscr{E}\mathbf{h}^T\mathbf{X} = N^{-\frac{1}{2}}\sum g_i \mathscr{E}n_i = N^{\frac{1}{2}}\sum g_i p_i$$
$$= N^{\frac{1}{2}}\sum g_i(q_i + \delta q_i^{(1)} + O(\delta^2)).$$

But

$$\sum g_i q_i = \sum h_i q_i^{\frac{1}{2}} = 0; \qquad c \sum g_i q_i^{(1)} = \sum h_i h_{2i} = 0.$$

Further $g_i = h_i q_i^{-\frac{1}{2}} \leqslant q_i^{-\frac{1}{2}}$, which is bounded since q_i is bounded away from zero and $|h_i| \leqslant 1$. $\mathscr{E}\,Y$ is thus $0(N^{\frac{1}{2}}\delta^2) = 0(N^{-\frac{1}{2}})$. Further

(5.7)
$$\begin{aligned} \text{var } Y &= \mathscr{E}(\mathbf{h}^T\mathbf{X})^2 - [\mathscr{E}(\mathbf{h}^T\mathbf{X})]^2 \\ &= \mathbf{h}^T(\mathscr{E}X_iX_j - \mathscr{E}X_i\mathscr{E}X_j)\mathbf{h} \\ &= \mathbf{g}^T(\mathscr{E}n_in_j - \mathscr{E}n_i\mathscr{E}n_j)\mathbf{g}/N \\ &= \mathbf{g}^T(p_i\delta_{ij} - p_ip_j)\mathbf{g}, \end{aligned}$$

a well known result. It has now to be shown that the final expression in (5.7) differs little from $\mathbf{g}^T(q_i\delta_{ij} - q_iq_j)\mathbf{g}$. First,

(5.8)
$$\mathbf{g}^T \operatorname{diag}(p_i - q_i)\mathbf{g} = \sum g_i^2(\delta q_i^{(1)} + 0(\delta^2)).$$

As before, $g_i^2 \leqslant q_i^{-1}$ and so the expression (5.8) is bounded above by $|\delta| \sum_{i=1}^{k} |q_i^{(1)}||q_i^{-1} + 0(\delta^2)$ and therefore is $0(\delta) = 0(N^{-\frac{1}{2}})$. Second,

(5.9)
$$p_ip_j - q_iq_j = \delta(q_iq_j^{(1)} + q_i^{(1)}q_j) + 0(\delta^2).$$

However,

(5.10)
$$\mathbf{g}^T(q_iq_j^{(1)})\mathbf{g} = \sum_i g_iq_i \sum_j g_jq_j^{(1)}$$

and these sums are both zero. Consequently,

(5.11)
$$\mathbf{g}^T((p_i - q_i)\delta_{ij} - (p_ip_j - q_iq_j))\mathbf{g} = 0(\delta).$$

(5.7) may thus be rewritten as

(5.12)
$$\begin{aligned} \text{var } Y &= \mathbf{g}^T(q_i\delta_{ij} - q_iq_j)\mathbf{g} + 0(\delta) \\ &= \mathbf{h}^T(\delta_{ij} - q_i^{\frac{1}{2}}q_j^{\frac{1}{2}})\mathbf{h} + 0(\delta). \end{aligned}$$

But

(5.13)
$$\mathbf{H}^T(\delta_{ij} - q_i^{\frac{1}{2}}q_j^{\frac{1}{2}})\mathbf{H} = \operatorname{diag}(0, 1^{k-1}),$$

and since $\mathbf{h}$ is orthogonal to $\mathbf{h}_1$, $\mathbf{h}^T(\delta_{ij} - q_i^{\frac{1}{2}}q_j^{\frac{1}{2}})\mathbf{h} = 1$. var Y is thus $1 + 0(\delta)$. In the analysis above we may write

(5.14)
$$Y = Y_u \cos \psi + Y_v \sin \psi, \qquad u > 2, \qquad v > 2$$

and equate the variances on the two sides. It follows that cov $(Y_u, Y_v) = 0(\delta)$.

It may be remarked that the results hold uniformly in an interval containing the true value of the parameter. Y_1 is a degenerate variable identically equal to zero and no statement has as yet been made on the distribution of Y_2.

However, for a fixed θ and hence $\{q_i\}$, the variables Y_i are all asymptotically normal, Y_1 being degenerate at zero.

$$(5.15) \qquad \mathbf{H}^T((n_i - Nq_i)/(Nq_i)^{\frac{1}{2}}) = \mathbf{Y} = \begin{bmatrix} Y_1 \\ Y_2 \\ \cdot \\ \cdot \\ \cdot \\ Y_k \end{bmatrix}$$

By the invariance rule for orthogonal transformations $\sum (n_i - Nq_i)^2/(Nq_i) = \mathbf{Y}^T\mathbf{Y}$. For any θ, we consider only those samples for which Y_2 is zero. Under this condition, $\sum (n_i - Nq_i)^2/(Nq_i) = Y_3^2 + Y_4^2 + \ldots + Y_k^2$, which is distributed approximately as χ^2 with $(k - 2)$ degrees of freedom. However, the requirement that Y_2 is zero is the condition,

$$0 = c \sum n_i q_i^{-1} q_i^{(1)} \equiv c \sum n_i \frac{1}{p_i(\theta)} \frac{\partial p_i(\theta)}{\partial \theta}$$

$$= c \frac{\partial}{\partial \theta} \log [p_1^{n_1} p_2^{n_2} \ldots p_k^{n_k}],$$

the maximum likelihood equation. If, therefore, maximum likelihood estimation is used, Y_2 vanishes and $Y_3^2 + Y_4^2 \ldots + Y_k^2$ is approximately distributed as χ^2 with $(k - 2)$ degrees of freedom. The case of several parameters is treated similarly. We consider the joint distribution of $Y_{s+2}, Y_{s+3}, \ldots, Y_k$ defined as in (5.6) where now $\mathbf{h}$ is to be orthogonal to each column of $\mathbf{B}$, where the elements of $\mathbf{B}$ are defined by

$$(5.15') \qquad \begin{aligned} b_{i1} &= q_i^{\frac{1}{2}} \\ b_{i,\alpha+1} &= q_i^{-\frac{1}{2}} q_i^{(\alpha 0)}. \end{aligned}$$

The first $(s + 1)$ columns of $\mathbf{H}$ are to be $\mathbf{B}(\mathbf{B}^T\mathbf{B})^{-\frac{1}{2}}$, which is an appropriate choice since $(\mathbf{B}^T\mathbf{B})^{-\frac{1}{2}}\mathbf{B}^T\mathbf{B}(\mathbf{B} \ \mathbf{B})^{-\frac{1}{2}} = \mathbf{1}_{s+1}$. If $\mathbf{h}$ is orthogonal to each column of $\mathbf{B}$ it is orthogonal to each column of $\mathbf{B}(\mathbf{B}^T\mathbf{B})^{-\frac{1}{2}}$ and conversely since this latter matrix and $\mathbf{B}$ "span the same space". It is now supposed that we select a $\hat{\theta}$ such that the norm, $\|\theta - \hat{\theta}\| = \delta = O(N^{-\frac{1}{2}})$, $\delta \geqslant 0$. As before, we find var Y and $\mathscr{E}Y$ are approximately unity and zero and that for $u > s + 1$, $v > s + 1$, corr (Y_u, Y_v) is $O(N^{-\frac{1}{2}})$ if $u \neq v$. Modifications have to be made thus (5.3) is replaced by

$$(5.16) \qquad p_i = q_i + \sum_\alpha \delta_\alpha q_i^{(\alpha 0)} + O(\delta^2), \qquad \delta^2 = \sum_{\alpha=1}^s \delta_\alpha^2.$$

(5.8) is similarly modified and the upper bound for (5.8) becomes

$$\sum_\alpha \sum_{i=1}^k \delta_\alpha |q_i^{(\alpha 0)}| q_i^{-1} + O(\delta^2).$$

(5.10) is replaced by a sum,

$$\sum_\alpha \sum_i g_i q_i \sum_j \delta_\alpha g_j q_j^{(\alpha 0)},$$

and so once again var Y is given by (5.12) and the device for obtaining cov (Y_u, Y_v) can again be used. Once again the results hold uniformly in a region containing the true parameter point, $\boldsymbol{\theta}$. For a fixed $\hat{\boldsymbol{\theta}}$, such that $\|\boldsymbol{\theta} - \hat{\boldsymbol{\theta}}\|$ is $O(N^{-\frac{1}{2}})$, Y_1 is again degenerate at zero, Y_i, $i > 1$, is asymptotically normal and the Y_i jointly asymptotically normal. We consider now for a fixed $\hat{\boldsymbol{\theta}}$ only those samples for which $Y_2, Y_3, \ldots, Y_{s+1}$ are all zeroes, in which case $\hat{\boldsymbol{\theta}}$ is the maximum likelihood estimate of $\boldsymbol{\theta}$. Further $\|\boldsymbol{\theta} - \hat{\boldsymbol{\theta}}\|$ is $O(N^{-\frac{1}{2}})$ by the general theory of maximum likelihood estimation. We can now summarise the results.

Theorem 5.1. *Suppose a sample of size N is drawn from a multinomial distribution with class frequencies, $p_i(\boldsymbol{\theta})$, depending on s unknown parameters, $\theta_1, \theta_2, \ldots, \theta_s$. Suppose further that $\partial p_i/\partial \theta_\alpha$, $\partial^2 p_i/\partial \theta_\alpha \, \partial \theta_\beta$ are bounded for all values of θ_α and θ_β in an open interval containing $\boldsymbol{\theta}$ and $\hat{\boldsymbol{\theta}}$. Then if $\hat{\boldsymbol{\theta}}$ be estimated from the data by means of the method of maximum likelihood, the Pearson variable,*

$$\sum_{i=1}^k (a_i - N q_i)^2/(N q_i), \qquad q_i = p_i(\hat{\boldsymbol{\theta}}),$$

is distributed asymptotically as χ^2 with $k - s - 1$ degrees of freedom.

6. ESTIMATED PARAMETERS AND ORTHONORMAL THEORY

We use the same notation as in Section 5 and suppose that class probabilities, $q_i \equiv p_i(\hat{\theta})$, are computed for each i, $i = 1(1)k$. Since $p_i < 1$ and $q_i > 0$, $\sum p_i^2/q_i < \infty$ so that the unknown distribution is ϕ^2-bounded with respect to the q_i if $\{p_i\}$ and $\{q_i\}$ are the probabilities with parameter values θ and $\hat{\theta}$. If the distributions at θ and $\hat{\theta}$ are of general form, then the condensed distribution, $\{p_i\}$, is ϕ^2-bounded with respect to the condensed distribution, $\{q_i\}$ even though this may not hold for the "uncondensed" distributions. As before we consider values of $\hat{\theta}$ or $\hat{\boldsymbol{\theta}}$ close to θ or $\boldsymbol{\theta}$ respectively. p_i can be expanded in terms of q_i and an orthonormal set on the distribution $\{q_i\}$, so that

$$(6.1) \qquad p_i = q_i\left(1 + \sum_1^\infty a_j x^{(j)}\right).$$

The a_i will depend on the unknown parameter, $\boldsymbol{\theta}$. In the special case, where the assumptions of the last section can be made, there will be a good approximation available

$$(6.2) \qquad p_i = q_i + \sum_{\alpha=1}^s \delta_\alpha \frac{\partial q_i}{\partial \theta_\alpha} + O(\delta^2).$$

This may be written, if the last term be neglected,

$$(6.3) \qquad p_i = q_i \left(1 + \sum_{\alpha=1}^{s} \frac{\delta_\alpha}{q_i} \frac{\partial q_i}{\partial \theta_\alpha}\right).$$

If now the terms are orthonormalized it follows that

$$(6.4) \qquad p_i = q_i \left(1 + \sum_{1}^{s} b_j t_i^{(j)}\right), \qquad \text{each } |b_j| < \phi,$$

where $t_i^{(j)}$ is the value of the jth orthonormal function at the class, i. The b_j are functions of θ. Now consider the joint distribution of $N^{-\frac{1}{2}}\mathscr{S}t^{(v)}$.

$$(6.5) \qquad \mathscr{E}t^{(v)} = \sum t_i^{(v)} p_i = \sum t_i^{(v)} \left(1 + \sum_{1}^{s} b_j t_i^{(j)}\right) q_i$$
$$= 0, \qquad \text{if } v > s$$

and

$$= b_k \quad \text{if} \quad v \leqslant s.$$

However, $t^{(v)^4} < \infty$ since the distributions $\{p_i\}$ and $\{q_i\}$ are condensed and hence possess only finitely many points. So that $t^{(v)^2}$ and also $t^{(u)}t^{(v)}$ can be expressed as Fourier series.

$$(6.6) \qquad t^{(v)^2} = 1 + \sum_{1}^{k-1} v_\alpha t^{(\alpha)}, \text{ say. Then,}$$

$$(6.7) \qquad \mathscr{E}t^{(v)^2} = \mathscr{E}\left(1 + \sum_{1}^{k-1} v_\alpha t^{(\alpha)}\right)$$
$$= \sum_{i=1}^{k} \left(1 + \sum_{1}^{k-1} v_\alpha t^{(\alpha)}\right) p_i$$
$$= \sum_{i=1}^{k} \left(1 + \sum_{1}^{k-1} v_\alpha t_i^{(\alpha)}\right)\left(1 + \sum_{1}^{s} b_j t_i^{(j)}\right) q_i$$
$$= 1 + \sum_{1}^{s} b_j v_\alpha, \qquad |b_j| < \phi.$$

The second term is in absolute value less than ϕ multiplied by a bounded sum and if $\{q_i\}$ differs little from $\{p_i\}$, it will be negligible. Similarly the covariances will be negligible. Now for sample size, N, and fixed $\{q_i\}$, the mean and variance of the sample functions, $N^{-\frac{1}{2}}\mathscr{S}t^{(j)}$, for $j > s$, will be approximately zero and unity, respectively and the correlation between any two such functions will be negligible. For $j \leqslant s$, the expectation of $N^{-\frac{1}{2}}\mathscr{S}t^{(j)}$ will be $N^{\frac{1}{2}}b_j$, so that $[N^{-\frac{1}{2}}\mathscr{S}t^{(j)}]^2$ is approximately a non-central χ^2 with one degree of freedom and non-centrality parameter equal to Nb_j^2 which is not negligible. If now we wish to minimize the expectation of χ^2, our procedure will be to estimate θ by ensuring that the sample value of $N^{-\frac{1}{2}}\mathscr{S}t^{(j)} = 0$ for $j = 1$,

$2, \ldots, s$. We thus have s equations to estimate the s parameters. The Pearson χ^2, however, on a discrete distribution is equal to $N^{-1} \sum_{1}^{k-1} [\mathscr{S}t^{(j)}]^2$, where $\{t^{(j)}\}$ is any orthonormal set. So that the estimation process in effect sets equal to zero or deletes s of the $(k-1)$ squares and has little effect on the distribution of the remaining $(k-1-s)$ variables, $N^{-\frac{1}{2}}\mathscr{S}t^{(j)}, j > s$.

The procedure may be regarded from another point of view. We consider the distribution of $\sum_{1}^{k} (N^{-\frac{1}{2}}\mathscr{S}t^{(j)})^2$ at all points, θ, in the parameter space, and consider the distribution of the last $(k-1-s)$ squares conditional on the first being zero.

The reasoning of this section and of the preceding can be generalized from the Pearson case to more general χ^2-analogues. The Pearson χ^2 is based on the orthonormalized indicator functions; other orthonormal systems are possible of which the Legendre and Hermite systems may be especially mentioned.

The proof is founded on the principle that if an arbitrary linear form in a set of variables is normal then the joint distribution is jointly normal (i.e. the variables are distributed in a normal correlation in the words of Russian authors).

7. THE TEST OF GOODNESS OF FIT

A common use of the Pearson χ^2 is to determine whether a given theoretical distribution fits an observed empirical distribution. If the parameters are given theoretically, there is little difficulty as

(7.1) $Q = \sum (\text{observed} - \text{expected})^2/(\text{expected})$,

with summation over the n classes, is distributed asymptotically as χ^2 with $(n-1)$ degrees of freedom as we have seen in Chapter V. In this section, some tests of goodness of fit in the common univariate distributions are considered, when the parameters have been estimated from the data, the discussion of contingency tables being deferred to Chapters XI and XII.

(i) The Poisson Distribution

Only the single parameter, λ say, has to be estimated. N independent observations are made on a Poisson variable, i occurring N_i times. Then the method of moments, or of maximum likelihood, yields the estimate,

(7.2) $\hat{\lambda} = \sum iN_i/N.$

As a rule the observations will have been ungrouped and there are no difficulties. This problem has been discussed, in particular, by "Student"

(1907 and 1919) and by Fisher (1922c; *Statistical Methods*). If the observations in the tails have been pooled then a provisional value, γ_0 say, can be given to the mean value to be assigned to the observations in the "pooled" classes; for example, γ_0 might be set equal to the lowest count in the upper tail. $\hat{\lambda}_0$ will be the estimate of λ by (7.2). However, with the aid of $\hat{\lambda}_0$ a better mean value, γ_1 say, can be obtained and this enables an improved estimate, $\hat{\lambda}_1$, to be made. This process may be repeated until

$$(7.3) \qquad \hat{\lambda}_j = \left(\sum_{i=0}^{k-1} i N_i + \gamma_j N_k^* \right) \Big/ N = \hat{\lambda}_{j-1} \quad \text{approximately,}$$

where N_k^* is the number of observations in the tail class, namely where the random variable has taken the value $k, k + 1, \ldots$ Usually no more than one or two iterations is required.

Estimation by minimum χ^2 is not simple and is not recommended, for single extreme values can have an undue influence on the estimation of λ; for example if a single observation has been made in the extreme class the contribution to the Pearson χ^2 will be p_k^{-1} and this may well be a large number and the estimate of λ may be unduly affected by the presence of such a single observation in the tail.

In the Poisson distribution, the Pearson χ^2, Q, with parameter estimated from the data will be distributed asymptotically as χ^2 with $(\nu - 2)$ degrees of freedom, ν being the number of classes. "Pooling" in the tails when the parameter has been estimated from the data does not result in the deletion of a whole degree of freedom but this effect can usually be ignored.

The Pearson χ^2 method may be unsuitable when the parameter is small, for example below unity, because in samples of even moderate size, such as 100–1000, there may be only a few classes with expectations greater than 5. Combinatorial tests are available in such cases, for example, as in Fisher (1950b).

(ii) The Binomial Distribution: p to be Estimated, the Index n given

Maximum likelihood methods (Fisher, 1922c) yield the estimate

$$(7.4) \qquad \hat{p} = \sum i N_i / N.$$

Q will be distributed asymptotically as χ^2 with $(\nu - 2)$ d.f. If there is no pooling, $\nu = n + 1$ and so the degrees of freedom are $\nu - 1$. Weldon's dice throwing experiments, given by Pearson (1900a) and Fisher in his *Statistical Methods* are classic. Another widely quoted example is Geissler's table of the number of boys in sibships of eight. A very detailed and penetrating analysis of this table has been given by R. A. Fisher in his *Statistical Methods*. Geissler's data have been reconsidered by Lancaster (1950d) who took Geissler's estimate of p, the probability of a child being a boy, as known theoretically. Q under such circumstances is distributed approximately as χ^2 with $(\nu - 2)$ degrees of freedom $+ \lambda\chi_1^2$, where χ_1^2 is distributed as χ^2 with 1 d.f. and is independent of the first component; $0 \leqslant \lambda \leqslant 1$ and the actual value of λ depends on the degree to which the particular observations (here the families of eight) determine the estimate. $\lambda = 0$ if no other observations are used, $\lambda = 1$ if the p is calculated from independently observed data alone.

In the test of goodness of fit, since a sufficient statistic is available, combinatorial methods could be used, as in Cochran (1936a).

If n is large and np or $n(1 - p)$ is small, the distribution may be approximately tested as a Poisson distribution.

(iii)   The Negative Binomial and Binomial with Index and p to be Estimated

The probability of the random variable, Y, assuming the value, y, in the negative binomial distribution can be written

(7.5)
$$p = \binom{y + k - 1}{k - 1} p^k (1 - p)^y, \qquad y = 0, 1, 2, \ldots,$$

where k is termed the exponent.

(7.6)
$$\mathscr{E} Y = k(1 - p)/p, \qquad \text{var } Y = k(1 - p)/p^2.$$

It is supposed that N independent observations are made on Y and n_y is the number of times y has occurred. For a single observation, the variance is greater than the mean. Two commonly used methods of estimation equate the observed mean to the value given by (7.6) so that

(7.7)
$$\bar{y} = \sum_y y n_y / N = \hat{k}(1 - \hat{p})/\hat{p}.$$

If k is given by hypothesis, $\hat{p}$ can be estimated from (7.7) with $\hat{k}$ replaced by k. The method of moments obtains $\hat{p}$ and $\hat{k}$ by (7.7) together with

(7.8)
$$\sum (\bar{y} - y)^2 n_y/(N - 1) = \hat{k}(1 - \hat{p})/\hat{p}^2$$

Division of the right side of (7.7) by the right side of (7.8) yields an estimate of $\hat{p}$ and substitution of this result in (7.7) yields $\hat{k}$.

The maximum likelihood method yields the equations (7.7) and

(7.9)

$$N \log (1 + y/k) = \sum_1^m n_y/k + \sum_2^m n_y/(k + 1) + \cdots + n_m/(m + k - 1),$$

in which m is the largest integer for which $n_m > 0$. A discussion of the fitting is available in Anscombe (1950) and Williamson and Bretherton (1963).

(iv) The Logarithmic Series

The reader is referred to Anscombe (1950) for the definition of this distribution and the estimation of the parameters.

(v) The Normal Curve

Examples of the fitting of the normal distribution to empirical data are numerous. We may cite the example given in Kendall and Stuart (1963), namely of the heights of men in groups. The mean is estimated as the sample mean and the variance is estimated as the sample variance, calculated with the aid of Sheppard's corrections. Another example of the fitting of the normal curve is given in Cramér (1946). He estimates the mean and variance in the same way and then uses further approximations, expanding the fitting distribution in the form,

$$(7.10) \qquad \hat{\hat{f}}(x) = \hat{f}(x)(1 - g_1 H_3(x)/3! + g_2 H_4(x)/4! + 10g_1^2 H_6(x)/6!)$$

in which $H_j(x)$ is the Hermite polynomial of degree, j, and g_1 and g_2 are the estimates of γ_1 and γ_2, the coefficients of skewness and excess, $\gamma_1 = \mu_3/\sigma^3$ and $\gamma_2 = (\mu_4 - 3\sigma^2)/\sigma^4$, g_1 and g_2 being estimated by replacing the theoretical value by the sample values. The Pearson χ^2 is then computed with the integral of (7.10) providing the expected values. That Cramér's use of χ^2 to test whether the series (7.10) gives a good fit is a reasonable solution is open to question. The coefficients have been obtained by minimising a certain sum of squares not the Pearson χ^2. It is known that the method may yield negative and hence also zero frequencies. A more appropriate test of normality would be as follows. If the data are ungrouped, estimate the mean and variance by means of the sample values for them. Now calculate the standardized sums of the standardized Hermite polynomials, $x^{(j)}$ say. Then

$$(7.11) \qquad Z_i = N^{-\frac{1}{2}} \mathscr{S} x^{(i)}$$

is a standardized asymptotically normal variable. Under hypotheses made and because of the estimation of the parameters, $Z_1 \equiv 0$ and $Z_2 \equiv 0$, and $Z_3, Z_4, \ldots$ are asymptotically normal $(0, 1)$ and the sum of squares of a finite number of them is distributed approximately as a χ^2 with the same number of degrees of freedom. In the case of ungrouped observations, this procedure seems preferable to the classic treatment as outlined by Cramér (1928, 1946). It also avoids the difficulties introduced by estimating the parameters on the ungrouped data and applying the resulting distribution to the grouped data. The values of the orthogonal polynomials for each observation are readily computed by a recurrence relation on the automatic computers.

(vi)　The Pearson and other Systems of Curves

The normal distribution belongs to the system of curves, first described by K. Pearson (1895). The fitting of the curves in this system has been described by Elderton (1927), Elderton and Hansmann (1934) and in pages cxxi–cxliii, clxxxvii–ccvii, et cetera of Tables for Statisticians and Biometricians, Part II of K. Pearson (1931) and in Kendall and Stuart (1963). However, the fitting by the method of moments of the Pearson system of curves has been severely criticised by R. A. Fisher and his school. If the Pearson curves are fitted by their first four moments, Fisher (1922c) has shown that, only when the coefficients of symmetry and kurtosis are small, can accurate results be obtained. The Pearson method of estimation is inefficient elsewhere and more efficient methods, especially that of maximum likelihood, should be used. For some members of this system and for more general curves, the computation of the maximum likelihood or other efficient estimates often requires special methods. These applications may be left to the reader who may refer to the references listed in the bibliography. For the general theory, he is referred especially to Fisher (1922c, 1924 and Statistical Methods), Neyman (1949), Neyman and Pearson (1928), K. Pearson (1916b) and Rao (1965).

EXERCISES AND COMPLEMENTS

Least Squares

1. Suppose that there are n points (x_i, y_i) given in the plane. Show how to choose a straight line, $Y_x = mx + c$, so that $\sum (Y_x - y)^2$ is as small as possible. Determine the straight line which minimizes $\sum (X_y - x)^2$, summation being over the values of Y_x, X_y, x and y.

2. Generalize the problem of Exercise 1 to a discrete bivariate distribution of points and finally to probability distributions of general nature.

3. What is the condition that the two least square regression lines should coincide?

4. Comment on the statement, "where the regression appears to be linear, Bravais' formula may be used at once without troubling to investigate the normality of the distribution". [Note. Bravais had derived the regression formula $y = \rho x \sigma_2 / \sigma_1$ under the assumption of joint normality. Yule points out that the regression coefficient can be obtained by minimizing $\sum (y - b_1 x)^2$.] (Yule, 1897a and b.)

5. Let $\beta^* = Ay$ be an estimator of β in the model given by (2.12). If β^* is unbiased, prove that $AX\beta = \beta$ and hence if the estimator is to be unbiased for all variations in β, $AX = 1$. Further prove that if A_1y and A_2y are two unbiased estimators, then $(A_1 - A_2)X = 0$. Hence prove that any unbiased estimator is of

the form $A = S^{-1}X^T + B$, where $BX = 0$, the $s \times s$ matrix of zeroes. Derive the covariance matrix for the estimator and prove that of all unbiased estimators, $S^{-1}X^TY$ has the minimum variance.

6. In the model $Y = X\beta + \epsilon$ of (2.12), suppose that $\mathscr{E}\epsilon = 0$, show that the properties of unbiasedness and minimum variance are independent of the assumption of normality.

7. Prove the equivalence of (2.7) and (2.18).

8. Let $u_1, u_2, \ldots, u_s$ be linearly independent vectors, and u_{s+1} be orthogonal to each of them. Then the set of $(s + 1)$ vectors is a linearly independent set. Prove that B^TB of (5.15') is of the form $1 + C$. [Hint. Recall that the first column of B is orthogonal to each of the others.]

9. Prove that $XS^{-\frac{1}{2}}$, where $S = X^TX$, can form the first p columns of an orthogonal matrix.

10. Show that the number of linearly independent linear forms in the y_i which have zero expectation under the model of (2.12) is $N - s$.

11. Show that the generalization of (2.19) and (2.20) is equivalent to the minimization of the quadratic form, $e^TV^{-1}e$.

12. Let $X_1, X_2, \ldots, X_v$ be normal $(0, 1)$ and mutually independent. Prove that the distribution of $\sum_1^v X_i^2$, subject to the condition $\sum p_i X_i = 0$, is that of χ^2_{v-1}. (Fisher, 1922a; Neyman and Pearson, 1928; Pearson, 1916b.)

13. With the notation of Exercise 13, suppose that the X_i are subject to linear restrictions, $AX = a$, A being $k \times v$ and of rank, k and a being a vector of constants. If a is the zero vector, the restrictions may be termed homogeneous, otherwise non-homogeneous. Prove that $\sum X_i^2$ is distributed as χ^2_{v-k} in the homogeneous case and as χ^2_{v-k} + a constant in the non-homogeneous case. Evaluate the constant. (Pearson, 1916b; Neyman and Pearson, 1928; Patnaik, 1949; Bateman, 1949.)

14. If $T_1, T_2, \ldots, T_N$ are unbiased estimators of a parameter, θ, prove that any linear form $\sum c_i T_i$ is also unbiased. If the T_i are uncorrelated determine the unbiased estimator with minimum variance and the minimum variance subject to $\sum c_i = 1$. [Note. $\mathscr{E}T_i = \theta, \mathscr{E}(T_i - \theta_i)^2 = \sigma_i^2$. $\mathscr{E}c^TT = \sum c_i\mathscr{E}T_i = \theta$. var $c^TT = \sum c_i^2\sigma_i^2$ and this is a minimum when $c_i = \sigma_i^{-2}/\sum \sigma_i^{-2}$. The weight should be inversely proportional to the variances.]

15. If $T_1, T_2, \ldots, T_N$ are unbiased estimators of a parameter, θ, having covariance matrix, V, find the linear form, c^TT, which is unbiased and has minimum variance. Prove that every c^TT, subject to $\sum c_i = 1$, is unbiased.

16. Show that if an inconsistent set of estimators is used in fitting a distribution, then the expectation of χ^2 is unbounded as $N \to \infty$. (Fisher, 1924.)

17. "In the theory of large samples, we may speak of χ^2 as made up of two parts, one of which is due to errors of estimation and vanishes when estimation is efficient, while the other is distributed in random samples as is the sum of squares of $s - r - 1$ quantities each" normally distributed about zero with unit standard deviation. (Fisher, 1928.)

18. If s theoretical parameters are provided for by hypothesis calculation of the probabilities of k classes, then χ^2 is distributed with $(k - 1)$ d.f. If the s parameters are estimated from the data, then the resulting χ^2 is distributed with $(k - s - 1)$ d.f. The correlation between these two test criteria is $(k - s - 1)^{\frac{1}{2}}(k - 1)^{-\frac{1}{2}}$. Deduce that this is compatible with the hypothesis that one of these test criteria is a component of the other. The χ^2 of a contingency table with fixed marginal totals has a correlation with the χ^2 of the same table with row and column parameters given by hypothesis equal to

$$[(m - 1)(n - 1)/(mn - 1)]^{\frac{1}{2}}.$$

Show this is again compatible with the hypothesis of a component χ^2. (Neyman and Pearson, 1928; Lancaster, 1954.)

19. Prove that χ^2/N converges in probability to the theoretical ϕ^2. [*Hint.* For a pair of distributions specified by H_0 and H_1, $F(x)$ and $G(x)$, there is a finite partition so that $\int \Omega^{*2}\, dF^* > 1 + \phi^2(1 - \epsilon)$ for arbitrary $\epsilon > 0$ and where F^* and G^* are the distributions functions on the finite partition and $\Omega^* = dG^*/dF^*$ the Radon–Nikodym derivative on the partitioned space. Now consider the expectation of χ^2 under H_1 and use the law of large numbers.]

20. Interpret χ^2/N as the sum of squares of the Fourier coefficients of the expansion of $\Omega(x)$ in an orthonormal series. (Ali and Silvey, 1965a and b; Lancaster, 1960b; E. S. Pearson, 1932; K. Pearson, 1904; von Mises, 1931.)

21. Fisher's proof given in VIII.4 assumes that the difference of two χ^2 variables has also the χ^2 distribution. [This is so if and only if the subtrahend is a component of the other χ^2 variable in the sense that both χ^2 variables and their difference are quadratic forms in an underlying set of independent normal variables. See Theorem II.5.5.] Verify that a counter example to Fisher's assumption is provided by

$$p(\xi, \eta) = (2\pi)^{-1}\xi^{-\frac{1}{2}}\eta^{-\frac{1}{2}}e^{-\frac{1}{2}(\xi+\eta)} + \epsilon(\xi - \eta)(\xi\eta - \xi - \eta + 2)e^{-(\xi+\eta)},$$

where ξ, $\eta \geqslant 0$ and ϵ is a small positive constant. Verify that this a properly defined probability distribution and that the m.g.f. is

$$M_{\xi,\eta}(t, u) = (1 - 2t)^{-\frac{1}{2}}(1 - 2u)^{-\frac{1}{2}} + 2\epsilon tu(t - u)(1 - t)^{-3}(1 - u)^{-3}$$

valid for $|t|, |u| < \frac{1}{2}$. (Fisher, 1924a; James, 1952; Lancaster, 1954; Neyman and Pearson, 1928.)

22. Prove that the linear forms, Y_i, in the notation of (5.15) which are not annihilated in the process of estimation cannot be independent of the values, θ_i, assigned to the unknown parameter, since all the Y_i are linear forms in the class frequencies, a_j. These class frequencies can be regarded as independent Poisson variables. Mutual independence of the Y's would characterize the normal distribution and a contradiction would result. Use this proposition to explain why independence between linear or quadratic forms in the cell frequencies can only be justified as an asymptotic result as the sample size becomes very large.

23. Let X be distributed as a rectangular variable on $(0, \theta)$, θ unknown. The sufficient estimator of θ is

$$\hat{\theta} = x_{max},$$

where x_{max} is the greatest observed value of X in N independent observations. Prove that the joint distribution of the other X's given x_{max} is rectangular on $(0, x_{max})$. Hence a χ^2 test for rectangularity of the original distribution can be devised by breaking up the interval $(0, x_{max})$ and setting the expected number of observations in any interval proportional to its length. Note that this estimate reduces the sample size by unity but not the degrees of freedom. [See Kendall and Stuart, 1961, Ex. 17.16.]

24. Let n_s be the observed class frequencies, $\tilde{m}_s$ the expected frequencies and $x_s = n_s - \tilde{m}_s$, $s = 1, 2, \ldots, k$.

$$x_1 + x_2 + \ldots + x_k = 0.$$

If the parameters are known then the density is given by

$$D = D_0 \exp -\tfrac{1}{2}\chi^2, \qquad \chi^2 = x_1^2/\tilde{m}_1 + x_2^2/\tilde{m}_2 + \ldots + x_k^2/\tilde{m}_k.$$

Let a frequency law of prescribed form be fitted. The resulting frequencies m_1, $m_2, \ldots, m_k$ will be represented by a point given by

$$X_1 = m_1 - \tilde{m}_1, \ldots, X_k = m_k - \tilde{m}_k$$

which must lie on the population locus

$$X_s = f_s(\alpha_1, \alpha_2, \ldots, \alpha_c), \qquad s = 1, 2, \ldots, (k-1);$$
$$X_1 + X_2 + \ldots + X_k = 0.$$

The deviations from the estimated frequencies are thus $n_s - m_s = n_s - \tilde{m}_s - (m_s - \tilde{m}_s) = x_s - X_s$ and the value of the Pearson χ^2 with estimated parameters is now

$$\chi_1^2 = \frac{(x_1 - X_1)^2}{\tilde{m}_1 + X_1} + \ldots + \frac{(x_k - X_k)^2}{\tilde{m}_k + X_k}.$$

Let now the simplifying assumption be made that X_s is negligible in the denominators on the right and let

$$z_s = \frac{x_s - X_s}{\tilde{m}_s^{\frac{1}{2}}}, \quad \text{then} \quad \chi_1^2 = \sum z_s^2.$$

Suppose that $\{X_s\}$ has been determined by c linear relations on the x_s. Let us write these out. Then any other linearly independent form in the $\{z_s\}$ does not contain X_s since this would be a further condition on the X_s. If $\sum z_s^2$ is to be minimised by choice of $\{X_s\}$ we have

$$\mathbf{v} = \begin{bmatrix} \mathbf{H}_1 \\ \mathbf{H}_2 \end{bmatrix} \mathbf{z}$$

where the X_s appear only in the first $(c+1)$ variables v_i. The $k - c - 1$ other variables are independent of the value of $\{X_s\}$. The first row of $\mathbf{H}_1$ may be taken to have elements proportional to $\tilde{m}_s$ and the other c rows can be written in ortho-normalized form. $\mathbf{H}_2$ consists of the additional rows necessary to complete the orthogonal matrix. The minimising procedure annihilates $H_1 z$. On the hypotheses made $(H_2 z)^T (H_2 z) = \chi_1^2$ is distributed as χ^2 with $k - c - 1$ degrees of freedom. But $H_2 z = H_2 x$, so that

$$x^T x = (H_2 x)^T H_2 x + (H_1 x)^T (H_1 x)$$
$$= \chi_1^2 + (\chi^2 - \chi_1^2)$$

(Neyman and Pearson, 1931).

25. Of 26,306 throws with 12 dice, a success is said to have been scored if 5 or 6 is observed on an individual die. The number of possible successes were 0(1)12 and the observed frequencies were 185, 1149, 3265, 5475, 6114; 5194, 3067, 1331, 403, 105; 14, 4, 0. Fit a binomial, $N(\frac{2}{3} + \frac{1}{3})^{12}$, to the data and verify that χ^2 is 35·49 with 10 d.f. if the last three classes are pooled. Under the hypothesis that the mean number of successes is 0·3377, verify that the χ^2 is 8·179 with 9 d.f. Compare the distributions with all dice equally biased or with all the bias concentrated on one die so that $p = \frac{1}{3}$ for 11 of the dice and $p = 0·40524$ for the remaining die. (Weldon, cited by Pearson 1900a; Fisher, *Statistical Methods*.)

26. 1000 observations have been made at Greenwich of the right ascension of the Polaris star. Let x denote deviations from the mean measured to the nearest half minute of seconds of time. Let the observed frequencies be calculated as being from a normal distribution with the same mean and variance as the sample. The results are as follows,

x	Observed Frequency	Calculated Frequency
−3·5	2	4
−3·0	12	10
−2·5	25	22
−2·0	43	46
−1·5	74	82
−1·0	126	121
−0·5	150	152
0	168	163
0·5	148	147
1·0	129	112
1·5	78	72
2·0	33	40
2·5	10	19
3·0	2	10

Test the fit of a normal distribution. (Whittaker and Robinson, 1944)

27. Merriman (1891) gave on page 14 of his *A Textbook of the Theory of Least Squares* the following observed and expected number of hits in various belts (or zones) on a target, the expected frequencies being computed on the assumption of normality and estimation of the mean and variance.

Belt	Observed Frequency	Expected Frequency
1	1	1
2	4	6
3	10	27
4	89	67
5	190	162
6	212	242
7	204	240
8	193	157
9	79	70
10	16	26
11	2	2
Totals	1000	1000

Comment on whether Merriman was correct in asserting that the normal gives a good fit to the observations. (Pearson, 1900a.)

28. The times of onset of labour of selected cases in Birmingham, 1950–1951 were classified by hours beginning at midnight with frequencies 52, 73, 89, 88, 68, 47; 58, 47, 48, 53, 47, 34; 21, 31, 40, 24, 37, 31; 47, 34, 36, 44, 78, 59. Verify that if the null hypothesis specifies a uniform distribution over the 24 hourly periods, $\chi^2 = 162 \cdot 77$ for 23 d.f. (E. Charles, *Brit. J. Soc. Med.* 7 (1953), 43–59.)

29. It was observed from the Royal Society Mathematical Tables that for the 1000 instances, $n = 1, 2, \ldots, 1000$, $p(n)$ was odd on 528 and even on 472 occasions. Does this indicate that $p(n)$ is more frequently odd than even? According to calculations of Miss R. Wood, for $n = 1, 2, 3, \ldots, 5,000$, $p(n)$ is odd on 2541 occasions. Comment. (Pathria, 1961.)

30. Yule constructed fourfold tables with the aid of random sampling numbers and then for each table calculated χ^2 with fixed marginal totals. He computed the probability that χ^2 with one degree of freedom should fall into the probability classes and constructed the following table

Value of χ^2	Number Expected	Number Observed
0–0·25	134·02	122
0·25–0·50	48·15	54
0·50–0·75	32·56	41
0·75–1·00	24·21	24
1·00–2·00	56·00	62
2·00–3·00	25·91	18
3·00–4·00	13·22	13
4·00–5·00	7·05	6
5·00–6·00	3·86	5
6·00 +	5·01	5

Is the table in favour of the hypothesis that χ^2 has indeed 1 d.f.? (Yule, 1922; Fisher, 1923.)

31. The absolute deviations from the true mean in measurements of right ascension were given by K. Smith as follows

Deviations in Tenths of a Second	Observed	Gaussian Curve by Moments	Gaussian Curve Improved by Minimum χ^2
0–1	114	101·61	98·63
1–2	84	84·12	82·59
2–3	53	57·65	57·91
3–4	24	32·71	34·00
4–5	14	15·36	16·72
5–6	6	5·974	6·881
6–7	3	1·923	2·372
7–8	1	0·5122	0·6843
8–9	1	0·1370	0·2053

Smith (1916) found that χ^2 was 10·833 when fitted by moments and 9·720 when fitted by minimum χ^2. The estimate by minimum χ^2 of the standard deviation was 2·355860. Fisher (1922c) found that the standard deviation estimated by maximum likelihood was 2·264214. (K. Smith, 1916; Fisher, 1922c.)

32. Adapt the proof given for the distribution of the Pearson χ^2 with parameters fitted from the data to the smooth test of Neyman. (Barton, 1953, 1954, 1955, 1956; Neyman, 1937, 1949.)

33. If parameters are estimated from grouped data, then Q calculated after pooling of classes (or groups) need not have the presumed asymptotic distribution. (Neyman and Pearson, 1931.)

34. Fit a Poisson distribution to the following frequency distribution for counts of red cells on small haemocytometer squares.

Cell Count	Frequency	Cell Count	Frequency
0	11	6	38
1	36	7	17
2	76	8	6
3	80	9	3
4	74	11	1
5	58	Total	400

Mean $= 3·6175 = \hat{\lambda}$. χ^2 for 8 d.f. $= 2·340$. $P = 0·97$. (Lancaster, 1950b.)

CHAPTER IX

Problems of Inference

1. INTRODUCTORY

The theory of statistical hypothesis testing was rudimentary at the time of the introduction of the Pearson system of frequency curves in 1895. Before that time, there could be little point in such a theory as there were few distributions alternative to the normal. Goodness of fit tests were developed only when there was need to compare the fits of different theoretical distributions to empirical data. Pearson (1895) suggested that the goodness of fit might be tested by the percentage deviation of the observed values from the expected values and that binomial theory might be applied for the goodness of fit in individual classes. However, there is correlation between such binomial variables and so there is a need to consider the joint distribution of the deviations. Pearson (1900a) considered the distribution of a quadratic form, which was shown to be a remarkably simple function of the observed and expected frequencies, the Pearson χ^2.

Further developments came later when R. A. Fisher saw the need to test clearly stated hypotheses and introduced the notion of the "null" hypothesis. Fisher also recognised the need for correct design; in other words, a question must be asked in precise terms and the experiment designed in such a way as to give a precise answer as economically as possible.

Neyman and Pearson extended this theory by showing that consideration should be given to possible alternative hypotheses. They gave criteria for powerful tests. We can justify many of the χ^2 tests by reference to the Neyman-Pearson theory.

However, it should be noted that a χ^2 test is often useful because it is not specialised to test a small class of alternative hypotheses against the null hypothesis. There is a close analogy here with the use of the F-test by Fisher in the analysis of variance. It is evident that the problems of multiple comparisons are as difficult in the χ^2 theory as they are in the theory of analysis of variance. It is also common for a χ^2 test to be used for the analysis of data which have not resulted from a planned experiment. Many examples of this could be cited from vital statistics and the social sciences. In such cases χ^2 is being used as a general-purpose preliminary screening test. Often observations are available on two variables which enable the null hypothesis of independence to be tested. If it is rejected, then usually fresh observations will be sought and these may be unplanned or perhaps take the form of a sampling survey or experiment.

2. TESTS OF HYPOTHESES

In many practical situations, an observation on a specified random variable, Z, is used to test the truth of an hypothesis. Z may be a single observation but it may also be some statistic derived from a set of observations on one or more random variables. Z is supposed to have a distribution belonging to some class $\{P_\theta\}$, where $\theta \in \Omega$. θ may represent a single parameter or it may represent a set of parameters. Ω is the *parameter space* or the *space of admissible parameter values*. The parameter space can be partitioned so that $\Omega = \Omega_H \cup \Omega_K$; and it is convenient to think of Ω_H as being determined by the null hypothesis, $H_0 \equiv H$, and of Ω_K as being determined by the alternative hypothesis, $H_1 \equiv K$. If $\theta \in \Omega_H$, then the null hypothesis is true; if $\theta \in \Omega_K$, then some alternative hypothesis is true.

The test of hypothesis will take the form that if $Z \in S_0$, the hypothesis is to be accepted; whereas, if $Z \in S_1$, the hypothesis is to be rejected. Alternatively, if $Z \in S_0$, $\theta \in \Omega_H$ is judged to be the better decision and if $Z \in S_1$, $\theta \in \Omega_K$ is the better decision. $S_0 \cup S_1$ is the whole sample space of Z or $\mathscr{P}(Z \in S_0 \cup S_1) = 1$. S_1 is called the *critical region*. It is recognised in the Neyman-Pearson theory that the best test of an hypothesis H depends on the nature of the alternative hypothesis K.

The testing procedure may lead to one of two types of errors,

(i) an *error of the first kind*, the rejection of a true hypothesis, or
(ii) an *error of the second kind*, the acceptance of a false hypothesis.

Errors of the first kind are controlled by making small the probability of rejecting the null hypothesis, when it is true. The critical region is chosen so as to have $\mathscr{P}(Z \in S_0) = \alpha, 0 < \alpha < 1$ after α has been fixed at some arbitrary low figure, commonly 0·05 or 0·01. Among all possible critical regions, it is sometimes possible to choose a region which minimises the probability of acceptance of the hypothesis when it is false, that is, the probability of an error of the second kind. The probability of rejection for a given critical region and given alternative hypothesis is known as the *power* and written $\beta(\theta)$, where $\theta \in \Omega_K$. The probability of an error of the second kind is thus $1 - \beta(\theta)$, where $\theta \in \Omega_K$. It is convenient to calculate also the *critical level*, $\bar{\alpha} = \bar{\alpha}(Z)$, the smallest level of significance at which the hypothesis would be rejected. The *critical level* is often referred to as the *probability of the experiment.* The critical level is a useful guide as to how strong the evidence is against the null hypothesis.

The Pearson χ^2 is used to test hypotheses in discrete distributions and so some modification has to be made to the general theory. In Chapter III, we have described techniques to overcome this difficulty. We have given reasons why the "corrections for continuity" and the "auxiliary random sampling" methods may be undesirable. As Professor E. J. G. Pitman has pointed out to me, although the device of the auxiliary random experiment may not be desirable on practical grounds, it is useful from a theoretical point of view for it enables us to treat the discrete χ^2 as though it were a continuous variable. For example, in using the normal approximation to the binomial distribution, the randomising procedure converts the distribution with weights at $(n + 1)$ points into a density distributed over an interval. The methods of the differential calculus are then applicable.

A version of the Neyman-Pearson fundamental lemma is as follows. Let $H \equiv H_0$ and $K \equiv H_1$ be hypotheses specifying probability distributions, P_0 and P_1, possessing densities, p_0 and p_1, with respect to some fundamental measure, μ. The measure μ always exists since $\mu \equiv P_0 + P_1$ is such a measure, since for any set, A, $\mathscr{P}_0(A) \leqslant \mu(A)$ and $\mathscr{P}_1(A) \leqslant \mu(A)$ and hence P_0 and P_1 are absolutely continuous with respect to μ. The densities, p_0 and p_1, are then the Radon-Nikodym derivatives of P_0 and P_1 with respect to μ and $0 \leqslant p_0$, $p_1 \leqslant 1$. Let α be some fixed number, such that $0 < \alpha < 1$. Then there always exists a significance test which has a size α under H and has a maximum power under K and this test is of the form "reject the hypothesis if $Z = z$ and $p_1(z) > kp_0(z)$ and accept it if $p_1(z) < kp_0(z)$". It is assumed that if there are values $z \in M$ such that $p_1(z) = kp_0(z)$ then $\mathscr{P}(Z \in M) = 0$. If a test is most powerful at level α for testing H against K, then for some k, it is of the form, reject if $p_1(z) > kp_0(z)$. A test having a size α under H can always be found, unless there is a test of size less than α with power 1 under K. In this last case, the whole of the distribution under K is concentrated on a set which has measure less than α under H. There is a corollary to these statements. If β is the power of the most powerful level α test $(0 < \alpha < 1)$ of H against K, then $\alpha < \beta$ unless $P_0 = P_1$. Proofs of all these statements may be found in Lehmann (1959).

Difficulties will arise when there is a marginal set M such that $\mathscr{P}(Z \in M) > 0$ and there can be no S_1 with $\mathscr{P}(Z \in S_1) = \alpha$. According to Fisher, in such cases, S_1 must be chosen so that $\mathscr{P}(Z \in S_1) \leqslant \alpha$ so that the exact test or an approximation to it must be used. The Neyman-Pearson theory requires that there shall be an auxiliary random experiment so that if $Z \in M$, sometimes Z is said to be in S_0 and other times Z is said to be in S_1. This random experiment is arranged so as to make the size equal to α precisely. In Chapter III, under some very plausible assumptions, we have shown that our suggested procedures will agree with the Neyman-Pearson procedure in all cases in which $Z \notin M$ and in about 75% of cases in which $Z \in M$. Some additional advantages in the applications were also pointed out there. It will be found that the methods of Chapter III often result in some test statistic Z being assumed to have the normal or χ^2 distributions in the null case. In many such cases, the Neyman-Pearson theory leads directly to the test to be applied. It has been the common custom following Wilks (1938 and 1962) and Wald (1941a and b) to set up the likelihood ratio criterion, λ, and then to apply asymptotic theory, whereby $-2 \log \lambda$ is shown to have the distribution of χ^2. However, it is possible to apply the normal approximation and then to use the theory of hypothesis testing as it has been developed in the univariate and multivariate normal distributions. The equivalence of these two approaches is proved in Section 4 below. Such a procedure enables much use to be made of the distribution of the non-central χ^2 and hence to determine the choice of tests and the powers of the tests used. The exact distribution of the likelihood ratio is unknown. The distribution, conditional on restraints, can only be determined by the use of the theory of the asymptotic χ^2 distribution. It is further possible that if the distribution, conditional on specified linear forms taking fixed values, is independent of these fixed values, then there is a characterisation of the normal distribution and the statement can only be true asymptotically. In Chapter VII, it has been shown that the likelihood ratio plays an important role in the choice of test criteria and that the parameter of non-centrality of test functions depends on the correlation of the particular criterion chosen and the likelihood ratio.

3. SIGNIFICANCE LEVELS

The choice of significance level is to some extent arbitrary and will often depend on the use which is to be made of the observations. For example, if the observations are exploratory, it will not be advisable to choose too small a significance level (or critical region) for then small but promising departures from the null hypothesis may be overlooked. If firm decisions have to be made, it might well be that the significance level will have to be chosen quite small and the experiment planned to be large so that the power also is large. It is often claimed that the performance of a statistical test is equivalent to an estimation of a parameter. While this is often so, the original observations may be quite unsuitable for estimation. The original test may be of independence of two random variables; if the test is judged non-significant then nothing further may be done, while if it is judged significant there may be far more suitable methods of determining some measure of the dependence and other relevant features, such as the reasons for the dependence or the form of dependence.

4. THE LIKELIHOOD RATIO TEST AND χ^2

There is a close relationship between theories of χ^2 and of the likelihood ratio. In designing tests of significance, one can select a likelihood ratio test and then make the transition to normal theory by Stirling's approximation or another equivalent method such as the use of the central limit theorem; alternatively, one can proceed to the asymptotic theory immediately by the application of Stirling's approximation or other equivalent procedure and then justify the test chosen by the likelihood ratio test theory. In this monograph, we have chosen the second method. We may mention that many proofs are available in the literature of the approximate asymptotic equivalence of the two methods. Moreover, direct enumeration has shown that the probabilities in the multinomial distribution tend to be ordered by the likelihood ratio and by the approximate Stirling's formula in the same way.

Example Let H_0 and H_1 specify that Z should have the Poisson distribution with parameters, ξ and η respectively, $\eta > \xi$.

The likelihood ratio for a given value $Z = z$ is

$$(4.1) \qquad e^{-\eta}\eta^z/(e^{-\xi}\xi^z) = e^{-(\eta-\xi)}\eta^z\xi^{-z}$$

so that the critical region contains the large values of Z, the upper tail of the distribution. The logarithm of the likelihood is $z \, (\log \eta - \log \xi)$, and this is to be the test criterion. Except for a multiplier, this is simply Z. But Z is a Poisson variable and so if ξ is large, the test function can be taken to be $(Z - \xi)\xi^{-\frac{1}{2}}$ and this is asymptotically a standard normal variable under H_0. H_1 specifies that $(Z - \eta)\eta^{-\frac{1}{2}}$ is a standard normal variable, so that, $(Z - \xi)\xi^{-\frac{1}{2}}$ is normal with expectation $(\eta - \xi)\xi^{-\frac{1}{2}}$ and variance $\eta\xi^{-1}$. The likelihood theory for the test of the displacement of the mean of a normal distribution makes it clear that $(Z - \xi)\xi^{-\frac{1}{2}}$ is the appropriate test function. The displacement of the mean is $(\eta - \xi)\xi^{-\frac{1}{2}}$ standard deviations. The power functions can then be computed. If there is interest only in the 50% point for the power, the differences in variances are irrelevant. In many cases the variance under H_1 will be close to unity so that $N(\eta - \xi)^2/\xi$ can be treated as the usual parameter of non-centrality for χ^2 with one degree of freedom; values of the power can be read from the tables of the non-central χ^2 distribution. If $(\eta - \xi)$ is larger, the mean and variance under H_1 of the test function can be computed.

Similar arguments can be used in more complicated cases as when the means, rather than the variable itself, are being considered or when the goodness of fit is being tested in the multinomial distribution, in contingency tables and so on. A formal proof of the equivalence of the χ^2 and the likelihood ratio tests, when parameters are given by hypothesis, is as follows.

Let H_0 specify the class probabilities, p_{0i}, of a multinomial distribution and let H_1 specify class probabilities, p_{1i}, $i = 1, 2, \ldots, n$. Let the observed numbers in the classes be N_i, $\sum N_i = N$. Then

(4.2)
$$\mathscr{P}(\{N_i\} \mid H_0) = \mathscr{P}(\{N_i\} \mid \{p_{0i}\}) = N! \prod_{i=1}^{n} [p_{0i}^{N_i}/N_i!]$$

$$\mathscr{P}(\{N_i\} \mid H_1) = N! \prod_{i=1}^{n} [p_{1i}^{N_i}/N_i!].$$

The likelihood ratio is then given by

(4.3) $$l = \mathscr{P}(\{N_i\} \mid H_0)/\mathscr{P}(\{N_i\} \mid H_1) = \prod_{i=1}^{n} (p_{0i}/p_{1i})^{N_i}.$$

Now the $\{p_{1i}\}$ are not known but may be estimated from the data by maximum likelihood methods. The estimates are

(4.4) $$\hat{p}_{1i} = N_i/N.$$

It follows that

(4.5)
$$-2 \log l = -2 \sum N_i [\log p_{0i} - \log \hat{p}_{1i}]$$
$$= -2 \sum N_i [\log p_{0i} - \log N_i + \log N]$$
$$= -2 \sum N_i \log p_{0i} + 2 \sum N_i \log N_i - 2N \log N.$$

Define differences

(4.6)
$$\Delta_i = Np_{0i} - N_i, \qquad N_i = Np_{0i} - \Delta_i$$
$$\sum \Delta_i = 0.$$

Then approximately Δ_i is of order $(Np_{0i})^{\frac{1}{2}}$ under H_0. When (4.6) is substituted in (4.5) it follows that,

(4.7)
$$-2 \log l = 2 \sum (Np_{0i} - \Delta_i) \log (1 - \Delta_i/Np_{0i})$$
$$\simeq \sum \Delta_i^2/(Np_{0i}), \quad \text{the Pearson } \chi^2.$$

It is of interest to note what happens to this computation when Stirling's approximation or other method of Chapter V is used to approximate to the probability

(4.8)
$$\mathscr{P}(\{N_i\} \mid H_0) \simeq (2\pi N)^{-\frac{1}{2}(n-1)} \prod_{i=1}^{n} p_{0i}^{-\frac{1}{2}} \exp -\tfrac{1}{2}\Delta^T V^{-1} \Delta$$

$$\mathscr{P}(\{N_i\} \mid H_1) \simeq (2\pi N)^{-\frac{1}{2}(n-1)} \prod_{i=1}^{n} p_{1i}^{-\frac{1}{2}} \exp -\tfrac{1}{2}\Delta^{*T} V^{-1} \Delta^*$$

where Δ has $(n - 1)$ terms of the form $N_i - Np_{0i}$ and Δ^* has $n - 1$ terms of the form $N_i - Np_{1i}$. An approximate maximum for $\mathscr{P}(\{N_i\} \mid H_1)$ is obtained by setting $\Delta^* = 0$. $\Pi p_{0i}^{-\frac{1}{2}} \simeq \Pi p_{1i}^{-\frac{1}{2}}$, since $|p_{0i} - p_{1i}|$ is $O(N^{-\frac{1}{2}})$ and then,

(4.9) $$-2 \log_e l = \Delta^T V^{-1} \Delta = \sum \Delta_i^2/(Np_{0i}),$$

the Pearson χ^2.

5. MULTIPLE COMPARISONS

It is not proposed to give here a general review of the theory of multiple comparisons, for which the reader may be referred to the March, 1965 issue of *Technometrics*. Instead a few special problems are mentioned.

Perhaps the simplest case of the multiple decision problem is as follows. Let Z_1 and Z_2 be mutually independent standard normal variables. Then, if ψ is specified independently of Z_1 and Z_2, every linear form, $Z_1 \cos \psi + Z_2 \sin \psi$, is a standard normal variable. If, however, ψ is dependent on the values assumed by Z_1 and Z_2, then the form cannot be treated as a linear form. In particular, if $\psi = \tan^{-1}(Z_2/Z_1)$, the form has the distribution of $(Z_1^2 + Z_2^2)^{\frac{1}{2}}$, the square root of χ^2 with 2 d.f. Suppose now that $Z_1, Z_2, \ldots, Z_n$ are a set of n mutually independent standard normal variables. Then for all unit vectors $\mathbf{a}$ and all sets of values $\mathbf{z}$, $(\mathbf{a}^T \mathbf{z})^2 \leqslant \mathbf{z}^T \mathbf{z}$ by the Schwarz inequality. χ^2 can, therefore, be characterized as the greatest value of $(\mathbf{a}^T \mathbf{z})^2$ for all choices of the unit vector, $\mathbf{a}$. Such a specially chosen "linear form" evidently cannot be tested as a standard normal variable.

A non-significant value of χ^2 can be interpreted in the same way as a non-significant value of F, namely that the observations do not give evidence for a departure from the null hypothesis. It is, however, possible for χ^2 to be non-significant and some comparison or linear form to be significant. Thus, it may be specified, before the observations are made, that Z_1 and Z_2 are to be tested separately, in which case the observed Z_1 may be judged significant at the α-level whereas Z_2 may be small and the combined variable, $Z_1^2 + Z_2^2$, can be less than the value at the α-level of significance of χ^2 with two degrees of freedom. With linear orthogonal forms, $Y_1 = 2^{-\frac{1}{2}}(Z_1 + Z_2)$ and $Y_2 = 2^{-\frac{1}{2}}(Z_1 - Z_2)$, the same kind of results could also occur. The theory of multiple comparisons is not yet complete but it suggests that, wherever possible, a definite hypothesis of interest should be stated before making the observations and then tested by them by the most powerful test available.

A good example of this last statement is the testing of the mutual independence of the marginal variables, Z_1 and Z_2, in a general $m \times n$ contingency table. If nothing further is given, then a χ^2 variable with $(m-1) \times (n-1)$ degrees of freedom tests all departures from independence simultaneously. However, if it is also given that the distribution is jointly normal, then for such partitions of the marginal spaces as are likely to be met in practice, the most powerful test will be the test of the correlation between ζ_1 and ζ_2, where ζ_i is the orthonormal function on the condensed marginal distribution which has maximum correlation with the marginal variable, Z_i, as is pointed out in Chapter X. The standardized function, $N^{-\frac{1}{2}} \mathscr{S}(\zeta_1 \zeta_2)$, can be tested as a standard normal variable with one degree of freedom. A less obvious example of the multiple comparison problem is when $N R_1^2$, where R_1

is the first canonical correlation in an $m \times n$ contingency table, is treated as χ^2 with $(m + n - 3)$ degrees of freedom, which has been suggested by some authors. Lancaster (1963c) showed that NR_1^2 is stochastically larger than χ^2 with $(m + n - 3)$ degrees of freedom and that the distribution of other NR_i^2 did not approximate well to the distribution of χ^2 with $(m + n - 2i - 1)$ degrees of freedom, R_i being the ith canonical correlation in the sense of Fisher (Maung, 1942).

A form of the multiple comparison problem that has often not been recognised as such in the past, is as follows. An observer wishes to detect those persons who can guess the face values of cards without viewing them. N_0 persons are submitted to the test of, say, guessing for n cards. Those with the N_1 best scores are re-examined at another trial. Of these N_2 are selected to undergo a third trial. Finally after r trials there will be N_r persons each of whom will have had high scores at each trial. These N_r persons will have total scores differing significantly from the mean, if each person is considered in isolation. However, it is obvious that even if the null hypothesis of no variation in ability were true, this would still almost certainly be so, for the N_r are those selected from the N original persons. Some form of order statistics will be needed to devise an experiment along these lines. Alternatively the past results could be discarded at each trial and some confirmatory experiments carried out after their completion to test anew whether the subjects did have the ability.

In conclusion, the use of the χ^2 test in many situations, where a precise alternative hypothesis cannot be formulated, can be justified as follows. If the expectation of a quadratic form is to increase with N under all possible H_1 specifying n mutually independent $(c_i, 1)$ variables, $X_i, i = 1, 2, \ldots, n$, then the quadratic form must be non-negative and of full rank. Moreover, if the quadratic form is to be equally sensitive to all non-centrality parameters, $\sum c_i^2 = \lambda$, then it must be of the form $\sum X_i^2$ by symmetry considerations.

6. GROUPING, OR CHOICE OF PARTITIONS OF THE MEASURE SPACE

It can be said that there is no general theory of the choice of partitions of the sample space. Some authors have discussed how the space should be partitioned if N ungrouped observations are given. If the null hypothesis specifies the normal distribution and the alternative hypothesis specifies the normal distribution with a shift in mean and variance, then there will be more powerful tests than the ordinary χ^2, so that the problem should not arise. It may also be more appropriate to use a χ^2-analogue based on the Hermite series rather than the indicator variables. Other special cases will come readily to mind.

One possibility has been dealt with in detail by Mann and Wald (1942), namely, there is a difference between the cumulative probability functions $F(z)$ of the null hypothesis and $G(z)$ of the alternative so that

$$|F(Z_0) - G(Z_0)| \geqslant \Delta$$

at some point. Mann and Wald (1942) gave as a plausible procedure the rule

(6.1) $$k_N = 4[2(N-1)^2 C^{-2}]^{\frac{1}{5}}$$

where k_N is the number of groups, N the sample size and $\mathscr{P}\{Z > C\} = \alpha$, if Z is a standard normal variable. They further gave the expected power of 0.5 for $\Delta_N = 5k_N^{-1} - 4k_N^{-2}$. Practical aspects of the problem were set out by Williams (1950).

It seems that the Mann and Wald procedure may recommend too many classes for some forms of alternative hypothesis. An easily computed example is the shift of the mean of a normal distribution. Hamdan (1963) found that for $N = 1,000$ and a shift of 0.1 standard deviations 6 classes gave a parameter of non-centrality of 8.978 whereas the greatest possible value would be 10.05. It appeared from his work that in this particular case nothing could be gained by increasing the number of classes since it would add little to the parameter of non-centrality but much to the degrees of freedom and hence to the least critical value of χ^2. There would in general be a loss of power in using more than six classes.

In conclusion, the recommendation of Mann and Wald (1942) can only be held to apply in special cases. In other cases, quite different solutions may be appropriate. It seems likely that the solution of Hamdan (1963) is capable of generalization in all those cases, where there are definite null and alternative hypotheses.

7. LARGE VALUES OF χ^2

To avoid confusion, the usual Pearson χ^2 as observed will be termed Q. As a rule, the critical region for Q will be chosen so as to contain the large values and it is assumed in this section that this is so and that H_0 is to be rejected if Q is large. The interpretation of abnormal, chiefly large, values of Q may be carried out with the aid of principles, (i) to (iv) and (viii), laid down by Fisher (1924). The conditions, under which Q will not have a distribution approximating to the theoretical χ^2, may be detailed as follows:

  (i) the degrees of freedom have not been calculated correctly;
  (ii) the hypothesis tested is not in fact true;
  (iii) the method of estimation employed is inconsistent;
  (iv) the method of estimation is inefficient;
  (v) estimation has been carried out in a different space;

(vi) the experimental results, to which the test is applied, have been selected;

(vii) an incorrect mathematical model has been used;

(viii) the necessary conditions for asymptotic normality have not held;

(ix) an event can occur with positive probability under an alternative hypothesis, which has zero probability under the null hypothesis (this is really a special case of (ii)).

(i) *Degrees of Freedom.* The problem of the correct number of degrees of freedom was solved by Fisher in his papers of the years, 1922–1924. It seems necessary, however, to add, that K. Pearson (1900a) was well aware of such a problem and that the paper of Pearson (1916b) goes far towards giving a general solution, since he considered the effects of homogeneous and non-homogeneous linear constraints on the cell contents. He failed, however, to recognize that certain procedures, such as the fixing of the marginal totals in a contingency table imposed such restrictions and this failure led to much argument. His reckoning for the degrees of freedom tended to yield too high a probability to the value of Q derived from the observations. Perhaps Fisher's point here is of historical interest only; for since the general adoption of his ideas, there is likely to be less disagreement as to the calculation of the degrees of freedom than in the early years after the introduction of the Pearson χ^2.

(ii) *False Hypotheses.* Fisher (1924) observed that if the hypothesis tested is not true, then the calculated Q becomes indefinitely large if the sample size is large. For if the sample size is large the class numbers a_i differ from the expectations, Ng_i, under the alternative hypothesis by quantities of order, $N^{\frac{1}{2}}$. The expected class frequencies, however, are Nf_i. Q/N thus tends to equal $\phi^2 = \sum g_i^2/f_i - 1$ and if $g_i \neq f_i$ for any class, $\phi^2 > 0$. $\mathscr{E}Q$ thus increases as $N\phi^2$, as $N \to \infty$.

(iii) *Inconsistent Estimation.* Fisher (1924) gives a detailed discussion of some experimental work of Brownlee (1924). Fisher (1924) points out that if the distribution is really the binomial, $(\frac{1}{2} + \frac{1}{2})^4$, but the class frequencies are computed using certain areas under a normal curve with the same mean and standard deviation, then the quantity, Q, will not have the correct distribution since the class probabilities will be the f_i of the binomial not the f_i^*, of the normal distribution with the same mean and variance. The apparent parameter of non-centrality will be $N[\sum f_i^2/f_i^* - 1]$ and not zero as it should be under a true null hypothesis. The expected value of Q/N will not be equal to the degrees of freedom with such a procedure.

(iv) *Inefficient Estimation.* If parameters are estimated inefficiently, the distribution of Q will no longer approximate to the theoretical. Efficient estimation annihilates linear forms equal in number to the parameters estimated.

Inefficient estimation will result in some linear forms in the frequencies, which should have been annihilated, having distributions not degenerate at zero. These will contribute amounts proportional to N to the expectation of Q under H_1.

(v) *Estimation in the Correct Space.* Let U be an arbitrary random variable. Then the Pearson χ^2 test is not always carried out on the observations on U but on a condensed variable, U^* say. U^* takes only a finite set of values. A_i is the number of times U^* assumes the value U_i^*, say. This corresponds to an event, $U \in B_i$ say. Now the least square or maximum likelihood estimates should be made using the space of U^* and not the space of U. It is shown in Chapter VIII that anomalies will occur if estimation is carried out in the space of U. In fact, some parameters in the space of U may be irrelevant in the space of U^*.

(vi) *Multiple Comparisons.* It may be briefly noted that if k experimenters carry out an experiment and only the one with the most striking findings publishes his results, then the ordinary significance levels cannot be used. Other aspects of this problem have been treated in Section 5.

(vii) *Incorrect Mathematical Models.* Care must be taken to see that the mathematical model is appropriate. Thus suppose s seeds are planted in each of the plots laid out in the form of an $m \times n$ rectangle. If a_{ij} are the number of plants developing, then the $\{a_{ij}\}$ cannot be considered to be an array of the observed numbers in a contingency table of two dimensions since if so then $\{s - a_{ij}\}$ is also such an array. The "χ^2" of such contingency tables will not agree. The true model in this type of experiment is an $m \times n \times 2$ contingency table, which is of the type, which we have termed 2B in Section XII.3.

(viii) *Asymptotic Conditions.* Q is computed on the assumption that every cell has an expected number not too small. This has been discussed in Chapter V. Without undue risk of disturbing the distribution, the expectation in one or two cells out of, say, 6 or more may be allowed to be in the interval $(1, 2)$; but it is unwise to have any cell expectations lower than unity. It may be that the experimental conditions require many small expectations, perhaps less than unity. In this case it may be possible to obtain, either directly by an exhaustive analysis of the possibilities or by simulation, a null distribution of the criterion, Q, and even some distributions under H_1.

(ix) *Infinite Values of Q.* The possibility of infinite values of Q was mentioned by Pearl (1906) as a criticism of the Pearson methods. Pearson and Heron (1913) gave what must be the final answer. The null hypothesis must be rejected if Q is infinite. Infinite values of Q correspond to a positive probability under the alternative hypothesis of an event which has zero probability under the null hypothesis. One such observation is sufficient to disprove the null hypothesis.

8. SMALL VALUES OF χ^2

From the earliest times, there have been misapprehensions as to the distribution under a true null hypothesis of the Pearson χ^2, which we again write as Q. Authors have often appeared to expect a satisfactory hypothesis to correspond with a high value of the probability,

$$(8.1) \qquad \mathcal{P}(Q) = \int_Q^\infty \psi(q)\,\mathrm{d}q,$$

where $\psi(q)$ is the density function of χ^2 with the appropriate number of degrees of freedom. However, $\mathcal{P}(Q)$ under H_0 is rectangularly distributed on the unit interval, so that when H_0 is true $\mathcal{P}(Q)$ may be expected to take any value in the unit interval. Under certain alternative hypotheses, Q may have the distribution of the non-central χ^2 and then it will be appropriate to choose the critical region as $\{Q : Q > \chi_\alpha^2\}$. However, there are important cases where the alternative hypothesis may well specify low values of Q. Fisher, Thornton and MacKenzie (1922), who made (8.1) the basis of a control chart for the counting of replicate plates of bacterial colonies, were among the first to suggest that a low value of Q might also cast doubts on the null hypothesis. There are physical reasons in some instances for regarding low values of Q as abnormal. Thus the growth of some colonies may inhibit the growth of other bacterial colonies and they may remain unobservable with the usual techniques. In blood counting, there is a "crowding effect" which diminishes the variance between the counts on adjacent small areas of the haemocytometer.

A similar situation may arise with the distribution of the continuous χ^2 variable. For example let us suppose that Z_i is a set of metrical observations, so that we have a set of n independently distributed and standardized normal variables, $Z_1, Z_2, \ldots, Z_n$. If Z_1 were to take a small value, it would occasion no surprise since a value of Z in the neighbourhood of zero is the most likely event under the null hypothesis. However, if each of $Z_1, Z_2, \ldots, Z_n$ were to have a small value, then the product of the independent probabilities, $\mathcal{P}\{|Z_i| \leqslant \epsilon\}$, might be quite small if ϵ is some small positive quantity. Similarly the probability that the sum of squares, $\sum Z_i^2$, should be less than a small positive quantity, η, is also small. In either case, the cumulative probability of obtaining a value greater than the observed value would be close to unity. It would, for example, be a cause for concern if some observer gave results with standard errors very much smaller than might have been anticipated with his equipment.

Example (*i*) *Diminished variance.* In blood counting we may have in the haemocytometer chamber a distribution of red cells which approximates closely to hypothesis, so that the number of cells per unit volume is a Poisson variable, and the numbers in disjunct regions are mutually independent. If counts could be made at this time, the test criterion of consistency, $Q = \sum (Z_i - \bar{Z})^2/\bar{Z}$, has a distribution close to the theoretical. At higher densities this will not be true if the counts are made after the cells have settled on the haemocytometer grid. Q has an expectation less than the degrees of freedom, $n - 1$.

Q may be small because of the introduction of subjective errors.

Example (*ii*) *Diminished variance due to subjective choice of haemocytometer squares.* Some technicians may observe a few typical squares on a haemocytometer chamber and then proceed to count only on groups of squares which have about the same density of cells. A diminished variance results but also a quite unreliable mean. It is difficult to avoid the conclusion that even grosser forms of sophistication or correction of results occur.

Example (*iii*) *Sophistication of parallel counts of bacterial density.* Suppose γ is some integer. Then it is not unusual to find sets of counts reported on parallel estimations of bacterial density of the form $(\gamma, \gamma, \ldots, \gamma)$. There is often an extremely small probability of such an event. In fact, the probability of such a distribution, conditional on the total of $n\gamma$ for the n counts, is given by a term of the symmetrical multinomial distribution.

$$8.2) \qquad \mathscr{P}\{(\gamma, \gamma, \ldots, \gamma) \mid n\gamma, \text{ Poisson}\} = n^{-n\gamma}(n\gamma)!/(\gamma!)^n$$

With $\gamma = 14$ and $n = 5$, the probability is found approximately to equal $(140\pi)^{-2}52^{2\cdot5}$ or about 1 in 12×10^6, with the aid of the Stirling approximation.

Spuriously low values of Q can be introduced by the statistician using the corrected χ^2 or $-2 \log_e P$ transformations to combine the probabilities from different experiments; for an example of this use of a test function incorrectly applied in discrete distributions see Exercise VII.32.

In summary, low values of Q can be due to chance, to physical conditions, to subjective error or sophistication and to faulty statistical analysis.

9. HIDDEN PARAMETERS

An unsolved problem concerns the choice of distribution to be fitted from some class of distribution. If the class is, say, the class of normal distributions $N(\alpha, \sigma^2)$ where the mean α, and variance, σ^2, are to be fitted from the data, then the solution is well known. The class of distributions might be the Γ-variables; here again the theory is well known since there is a parameter to be estimated. Somewhat less definite is the solution to the problem suggested by the discussion of Student (1919) in reply to criticism by L. Whitaker (1914) of Student (1907). If it is uncertain *a priori* as to whether a binomial, a negative binomial or a Poisson distribution should be fitted, how many parameters should be considered as having been fitted for the Poisson distribution? As Student (1919) points out, if the variance is greater than the mean the negative binomial is possibly the appropriate distribution; if the variance is less than the mean the ordinary binomial will be appropriate; whereas if the mean equals the variance, the Poisson distribution will be appropriate. Now the Poisson distribution will be closely approximated by binomial distributions from either side as $n \to \infty$ and $np \to \lambda$. There will obviously be a discontinuity in the way we treat the degrees of freedom. No such problem arises if we agree beforehand to fit a distribution of given type. Similarly, we may have a choice of fitting a Γ-distribution, a normal distribution, another Pearson type and so on. How should this choice be allowed for? Similar unsolved problems arise in regression analysis where the choice of a logistic, the cumulative normal curve, the cumulative Γ-distribution and so on may be made. Possibly, the best solution is to make an opinion on what curve or distribution should be fitted by a consideration of collateral information *e.g.* the negative binomial in the decay of a radio-active source.

10. χ^2 AND THE SAMPLE SIZE

Let us consider the fitting of the normal curve to empirical data, for example, the heights of men. Satisfactory fits, as measured by χ^2, have been reported in the literature for sample sizes of the order of 100,000. Yet is it evident that if the sample size were sufficiently large the observed χ^2 would be expected to be very large and the null hypothesis used in the test of fit, namely, that the class frequencies are such as would be generated by a normal curve with the same mean and variance as the sample, would be rejected. The non-significant χ^2 at moderate sample size must be interpreted as meaning that at this sample size, it is possible without loss of information to regard the data as having been generated by sampling from a normal population; in particular, the Pearson ϕ^2 comparing the unknown distribution with the normal distribution is at most of order, N^{-1}, when the comparison is made with the given grouping or partition of the space. The null hypothesis in the example of heights is not an exact one. It has been said by some critics of χ^2 that χ^2 will always be large if the sample size is large. This need not be the case; for example, the hypothesis that almost every number on the unit interval is "normal" in the scale of two or ten or in any other scale is exact; where large computations of irrational numbers such as e or π are available, the "normality" hypothesis yields χ^2 in accordance with the theory. It may be noted in passing that many of the published tests of this hypothesis have been marred by the use of χ^2's which are not mutually independent. A χ^2 has often been computed of the form, $10N^{-1} \sum_{i=1}^{10} (a_i - N/10)^2$, for $N = N_1$. The results are retained until

$N_2 - N_1$ further digits of the irrational number have been examined. The cumulated class totals are again tested with $N = N_2$. There is, of course, a correlation of $N_1^{\frac{1}{2}}/N_2^{\frac{1}{2}}$ between these successive χ^2's.

11. SMALL CLASS FREQUENCIES

The distribution of the Pearson χ^2 has been computed on the basis that no class expectation is small. It used to be feared by authors that the test would not be accurate if carried out when any class expectation was less than 10. These fears have proved unjustified by the investigations of authors using simulation methods or complete enumeration of all possibilities in given discrete distributions. Some general conclusions can be made—it is probably desirable not to have any expectation less than unity; with several degrees of freedom, for class frequencies of 5 or more, the distributions of the Pearson χ^2 approximate satisfactorily to the asymptotic or theoretical χ^2 distributions. If there are a number of classes, perhaps a third or a quarter of them can have expectations in the interval, 1 to 5, without causing serious departures of the distribution of the Pearson χ^2 from the theoretical. These considerations are of importance in testing the null hypothesis that sampling is from a Poisson distribution, for the parameter of non-centrality of the χ^2 test may be greatly diminished if too much "pooling" is carried out, for example, in the positive tail of the distribution, when the parameter under H_1 is greater than the parameter under H_0.

12. THE PARTITION OF χ^2

The objection, that χ^2 is an overall or omnibus test, has been partly answered by K. Pearson in his Editorial (1917). His discussion may be amplified with the aid of the theory of the partition of χ^2 by orthogonal transformations. Suppose that N observations are given on a set of the n indicator variables of the multinomial distribution. Only $(n - 1)$ of these indicator variables are linearly independent and so if a subset of $(n - 1)$ of them is chosen, the remaining variable is determined by their values. Equivalently, any set of $(n - 1)$ orthonormal functions, $\{U^{*(i)}\}$, may be considered and their standardized sums,

(12.1) $$U^{(i)} = N^{-\frac{1}{2}}\mathscr{S}U_j^{*(i)}, \quad j = 1(1)N; \quad i = 1(1)(n - 1).$$

The usual χ^2 is $\sum U^{(i)^2}$ and is invariant for any choice of the set $\{U^{*(i)}\}$. Thus we have given examples where the orthonormal sets are the orthonormalized indicator variables, the Walsh functions, the orthogonal polynomials on n points with equal weights, the Kravčuk polynomials and so on. For example, the "smooth test" of goodness of fit as applied to discrete distributions by Barton (1955) becomes identical with the χ^2 test when $\Psi_{n-1}^{'2}$ is used.

H_1 may specify particular changes, in which case the $U^{(i)}$ may be given special forms or alternatively, if the $U^{(i)}$ are a fixed set, orthonormal linear forms in them may be chosen so that the parameter of non-centrality is as large as possible for a few degrees of freedom as possible. In the two following examples it is supposed that H_0 specifies the rectangular density function, $f(u) = 1$ in $(0, 1]$, $f(u) = 0$ elsewhere.

Example (i) Let H_1 specify the density, $g(u) = 1 + \alpha$, on $(0, \frac{1}{2}]$; $g(u) = 1 - \alpha$, on $(\frac{1}{2}, 1]$; $g(u) = 0$ elsewhere. Suppose that N observations are available on the indicator variables for n classes, n even. Q has a parameter of centrality equal to $N\alpha^2$ and $(n - 1)$ degrees of freedom. Supposing that R is the first Rademacher function, $R = +1$ if $u < \frac{1}{2}$, $= -1$ if $u \geqslant \frac{1}{2}$, and that a is the number of observations greater than $\frac{1}{2}$, the standardized sum, $N^{-\frac{1}{2}}\mathscr{S}R = N^{-\frac{1}{2}}(N - 2a)$ is asymptotically normal, $N(0, 1)$, under H_0 and its square is asymptotically non-central χ^2 with 1 d.f. and parameter of non-centrality equal to $N\alpha^2$. If it is desired merely to test $H_1: \alpha \neq 0$, against H_0, clearly this test function is preferable to the Pearson χ^2 which contains it as a component.

Example (ii) Let H_1 specify a density function, $g(u) = 1 + \alpha(u - \frac{1}{2})\sqrt{12}$, and suppose the unit interval is partitioned into n equal intervals. For the original distribution, the value of ϕ^2 is α^2 and for the condensed distribution ϕ^2 is $\alpha^2(1 - n^{-2})$. The variable taking values, $V = -\frac{1}{2} + (2k + 1)/(2n)$, $k = 0, 1, 2, \ldots, n - 1$ is associated with a ϕ^2 of $\alpha^2(1 - n^{-2})$ and no variable defined on the condensed distribution and orthogonal to V is associated with a non-zero ϕ^2. The standardized sum, $N^{-\frac{1}{2}}\mathscr{S}V/\sqrt{(\text{var } V)}$, is thus asymptotically normal $(0, 1)$ under H_0 and its square is χ^2 with 1 d.f. Under H_1 it is a non-central normal variable.

Example (iii) Variations can be made on Example (i). Let H_1 specify $g(u) = 1 + \alpha R_1 + \beta R_2 + \gamma R_1 R_2$, where R_1 and R_2 are the first and second Rademacher functions. The squares of the standardized sums, $N^{-\frac{1}{2}}\mathscr{S}R_1$, $N^{-\frac{1}{2}}\mathscr{S}R_2$ and $N^{-\frac{1}{2}}\mathscr{S}R_1R_2$, are associated with parameters of non-centrality, $N\alpha^2$, $N\beta^2$ and $N\gamma^2$ respectively. If the unit interval is subdivided into n intervals, $n = 4k$, then the subdivision corresponding to $n = 4$ is a sufficient partition. If the hypothesis, $\alpha = 0$, $\beta = 0$, $\gamma = 0$ is to be tested, the Pearson χ^2 with 3 d.f. has parameter of non-centrality $N(\alpha^2 + \beta^2 + \gamma^2)$ and is the best test function if nothing further is specified. It may be, however, desirable to test individually whether α, β or γ is zero and this is readily done with the use of the standardized sums, $N^{-\frac{1}{2}}\mathscr{S}R_1$, $N^{-\frac{1}{2}}\mathscr{S}R_2$, $N^{-\frac{1}{2}}\mathscr{S}R_1R_2$. This example can be readily given a genetic interpretation with individuals from crosses of the form, $AaBb \times aabb$, being the observations. α could then be the parameter of departure of the population from equality of the classes Aa and

aa in the F_1 generation, β the similar parameter for the B-classification and γ the interaction parameter between the A-factor and the B-factor.

Examples of the use of the partition of χ^2 are frequent in our later chapters on contingency tables.

13. MISCLASSIFICATION AND MISSING VALUES

In agricultural experimentation, it has been found necessary to make provision for the loss of observations by accident in some of the experimental units. It has been held necessary by some to make corresponding provision in the multinomial distribution and in the contingency tables. For example, in the multinomial, the observations likely to fall into class i may fall into classes $i \neq j$ with some probabilities, s_{ij}. The observable distribution will then depend on $\{p_{0i}\}$ specified by H_0, $\{p_{1i}\}$ specified by H_1 if the $s_{ij} = 0$. If the s_{ij} are not all zero but are known, then modified sets will be available, $\{p'_{0i}\}$ and $\{p'_{1i}\}$, say. If the s_{ij} are not zero and are not known, then the effects of misclassification on the test are a matter for speculation. As a rule misclassifications reduce the power of the tests (see Mote and Anderson, 1965). In contingency tables there may be confusion as to which of two cells a certain cell number belongs. For example, two entries in a contingency table are given and it is not known which belongs to the cell at position $(1, 1)$ and which belongs to $(1, 2)$. Watson (1956) has given a solution to this problem which is equivalent to partitioning χ^2 of the contingency table so that a_{11} and a_{12} appear in only one comparison with opposite sign and the contribution from this comparison is deleted from the sum of squares. Consequently the degrees of freedom are reduced by unity. χ^2 is computed as

$$(a_{11} + a_{12})^2/(\bar{a}_{11} + \bar{a}_{12}) + \sum_2^m a_{i1}^2/\bar{a}_{i1}$$

$$+ \sum_2^m a_{i2}^2/\bar{a}_{i2} - (a_{.1} + a_{.2} - a_{11} - a_{12})^2/(a_{.1} + a_{.2} - \bar{a}_{11} - \bar{a}_{12})$$

$$+ \sum_{i=1}^m \sum_{j=3}^n a_{ij}^2/\bar{a}_{ij} - N.$$

The fourth expression is introduced because pooling two cells does not delete the whole of the χ^2 corresponding to one degree of freedom. A simpler prescription is to pool the entries in the two first columns and the χ^2 of this table has $(n - 2)(m - 1)$ degrees of freedom. A homogeneity test can then be carried out on the entries $a_{21}, a_{22}, a_{31}, a_{32} \ldots a_{m1}, a_{m2}$ treated as an $(m - 1) \times 2$ contingency table. The two χ^2's may be summed to give a total χ^2 of $(n - 2)(m - 1) + (m - 2) = (n - 1)(m - 1) - 1$ degrees of freedom. We leave further study of the problem to the exercises and give several references to the problem in the bibliography.

14. THE RECONCILIATION OF χ^2

Often it is not realised that two apparently different tests are either identical or differ only slightly. As an example of this latter, the tests of a binomial variable by the use of median probability or the standardized normal deviate as test functions give substantially equivalent results. Similarly, the corrected normal deviate and the exact test yield substantially equivalent results. In the fourfold table under Model 2, the usual χ^2 with fixed marginal totals can readily be shown equal to the test of the proportions in the two rows (whose totals have been determined) and this was used by Fisher (1922a) to establish that the χ^2 of the fourfold table had but one degree of freedom.

We have earlier shown the equivalence of many likelihood tests and the Pearson χ^2 test. Indeed, the partition of the logarithm of the likelihood has a sound theoretical basis only when the identification is made.

If the χ^2 is considered in the form $\sum_{i=1}^{n-1} [N^{-\frac{1}{2}}\mathscr{S}x^{(i)}]^2$ in the multinomial, and if $\{x^{(i)}\}$ together with the constant function is an orthonormal basis on the n cells of the multinomial, then all such tests are the same function, whether they be orthogonalised indicator variables or the set of orthogonal polynomials. Similar considerations apply to contingency tables in which in place of $\{x^{(i)}\}$, we have a product set $\{x^{(i)}\} \times \{y^{(i)}\}$.

15. MISCELLANEOUS INFERENCE

This Section has been included for bibliographical convenience only and contains references to some topics marginal to the purpose of the monograph or insufficiently treated in it. See the *Index to the Bibliography*.

EXERCISES AND COMPLEMENTS

1. Suppose that H_0 and H_1 specify class probabilities, f_i and g_i respectively, $i = 1, 2, \ldots, n$. Use the law of large numbers to deduce the asymptotic values of the observed proportions, a_i/N, when N independent observations are made and a_i observations fall into the ith class, supposing that n is fixed and $n = o(N)$, N being indefinitely large. Hence verify that the Pearson χ^2, Q, is consistent in the sense that if $g_i \neq f_i$ for at least one class, Q will increase without limit as $N \to \infty$ and hence will reject H_0.

2. Use Exercise 1, to obtain

$$Q/N \simeq \phi^2, \qquad \phi^2 = \sum g_i^2/f_i - 1.$$

Obtain the asymptotic distribution of Q under H_1, by considering the expectations and covariance matrix of the variables $a_i/(Nf_i)^{\frac{1}{2}}$. (Eisenhart, 1938; Patnaik, 1949.)

3. Eleven authors have given their goodness of fit results in the form of a percentage averaged over the n classes of a multinomial distribution,

$$100n^{-1} \sum |\text{observed} - \text{expected}|/(\text{expected})$$

Pearson (1895) cites averages of 9%, 13·5%, 7 to 15% and so on. Discuss how this criterion is likely to vary with N. How would this average behave under H_0 and H_1? How would it be affected by the choice of groupings (partitions of the space) with small values of some f_i or g_i.

4. For the point binomial, $(\frac{1}{2} + \frac{1}{2})^n$, the expected frequencies are given by $p_r = \binom{n}{r} 2^{-n}$. Plot them as a frequency polygon. The slope of the line joining the ordinates at the points, r and $r + 1$, is $p_{r+1} - p_r$. Set $\sigma^2 = (n + 1)\frac{1}{2}\frac{1}{2}$. Prove that

$$\frac{\text{slope of polygon}}{\text{mean ordinate}} = -\frac{2 \times \text{mean abscissa}}{2\sigma^2}$$

where the mean abscissa is defined as $\frac{1}{2}(2r + 1) - \frac{1}{2}n$.

$\hat{\sigma}^2 = (n + 1)/4$ is preferable to $\hat{\sigma}^2 = n/4$ as an estimator of the variance of the binomial if it is to be used to fit the normal curve to the symmetric binomial distribution. (Pearson, 1895, Feller, 1945.)

5. Let $\mathbf{X}$ be a normal vector with n mutually independent, standardized components. Show that the region $\chi^2 = \mathbf{X}^T\mathbf{X} < t^2$ is the same as the region such that $\mathbf{a}^T\mathbf{X} < t$ for every unit vector $\mathbf{a}$.

6. Let $X_1, X_2, \ldots, X_n$ be mutually independent and, under H_0, normal $N(0, 1)$. Let H_1 specify that $\mathscr{E}X_i = c_i$, where $\sum c_i \neq 0$. Then all positive sets of c_i^2 can be tested by allowing $\mathbf{a}$, a unit vector to take all possible values. $\mathbf{a}^T\mathbf{X}$ can be tested as a standardized normal variable if $\mathbf{a}$ be chosen before the observations have been made. If $\mathbf{a}$ be allowed to take any arbitrary value show that $(\mathbf{a}^T\mathbf{X})^2 \leqslant \sum X_i^2$. Hence justify the test of H_0 that $\sum c_i^2 = 0$ by the test function, $\sum X_i^2$ and the choice of the critical region $\sum X_i^2 > \chi_{0.50}^2$, the 50% point of χ^2 with n degrees of freedom. (Bravais, 1846; Czuber, 1898; Pearson and Lee, 1908.)

7. The χ^2 variable can be considered as the square of a generalized probable error. It can be displayed as $\chi^2 = \mathbf{Y}^T\mathbf{V}^{-1}\mathbf{Y}$, where the Y's are jointly normal with covariance matrix, $\mathbf{V}$. "The ellipsoid in n-fold space can be reduced to a sphere by proper squeezes in the directions of its proper axes, and accordingly its volume can be found from that of a sphere in n-fold space". Verify that the square of Pearson's probable errors on page 62 of Pearson and Lee (1908) are the 50% points of χ^2.

8. Verify that Pearson's use of the generalized probable error and the use of similar measures for $n = 2$ and $n = 3$ by Bravais and Czuber are equivalent to the test of the departure of the means of $X_1, X_2, \ldots, X_n$ from zero, when H_0 specifies that the X_i form a mutually independent set of normal (0, 1) variables.

9. Let $\xi_1, \xi_2, \ldots, \xi_n$ be a set of random variables with zero means. Suppose that a subset of m linearly independent variables can be found such that the covariance matrix is positive definite but that no subset of $(m + 1)$ variables has a positive definite covariance matrix. Suppose further that the linearly independent variables

are jointly normal. Then if $\epsilon = A\xi$, and $\eta = B\xi$ are two $m \times 1$ vectors such that $\mathscr{E}\epsilon\epsilon^T = F$ and $\mathscr{E}\eta\eta^T = V$ are positive definite, $\epsilon^T F^{-1}\epsilon \equiv \eta^T V^{-1}\eta$ and is distributed as χ^2 with m degrees of freedom (Pearson, 1900a; Pearson and Lee, 1908; Sheppard, 1929.)

10. Prove that the Pearson χ^2 test of the multinomial distribution is unbiased and consistent. (Kendall and Stuart, 1961; Mann and Wald, 1942; Neyman, 1949.)

11. Let $\mathbf{a}$ be a fixed unit vector. Let X be a vector of mutually independent normal variables with unit variance. Let H_0 specify that the means are all zeroes and H_1 that the means are λa_i, $\lambda > 0$. Determine by the use of the Neyman–Pearson theory, the most powerful test of H_0 against H_1. [The test function will be of the form $\mathbf{a}^T X$ with parameter of non-centrality, λ.] If H_1 specifies that $\lambda \neq 0$, then the test-function will be $\mathbf{a}^T X$ and the critical region $|\mathbf{a}^T X| > k$ or $(\mathbf{a}^T X)^2 > k^2$. This test function is χ^2 with 1 d.f. Prove that any other χ^2 variable cannot have a parameter of non-centrality greater than λ.

12. The computations of various authors can be examined to see whether the χ^2 approximation yields reasonable approximations to the size of the test. The general conclusion is that the approximations are good. (Bennett, 1962; Cochran, 1936, 1942; el Shanawany, 1936; Kathirgamatamby, 1953; Lancaster, 1952, 1961a; Lancaster and Brown, 1965; Neyman and Pearson, 1931; Pearson, 1947; Plackett, 1964; Sukhatme, 1937a,b, 1938; Uppuluri and Bowman, 1966; Wise, 1963, 1964; Yates, 1934.)

13. Indicate how the size of the Pearson χ^2 in the multinomial will be affected by

(i) symmetries such as $p_i = p_j$ for some pairs (i, j).
(ii) the possibility that in the symmetrical multinomial, $\sum a_i^2 = \sum b_i^2$ when the $\{a_i\}$ and $\{b_i\}$ correspond to distinct partitions of N, the sample size.
(iii) the possibilities of identities between the χ^2 at distinct points, $(i_1, i_2, \ldots, i_n)$ even in the absence of symmetries among the class probabilities.
(iv) the choice of sets of $\{p_i\}$ for computing convenience. The choice of $\{0 \cdot 2, 0 \cdot 3, 0 \cdot 5\}$ leads to the possibility that certain distinct points will have the same values of χ^2.

Note that, since the logarithms of these rational p_i are not linearly dependent with respect to the positive integers, for the same set of p_i, the likelihood ratios at all points may be distinct. Give a general example where this may not be so. [*Hint.* Find positive p_i such that,

$$a \log p_1 + b \log p_2 + c \log p_3 = 0$$

$$d \log p_1 + e \log p_2 + f \log p_3 = 0, \quad p_1 + p_2 + p_3 = 1,$$

for integers a, b, c, d, e, f.]

14. In the multinomial distribution with k classes and estimated parameters, write

$$(*) \quad f(p_1, p_2, \ldots, p_{k-1}) = (\tfrac{1}{2}n\pi)^{-\frac{1}{2}(k-1)} \prod_1^k p_i^{-\frac{1}{2}}$$

$$\times \exp -\tfrac{1}{2}N \sum_{i,j} (\delta_{ij}p_i^{-1} + p_k^{-1})(\hat{p}_i - p_i)(\hat{p}_j - p_j)$$

and the likelihood ratio,

(**)
$$\lambda = \frac{N^N \prod_j n_j!}{\prod_i n_i^{n_i} N!} \frac{N!}{\prod_j n_j!} \prod_i p_i^{n_i}.$$

Verify that if the variables in (*) are changed from $\{p_i\}$ to $\{n_i\}$, then the function will remain unchanged except for a factor $N^{\frac{1}{2}(k-1)}$, which disappears because $N\,dp_i = dn_i$. An application of Stirling's approximation to the second factor of (**) shows the equivalence of (*) and (**). (See p. 215, Mood, 1950; Wilks, 1938.)

The three following examples are designed to show that it is immaterial for the usual applications of χ^2 whether the likelihood ratio test be devised and then the normal approximation be applied to obtain a χ^2 distribution or the normal approximation be first applied and then tests of significance be devised with the aid of likelihood ratio theory.

15. Let H_0 specify p_0 and H_1 specify p_1, $p_0 + q_0 = 1$, $p_1 + q_1 = 1$, $p_0 < p_1$. Suppose now that an observation of m successes has been made from a binomial population with index, N. Derive the likelihood ratio test of H_1 against H_0 and a critical region of the form, $\{x : x > k\}$. Show that the asymptotic distribution of the variable, $X = (m - Np_0)/(Np_0q_0)^{\frac{1}{2}}$, is $N(0, 1)$ under H_0 and

$$N(N^{\frac{1}{2}}(p_1 - p_0), (p_0q_0)^{-1}p_1q_1)$$

under H_1.

Proceed alternatively and consider X as asymptotically normal under H_0 and H_1. Derive by likelihood ratio considerations the critical region for X.

16. Treat the Poisson distribution in the manner suggested by Exercise 15.

17. Let H_0 specify a hypergeometric distribution generated by sampling from a 2×2 distribution with independently distributed marginal variables. Let H_1 specify that $\psi = p_{11}p_{22}/(p_{12}p_{21})$ is not unity. Use the likelihood ratio method to determine the best test of H_0 against H_1. Obtain the asymptotic distribution and parameter of non-centrality of the approximating normal variable. Also give the normal approximating distributions under H_0 and H_1 and derive the most powerful test criterion.

18. Re-examine the approximations given by the parameter of non-centrality in the following case. H_0 specifies p_{0i} and H_1, p_{1i}. Each $|p_{0i} - p_{1i}|$ is small. What is the mean value of χ^2 under H_0 and H_1 using normal approximation theory?

19. Berkson states that there is never any valid reason for the rejection of the null hypothesis except on the willingness to embrace an alternative. Consider this remark in relation to (i) the test of counting experiments for consistency, where there is reason to believe that large values of χ^2 are associated with defective technique and small values with the selection or sophistication of the data; (ii) the observation of an unusual deal in bridge, one hand being dealt all thirteen spades. (Berkson, 1938; Fry, 1938.)

20. Suppose that a normal curve is fitted to a frequency table of the heights of men. With sample sizes of the order of tens of thousands, acceptable values of χ^2 are usually obtained. Yet it is certain that if more extended observations could be

made that the values of χ^2 would be high. [The hypothesis of normality is only acceptable as an approximation when the sample size is not too large. The ϕ^2 of the observed distribution with respect to the condensed normal is necessarily positive.] By the consideration of counting experiments or the counts of the frequencies of digits in randomly chosen real numbers, show that χ^2 need not tend to increase without limit as the number of observations is increased without limit. In this last case, note that the hypothesis is exact. (page 526, Berkson, 1938.)

21. Berkson (1938) found that although the variance of the red cell count on small squares of the haemocytometer was diminished below the mean, the over-all χ^2 test of goodness of fit failed to reveal a significant departure from the null hypothesis that Poisson conditions held. Comment on Berkson's remark, taking account of the following:

(i) the pooling in the tails of the distribution diminishes the expected value of χ^2 by more than the number of degrees of freedom lost, since a plausible alternative hypothesis is not likely to assign the same ratio between the probabilities under null and alternative hypotheses in all the classes pooled.
(ii) the variance may be tested by treating $\sum (x - \bar{x})^2/\bar{x}$ as though it were χ^2 with $(n - 1)$ degrees of freedom.
(iii) the observed value of the variance may be tested as a component of the overall χ^2.

22. 134 sets of three parallel bacterial counts have been carried out and the sets classified by the observed value of the χ^2 as under.

Value of χ^2	Frequency	Value of χ^2	Frequency
0–1	43	5–6	4
1–2	30	6–7	4
2–3	24	7–8	2
3–4	12	8+	5
4–5	10	Total	134

Compute the expected frequency for each class with the aid of the tables of Pearson (1900a) or the *Biometrika Tables*. (Fisher, Thornton and MacKenzie, 1922.)

23. Compare the following statement with the justification of statistical control of counting experiments of Exercise 22. "In recent times one often repeated exposition of the tests of significance, by J. Neyman, a writer not closely associated with the development of these tests, seems liable to lead mathematical readers astray, through laying down axiomatically, what is not agreed or generally true, that the level of significance must be equal to the frequency with which the hypothesis is rejected in repeated sampling of any fixed population allowed by hypothesis. This intrusive axiom, which is foreign to the reasonings on which the tests of significance were in fact based, seems to be a real bar to progress." (Fisher, 1945a.)

24. Brownlee (1924) obtained 32 samples of 256 tosses of a coin, so that his expected values were in the ratios 1, 4, 6, 4, 1. He found that the χ^2 with 4 d.f.

appeared to be of reasonable size. If he fitted a normal curve calculated with the use of Sheppard's correction for the variance, unduly high values of χ^2 were obtained. Fisher (1924) points out that the method of estimation is inconsistent since the fractions in the 5 classes are

Number of Heads	0 or 4	1 or 3	2
Binomial	0·0625	0·25	0·375
Fitted normal	0·0668072	0·2417303	0·3829250

and hence as the sample size increases χ^2 tends to infinity.

25. Examine the Exercises of Chapter 8 to obtain examples of high Q due to inefficient fitting. (Brownlee, 1924; Fisher, 1924; Koshal, 1933, 1935, 1939; Koshal and Turner, 1930.)

26. Verify by a consideration of tables of the theoretical χ^2 that incorrect calculation of the degrees of freedom are especially important in the case of the fourfold table, where controversy once centred on whether the d.f. were 1 or 3. (See Fisher, 1922a for the fourfold table.)

27. In the symmetrical multinomial distribution with n cells, is it proper to select the cell with the highest count and treat this count as a binomial variable? Give a brief argument to show that for α not greater than, say, 0·05, and $n = 4$ or 5, a modification can be made as follows: let m_α be the value such that $\mathscr{P}(m > m_\alpha) = \alpha/n$. Then the test, $\{Y - Nn^{-1}\}/(N(n-1)n^{-2})^{\frac{1}{2}} > m_\alpha$ has size approximately equal to α if Y is the largest entry in any of the n cells.

28. In sets of parallel counts assembled either by simulation or in experiments carried out under favorable conditions, there is little correlation between the mean and the variance. The affixing of very small standard errors to means obtained in counts with low variances is thus an incorrect use of the usual confidence limits technique. (Lancaster 1950b, 1952, 1953b.)

29. In a contingency table show that NR_1^2 is stochastically larger than χ^2 with $m + n - 3$ degrees of freedom by the following device. Let E be the $(m-1) \times (n-1)$ matrix of standardized asymptotically normal variables, make an orthogonal transformation,

$$G = RE, \quad \text{where} \quad r_{1j} = e_{j1}\left(\sum_1^{m-1} e_{j1}^2\right)^{-\frac{1}{2}}$$

The leading element of this matrix is the square root of a χ^2 variable with $(m-1)$ degrees of freedom. Now make an orthogonal transformation

$$H = GS, \quad \text{where} \quad s_{j1} = g_{1j}\left(\sum_1^{n-1} g_{1i}^2\right)^{-\frac{1}{2}}$$

h_{11}^2 is χ^2 with $n + m - 3$ d.f. It is, however, less than NR_1^2 with probability one. (Kendall and Stuart, 1961; Lancaster, 1963a; Wijsman, 1958; Williams, 1952.)

30. (i) Compute the unconditional probability of a difference between two Poisson variables each with parameter, λ. (Irwin, 1937.)

(ii) Show that for large λ, the difference between Poisson variables is asymptotically normal with zero mean and variance, 2λ, and can be used as a test function if λ be given.

(iii) Show that if X_1 and X_2 are independently distributed Poisson variables with parameter, λ, then the distribution of $X_1 - X_2$ for given values N of their sum is that of twice a variable, namely $2X_1 - N = 2(X_1 - \frac{1}{2}N)$, from a binomial distribution with $p = \frac{1}{2}$ and parameter, N.

Observe that in this last case, the test of "homogeneity" of X_1 and X_2 can be made purely combinatorial without reference to an unknown parameter, λ.

31. From Geissler's data on the sex ratio of sibships (families), sibships of eight can be constructed from his sibships of nine by subtracting the most recent child, the sex of whom is given. Geissler gave a value $\hat{p} = 0.5147676$ for the probability of a child born being a male. With this value of $\hat{p}$, the expected number of sibships can be computed.

No. of Males	Observed No. of Sibships	Expected No. of Sibships
8	264	189·78
7	1,655	1,431·28
6	4,948	4,722·03
5	8,498	8,902·20
4	10,263	10,489·31
3	7,603	7,909·99
2	3,951	3,728·09
1	1,152	1,004·03
0	161	118·30
Total	38,495	38,495·01

Calculate (observed-expected) for each size of sibship. Compare this result with Table 11 of R. A. Fisher's *Statistical Methods*. Compute χ^2. (Gini, 1951; Lancaster, 1950d.)

32. In the previous example, Geissler's estimate $\hat{p}$ was not independent of the table to which it was fitted. In fact, his estimate in M births includes the N births in the table. Verify that the number of males in the table is not a binomial, but a hypergeometric variable if $\hat{p}$ is so estimated. This has the effect of reducing the variance of the corresponding component of χ^2, namely $(m - N\hat{p})(N\hat{p}\hat{q})^{-\frac{1}{2}}$, since the variance of male births would be $N\hat{p}N\hat{q}(M - N)(N)/N^3$ or $\hat{p}\hat{q}(M - N)$ rather than $M\hat{p}\hat{q}$. (See Exercise XI.2.) If N is not a large part of M little disturbance will be caused to the distribution of the Q.

33. How would identical twinning disturb the distribution of sex ratios in families? (Discussion on Geissler's data in R. A. Fisher's *Statistical Methods*.)

34. "If a curve is a good fit to a sample, to the same fineness of grouping it may be used to describe other samples from the same general population. If it is a bad fit, then this curve cannot serve to the same fineness of grouping to describe other samples from the same population.

We thus seem in a position to determine whether a given form of frequency curve will effectively describe the samples drawn from a given population to a certain degree of fineness of grouping.

If it serves to this degree, it will serve for all rougher groupings, but it does not follow that it will suffice for still finer groupings. Nor again does it appear to follow that if the number in the sample be largely increased the same curve will still be a good fit. Roughly the χ^2's of two samples appear to vary for the same grouping as the r total contents. Hence if a curve be a good fit for a large sample it will be good for a small one, but the converse is not true, and a larger sample may show that our theoretical frequency gives only an approximate law for samples of a certain size."

Comment on this statement with the aid of the partition of χ^2 and the asymptotic formula $\chi^2/N \simeq \phi^2$. (Pearson, 1900a.)

35. Let X be rectangularly distributed on the unit interval. Suppose a realised value of X is expanded as a decimal expansion, $0 \cdot X_1 X_2 X_3 \ldots X_n \ldots$. Prove that any finite set of the X_i is a mutually independent set and that each X_i is uniformly distributed on the 10 points, $0, 1, 2, \ldots, 9$. Prove that any subset of $\{X_{2i-1} + X_{2i}\}$, $i = 1, 2, 3, \ldots$, has also the property of independence. Prove that each $Y = X_{2i-1} + X_{2i}$ takes values $0(1)18$ with probabilities $0 \cdot 01(0 \cdot 01)0 \cdot 10$ and that $\mathscr{P}(Y_i = s) = \mathscr{P}(Y_i = 18 - s)$. Prove also that Y_i mod 10 is uniformly distributed on the points $0(1)9$. The frequencies of the values of Y have been found to occur as follows $0, 24$; $1, 45$; $2, 78$; $2, 95$; $2, 112$; $5, 160$; $6, 150$; $7, 217$; $8, 220$; $9, 281$; $10, 215$; $11, 190$; $12, 172$; $13, 135$; $14, 135$; $15, 117$; $16, 77$; $17, 51$; $18, 26$. Test these frequencies against the hypothesis suggested. If the Y's are defined by $Y_i = X_i + X_{i+1}$, are the Y_i's mutually independent? Are the $Z_i = Y_i$ (mod 10) in this case mutually independent? Test whether the Z_i are uniformly distributed on the points $0(1)9$. (For the original data, see Pathria, 1961.)

36. In a breeding experiment, a cross $AaBbCcDdEe \times aabbccddee$ yielded classes with 0, 1, 2, 3, 4, 5 dominants respectively having frequencies 17, 81, 152, 180, 104 and 17. Is the fit to the observed distribution by a binomial, $551 (\frac{1}{2} + \frac{1}{2})^5$, satisfactory? Calculate χ^2 with expected numbers proportional to 1, 5, 10, 10, 5, 1. (Roberts, Dawson and Madden, 1939; Haldane, 1939a; Lancaster, 1965a.)

37. Test a similar hypothesis for a breeding experiment with crosses of the form $AaBbCc \times aabbcc$. Test the observed numbers having 0, 1, 2, 3 dominants equal to 181, 484, 570, 192 against the binomial hypothesis which assigns relative frequencies proportional to 1, 3, 3, 1. Calculate χ^2 as the test of hypothesis [data from Fisher *Statistical Methods*].

38. Examine the cited papers for examples, where the sample size is large, but the values of Q are moderate. (Pathria 1961, 1962, 1964.)

39. Comment on the following. "First, if according to Mendelian theory there must be zero frequency in any particular class, then the improbability of an occurrence in that class is zero. Secondly, the χ^2 test can be applied to any number whatever of Mendelian classes, to s classes individually and all the remainder, or in a particular case to *one* class and all the remainder." Justify the statements made by the partition of χ^2 and indicate how asymptotically independent sets of comparisons can be made. (Editorial (of K. Pearson), 1917; Pearson and Heron, 1913.)

40. Suppose that X and Y are jointly normal and standardized and that H_0 specifies independence and H_1 specifies a non-zero coefficient of correlation, ρ. Prove that the maximum parameter of non-centrality is $N\rho^2/(1 - \rho^2)$ for any χ^2 test. Suppose that the marginal distributions have been partitioned into m and n sets, respectively so that there are orthonormal functions, ξ and η, with correlations greater than $1 - \epsilon$ with X and Y respectively. Then under H_1, $\mathscr{E}\xi\eta = \rho + O(\epsilon^2)$. The parameter of non-centrality, with N independently made observations of the test function $(N^{-\frac{1}{2}}\mathscr{S}\xi\eta)^2$ is thus approximately $N\rho^2$ if ϵ be small. If ρ^2 be negligible this parameter of non-centrality is not greatly different from that of the χ^2 of the contingency table. The test function $N^{-\frac{1}{2}}\mathscr{S}\xi\eta$ is thus a more powerful test than the overall χ^2 of the hypothesis of independence.

41. Berkson (1938 and 1940) considers the variance of the blood count. $\sum (x - \bar{x})^2/\bar{x}$ is χ^2 with 399 d.f. It is possible to obtain this in the form,

(*) $$N^{-1}\{N\sum x^2f_x - (\sum xf_x)^2\}/(\text{variance}),$$

where

(**) $$\sum (x - \bar{x})^2 = \sum x^2f_x - (\sum xf_x)^2/N$$

the summation on the left being over counts, the summation on the right being over classes. The expression on the right being standardized in the expression above (**) can be considered as the contribution of the second Gram–Charlier polynomial to the overall χ^2 of the fitting of the Poisson distribution to the frequency distribution of the counts. The expression $\sum (x - \bar{x})^2/\bar{x}$ can be regarded as a standardized variance.

42. In counting experiments the alternative hypothesis may specify that the variance is high (defective technique) or low (crowding effect). (Berkson, 1938, 1940; Fisher, Thornton and MacKenzie, 1922; Lancaster, 1950b.)

43. An author gives a set of six parallel counts of 13 colonies per plate. Compute the probability that with a total of 78 colonies on 6 plates the distribution will be 13, 13, 13, 13, 13, 13. Comment on the low χ^2. (Bodmer, 1959; Camp, 1938 and 1940; Fisher, 1936; Fisher, Thornton and MacKenzie, 1922; Keynes, 1921.)

44. Smith, C. (1931) "Normal variations in erythrocyte and haemoglobin values in women", *Arch. Internal Med.*, **47**, 206–229, gives the results of training technicians. Technician A gave repeated counts as follows: 437, 444, 465, 435, 431, 442, 422, 417, 420, 417, 441, 427, 458, 431, 429, 483, 443, 412, 431 and 425. At a later date he returned parallel counts as follows: 432, 424, 440, 434, 434, 419, 413, 439, 428, 419, 418, 429, 413, 423, 423, 426, 426, 424, 422, 420. Show that the χ^2 for consistency with the multinomial hypothesis is 13·27 with 19 d.f. for the first set and in

his later count χ^2 was 2·54 with 19 d.f. Technician F gave counts as follows: 393, 393, 394, 390, 391, 392, 399, 392, 391, 393 and later 393, 394, 392, 393, 394, 393, 393, 393, 392, 394. His first χ^2 is 0·14 with 9 d.f., his second is 0·01 with 9 d.f. Comment on these values of χ^2.

45. Calculate ϕ^2 when H_0 and H_1 specify Poisson distributions with parameters, λ_0 and λ_1. With $\lambda_0 = 6$ and $\lambda_1 = 7$ evaluate numerically ϕ^2 when the classes of the positive tail are pooled (i.e. condensed) for counts, k to ∞, $k = 10, 11, 12, \ldots$. With 400 observations, what would be a resonable value of k when $\lambda_0 = 6$.

46. Show that the process of pooling two classes deletes a component χ^2 of 1 d.f. if parameters are given theoretically. Prove that this is no longer so if the parameters are estimated from the data. Use, as an example, the pooling of two cells in a fourfold table with fixed marginal totals. (Neyman and Pearson, 1931; Lancaster, 1950a.)

47. "χ^2 becomes 1·297 [with 3 d.f.] instead of 7·035, so that the corresponding value of P is 0·733. This, of course, denotes a good fit but one which is still very inferior to the Pearsonian Type II curve for which $P = 0·999$." Comment on this as an example of an author expecting that χ^2 should be very small. (Isserlis, 1917.)

CHAPTER X

Normal Correlation

1. INTRODUCTORY

The bivariate and multivariate normal distributions, the normal correlations of Russian authors, have played an important part in the development of the theory of χ^2 by K. Pearson and his school. Moreover, there is a large literature dealing with the Pearson tetrachoric and polychoric correlations. Some aspects of these correlations are treated in this chapter with the aid of the orthonormal theory. Let the random variables, $Z_1, Z_2, \ldots, Z_n$ be written as the elements of a vector, $\mathbf{Z}$ and suppose that $n = p + q$. It is convenient to denote two subsets of $\{Z_i\}$ by $\{X_i\}$ and $\{Y_j\}$ and to write them as elements of vectors adopting the convention,

$$(1.1) \qquad \mathbf{Z} = \mathbf{X} \dotplus \mathbf{Y},$$

so that $Z_i = X_i$ for $i = 1, 2, \ldots, p$; $Z_i = Y_{i-p}$ for $i = p + 1, p + 2, \ldots, n$. We suppose first that $\mathbf{V}$ is an $n \times n$ positive definite matrix. It will appear from the later analysis that

$$(1.2) \qquad \mathbf{V} = \mathscr{E}\mathbf{Z}\mathbf{Z}^T,$$

is the covariance matrix. A set of variables $\{Z_i\}$ will be said to be distributed in a normal correlation (i.e. in the joint normal distribution) if the joint density function is given by

$$(1.3) \qquad f(\{z_i\}; \mathbf{V}) = (2\pi)^{-\frac{1}{2}n} |\mathbf{V}|^{-\frac{1}{2}} \exp -\tfrac{1}{2}\mathbf{z}^T\mathbf{V}^{-1}\mathbf{z},$$

where $\mathbf{V}$ is positive definite. Let $\mathbf{V}$ be partitioned, so that

$$(1.4) \qquad \mathbf{V} = \begin{bmatrix} \mathbf{V}_{11} & \mathbf{V}_{12} \\ \mathbf{V}_{21} & \mathbf{V}_{22} \end{bmatrix}, \qquad \mathbf{V}_{12} = \mathbf{V}_{21}^T$$

where $\mathbf{V}_{11}$ is $p \times p$.

The marginal distribution of the first p variables can now be obtained. Let new variables be defined by

$$(1.5) \qquad \mathbf{X} \dotplus \mathbf{T} = \mathbf{S} = \begin{bmatrix} \mathbf{1}_p & \mathbf{0} \\ -\mathbf{V}_{21}\mathbf{V}_{11}^{-1} & \mathbf{1}_q \end{bmatrix}(\mathbf{X} \dotplus \mathbf{Y}) = \mathbf{CZ}.$$

Then a computation shows that $\mathbf{CVC}^T$ is the direct sum of two matrices,

$$(1.6) \qquad \mathbf{CVC}^T = \mathbf{V}_{11} \dotplus (\mathbf{V}_{22} - \mathbf{V}_{21}\mathbf{V}_{11}^{-1}\mathbf{V}_{12}) = \mathbf{V}_{11} \dotplus \mathbf{B}, \quad \text{say.}$$

The joint distribution of the S_i can now be obtained by applying the transformation (1.5) to (1.3). The Jacobian of this transformation is unity. Further

$$(1.7) \qquad \mathbf{z}^T\mathbf{V}^{-1}\mathbf{z} = (\mathbf{C}^{-1}\mathbf{s})^T\mathbf{V}^{-1}\mathbf{C}^{-1}\mathbf{s} = \mathbf{s}^T(\mathbf{CVC}^T)^{-1}\mathbf{s}$$
$$= \mathbf{s}^T(\mathbf{V}_{11}^{-1} \dotplus \mathbf{B}^{-1})\mathbf{s} = \mathbf{x}^T\mathbf{V}_{11}^{-1}\mathbf{x} + \mathbf{t}^T\mathbf{B}^{-1}\mathbf{t};$$

and also $|\mathbf{V}| = |\mathbf{V}_{11}||\mathbf{B}|$. It follows that the joint density function of the $\{S_i\}$ given by

$$(1.8) \quad g(\{s_i\}) = (2\pi)^{-\frac{1}{2}n} |\mathbf{V}|^{-\frac{1}{2}} \exp -\tfrac{1}{2}\mathbf{s}^T(\mathbf{V}_{11}^{-1} + \mathbf{B}^{-1})\mathbf{s}$$
$$= (2\pi)^{-\frac{1}{2}p} |\mathbf{V}_{11}|^{-\frac{1}{2}} \exp -\tfrac{1}{2}\mathbf{x}^T\mathbf{V}_{11}^{-1}\mathbf{x} \, (2\pi)^{-\frac{1}{2}q} |\mathbf{B}|^{-\frac{1}{2}} \exp -\tfrac{1}{2}\mathbf{t}^T\mathbf{B}^{-1}\mathbf{t},$$

factorizes into the product of two independent joint density functions of the normal correlation. The variables, $t_1, t_2, \ldots, t_q$, can now be integrated out to give

$$(1.9) \qquad\qquad p(\{x_i\}) = (2\pi)^{-\frac{1}{2}p} \exp -\tfrac{1}{2}\mathbf{x}^T\mathbf{V}_{11}^{-1}\mathbf{x}.$$

In particular, the density of X_1 is normal with variance, v_{11}. Further, X_1 and X_2 have a joint normal distribution with density function,

$$(1.10) \qquad h(x_1, x_2) = (2\pi)^{-1} |\mathbf{V}_{11}|^{-\frac{1}{2}} \exp -\tfrac{1}{2}[x_1 x_2]^T\mathbf{V}_{11}^{-1}[x_1 x_2],$$

where $[x_1 x_2]$ is a column vector. This is the density function of the usual bivariate normal correlation. v_{12} is thus shown to be the covariance between X_1 and X_2. It is clear that $\mathbf{V}$ is the covariance matrix of the variables, Z_i, since what has been proved for X_1, and X_1 and X_2 jointly could be proved for any X_j or pair, X_j and $X_{j'}$.

2. THE PARTIAL CORRELATIONS

The partial correlations can also be determined by the computations given above. The variances and covariances of $Y_1, Y_2, \ldots, Y_q$ have to be determined conditional on certain fixed values of $X_1, X_2, \ldots, X_p$. The variables, $T_1, T_2, \ldots, T_q$, are jointly independent of the X's, so that their joint distribution conditional on $\{X_i\} = \{x_i\}$ is given by

$$(2.1) \qquad\qquad k(\{t_i\}) = (2\pi)^{-\frac{1}{2}q} |\mathbf{B}|^{-\frac{1}{2}} \exp -\tfrac{1}{2}\mathbf{t}^T\mathbf{B}^{-1}\mathbf{t}.$$

However, from (1.5), $\mathbf{Y} = \mathbf{T} \dotplus \mathbf{V}_{21}\mathbf{V}_{11}^{-1}\mathbf{x}$. So that the mean values of the Y's are given by $\mathscr{E}\mathbf{Y} = \mathbf{V}_{21}\mathbf{V}_{11}^{-1}\mathbf{x}$. The covariance matrix of the Y's is the same as that of the T's, because the Y's differ from the T's only in having different means, and it is given by

(2.2) $$\mathscr{E}(YY^T \mid X = x) = V_{22} - V_{21}V_{11}^{-1}V_{12}.$$

In particular, for $p = 1$, if the covariance matrix is the correlation matrix, we obtain of $\{Z_{i.1}\}$, $i = 1, 2, \ldots, n$

(2.3)
$$\sigma_{i.1}^2 = 1 - \rho_{1i}^2, \qquad \text{for the variances}$$

$$\sigma_{ij.1} = \rho_{ij} - \rho_{1i}\rho_{1j}, \qquad \text{for the covariances,}$$

conditional on a fixed value of Z_1, and

(2.4) $$\rho_{ij.1} = (\rho_{ij} - \rho_{1i}\rho_{1j})(1 - \rho_{1i}^2)^{-\frac{1}{2}}(1 - \rho_{1j}^2)^{-\frac{1}{2}}$$

for the correlation.

Other partial correlations can be defined of the form, $\rho_{ij.12}, \rho_{ij.123}, \ldots$. These can be obtained in one step from (2.2) or calculated recursively from the formulas (2.3) and (2.4).

3. THE CANONICAL CORRELATIONS OF HOTELLING

Hotelling (1936) sought the linear functions of the X_i and Y_j respectively which had maximum correlation, namely

(3.1) $$\xi_1 = \sum a_i X_i, \qquad \eta_1 = \sum b_i Y_i$$

such that the correlation is maximized. If the coefficients a_i and b_i are suitably standardized, ξ_1 and η_1 have unit variance and the problem is then to maximize $\mathscr{E}\xi_1\eta_1$. A further pair of variables, ξ_2 and η_2, is then to be defined after the manner of (3.1) and the correlation maximized. Hotelling (1936) carried his proof through with the aid of Lagrange multipliers. We give an alternative method after the style of Lancaster (1966b). Two lemmas are first proved.

Lemma 3.1. Let $P_0, P_1, P_2, \ldots, P_{n_1}$ and $Q_0, Q_1, Q_2, \ldots, Q_{n_2}$ be two sets of orthonormal functions defined on a probability measure, with $P_0 = Q_0 = 1$; so that

(3.2) $\mathscr{E}P_iP_j = \mathscr{E}Q_iQ_j = \delta_{ij}, i = 0, 1, 2, \ldots, n_1; \quad j = 0, 1, 2, \ldots, n_2.$
$\mathscr{E}P_iQ_j = \delta_{ij}\rho_i, \rho_1 \geqslant \rho_2 \ldots \geqslant 0.$
Then $\mathscr{E}(\sum a_iP_i \sum b_jQ_j)$, subject to $\sum a_i^2 = \sum b_j^2 = 1$, is maximized by taking $a_1 = b_1 = 1$. ($\mathscr{E}$ is the expectation operator.)

Proof.

$$\mathscr{E}(\sum a_iP_i \sum b_jQ_j) = \sum_{i=1}^{n_1} \mathscr{E}a_ib_iP_iQ_i = \sum a_ib_i\rho_i \leqslant \tfrac{1}{2}\sum(a_i^2 + b_i^2)\rho_i,$$

and this sum is obviously maximized by taking $a_1 = b_1 = 1$.

Lemma 3.2. *For an arbitrary real $p \times q$ matrix, there exist orthogonal matrices, $\mathbf{P}$ and $\mathbf{Q}$, of sizes, p and q, such that*

(3.3) $$\mathbf{P}^T \mathbf{A} \mathbf{Q} = \mathbf{C} = [\mathrm{diag}\,(c_{ii}), \mathbf{0}_{p,q-p}].$$

Proof. There is an orthogonal $\mathbf{P}$ such that $\mathbf{P}^T(\mathbf{A}\mathbf{A}^T)\mathbf{P} = \mathrm{diag}\,(c_{ii}^2)$, where $c_{11} \geqslant c_{22} \geqslant \ldots \geqslant 0$. The r non-zero columns of $\mathbf{A}^T\mathbf{P}$ are thus mutually orthogonal and may be standardized by division by $c_{11}, c_{22}, \ldots, c_{rr}$. The first r columns of an orthogonal matrix are thus obtained. The matrix may then be completed by adjoining $(n - r)$ columns with due regard to the orthogonality conditions to yield an orthogonal matrix, $\mathbf{Q}$. $\mathbf{P}^T\mathbf{A}\mathbf{Q}$ now has the stated form.

For the proof of this lemma, one may also refer to Schwerdtfeger (1958) and Lancaster (1958).

To solve Hotelling's problem, new variables are defined by

(3.4) $$\mathbf{U} + \mathbf{W} = (\mathbf{P}^T + \mathbf{Q}^T)(\mathbf{V}_{11}^{-\frac{1}{2}} + \mathbf{V}_{22}^{-\frac{1}{2}})(\mathbf{X} + \mathbf{Y}),$$

where $\mathbf{P}$ and $\mathbf{Q}$ are orthogonal matrices obeying the conditions of Lemma 3.2 with $\mathbf{A} = \mathbf{V}_{11}^{-\frac{1}{2}}\mathbf{V}_{12}\mathbf{V}_{22}^{-\frac{1}{2}}$. It follows that the covariance matrix now takes the form,

(3.5) $$\mathscr{E}(\mathbf{U} + \mathbf{W})(\mathbf{U} + \mathbf{W})^T = \begin{bmatrix} \mathbf{1}_p & \mathbf{C} \\ \mathbf{C}^T & \mathbf{1}_q \end{bmatrix}$$

There is a biunique relation between the sets, $\{X_i\}$ and $\{U_i\}$, and between the sets, $\{Y_i\}$ and $\{W_i\}$. A form linear in the X_i is linear in the U_i and conversely; similarly for Y_i and W_i. The Hotelling problem now reduces to finding linear functions, $\sum a_i U_i$ and $\sum b_i W_i$, such that the expectation of their product is maximal subject to the condition, $\sum a_i^2 = \sum b_i^2 = 1$. Lemma 3.1 shows that U_1 and W_1 form the first pair of canonical variables and the first canonical (maximal) coefficient of correlation is $\rho_1 = c_{11}$. We look for the second pair. The orthogonality to ξ_1 forces ξ_2 to be of the form, $\sum_2^p c_i U_i$, and similarly η_2 is of the form $\sum_2^q d_i V_i$. $\sum c_i^2 = \sum d_i^2 = 1$. The second pair is U_2, W_2 and the second canonical correlation is $\rho_2 = c_{22}$. The canonical variables and coefficients of correlation are thus (U_i, W_i) and ρ_i, $i = 1, 2, \ldots, r$; $r \leqslant p$. If the Hotelling problem is specialized so that $p = 1$, there is one canonical coefficient and this is the coefficient of multiple correlation, which sometimes is defined as the maximum correlation possible between the regressand X, and the regressors, Y.

4. KOLMOGOROV'S CANONICAL PROBLEM

With the notation of Section 3, we may seek the pair of square summable functions in the $\{X_i\}$ and $\{Y_i\}$ respectively which has maximal correlation.

The solution was stated by Kolmogorov (cited by Sarmanov and Zaharov, 1960) to be U_1, W_1 in our notation.

Because of the biunique relation (3.4), we may look for functions in the $\{U_i\}$ and $\{W_i\}$. $\{U_i\}$ is a mutually independent set and so the product set of standardized Hermite-Chebyshev polynomials,

$$\{u_1^{(i_1)}\} \times \{u_2^{(i_2)}\} \times \ldots \times \{u_p^{(i_p)}\}$$

is complete on the product measure of $\{U_i\}$ with respect to square summable functions $\xi(U_1, U_2, \ldots, U_p)$. It is also complete with respect to these functions on any measure space for which $\{U_i\}$ is a set of marginal variables since both $\xi(U_1, U_2, \ldots, U_p)$ and the approximating series are constant for a given set of values $\{u_1, u_2, \ldots, u_p\}$. Similarly the product set $\{w_1^{(j_1)}\} \times \{w_2^{(j_2)}\} \times \ldots \times \{w_q^{(j_q)}\}$ is complete with respect to functions $\eta(W_1, W_2, \ldots, W_q)$. Let us write ξ and η as linear forms in their respective orthonormal sets. We may assume that ξ and η are standardized so that there is no constant term, and further the sum of squares of the Fourier coefficients of ξ and η will each be unity. Let us write $\xi = \sum a_i P_i$, where the P_i are the elements of the product set in some order. Now each P_i will have the form $\prod_{1-j}^{p} u_j^{(k_j)}$, where $k_j = 0, 1, 2, \ldots$. There will be only one element of the product set of the $w_j^{(k_j)}$ product set for which the correlation is non-zero, namely $\prod_{j=1}^{p} w_j^{(k_j)}$, which we may term $Q_{\tau(i)}$. Then we have to maximize $\mathscr{E}\xi\eta = \mathscr{E}(\sum a_i P_i \sum b_j Q_j)$ and by Lemma 3.1, we may take the canonical variables to be the pairs, P_i, $Q_{\tau(i)}$, and the canonical correlations are in some order all the non-zero products, $\rho_1^{k_1} \rho_2^{k_2} \ldots \rho_r^{k_r}$, which can be arranged in descending order of magnitude. The largest is evidently ρ_1 and so the first pair of canonical variables is (U_1, W_1).

The above reasoning can be justified as follows. The multivariate density can be written in the following form,

$$(4.1) \qquad g(\{u_i\}, \{w_j\}; \{\rho_i\}) = \prod_{i=1}^{r} f(u_i, w_i; \rho_i) \prod_{r+1}^{p} f(u_i) \prod_{r+1}^{q} f(w_i)$$

where $f(u)$ is the univariate standardized normal density function and $f(u_i, w_i; \rho_i)$ is the bivariate standardized normal density function with correlation coefficient equal to ρ_i. So that if an expression contains any non-constant term in $u_k^{(j)} w_k^{(j')}$ it will be zero if $j \neq j'$. Therefore, an expectation of products of the form, $P_i Q_i$, will be zero unless the product is of the form

$$(4.2) \qquad \mathscr{E}(PQ) = \prod_{k=1}^{p} u^{(j_k)} w^{(j_k)}.$$

A discussion of the Kolmogorov generalization is given in Lancaster (1966b).

5. MULTIVARIATE NORMALITY

It is often unduly restrictive to assume that the marginal variables have a positive definite covariance matrix. For example, in the Hotelling problem there may be a linear form in the first set of variables which has unit correlation with some linear form in the second set; in such a case, the frequency function cannot be written out in the standard form (1.3), for the covariance matrix is singular. If such cases are not to be excluded from consideration, an alternative definition of a joint normal distribution is required.

A set of random variables is said to be *jointly normal* if every linear form in the variables is normal. It is easily shown by the use of generating functions or orthogonal transformations, that a distribution jointly normal by the definition of Section 1, namely equation (1.3), is jointly normal by the present definition. A theorem going in the other direction is as follows:

Theorem 5.1. *Let* $Z_1, Z_2, \ldots, Z_n$ *be a set of random variables, forming the elements of the vector,* $\mathbf{Z}$, *such that, for arbitrary real* $\mathbf{b}$, $\mathbf{b}^T \mathbf{Z}$ *is normal. Then the joint distribution can be exhibited as a joint normal distribution of* $r \leqslant n$ *variables,* $W_1, W_2, \ldots, W_r$, *of the form*

$$(5.1) \qquad f(w_1, w_2, \ldots, w_r) = (2\pi)^{-\frac{1}{2}r} |\mathbf{B}|^{-\frac{1}{2}} \exp -\tfrac{1}{2} \mathbf{y}^T \mathbf{B}^{-1} \mathbf{y}$$

and a set of $(n - r)$ *sure variables,* $W_{r+1}, W_{r+2}, \ldots, W_n$. *A transformation can always be found so that* $\mathbf{B}$ *is the unit matrix of size,* r.

Proof. Every Z_i is normal since $\mathbf{b}$ can be specialised so as to have the ith element equal to unity and all others equal to zero. Every Z_i has finite variance and consequently the covariances are all finite. It can be supposed without loss of generality that all means are zeroes. Let $\mathbf{V}$ be the covariance matrix of $\mathbf{Z}$, then there exists an $\mathbf{H}$ such that

$$(5.2) \quad \mathbf{H}^T \mathbf{V} \mathbf{H} = \operatorname{diag}(v_1, v_2, \ldots, v_r, 0^{n-r}), \qquad v_i > 0, \quad i = 1, 2, \ldots, r.$$

Write $\mathbf{W} = \operatorname{diag}(v_1^{-\frac{1}{2}}, v_2^{-\frac{1}{2}}, \ldots, v_r^{-\frac{1}{2}}, 1^{n-r}) \mathbf{H}^T \mathbf{X}$. Then

$$(5.3) \qquad\qquad \mathscr{E} \mathbf{W} \mathbf{W}^T = \operatorname{diag}(1^r, 0^{n-r}).$$

The last $(n - r)$ of the W's are degenerate. If $\mathbf{V}$ is of rank, n, then the set of degenerate variables is empty. For $i = 1, 2, \ldots, r$, $U_i = W_i$ is a linear form in the Z's and so is a standardized normal variable after (5.3). Further, $\mathbf{a}^T \mathbf{U}$ is normal for every real vector, $\mathbf{a}$.

$$(5.4) \qquad\qquad \mathscr{E} \exp t\mathbf{a}^T \mathbf{U} = \exp \tfrac{1}{2} t^2 A,$$

where A is some constant and t is a complex variable. But A must be the variance of $\mathbf{a}^T\mathbf{U}$ and hence $A = \mathbf{a}^T\mathbf{a} = \sum_1^r a_i^2$. If $ta_i = c_i$, we have for an arbitrary set of real numbers, c_i,

(5.5)
$$\mathscr{E}\exp\sum c_iU_i = \prod_{i=1}^r \exp \tfrac{1}{2}c_i^2,$$

but this is the condition that $U_1, U_2, \ldots, U_r$ form a mutually independent set of standardized variables. A linear transformation will give a distribution of more general type (5.1). An alternative proof of the mutual independence is given as an exercise.

The Hotelling and Kolmogorov problems can thus be stated and proved without the assumption that the covariance matrices of the first set, of the second set and of the combined set are positive definite. The broader definition has been developed out of work by Cramér and Wold (1936), Mourier (1953), Fréchet (1951) and Basu (1956) and has been used by Rao (1965).

6. THE CANONICAL CORRELATIONS

Fisher (Fisher, 1940 and in Maung, 1942) analysed contingency tables from the point of view of the theory of Hotelling (1936). This work can be said in a sense to be a development of the Pearson idea that, if there is an underlying joint normal distribution, it is often possible to estimate ρ, the coefficient of correlation, without being given any metrical data on the marginal variables.

First, a bivariate normal distribution may be given theoretically. Let complete orthonormal sets be given on the marginal distributions, taken without loss of generality to be the standardized Hermite polynomials $\{x^{(i)}\}$ and $\{y^{(j)}\}$. Let the functions, $x^{(1)}, x^{(2)}, x^{(3)}, \ldots$ and $y^{(1)}, y^{(2)}, y^{(3)}, \ldots$ be written as elements of vectors, $\mathbf{x}$ and $\mathbf{y}$; then corresponding to (1.4), there follow

(6.1)
$$\mathbf{V}_{11} = \mathscr{E}\mathbf{x}\mathbf{x}^T = \mathbf{1}$$
$$\mathbf{V}_{22} = \mathscr{E}\mathbf{y}\mathbf{y}^T = \mathbf{1},$$

where each $\mathbf{1}$ is an infinite unit matrix.

(6.2)
$$\mathbf{V}_{12} = \mathbf{V}_{21}^T = \mathscr{E}\mathbf{x}\mathbf{y}^T = \text{diag}\,(\rho^i),$$

an infinite diagonal matrix. From Lemma 3.1, it follows that the pairs of canonical variables are $(x^{(i)}, y^{(i)})$ and the canonical correlations are $|\rho|^i$, a result due to Maung (1942) for $i = 1$ and to Lancaster (1957) for $i = 2, 3, \ldots$ It is clear that if we had begun with any other complete sets, then the same result could have been obtained by an orthogonal transformation to the Hermite series. We have proved

Theorem 6.1. *In the bivariate normal distribution, the pairs of canonical variables are $\{x^{(i)}, y^{(i)}\}$, the standardized Hermite polynomials of the marginal variables, and the canonical correlations are $|\rho|^i$.*

Proof. If ρ is negative the odd members of $\{y^{(i)}\}$ are to be taken with the negative sign. Let us now suppose that the marginal spaces have been condensed so that the condensed X variable takes m distinct values and the Y variable takes n distinct values. Let us write $g^{(0)} = 1 = h^{(0)}$ and suppose that orthonormal bases are formed by adjoining $(m - 1)$ additional functions, subject to the orthonormal conditions, on the condensed X distribution to form the elements of a vector, $\mathbf{g}$, having $(m - 1) = p$ elements. Similarly we define $\mathbf{h}$ on the condensed Y distribution. It is easily verified that the $(m - 1)$ elements of $\mathbf{g}$ together with $g^{(0)}$ form an orthonormal set on the original X distribution. Similar remarks apply to the elements of $\mathbf{h}$. By the general orthonormal theory we have

$$(6.3) \qquad \mathbf{g} = \mathbf{A}\mathbf{x}, \qquad \mathbf{h} = \mathbf{B}\mathbf{y},$$

where the rows of $\mathbf{A}$ have unit norm and are mutually orthogonal. We define also

$$(6.4) \qquad \mathbf{G} = \mathbf{A}^T\mathbf{A} \quad \text{and} \quad \mathbf{H} = \mathbf{B}^T\mathbf{B}.$$

$g_{ii} = \sum_{k=1}^{p} a_{ki}^2$ may be termed the proportional representation of $x^{(i)}$ in the set $\{g^{(k)}\}$ since it is the sum of squares of the Fourier coefficients of $x^{(i)}$ with respect to the members of the set $\{g^{(k)}\}$. A lemma due to Sarmanov and Zaharov (1959) is now cited.

Lemma 6.1. *Let $\mathbf{F}_{11}$ be a submatrix of $\mathbf{F}$, and let $\mathbf{F}$ be a matrix of correlations of two orthonormal sets, the elements of vectors, $\mathbf{u}$ and $\mathbf{v}$, so that $\mathbf{F} = \mathscr{E}\mathbf{u}\mathbf{v}^T$, $\mathscr{E}\mathbf{u}\mathbf{u}^T = \mathbf{1}_p$, $\mathscr{E}\mathbf{v}\mathbf{v}^T = \mathbf{1}_q$. Then the kth canonical correlation obtained from $\mathbf{F}_{11}$ is not greater than the kth canonical correlation obtained from $\mathbf{F}$. Equivalently the kth latent root of $\mathbf{F}_{11}\mathbf{F}_{11}^T$ is not greater than the kth latent root of $\mathbf{F}\mathbf{F}^T$.*

From the lemma it follows that no pair of linear forms, $\mathbf{a}^T\mathbf{u}$ and $\mathbf{b}^T\mathbf{v}$, can have a correlation greater than $|\rho|$. Let there be defined

$$(6.5) \qquad g^* = \mathbf{a}^T\mathbf{g} = g_{11}^{-\frac{1}{2}}\sum_{k=1}^{p} a_{k1}g^{(k)}, \qquad h^* = \mathbf{b}^T\mathbf{h} = h_{11}^{-\frac{1}{2}}\sum_{k'=1}^{q} b_{k'1}h^{(k')}.$$

g^* and h^* are standardized so that the correlation between them is given by

$$(6.6) \qquad \mathscr{E}g^*h^* = \mathscr{E}\mathbf{a}^T\mathbf{g}\mathbf{h}^T\mathbf{b} = \mathscr{E}\mathbf{a}^T\mathbf{A}\mathbf{x}\mathbf{y}^T\mathbf{B}^T\mathbf{b}$$
$$= \mathbf{a}^T\mathbf{A}\,\text{diag}\,(\rho^i)\mathbf{B}^T\mathbf{b}.$$

The coefficient of ρ in this expansion is $(g_{11}h_{11})^{\frac{1}{2}}$ since the first elements of $\mathbf{a}^T\mathbf{A}$ and $\mathbf{b}^T\mathbf{B}$ are $g_{11}^{-\frac{1}{2}}\sum_{k=1}^{p} a_{k1}^2 = g_{11}^{\frac{1}{2}}$ and $h_{11}^{\frac{1}{2}}$ respectively. If g_{11} is close to unity, then the other elements of $\mathbf{a}^T\mathbf{A}$ are negligible. Similar considerations apply to $\mathbf{b}^T\mathbf{B}$. So that it can be stated that if the representations of $x^{(1)}$ and $y^{(1)}$ in the sets $\{g^{(i)}\}$ and $\{h^{(j)}\}$ are large, then the first canonical correlation is close to $|\rho|$ and g^* and h^* are highly correlated with the members of the first pair of canonical variables. It may be noted that the correlation of g^* with $x^{(1)}$ is $g_{11}^{\frac{1}{2}}$. It is easy to show that the representation of $x^{(1)}$ in the set $\{g^{(i)}\}$ may be zero. As an example let X and Y be jointly normal and let X be replaced by a variable, X_s, so that $\mathscr{P}(X_s = x) = \mathscr{P}(X = x) + \mathscr{P}(X = -x)$, $x > 0$ and $\mathscr{P}(X_s \leqslant 0) = 0$. Then the Hermite polynomials of even degree are complete on the distribution of X_s, which have been obtained by a process of condensation from X. In the joint distribution of X_s and Y, the first canonical correlation is ρ^2. It is evident that no rule can be given other than the lemma cited unless something is known of the matrices, $\mathbf{A}$ and $\mathbf{B}$. We give as an example without proof the following theorem.

Theorem 6.2. *In a distribution obtained by condensation from the bivariate normal distribution, the canonical correlations can be computed from the matrix* $\mathbf{A}$ diag $(\rho^i)\mathbf{B}^T$. *The squares of the canonical correlations can be obtained from the solution of the equation,*

(6.7)
$$|\lambda\mathbf{1} - \mathbf{A}\,\text{diag}\,(\rho^i)\mathbf{B}^T\mathbf{B}\,\text{diag}\,(\rho^i)\mathbf{A}^T| = 0.$$

Suppose now that we have given a condensed distribution and we have repeated sampling so that an $m \times n$ contingency table results with sample size, N. The matrix of standardized asymptotically normal and mutually independent component χ's can be displayed as a $p \times q$ matrix, with entry at the i, j position equal to $N^{-\frac{1}{2}}\mathscr{S}g^{(i)}h^{(j)}$. Alternatively, at the same position, we could have the observed correlation, $N^{-1}\mathscr{S}g^{(i)}h^{(j)}$, between $g^{(i)}$ and $h^{(j)}$. The analysis of (6.1) and (6.2) is then appropriate with observed covariances conditional on the observed covariance matrices, $\mathbf{V}_{11}$ and $\mathbf{V}_{22}$, being unit matrices. If the null hypothesis specifies that X is independent of Y, the general theory of canonical variables may be invoked. Briefly, let Z_{ij}, $i = 1(1)p$, $j = 1(1)q$, be a set of mutually independent standardized normal variables. Then the distribution of the canonical variables, the diagonal elements of $\mathbf{P}^T\mathbf{Z}\mathbf{Q}$, can be determined where $\mathbf{P}$ and $\mathbf{Q}$ are orthogonal matrices and $\mathbf{P}^T\mathbf{Z}\mathbf{Q}$ is diagonal. Alternatively, the squares of the canonical variables are the roots of the equation,

(6.8)
$$|\lambda\mathbf{1} - \mathbf{Z}\mathbf{Z}^T| = 0.$$

The distribution of the roots of this equation have been obtained by Fisher (1939), Girshick (1939), Hsu (1939), Roy (1939), Mood (1951) and James (1964). Few useful tests of significance have resulted from the theory. Marriott (1952) appears to be the most useful reference. Some authors such as Kendall and Stuart (1958) and Williams (1952) have attempted to avoid the difficulties of the distribution theory by assuming that NR_1^2 is distributed as χ^2 with $(m + n - 3)$ degrees of freedom, NR_2^2 is distributed as χ^2 with $(m + n - 5)$ degrees of freedom, ..., NR_{m-1}^2 is distributed as χ^2 with 1 d.f. where, R_i is the ith canonical correlation. Lancaster (1963a) showed that this assumption is not well founded. NR_1^2 is stochastically greater than χ_{m+n-3}^2 and NR_{m-1}^2 is stochastically smaller than χ_1^2 with 1 d.f. Thus in a sampling experiment with $m = n = 3$, NR_1^2 had a mean and variance, equal to 3·27 and 8·09 instead of the theoretical 3 and 6 while NR_2^2 had a mean and variance equal to 0·27 and 0·25 instead of the theoretical 1 and 2. Possibly in the joint normal case if independence is the hypothesis being tested, the best test function is of the correlation between X and Y or of the correlation of functions, approximating to X and Y. But if no class of alternative hypotheses is specified, some over-all test such as χ^2 will be necessary.

7. THE WISHART DISTRIBUTION

The Wishart distribution might appear to be relevant. The observed co-variance matrix of the $(m + n - 2) = p + q$ orthonormal marginal variables of Section 6 might be set up in the form of (6.1) and (6.2) and the matrix of standardized asymptotically normal variables might be considered to be covariances multiplied by a factor, $N^{\frac{1}{2}}$, conditional on the variables, $N^{-1}\mathcal{S}x^{(i)}x^{(j)}$ and $N^{-1}\mathcal{S}y^{(i)}y^{(j)}$, taking their theoretical values, δ_{ij}. No tests of significance using this principle appear to have been devised. We refer the reader to Anderson (1958) and Wishart (1948) for a discussion of the Wishart distribution.

8. TETRACHORIC CORRELATION

K. Pearson (1900c) estimated the coefficient of correlation, ρ, of a bivariate normal distribution from the entries in a fourfold or tetrachoric table by his tetrachoric coefficient. Let the bivariate normal density be written in the form of a canonical expansion, so that

$$(8.1) \qquad f(x, y; \rho) = (2\pi)^{-1} |V|^{-\frac{1}{2}} \exp -\tfrac{1}{2}x^T V^{-1} x$$

$$= f(x)f(y)\{1 + \rho x^{(1)}y^{(1)} + \rho^2 x^{(2)}y^{(2)} + \ldots\},$$

where $V = (c_{ij})$, $v_{11} = v_{22} = 1$, $v_{12} = v_{21} = \rho$, x is the vector $[x, y]$ and $f(x)$ is the frequency function of the standardized normal distribution and $x^{(i)}$ and $y^{(i)}$ are the Hermite polynomials of degree, i, in X and Y respectively. For $|\rho| < 1$ the series on the right converges absolutely *a.e.* with respect to the product measure $f(x)f(y)$. This can be proved as a special case of the proposition on page 63 of Alexits (1961) where it is shown that a general orthogonal series $\sum c_n \phi_n(x)$ converges absolutely a.e. if $\sum |c_n| < \infty$, summation in each case being over $0, 1, 2, \ldots$. Term by term integration of the series on the right side of (8.1) is therefore permissible.

The Hermite polynomials possess the property,

$$(8.2) \qquad H_j(x) \exp -\tfrac{1}{2}x^2 = -\frac{d}{dx}\,[H_{j-1}(x) \exp -\tfrac{1}{2}x^2],$$

which is sometimes used as the defining property of the series. It follows that,

$$(8.3) \qquad \frac{1}{\sqrt{(2\pi)}} \int_h^\infty x^{(j)} \exp -\tfrac{1}{2}x^2 \, dx = \frac{1}{\sqrt{(2\pi)}} \frac{1}{\sqrt{j!}} H_{j-1}(h) \exp -\tfrac{1}{2}h^2$$

$$= \tau_j(h),$$

in the notation of Pearson (1900c). To obtain the amount of distribution in the quadrant $\{h \leqslant x < \infty;\, k \leqslant x < \infty\}$ it is therefore necessary only to integrate the right side of (8.1) term by term and apply the result (8.3). Pearson's result follows

$$(8.4) \qquad \int_h^\infty \int_k^\infty f(x, y;\, \rho) \, dx \, dy = \sum_0^\infty \rho^j \tau_j(h) \tau_j(k),$$

for the theoretical amount of density.

In the observations, Pearson (1900c) assumes that there is an underlying joint normal correlation and that both marginal distributions have been dichotomized to give a fourfold table. The N observations have been classified into $(a + b)$ values of X less than the value at the point of dichotomy and $(a + c)$ values of Y less than the value at the point of dichotomy. A cross classification has yielded the fourfold table,

	Y less than y_0	Y greater than y_0	Total
X less than x_0	a	b	$a + b$
X greater than x_0	c	d	$c + d$
Total	$a + c$	$b + d$	$a + b + c + d = N$

The points x_0 and y_0 are not known but are estimated by equations of the form,

$$(8.5) \qquad \frac{a + b}{N} = \frac{1}{\sqrt{(2\pi)}} \int_{-\infty}^h \exp -\tfrac{1}{2}x^2 \, dx$$

and x_0 is assumed to be equal to h; similarly y_0 is estimated by k using

$$(8.6) \qquad \frac{a + c}{N} = \frac{1}{\sqrt{(2\pi)}} \int_{-\infty}^k \exp -\tfrac{1}{2}y^2 \, dy.$$

Applying (8.5) and (8.6) to (8.4) there follows

(8.7)
$$\frac{ad - bc}{N^2} = \sum_1^\infty \rho^j \tau_j(h) \tau_j(k).$$

Unless $|\rho|$ is near unity, relatively few terms of the right side of (8.7) may be used and the estimate r of ρ is given by the solution of an equation. The methods to be used may be found on pages, *xliv* to *li* of Part II of Pearson's (1931) *Tables for Statisticians and Biometricians*. The method appears to give reasonable estimates of ρ but unfortunately no sampling theory is available.

9. THE POLYCHORIC SERIES

The following analysis is based on the methods of Lancaster and Hamdan (1964). Let **x** and **y** be vectors, the elements of which are the standardized Hermite polynomials. Let X and Y be jointly normal and standardized. Suppose now that finite partitions of the marginal distributions have been made and that **g** and **h** are the vectors with elements the non-constant members of the orthonormal bases on the finite marginal distributions. Then there is an **A** and a **B**, such that

(9.1)
$$\mathbf{g} = \mathbf{Ax}, \quad \mathbf{h} = \mathbf{By},$$

where the rows of **A** are normed and mutually orthogonal and similarly for **B**. Let

(9.2)
$$\mathbf{G} = \mathbf{A}^T \mathbf{A} \quad \text{and} \quad \mathbf{H} = \mathbf{B}^T \mathbf{B}.$$

Then g_{ii} is the sum of squares of the Fourier coefficients of $x^{(i)}$ with respect to the elements of **g**. $g_{ii} \leqslant 1$. g_{ii} is a measure of the representation of $x^{(i)}$ in the set $\{g^{(j)}\}$, the elements of **g**. Similar considerations apply to **H**. The correlations in the finite distribution can now be computed

(9.3)
$$\mathbf{R} = \mathscr{E}\mathbf{g}\mathbf{h}^T = \mathscr{E}\mathbf{Axy}^T\mathbf{B}^T$$
$$= \mathbf{A}\mathscr{E}\mathbf{xy}^T\mathbf{B}^T$$
$$= \mathbf{A}\,\text{diag}\,(\rho^i)\mathbf{B}^T.$$

The asymptotically normal component χ's of the $m \times n$ contingency table are all of the form, $N^{-\frac{1}{2}}\mathscr{S}x^{(i)}y^{(j)}$, and each has an expectation of $N^{\frac{1}{2}}\mathscr{E}x^{(i)}y^{(j)}$. The parameter of non-centrality for a sample of N under the assumption of asymptotic normality can now be computed

(9.4)
$$\lambda(\rho) = N \sum_{i=1}^{m-1} \sum_{j=1}^{n-1} \mathscr{E}[\mathscr{S}g^{(i)}h^{(j)}]^2.$$

If ρ is not large $\mathscr{E}[g^{(i)}h^{(j)}]^2$ will be approximately $1 + [\mathscr{E}g^{(i)}h^{(j)}]^2$, for it can be assumed that the variance is close to unity. From (9.4) it then follows that

$$(9.5) \qquad N^{-1}\lambda(\rho) = \operatorname{tr}(\mathscr{E}\mathbf{gh}^T\mathscr{E}\mathbf{hg}^T)$$
$$= \operatorname{tr}[\mathbf{A}\operatorname{diag}(\rho^i)\mathbf{B}^T\mathbf{B}\operatorname{diag}(\rho^i)\mathbf{A}^T]$$
$$= \operatorname{tr}[\mathbf{G}\operatorname{diag}(\rho^i)\mathbf{H}\operatorname{diag}(\rho^i)]$$
$$= \sum_i \sum_j g_{ij}\rho^j h_{ji}\rho^i$$

The method suggested for the estimation of ρ, is that the observed χ^2 of the contingency table should be equated to the sum of the degrees of freedom and the parameter of non-centrality,

$$(9.6) \qquad \chi^2 = (m-1)(n-1) + N\lambda(\hat{\rho}).$$

Confidence limits for $\hat{\rho}$ may be given by replacing in (9.6) the degrees of freedom by the 5% and 95% values of χ^2 for the same number of degrees of freedom. In each case, an equation results of the form,

$$(9.7) \qquad c_0 = c_2\hat{\rho}^2 + c_3\hat{\rho}^3 + \dots,$$

where the c's are constants. c_0 is calculated as mentioned; $c_2, c_3, \dots$, can be determined from (9.5). $c_2 = g_{11}h_{11}$; $c_3 = 2g_{12}h_{12}$; $c_4 = g_{22}h_{22} + 2g_{13}h_{13}$; $c_5 = 2g_{14}h_{14} + 2g_{23}h_{23}$; $c_6 = g_{33}h_{33} + 2g_{15}h_{15} + 2g_{24}h_{24} \dots$. Lancaster and Hamdan (1964) give an equation for the estimation of ρ in some classic data of Pearson and Lee (1903) treated as a 3×3 table

$$(9.8) \quad 0{\cdot}1702 = 0{\cdot}6238\hat{\rho}^2 + 0{\cdot}0096\hat{\rho}^3 + 0{\cdot}1966\hat{\rho}^4 + 0{\cdot}0296\hat{\rho}^5 + 0{\cdot}1215\hat{\rho}^6,$$

with solution $\hat{\rho} = 0{\cdot}4973$ and compare it with the formula of Pearson (1903)

$$(9.9) \qquad \phi^2 = \rho^2(1 - \rho^2)^{-1} = \rho^2 + \rho^4 + \rho^6 + \dots$$

which yields $\hat{\rho} = 0{\cdot}3814$. If the margins are partitioned more finely $m = n = 8$, for example, the recommended equation and the Pearson equation approximate more closely

$$(9.10) \qquad 0{\cdot}2889 = 0{\cdot}9071\hat{\rho}^2 + 0{\cdot}7729\hat{\rho}^4 + 0{\cdot}5256\hat{\rho}^6$$

giving an estimate, $\hat{\rho} = 0{\cdot}5041$. With $m = n = 18$,

$$(9.11) \qquad 0{\cdot}3445 = 0{\cdot}9758\hat{\rho}^2 + 0{\cdot}9424\hat{\rho}^4 + 0{\cdot}8602\hat{\rho}^6$$

giving an estimate $\hat{\rho} = 0{\cdot}5170$ and 95% confidence limits $0{\cdot}4956$ and $0{\cdot}5355$ as compared with the product moment estimate $0{\cdot}5157$.

It appears that the c_{2k+1} are usually negligible. Ritchie-Scott (1918) had considered the estimation of ρ in the polychoric case and had suggested averaging the tetrachoric correlations obtained by varying the point of dichotomy. It is possible that this would give good estimates. Thus in

Table 3 of Lancaster and Hamdan (1964), with variable values of h and k in the Pearson tetrachoric method of estimation, there are 49 estimates with a mean of 0·5042 and standard deviation of 0·037, and so any averaging process would also give a good estimate. The table also supports the validity of the Pearson tetrachoric method.

10. THE CORRELATION RATIO

In some applied work, for example in psychology, although there may be grounds for postulating that the fundamental distribution is the joint normal, the marginal variables may not be known. The observable variables are some function of the unknown marginal normal variables. The functional relationship may not be known but it is usually supposed that it is monotonic, even though not strictly monotonic. So that there are given observations from a bivariate distribution, of which both marginal spaces have been obtained by a condensation from the original spaces of the normal variables. The tetrachoric and the polychoric coefficients of correlation can be interpreted as attempts to reparametrise the data. Either method solves the problem by equating the cumulative probability to that of a standardized normal variable and assigning the corresponding value to the end point of the interval. However, less may be known about the underlying variables than is given above. The condensed variables may not be monotonic functions of the original normal variables. For such cases K. Pearson devised the method of the correlation ratio.

With a sample size of N from a normal distribution, the variances of the arrays of the Y's for a fixed value of X is equal to $\sigma_2^2(1 - \rho^2)$. Suppose now that the values of Y are given but only class values of X. Then we may compute the variances of Y within X-classes and use the relation mentioned or alternatively we may compute the difference of the Y's within classes and the total sum of squares and then give it as a ratio to the total sum of squares.

$$(10.1) \qquad N\eta^2 = \sum N_p(\bar{y}_p - \bar{y})^2 / \sum (y - \bar{y})^2,$$

which is distributed approximately as $(1 - \rho^2)\chi^2$ with $(m - 1)$ degrees of freedom if there are m X-classes. If Y be assumed to be normal and to have normal correlation with X, the use of η^2 is equivalent to that of r^2. The result (10.1) is due to R. A. Fisher (*Statistical Methods*). It is possible, if the values of X are given, to separate out the sums of squares due to linear regression and to calculate an expression $(\eta^2 - r^2)\sigma_2^2$. This last expression is necessarily positive, for

$$(10.2) \qquad N(\eta^2 - r^2)\sigma_2^2 = \sum \{N_x(\bar{y}_x - Y_R)^2.$$

Pearson (1905b and 1909) and Blakeman (1905) tested for linearity of regression by comparing the expression (10.2) with its presumed standard error. Fisher gives $N(\eta^2 - r^2)/(1 - r^2)$ as χ^2 with $(m - 2)$ degrees of freedom, it being assumed that N is so large that the denominator is effectively equal to its expected value.

11.   BISERIAL η

Pearson (1909) considered correlation where one variable was metrical and the other was grouped into several classes. He states that "if the non-measurable, or at least unmeasured character, be classed into a considerable number of groups there is no doubt that the most satisfactory method to adopt is that of the correlation ratio". Tate (1955) has reconsidered the biserial η and we may follow where practicable his notation. Tate takes Z to be the random variable induced by the dichotomy of the space of Y. Pearson (1909) writes for the regression of X on Y,

$$(11.1) \qquad (X - \mathscr{E}X) = r\frac{\sigma_1}{\sigma_2}(Y - \mathscr{E}Y).$$

This line will cut any Y-section at a point $(\bar{p}, \bar{q})$. If the joint distribution were known, the correlation coefficient could be estimated by

$$(11.2) \qquad r = \frac{\bar{p}/\sigma_1}{\bar{q}/\sigma_2},$$

where σ_1 and σ_2 are the standard deviations of X and Y respectively, σ_1 can be determined since measurements have been made on the sample of values of X. Without loss of generality σ_2 can be set equal to unity. Suppose now that the space of Y is dichotomized. In either, say the first, of the classes, $\bar{p}$ can be computed from the contingency table. Pearson makes good the absence of knowledge of the Y variable by computing its mean value in the first class under the assumption that Y is normal,

$$(11.3) \qquad \bar{q} = \int_y^\infty tf(t)\,dt \bigg/ \int_y^\infty f(t)\,dt = f(y)\bigg/\int_y^\infty f(t)\,dt.$$

These values can be obtained from tables of the normal curve. Tate (1955) gives the "biserial r" in the form

$$(11.4) \qquad r^* = \frac{1}{n}\sum(X_i - \bar{X})(Z_i - \bar{Z})\bigg/\left\{\frac{1}{n}\sum(X_i - \bar{X})^2\right\}^{\frac{1}{2}}f(T)$$

$$= r\left\{\frac{1}{n}\sum(Z_i - \bar{Z})^2\right\}^{\frac{1}{2}}\bigg/f(T)$$

where $f(T) = (2\pi)^{-\frac{1}{2}} \exp(-\frac{1}{2}T^2)$ and T is the value of the normal variable at the point of dichotomy, so that

(11.5) $$\mathscr{P}(Z = 1) = \int_T^\infty f(t)\,dt.$$

Soper (1913) gave an asymptotic variance

(11.6) $$AV(r^*) = \frac{1}{n}\left\{\rho^4 + \rho^2\left[\frac{pq\omega^2}{f^2} + \frac{(2p-1)\omega}{f} - \frac{5}{2}\right] + \frac{pq}{f^2}\right\}$$

where f, p and q have arguments $\omega \equiv T$, the point of dichotomy and $p = \mathscr{P}(Z = 1) = 1 - q$. Tate (1955) gives a table of these values. r^* is a consistent estimator. On occasion, it may exceed unity in absolute value. Tate (1955) recommends setting the estimate as $+1$ or -1 in such cases. r^* is asymptotically most efficient when ρ is small but is inefficient when $|\rho|$ is close to unity. This theory can be linked with the orthonormal theory. Suppose the (X, Y) are jointly normal with means $(0, 0)$ and variances (σ_1^2, σ_2^2) and correlation, ρ. Then the orthonormal function on the Y-space will be of the form

$$Y^* = b_1 y^{(1)} + b_2 y^{(2)} + \cdots, \qquad \sum b_i^2 = 1.$$

We may write the X variable as

$$X^* = a_1 x^{(1)} + a_2 x^{(2)} + \cdots, \qquad \sum a_i^2 = 1$$

but since there are a number of classes, a_1^2 will approach unity and the other coefficients, a_i, will be small.

$$\text{corr}(X^*, Y^*) = \sum a_i b_i \rho^i.$$

Now if ρ is not large, this correlation will be very nearly $a_1 b_1 \rho$ or $b_1 \rho$. However, b_1 for a dichotomy has a maximum value $(2/\pi)^{\frac{1}{2}}$ so that the maximum for all choices of the point of dichotomy is $(2/\pi)^{\frac{1}{2}}\rho$. As usually carried out, the biserial correlation must be biased.

12. TESTS OF NORMALITY

13. VARIOUS MEASURES OF CORRELATION

For these two sections, see the *Index to the Bibliography*.

EXERCISES AND COMPLEMENTS

1. Determine the conditions under which $f(\mathbf{x}) = C \exp -\frac{1}{2}W$, $-\infty < x < \infty$, where C is a suitable constant and $W = ax_1^2 + bx_2^2 + cx_3^2 + 2hx_1x_2 + 2fx_2x_3 + 2gx_1x_3$, can be a probability density function of three random variables, X_1, X_2 and X_3. For a fixed value of X_3 determine the joint distribution of X_1 and X_2. (Bravais, 1846.)

2. In the distribution of Exercise 1, determine the covariance matrix of the variables. (Bravais, 1846; Edgeworth, 1892.)

3. In the bivariate normal distribution determine the line of regression of X_2 on X_1 to minimise $\mathscr{E}(X_2 - \lambda X_1)^2$. Give this result in the special case when the variances of X_1 and X_2 are units. Determine also the regression of X_1 on X_2. When do these lines coincide? (Dixon, 1886; Galton, 1886.)

4. Prove that if X and Y have a joint distribution with unit variances, the mean of the X, conditional on a fixed value of $Y = Y_0$, is less distant from its mean than is Y_0. Give an exact result for the normal distribution in standard form. This is the famous regression to mediocrity of the sons of fathers of Galton. Tall fathers have sons whose means tend to be less than that of their fathers. Prove that the means of fathers conditional on fixed values of the sons also has the same property (which was overlooked by Galton). (Galton, 1886.)

5. Let a multivariate frequency function be written in the form

$$f(\{x_i\}) = c \exp [-\tfrac{1}{2}\mathbf{x}^T \mathbf{A} \mathbf{x}],$$

with x_i the elements of $\mathbf{x}$; $-\infty < x_i < \infty$ for $i = 1, 2, \ldots, n$. Show that $\mathbf{x}^T \mathbf{A} \mathbf{x}$ may be reduced to a sum of squares in which x_1 appears only in the first summand, x_2 appears only in the first two, ... For $n = 3$, show that $\mathbf{x}^T \mathbf{A} \mathbf{x}$ can be exhibited in the form, $\mathbf{x}^T \mathbf{A} \mathbf{x} = (b_1 x_1 + b_2 x_2 + b_3 x_3)^2 + (c_2 x_2 + c_3 x_3)^2 + (d_3 x_3)^2$. Also show that these three linear forms are mutually independent stochastically. Further determine the distribution of the second or third variables and their joint distribution conditional on fixed values of the first. (Bravais, 1846; Edgeworth, 1892; Pearson, 1896; p. 537 of Laplace, 1836.)

6. Prove that the section of a multivariate normal distribution by a hyperplane is again a multivariate normal distribution. (Edgeworth, 1892; Pearson, 1916a.)

7. If $f(\mathbf{x}) = C \exp -\tfrac{1}{2}\mathbf{x}^T \mathbf{A} \mathbf{x}$, then the covariance matrix of the variables X_1, $X_2, \ldots, X_n$ is $\mathbf{A}^{-1} = \mathbf{R}$, say. If the variables have been standardized to have unit variance, then the off-diagonal terms of $\mathbf{R}$ are the coefficients of correlation. (Edgeworth, 1892; Pearson, 1896.)

8. A NASC for the independence of the vectors $\mathbf{X}$ and $\mathbf{Y}$ of Section 1 is that $\mathscr{E}\mathbf{X}\mathbf{Y}^T = 0$. More generally a NASC for the independence of vectors, $\mathbf{C}_1\mathbf{Z}$ and $\mathbf{C}_2\mathbf{Z}$ is that $\mathscr{E}\mathbf{C}_1\mathbf{Z}\mathbf{Z}^T\mathbf{C}_2^T = \mathbf{C}_1\mathbf{V}\mathbf{C}_2^T = 0$. $\mathbf{Z}$ has n elements, $\mathbf{C}_1$ is $m_1 \times n$, $\mathbf{C}_2$ is $m_2 \times n$.

9. Let $X, Y_1, Y_2, \ldots, Y_q$ be jointly normal and standard normal. If, among all linear forms $\mathbf{b}^T\mathbf{Y}$, η has maximum correlation with X, then X is uncorrelated with any other linear form, $\mathbf{c}^T\mathbf{Y}$, which is uncorrelated with η. (The correlation of X with η is called the multiple correlation).

10. Prove that the result of Exercise 9 does not depend on normal theory but only on the finiteness of the variances. (The result is well known in the general theory of orthogonal functions.)

11. Define and give a general formula for the coefficients of partial correlation in a normal system. (Pearson, 1896; Yule, 1907; Kendall and Stuart, 1961.)

12. Prove that in a joint normal distribution, all distributions, conditional on fixed values of certain of the marginal variables, depend entirely on the elements of the covariance matrix and the set of fixed values. For example with variables

X, Y and Z, the joint distribution of X and Y for $Z = z_0$ is completely determined by the covariance matrix and z_0. Give an example to show that this proposition can be false in general distributions taking as your example, $p_{xyz} = (1 + \rho xyz)/8$. X, Y and Z each having the distribution of the form $X = 1$ and -1 with probabilities, $\frac{1}{2}$ and $\frac{1}{2}$.

13. Suppose that the n elements of $\mathbf{X}$ are standard normal and mutually independent. Let $\mathbf{A}$ be an $m \times n$ rectangular matrix, $m \leqslant n$, of rank m. Determine the distribution of the vector $\mathbf{Y} = \mathbf{AX}$. (Pearson, 1896.)

14. If $\mathbf{A}$ is of rank, $r < m$, prove that the distribution of the vector, $\mathbf{Y}$, is normal in the sense of Section 5.

15. In Exercise 13, make the transformation

$$Y_1 = (1 - \rho)^{\frac{1}{2}} X_1 + \rho^{\frac{1}{2}} X_2$$
$$Y_2 = \rho^{\frac{1}{2}} X_2 + (1 - \rho)^{\frac{1}{2}} X_3$$
$$Y_3 = X_3, \quad 0 < \rho < 1.$$

Verify that corr $(Y_1, Y_2) = \rho$. By integrating out Y_3, obtain a bivariate normal distribution.

16. In Exercise 15, obtain Mehler's identity (i) by a consideration of

$$\mathscr{E}[\exp(Y_1 t_1 - \tfrac{1}{2}t_1^2)\exp(Y_2 t_2 - \tfrac{1}{2}t_2^2)];$$

(ii) by writing out the Hermite polynomials with the use of Runge's identity of Exercise IV.57 in the form,

$$H_m(Y_1) = H_m((1 - \rho)^{\frac{1}{2}} X_1 + \rho^{\frac{1}{2}} X_2) = \sum (1 - \rho)^{\frac{1}{2}k}\rho^{\frac{1}{2}(m-k)}H_k(X_1)H_{m-k}(X_2)$$

and similarly for $H_m(Y_2)$. Verify that $\mathscr{E}H_m(Y_1)H_n(Y_2) = \rho^m \delta_{mn}$, which follows from the independence of the X's and the orthogonality of the elements of $\{H_m(X_2)\}$. Note that $\sum \rho^{2m}$ is convergent and that the Hermite polynomials are complete on the normal and then apply the theorem of Exercise IV.68 to obtain Mehler's identity.

17. Suppose that $\mathbf{X}$ is normal and $\mathscr{E}\mathbf{XX}^T = \mathbf{V}$. Find the linear form $\mathbf{a}^T\mathbf{X}$ which has maximum variance, subject to $\mathbf{a}^T\mathbf{a} = 1$. This is the principal component of Hotelling. (Pearson, 1901b; Hotelling, 1933.)

18. Let $H(t)$ be the g.f. for the Hermite polynomials. By a consideration of terms of the form, $t^i u^j v^k$ in the integral, $\int H(t)H(u)H(v)(2\pi)^{-\frac{1}{2}}\exp -\tfrac{1}{2}x^2\,dx$, obtain identities of the form

$$H_m^2(x) = a_{m0} + a_{m2}H_2(x) + \ldots + a_{m,2m}H_{2m}(x).$$

Hence obtain the variances of the functions, $x^{(i)}y^{(j)}$, in the bivariate normal distribution.

19. Suppose that the means are zero and the standard deviations of the jointly normal variables, X and Y, are σ_1 and σ_2 and are known. Prove that

$$\lambda = \mathscr{S}(xy)/(n\sigma_1\sigma_2)$$

is the maximum likelihood estimator of the coefficient of correlation. (Pearson, 1896.)

20. Give examples to show that V_{11} and V_{22} of (1.4) may be positive definite although V is not. In such a case prove that there is always a correlation of unit absolute value between some linear form in the first set and a linear form in the second set.

21. In the Hotelling problem of Section 3, if $p = 1$, ρ_1 is the multiple correlation between X_1 and the set $\{ Y_i \}$. If $p > 1$, ρ_1 is the greatest multiple correlation between linear forms $a^T X$ and the set $\{ Y_i \}$. (Hotelling, 1936.)

22. The canonical correlations can be given a geometric interpretation; $p + q$ variates determine two flat spaces of dimensions, p and q, intersecting at the origin. One invariant is the minimum angle between two lines in the two spaces and this is $\cos^{-1} \rho_1$. (Hotelling, 1936.)

23. An approximate value for the standard error of ρ_1 is $(1 - \rho_1^2) N^{-\frac{1}{2}}$. (Hotelling, 1936.)

24. Consider the classification of Aberdeen school children by hair and eye colours:

Eye Colour	Fair	Red	Hair Colour Medium	Dark	Black	Total
Blue	1368	170	1041	398	1	2978
Light	2577	474	2703	932	11	6697
Medium	1390	420	3826	1842	33	7511
Dark	454	255	1848	2506	112	5175
Total	5789	1319	9418	5678	157	22,361

This is the matrix $N\hat{F}$, where $\hat{F}$ is the sample value of the unknown matrix, F, of Theorem VI.2.1. Write $B = \text{diag} (f_{i.}^{-\frac{1}{2}}) \hat{F} \text{diag} (f_{.j}^{-\frac{1}{2}})$. Determine the M and N such that $M^T B N$ is in canonical form. Find the canonical variables and correlations by the method of Hotelling or of Section VI.2. The first pair of canonical variables assume values in the rows of $1 \cdot 1855$, $0 \cdot 9042$, $-0 \cdot 2111$, $-1 \cdot 5458$ and in the columns of $1 \cdot 3419$, $0 \cdot 2933$, $0 \cdot 0038$, $-1 \cdot 3643$, $-2 \cdot 8278$. Thus the first canonical variables assign values so that the rows and column variables are ordered in what appears to be the natural manner. (Maung, 1942; Lancaster, 1958.)

25. In the notation of Section 1, Wilks gave criteria, the vector correlation coefficient,

$$Q^2 = \frac{(-1)^p C}{|V_{11}| \, |V_{22}|}, \quad \text{where} \quad C = \begin{vmatrix} 0 & V_{12} \\ V_{21} & V_{22} \end{vmatrix}$$

and the *vector alienation coefficient*,

$$Z = \frac{|V|}{|V_{11}| \, |V_{22}|}.$$

Prove that both these indices are invariant under non-singular transformations of the form, $\mathbf{X} + \mathbf{Y} \to (\mathbf{G} + \mathbf{H})(\mathbf{X} + \mathbf{Y})$, if $\mathbf{G}$ and $\mathbf{H}$ are non-singular. Verify that they are generalizations of the coefficient of correlation, ρ, and the conditional variances of the bivariate normal correlation. (Wilks, 1932; Hotelling, 1936.)

26. Suppose that a square summable function in X is given. It is required to find the function of Y with which it has maximal correlation. Treat this problem (i) when X and Y are jointly normal (ii) when X and Y are the row and column variables of a bivariate distribution. [*Hint.* Let the function given be $\sum a_i x^{(i)}$; $\sum a_i^2 = 1$. It is required to maximise $\mathscr{E}[\sum a_i x^{(i)} \sum b_j y^{(j)}]$.] (Yates, 1948, Lancaster, 1958.)

27. The frequency function of the observed r is proportional to

$$(1 - \rho^2)^{\frac{1}{2}(n-1)}(1 - r^2)^{\frac{1}{2}(n-4)}\left(\frac{\partial}{\sin\theta\,\partial\theta}\right)^{n-1}\frac{\theta^2}{2}\,dr$$

$r = \mathscr{S}(x - \bar{x})(y - \bar{y})/\{\mathscr{S}(x - \bar{x})^2\mathscr{S}(y - \bar{y})^2\}^{\frac{1}{2}}$. (Fisher, 1915.)

28. If $\rho = 0$, the significance of r can be tested by the rule $t = (n - 2)^{\frac{1}{2}}r(1 - r^2)^{-\frac{1}{2}}$, where t is Student's t with $(n - 3)$ degrees of freedom. (Fisher, 1915.)

29. Let $\psi(t, u)$ be the bivariate characteristic function of the bivariate normal density, so that

$$\psi(t, u) = \iint \exp\,(itx + iuy)f(x, y; \rho)\,dx\,dy$$

$$= \exp\,[-\tfrac{1}{2}(t^2 + 2\rho tu + u^2)]\sum_{n=0}^{\infty}(-\rho)^n t^n u^n/n!.$$

Carry out the inversion on the left by multiplying by $\exp\,(-ixt - iyt)$ and integrating and do the same on the right term by term to obtain Mehler's identity. (A. C. Aitken, unpublished; Hardy, 1933; Kendall, 1941.)

30. Define a bivariate distribution as follows. Suppose that $\mathscr{P}(X = 1) = p$, $\mathscr{P}(X = 0) = q$, $p + q = 1$. Suppose that the distribution of Y is normal with variance, σ^2, and mean, μ_0 or μ_1, conditional on $Y = 0$ or $Y = 1$ respectively. Verify that $\mathscr{E}(X) = p$, var $(X) = pq$, $\mathscr{E}(Y) = q\mu_0 + p\mu_1$, var $(Y) = \sigma^2 + pq(\mu_1 - \mu_0)^2$ and corr $(X, Y) = (\mu_1 - \mu_0)(pq)^{\frac{1}{2}}/[\sigma^2 + pq(\mu_1 - \mu_0)^2]^{\frac{1}{2}}$. Note that the correlation increases with $(\mu_1 - \mu_0)^2$ and is asymptotically ± 1 as $(\mu_1 - \mu_0) \to \pm\infty$. Calculate the bivariate moments from the moment generating function,

$$q\exp\,(\mu_0 t + \tfrac{1}{2}\sigma^2 t^2) + p\exp\,(u + \mu_1 t + \tfrac{1}{2}\sigma^2 t^2).$$

(Tate, 1955.)

31. If ρ is not large, the leading term on the right of (6.7) namely $g_{11}h_{11}$, dominates the computation of the estimates. The orthonormal bases on the condensed marginal distributions is an orthonormal set on the marginal (normal) distribution. Suppose that this set is completed. Then the marginal variables can be expanded in the forms,

$$X = Z_1 \cos\theta + Z_2 \sin\theta$$
$$Y = Z_3 \cos\psi + Z_4 \sin\psi$$

where Z_1 and Z_3 are linear forms in the orthonormal bases of the condensed distribution and Z_2 and Z_4 are linear forms in their respective complements and where $\mathscr{E}Z_i = 0$, $\mathscr{E}Z_i^2 = 1$ for each i. Verify that $\cos^2 \theta = g_{11}$, $\cos^2 \psi = h_{11}$.

32. Suppose that, under the alternative hypothesis, X and Y are jointly normal with correlation, ρ, unknown but not large, say $\rho = 0\cdot25$. Suppose further that it is required to test the independence of X and Y by the χ^2 of a contingency table with sample size, N. Show that maximum power is attained with a small number of classes, the condensations being chosen so that the representations, g_{11} and h_{11} approach unity. The parameter of non-centrality is not greatly increased by finer partition of the marginal spaces.

33. Let a bivariate normal distribution be partitioned at the means. Prove that the variables have different signs with a probability, $p = \pi^{-1}\cos^{-1}\rho$; $\rho = \cos p\pi$.

[*Hint.* Let $\mathbf{B} = \begin{bmatrix} 1 & \rho \\ \rho & 1 \end{bmatrix}$, and suppose the variables are the elements of $\mathbf{X}$. Make the transformation, $\mathbf{U} = \mathbf{B}^{-\frac{1}{2}}\mathbf{X}$. Then the distribution of $\mathbf{U}$ is circularly symmetric. Find now the angle between the lines represented by the transforms of the two X-axes.] (Stieltjes, 1889; Sheppard, 1898b.)

34. Suppose a bivariate normal distribution is dichotomized at each mean. Show that the canonical correlation in the resulting four point distribution is $|1 - 2\pi^{-1}\cos^{-1}\rho| = 2\Pi^{-1}|\sin^{-1}\rho|$.

35. Let $\mathbf{Z} = (z_{ij})$ be a $p \times q$ matrix, the elements of which form a set of mutually independent standardized normal variables. Let the kth column be standardized and written as the first column of an orthogonal matrix, $\mathbf{P}_1$. Prove that the first element of the kth column of $\mathbf{P}_1^T\mathbf{Z}$ is the square root of a χ^2 with p degrees of freedom and that every other element in the kth column is zero. Prove that this non-zero element is distributed independently of every other element of $\mathbf{P}_1^T\mathbf{Z}$ and that every element of $\mathbf{P}_1^T\mathbf{Z}$ not in the kth column is standard normal and that these $p(q - 1)$ variables form a mutually independent set. (Wijsman, 1957.)

36. In Exercise 35, form an orthogonal $\mathbf{Q}_1$ whose first column is "parallel" to the first row of $\mathbf{P}_1^T\mathbf{Z}$. Prove that the first element of $\mathbf{P}_1^T\mathbf{Z}\mathbf{Q}_1$ is distributed as the square root of a χ^2 with $p + q - 1$ degrees of freedom. If the element in the leading position of $\mathbf{P}_1^T\mathbf{Z}\mathbf{Q}_1$ is isolated prove that it must be a diagonal element of $\mathbf{C} = \mathbf{P}^T\mathbf{Z}\mathbf{Q}$, where $\mathbf{C}$ has the canonical form, $c_{ij} = 0$ if $i \neq j$. If there are non-zero terms in the first column of $\mathbf{P}_1^T\mathbf{Z}\mathbf{Q}_1$, determine an orthogonal P_2 such that $\mathbf{P}_2^T\mathbf{P}_1^T\mathbf{Z}\mathbf{Q}_1$ has zeroes in every position in the first column except at the leading position. Now find a suitable $\mathbf{Q}_2$ and continue the iteration. Prove that the modulus of the term in the leading position tends to a limit, the square root of one of the latent roots of $\mathbf{Z}^T\mathbf{Z}$ or $\mathbf{Z}\mathbf{Z}^T$. (Lancaster, 1963a.)

37. Prove that if $R_1, R_2, \ldots, R_p$ are the canonical correlations formed from an $m \times n$ contingency table in the sense of Maung (1942) and the sample size is N, then NR_1^2 is stochastically larger than χ^2 with $(m + n - 3)$ degrees of freedom. (Lancaster, 1963a.)

38. Show that $\phi^2 + 1 = |(1 + \mathbf{P})(1 - \mathbf{P})|^{-\frac{1}{2}}$ in the multivariate normal with positive definite correlation matrix, $\mathbf{R} = 1 + \mathbf{P}$. (Pearson, 1904; Lancaster, 1957.)

39. Obtain the values of the generalised coefficients of correlation of the multi-variate normal distribution by a consideration of the integral of $\exp \left[\sum_i t_i x_i - \frac{1}{2}\sum t_i^2\right]$ over the distribution. Hence derive an expression for the expansion of the multivariate normal distribution generalising the Mehler identity. (Kendall, 1941; Moran, 1948.)

40. Expand $f(x, y; \rho)$ in a Taylor series in powers of ρ, writing $f(x, y; \rho) = (2\pi)^{-1} \exp \left[-\frac{1}{2}(x^2 + y^2)\right]\{u_0 + u_1 \rho/1! + u_2 \rho^2/2! + \ldots\}$. Differentiate $\log f(x, y; \rho)$ with respect to ρ to obtain

$$(1 - \rho^2)^2 \frac{\partial f}{\partial \rho} = [xy + \rho(1 - x^2 - y^2) + \rho^2 xy - \rho^3]f(x, y; \rho)$$

and then differentiate this expression n times with respect to ρ to obtain the recurrence formula

$$u_{n+1} = n(2n - 1 - x^2 - y^2)u_{n-1} - n(n-1)(n-2)u_{n-3} + xy\{u_n + n(n-1)u_{n-2}\}$$

Hence show that

$$u_0 = 1$$
$$u_1 = xy$$
$$u_2 = (x^2 - 1)(y^2 - 1)$$
$$\ldots$$
$$u_n = H_n(x)H_n(y),$$

where $H_n(.)$ is the Hermite polynomial of degree, n. (Pearson, 1900c.)

41. Prove that in the joint bivariate normal distribution the ratio, $\Omega(x, y) = f(x, y; \rho)/[g(x)h(y)]$, between the densities of the joint distribution and the product distribution can be approximated by the sum $S_n = \sum^n \rho^i x^{(i)} y^{(i)}$ by observing that

$$\iint \Omega^2(x, y)g(x)h(y)\, dx\, dy = (1 - \rho^2)^{-1}$$

and that the integral

$$\iint (\Omega - S_n)^2 g(x)h(y)\, dx\, dy$$

tends to zero as $n \to \infty$, since it is equal to $(1 - \rho^2)^{-1} - 1 - \rho^2 - \rho^4 \ldots - \rho^{2n}$.

42. Prove that the Mehler series converges absolutely a.e. since the series is of the form $(2\pi)^{-1} \exp \left[-\frac{1}{2}(x^2 + y^2)\right]\sum_0^\infty \rho^n \phi_n$, where $\sum \rho^n = (1 - \rho)^{-1}$ is absolutely convergent and $\phi_n = x^{(n)} y^{(n)}$, is an orthonormal set on the product measure. [This is a special case of a theorem on page 63 of Alexits, 1961.]

43. For random variables, $X_1, X_2, \ldots, X_p$ and $Y_1, Y_2, \ldots, Y_q$, suppose that $\mathscr{E}XX^T = 1_p$, $\mathscr{E}YY^T = 1_q$ and $\mathscr{E}XY^T = A$.

(a) Show that $\sum_i a_{ij}^2 \leqslant 1$,

(b) If $\sum_j a_{ij}^2 = 1$, $X_i = \sum a_{ij} Y_j$, a.e. (X_i is completely dependent on the set $\{Y_j\}$).

(c) If $\operatorname{tr} \mathbf{A}^T \mathbf{A} < 1$, then

$$\mathbf{V} = \begin{bmatrix} \mathbf{1}_p & \mathbf{A} \\ \mathbf{A}^T & \mathbf{1}_p \end{bmatrix}$$

is not singular.

(d) If a submatrix of $\mathbf{A}$ has unit rank and unit norm, then $\mathbf{V}$ is singular.

(e) If each X_i is linearly dependent on the set $\{Y_j\}$ and each Y_j is linearly dependent on the set $\{X_i\}$, then $\mathbf{A}$ is square and orthogonal.

CHAPTER XI

Two-way Contingency Tables

1. INTRODUCTORY

In this Chapter, we discuss contingency tables of two dimensions, leaving the more controversial topic of the contingency tables of higher dimensions to the next Chapter. We define a *contingency table* as an array resulting from an hierarchical classification of a collection of individuals, that is, the product space is partitioned into sets of the form, $A \times B \times \ldots \times C$. We do not insist that the individuals have been generated by a random process; for otherwise we would face a difficult problem as to whether an array is a contingency table or not before we have begun its analysis. Thus the collection may be an existent population or a subset of an existent population, for example, a specified heap of sand, the grains of which are to be classified by weight and largest diameter. After the classification has been made, it may be desired to give an interpretation of the findings. Thus the collection of the individual grains of sand may be supposed to have been generated by alternative methods from rock by weathering, fracturing by natural means or mechanical grinding by artificial means. The observed joint distribution of weight and diameter may well be a guide to the relevant hypothesis.

The classification by one character is termed a one-dimensional *marginal classification*. In this Chapter all marginal classifications are necessarily one-dimensional. In the higher-dimensional contingency tables, we may have marginal distributions of any order less than the number of dimensions of the table. In terms of the theory of probability measures we have one-dimensional marginal spaces, with an appropriate field of sets defined on each. X is a random variable defined on a space, A, with sets, $A_i \in \mathscr{A}$. Similarly Y is a random variable defined on a space, B, with sets $B_j \in \mathscr{B}$. For a classification, a system of sets $A_1, A_2, \ldots$ are chosen so that $A_1 \cup A_2 \cup \ldots = A$ and $A_i \cap A_j = \phi$ for $i \neq j$. The system of sets $\{A_i\}$ chosen is a partition of the space, A. Similarly, let us suppose that $\{B_i\}$ is a partition of B. The cardinalities of these two systems may be different and we do not assume at present that they are finite. We suppose now that to every individual to be classified, we can assign values of the random variables, X and Y. We do not exclude the possibility that the numbers of times $X \in A_i$ and/or $Y \in B_j$ may be sure variables. There are advantages in writing the sets in the hierarchial classification in the cross product notation of Cartesian product spaces, $A_i \times B_j$ rather than as $A_i \cap B_j$, for confusion may result in this second notation if A and B are the same spaces such as the real line belonging to different random variables, e.g. if the spaces of X and Y are the infinite real line. For each individual classified we will now have the result, $X_k \in A_i$ and $Y_k \in B_j$ or $(X_k, Y_k) \in A_i \times B_j$, $k = 1, 2, \ldots, N$. A contingency table corresponding to a two-way or two-dimensional classification of N individuals is the matrix or array of numbers, $(a_{ij}) = \mathbf{A}$, where a_{ij} is the number of times $(X_k, Y_k) \in A_i \times B_j$. a_{ij} can be written as a sum, $\mathscr{S}\psi_{A_i}(X_k)\psi_{B_j}(Y_k)$, the summation being over $k = 1(1)N$, and $\psi_{A_i}(x)$ is the indicator variable taking values 1 or 0, according as $x \in A_i$ or $x \notin A_i$. In practical examples, it is assumed that the classifications are exhaustive and mutually exclusive, although the result of the classification may not be known for some individuals, so that an additional class must often be adjoined to the system of sets giving the partition, the class "unknown" or "unclassified".

In many cases, the marginal variables can take an infinity or even a non-denumerable set of values, for example, the greatest diameters of grains of sand. Every classification of the grains by diameter then entails some loss of information. We are replacing a random variable, the diameter which is usually considered to take a continuum of values, by a variable which is the same for every individual falling into the same class. The original random variable is replaced by one assuming the finite set of values of a simple function; sometimes the value of this simple function in the set is the number of the row in the contingency table; at other times it may be the mid-point of, the end point of, or the average value of the random variable in a particular interval or set.

In summary a two-dimensional, or more briefly a two-way, contingency table is made when the individuals of a collection are classified by two characteristics and the number of individuals in the possible product sets are written as a_{ij}, the entries in a two dimensional array or matrix, $(a_{ij}) = \mathbf{A}$.

2. PROBABILITY MODELS IN A TWO-WAY CONTINGENCY TABLE

If probability methods of analysis are to be applied to contingency tables, the tables must be supposed to have been generated from some probability model. Following the investigations of such authors as E. S. Pearson (1947) Sheppard (1898), K. Pearson (1900, 1904 and *passim*), R. A. Fisher (1922 and *passim*), Bartlett (1935), Barnard (1947) and Roy and Mitra (1956), three models have been developed and given various names as follows:

Model 1. *Unrestricted bivariate sampling.* This is the two-variate case of Roy and Mitra (1956) and the generalization of the double dichotomy of Barnard (1947). It is supposed that there is a joint distribution, $\mathscr{P}(X \in A_i, Y \in B_j) = f_{ij}$, $i = 1(1)m$, $j = 1(1)n$. Each individual of the collection mentioned in the previous section is obtained by some random process assigning weights, f_{ij}, to the cell or product set, $A_i \times B_j$, $i = 1(1)m$, $j = 1(1)n$. N independent observations on the two-dimensional random variable (X, Y) are made and a contingency table with entries, a_{ij}, is formed. The probability of obtaining the set of frequencies $\{a_{ij}\}$ is given by the usual multinomial expression,

$$(2.1) \qquad \mathscr{P}\{\{a_{ij}\} \mid \{f_{ij}\}, N\} = N! \prod_{i,j} \{f_{ij}^{a_{ij}}/a_{ij}!\}.$$

Example (*i*) Two distinguishable coins are tossed independently of one another and the experiment is repeated N times. The results can be arranged in the form of a contingency table with two rows and two columns. Let the first suffix refer to the first coin, 1 for a head and 2 for a tail and a similar convention be used for the second coin. Then a_{11}, a_{12}, a_{21} and a_{22} are respectively the number of times HH, HT, TH and TT have been recorded. Each $f_{ij} = \frac{1}{4}$ if the coins are true. A genetic counterpart of this experiment is given in the next example.

Example (*ii*) Suppose that there are $a_{..}$ crosses of parents with genetic constitutions $AaBb$ and $aabb$ and that the genes at the A locus segregate independently of the genes at the B locus. There are four types of offspring $AaBb$, $Aabb$, $aaBb$ and $aabb$ and the numbers of each type can be arranged in the form of a 2×2 contingency table. $f_{ij} = \frac{1}{4}$ under the hypotheses made and the usual genetic theory.

Example (*iii*) A 6×6 table can be formed by tossing two dice independently of one another and making $a_{..}$ independent experiments.

Example (*iv*) The genetic example can be varied by assuming that the genes at the A locus do not segregate independently of the genes at the B locus. The loci may be on the same chromosome. A two-dimensional probability distribution results in which $f_{ij} \neq f_{i.}f_{.j}$. Similar modifications of the coin or dice examples would be rather artificial.

Example (*v*) Let X and Y have the joint normal distribution with coefficient of correlation, ρ. Let the marginal spaces of X and Y be partitioned into m and n classes respectively with boundaries of the form, $-\infty = a_0, a_1, \ldots,$ $a_m = +\infty$, $-\infty = b_0, b_1, \ldots, b_n = +\infty$. Then if N independent drawings are made from this bivariate population, an $m \times n$ contingency table of mn cells results.

Model 2.   *The comparative trial.* Let $a_{i.}$ observations be made in the class i; or in other words suppose $a_{i.}$ observations are made in the ith row of a contingency table. Suppose now that for each observation in the ith row, there is a probability, f_{ij}, that the column variable takes the value, j. Suppose further that each observation is independent of every other observation. It is thus assumed that

(2.2) $$\sum_{j=1}^{n} f_{ij} = 1, \qquad \text{for } i = 1, 2, \ldots, m.$$

and so the probability that the entries in the resulting contingency table take the values, a_{ij}, is a product of multinomial probabilities,

(2.3) $$\mathcal{P}\{\{a_{ij}\} \mid \{a_{i.}\}, \{f_{ij}\}\} = \prod_{i=1}^{m} \left[a_{i.}! \prod_{j=1}^{n} (f_{ij}^{a_{ij}}/a_{ij}!) \right].$$

Example   *(vi)* $a_{1.}$ animals are submitted to the first treatment, $a_{2.}$ to the second . . . They are then classified according to the results of treatment into n exhaustive and mutually exclusive classes. For example, the treatments may be for some disease and if $n = 2$, the two classes might be cure and failure.

Example   *(vii)* a_{i1} males and a_{i2} females are recorded among the children of $a_{i.}$ mothers in the ith age group. An $m \times 2$ table results.

This model is the comparative trial of Barnard (1947) and the "one variate, one classification" case of Roy and Mitra (1956) since the numbers in the rows but not the numbers in the columns are chosen arbitrarily. As suggested by the second example, many demographic observations are appropriately discussed by the use of this model.

Model 3.   *The permutation model.* Let $a_{i.} > 0$, and $a_{.j} > 0$ for $i = 1, 2, \ldots,$ m and $j = 1, 2, \ldots, n$ and $\sum_{i} a_{i.} = \sum_{j} a_{.j} = N$. From N distinguishable objects select $a_{1.}$ objects without replacement then $a_{2.}, \ldots, a_{m.}, \ldots$. Now by a random process without replacement select $a_{.1}$ objects from the whole table and assign them to column 1, $a_{.2}$ objects to column 2, $\ldots, a_{.n}$ objects to column n. Then the number of ways of obtaining a table with entries $\{a_{ij}\}$ is the product of terms of the form,

(2.4) $$a_{i.}! \Big/ \prod_{j} a_{ij}!$$

subject to the conditions, $\sum_{i} a_{ij} = a_{.j}$. The column totals can be chosen in

(2.5) $$N! \Big/ \prod_{j} a_{.j}!$$

ways. So that for fixed column totals, the probability of the set $\{a_{ij}\}$ is given by

$$(2.6) \qquad \mathscr{P}\{\{a_{ij}\} \mid \{a_{i.}\}, \{a_{.j}\}\} = \prod_i (a_{i.}!/\prod_j a_{ij}!) \Big/ \left(N!/\prod_j a_{.j}! \right)$$

$$= \prod_i a_{i.}! \prod_j a_{.j}! \Big/ \left(N! \prod_{i,j} a_{ij}! \right)$$

This permutation model is not much used but from it there can be derived a useful combinatorial formula, for fixed values of $a_{i.}$ and $a_{.j}$.

$$(2.7) \qquad \sum \prod_{i,j} (a_{ij}!)^{-1} = N! \Big/ \left(\prod_i a_{i.}! \prod_j a_{.j}! \right)$$

where summation is over all such values of a_{ij} that

$$\sum_i a_{ij} = a_{i.} \quad \text{and} \quad \sum_j a_{ij} = a_{.j}.$$

Example (*viii*) A set of four bridge hands may be considered. There are then 13 cards to be dealt to each player (row) and the total of each suit (column) is 13. The probability of any distribution of cards at a deal is then given by (2.6) with $a_{i.} = a_{.j} = 13$, $N = 52$. a_{ij} is the number of the jth suit dealt to the ith player. This example can be varied by considering an individual hand only. See page 45 of Feller (1958).

It is appropriate to give some examples here of two-way arrays of numbers, which do not constitute two-way contingency tables, generated by any of the three models given above.

Example (*ix*) rc sets of k seeds are submitted to two sets of treatments (rows and columns) and, a_{ij}, the number germinating from each pair of treatments, counted. This is a layer of an $r \times c \times 2$ three-dimensional table. This statement is confirmed by the consideration that the numbers of seeds not germinating, $\{k - a_{ij}\}$, constitute a complementary table. These two tables in fact are the two layers of an $r \times c \times 2$ three-way contingency table.

Example (*x*) Consider a pack of 52 cards with 16 court cards and 36 plain. Let there be N cards chosen without replacement and classified into red and black (rows) and court and plain (columns). This once again is not a two-dimensional contingency table of the permutation model type but a "layer" of a three-dimensional table.

Example (*xi*) Let two judges rank r items of M sets. Let a_{ij} be the number of times the first judge gave the ith rank and the second judge gave the jth rank to the same items. The appropriate analysis is given by R. L. Anderson (1959).

Theorem 2.1. *In all three models*

$$(2.8) \qquad \mathscr{P}\{\{a_{ij}\} \mid \{a_{i.}\}, \{a_{.j}\}\} = \prod_i a_{i.}! \prod_j a_{.j}! \bigg/ \bigg(N! \prod_{i,j} a_{ij}!\bigg).$$

Proof. This has already been proved for Model 3. It can be proved by dividing (2.1) by the product of the two marginal multinomial expressions in Model 1 and by dividing the product of expressions of the form, (2.3), by the column multinomial expression in Model 2. Theorem 2.1 can be expressed as—

Corollary 1. *The distribution of entries in a contingency table of two dimensions, under the hypothesis of independence of the marginal variables, does not involve the marginal parameters, $\{f_{i.}\}$ and $\{f_{.j}\}$ in Model 1, nor $\{f_j\}$ in Model 2, under the hypothesis of homogeneity, $f_{ij} = f_j$.*

The unconditional probability of the entries in a contingency table under Model 1 can be considered as a conditional probability, generated by a set of Poisson variates; so that, each a_{ij} is the observed value from a Poisson with parameter, $\lambda p_{i.} p_{.j}$, the Poisson distributions being mutually independent. The probability required is then the probability of obtaining a given set $\{a_{ij}\}$ conditional on $\sum_{i,j} a_{ij} = N$. The proof follows by straight out verification of the relation

$$(2.9) \qquad \mathscr{P}\{\{a_{ij}\} \mid \{f_{ij}\}, N\} = \prod_{i,j} \mathscr{P}\{a_{ij} \mid \{\lambda f_{ij}\}, H\}/\mathscr{P}\{N \mid \lambda, H\}$$

H being the hypothesis that each a_{ij} is a Poisson variable with parameter λf_{ij}. A converse of Theorem 2.1, due to Fisher (1935) and Wilson (1942) is as follows for the special case of the 2×2 contingency table.

Theorem 2.2. *If the distribution of entries in a 2×2 contingency table, conditional on the marginal totals, is independent of the marginal parameters, then*

$$(2.10) \qquad\qquad\qquad f_{11} f_{22} = f_{12} f_{21},$$

and

$$(2.11) \qquad\qquad\qquad \psi \equiv f_{11} f_{22}/(f_{12} f_{21}) = 1$$

Proof.

$$(2.12) \qquad \mathscr{P}\{\{a_{ij}\} \mid \{f_{ij}\}, N\} = f_{11}^{a_{11}} f_{12}^{a_{12}} f_{21}^{a_{21}} f_{22}^{a_{22}} N! \bigg/ \prod_{i,j} a_{ij}!$$

$$= f_{21}^{a_{.1}} f_{12}^{a_{1.}} f_{22}^{N - a_{1.} - a_{.1}} \psi^{a_{11}} N! \bigg/ \prod_{i,j} a_{ij}!$$

The right side of (2.12) is now in the form,

$$C_1 \psi^{a_{11}} N!/C_2,$$

where C_1 is constant for variations in a_{11} and C_2 only contains factorials in the four cell contents.

The probabilities of the different values of a_{11} will only be independent of the parameter if (2.10) or equivalently (2.11) holds. The conditional probability of a contingency table (2.6) can be computed for any given table but, except in the 2 × 2 table, there is usually difficulty in ordering the different possible sets of the observed numbers by their probability.

Theorem 2.3. *If the distribution of entries in an m × n contingency table, conditional on the marginal totals, is independent of the marginal parameters, then*

$$(2.13) \quad f_{ij}f_{i'j'} = f_{ij'}f_{i'j}, \quad i = 1, 2, \ldots, m, \quad j = 1, 2, \ldots, n.$$

Proof. The distribution of the $\{a_{ij}\}$ given $\{a_{i.}\}$ and $\{a_{.j}\}$ does not involve the marginal parameters. In particular, the distribution of a_{11} given $a_{1.}$ and $a_{.1}$ does not involve them. However, a_{11} can be regarded as the distribution of the entry in a cell of a 2 × 2 table. Theorem 2.2 can then be applied and $f_{11} = f_{1.}f_{.1}$. A similar process can be applied to the contents of any cell and $f_{ij} = f_{i.}f_{.j}$ follows and equivalently (2.13).

3. TESTS OF INDEPENDENCE

Often a bivariate distribution of general form is given by hypothesis. We may be interested in a contingency table formed from it by the process of condensation. Thus a marginal space may be partitioned into a finite number of sets. If these sets are numbered or if each is assigned a distinct value, a function taking a finite number of values is defined, a *simple function* in the terminology of measure theory. To every such partition of a marginal space there exists a simple function and conversely every simple function taking m distinct values on the marginal distribution determines a partition. If the simple function takes precisely two distinct values, it is a step function in the usual terminology of measure theory. Step functions may be defined on both marginal distributions. We define the Boas coefficient of correlation as the correlation between step functions, one defined on each marginal distribution.

Example. Suppose X and Y are jointly normal and each is standardized. Let step functions, ξ and η, be defined by $\xi = +1$ if $X \geqslant 0$, $\xi = -1$ if $X < 0$; $\eta = +1$ if $Y \geqslant 0$, $\eta = -1$ if $Y < 0$. Then the Boas coefficient of correlation is $(2/\pi) \sin^{-1} \rho$. Note that ξ and η each define a dichotomy. If a random sample of N observations be made, the observed coefficient of correlation will have a probability distribution dependent on ρ.

A Boas coefficient of correlation in a bivariate distribution on four points is uniquely determined, except for a convention as to sign, no matter what values are assigned to the step functions. For example, if X takes the values, 1 and 2, with probabilities, q and p, and we consider any arbitrary function, ξ, on the two points taking values, u and v; $u \neq v$, then $\mathscr{E}(\xi) = uq + vp$, $\xi - \mathscr{E}(\xi) = (u - v)p$ when $X = 1$, and $\xi - \mathscr{E}(\xi) = -(u - v)q$ when $X = 2$. The variance of ξ is $(u - v)^2 pq$. The standardized variable,

$$\{\xi - \mathscr{E}(\xi)\}/\{\text{var } \xi\}^{\frac{1}{2}},$$

then is independent of the values, u and v, and is uniquely determined by the values, p and q, except for a convention as to sign. In the notation of the following theorem we write without loss of generality,

$$\xi = \xi_1 = -f_{2.}^{\frac{1}{2}} f_{1.}^{-\frac{1}{2}}, \quad \text{when } X = 1$$

$$\xi = \xi_2 = +f_{1.}^{\frac{1}{2}} f_{2.}^{-\frac{1}{2}}, \quad \text{when } X = 2$$

(3.1)

$$\eta = \eta_1 = -f_{.2}^{\frac{1}{2}} f_{.1}^{-\frac{1}{2}}, \quad \text{when } Y = 1$$

$$\eta = \eta_2 = +f_{.1}^{\frac{1}{2}} f_{.2}^{-\frac{1}{2}}, \quad \text{when } Y = 2.$$

The correlation in such a table is $\mathscr{E}(\xi\eta)$.

Theorem 3.1. *In a 2-dimensional distribution on four points, with measures, $\{f_{ij}\}$, $i, j = 1, 2$, a NASC for independence is that the Boas coefficient of correlation, $\mathscr{E}(\xi\eta) = R$, should be zero.*

Proof. Let ξ and η be the standardized functions on the marginal distribution so that $R = \mathscr{E}(\xi\eta)$ and assume that no one of $f_{1.}, f_{2.}, f_{.1}$ and $f_{.2}$ is zero. Write $\Omega = \Omega(i,j) = f_{ij}/(f_{i.}f_{.j})$. Then Ω is finite and can be written as a linear form, $\Omega = a_{00} + a_{10}\xi + a_{01}\eta + a_{11}^*\xi\eta$. If Ω is multiplied by 1, ξ, η and $\xi\eta$, and summed with respect to the independence probabilities, $f_{i.}f_{.j}$, there follows that $a_{00} = 1$, $a_{11}^* = \mathscr{E}(\xi\eta) = R$, $a_{10} = a_{01} = 0$, where $\mathscr{E}(\xi\eta) = \sum \xi_i \eta_j f_{ij}$.

(3.2) $$f_{ij} = (1 + R\xi\eta)f_{i.}f_{.j}$$

$R = 0$ is then obviously a sufficient condition for independence; $R = 0$ is necessary because $\xi\eta$ is not identically null.

If we now consider any arbitrary bivariate distribution, we can generalize the previous theorem to

Theorem 3.2. *A NASC for the independence of two random variables, X and Y, is that the Boas coefficient of correlation should be zero for any arbitrary partition (dichotomy) of the marginal spaces.*

Proof. *Necessity.* This follows because

$$f_{ij} = \mathscr{P}\{X \in A, Y \in B\} = \mathscr{P}\{X \in A\}\mathscr{P}\{Y \in B\} = f_{i.}f_{.j}.$$

Sufficiency. For independence not to hold, the above relation must fail to hold for some non-trivial partition; then we should have $R \neq 0$ by the previous theorem and a contradiction would have resulted.

Theorem 3.3. *A NASC for the independence of two random variables, X and Y, is that every simple function on the distribution of X is uncorrelated with every simple function on the distribution of Y.*

Proof. This theorem can be deduced from Theorem 3.2.

It can be also be deduced by considering a finite bivariate distribution defined on mn points as follows. Let the step function ξ take m distinct values on the distribution of X, each with positive probability. Similarly suppose η takes n distinct values on the distribution of Y. Then the required theorem follows from Theorem VI.2.1. It can also be deduced from the following theorem, of which versions have been given by Sarmanov (1958), Rényi (1959) and Lancaster (1959, 1960c). Sarmanov's proof used the ϕ^2 boundedness condition.

Theorem 3.4. *A NASC for the independence of two random variables X and Y is that every function of X should be independent of every function of Y.*

Theorem 3.5. *A NASC for the independence of two random variables is that for complete orthonormal sets $\{x^{(i)}\}$ on the space of X and $\{y^{(j)}\}$ on the space of Y, the expectations $\mathscr{E}(x^{(i)}y^{(j)})$, $i \neq 0$, $j \neq 0$ should all be zero.*

Theorem 3.6. *A NASC condition for the independence of X and Y is that the maximum correlation between functions $\xi(x)$ and $\eta(y)$ should be zero.*

It follows from these theorems that the statistical test of independence must be equivalent to the statement that the $(m-1)(n-1)$ standardized variables of the form $N^{-\frac{1}{2}}\mathscr{S}x^{(i)}y^{(j)}$ do not differ significantly from zero. If there had been no hypothesis beforehand of the type of dependence, then any one of these variables could be tested against the null hypothesis that they were asymptotically normal $(0, 1)$. If no specific alternative hypothesis had been made, χ^2 is the overall test of independence.

We defer until Section 7 discussion on the procedure when more precise information on the alternative hypotheses is available.

4. THE FOURFOLD TABLE

Two-way tables are usually treated following Sheppard (1898a), Pearson (1904), and Fisher (1922a) with marginal totals fixed and parameters estimated from the data but this is by no means an essential feature of the analysis. We consider, as an example of Model 1, data from Table 1 of Roberts, Dawson and Madden (1939), which has been analysed in detail by Haldane (1939).

The numbers of offspring from a certain breeding experiment were recorded as follows:

Table 4.1

Factors	b	B	Totals
a	921	971	1892
A	907	935	1842
Totals	1828	1906	3734

Let us suppose that hypotheses are made as follows. The probabilities that an individual should be a or A are $f_{1.}$ and $f_{2.}$ respectively. Similarly the probabilities that he should be b or B are $f_{.1}$ and $f_{.2}$. The probability of an individual being ab is f_{11}. We suppose that a sample of N independently distributed individuals is made. A null hypothesis, H_0, of independence will be $f_{ij} = f_{i.}f_{.j}$, $i, j = 1, 2$. This null hypothesis would be appropriate if there were N matings of the type $AaBb \times aabb$ and if differential mortality were suspected. If no such differential mortality were suspected another natural hypothesis would be $f_{1.} = f_{2.} = f_{.1} = f_{.2} = \frac{1}{2}$. If independence between the A and B classifications is to be tested also, then the null hypothesis would be $f_{ij} = \frac{1}{4}$ for all four possibilities. The analysis of the table will then run as follows

$$(4.1) \qquad \mathscr{P}(\{a_{ij}\} \mid \{f_{ij}\}, H_0) = N! \prod_{i,j} [f_{ij}^{a_{ij}}/a_{ij}!]$$

However, the row totals and the column totals are independently distributed under the hypotheses made so that the probability of obtaining the marginal totals is

$$(4.2) \quad \mathscr{P}(\{a_{i.}\}, \{a_{.j}\} \mid \{f_{i.}\}, \{f_{.j}\}, H_0) = \frac{N!}{a_{1.}!a_{2.}!} f_{1.}^{a_{1.}} f_{2.}^{a_{2.}} \cdot \frac{N!}{a_{.1}!a_{.2}!} f_{.1}^{a_{.1}} f_{.2}^{a_{.2}}.$$

From (4.1) and (4.2), after substituting $f_{ij} = f_{i.}f_{.j}$ in (4.1), the conditional probability of the table is obtained

$$(4.3) \qquad \mathscr{P}(\{a_{ij}\} \mid \{a_{i.}\}, \{a_{.j}\}, H_0) = \frac{a_{1.}!a_{2.}!a_{.1}!a_{.2}!}{N!a_{11}!a_{12}!a_{21}!a_{22}!}.$$

In the example given this is a product of factorials,

$$\frac{1892!1842!1828!1906!}{3734!921!971!907!935!}.$$

In principle, all the terms of the distribution of the variable Z, the number of individuals falling into the cell at $(1, 1)$ position in the table, could be computed and an exact probability associated with each value of Z. Usually the normal approximation would be used.

Model 2 is often appropriate for the analysis of clinical or prophylactic field trials. A classical example is the analysis of the results of inoculation against cholera given by Greenwood and Yule (1915) as follows:—

Table 4.2

	Not Attacked	Attacked	Totals
Inoculated	276	3	279
Not inoculated	473	66	539
Totals	749	69	818

The null hypothesis is that treatment is ineffective, so that $f_{11} = f_{21} = p$, $f_{12} = f_{22} = q = 1 - p$. The row totals are fixed in this example, for they are the numbers of inoculated and uninoculated before the epidemic began. It is assumed that every individual is exposed to the same risk of infection. H_0 says that this risk (or attack rate) is not altered by the prophylactic inoculation. There follows

$$(4.4) \qquad \mathscr{P}(\{a_{ij}\} \mid p, \text{binomial}) = \frac{a_1.!}{a_{11}!a_{12}!} p^{a_{11}}q^{a_{12}} \frac{a_2.!}{a_{21}!a_{22}!} p^{a_{21}}q^{a_{22}}$$

p is unknown, so that this probability cannot be computed. However, a conditional probability $\mathscr{P}(\{a_{ij}\} \mid \{a_{i.}\}, \{a_{.j}\})$ can be obtained as the ratio, $\mathscr{P}(\{a_{ij}\} \mid \{a_{i.}\}, \text{binomial})/\mathscr{P}(\{a_{.j}\} \mid \{a_{i.}\}, \text{binomial})$. The conditional probability is (4.4) divided by the probability that $a_{.1}$ individuals fall into the first column. Once again the same expression as in (4.3) is obtained.

Although this model was mentioned by Pearson (1916b) it was Fisher (1922a) who clarified with its aid some outstanding problems on the distribution of χ^2 with the following reasoning. Let it be assumed that the asymptotic normal theory is appropriate for the binomial. With k observations, the standard error of the proportion observed is $\sqrt{(pq/k)}$. The observed proportion in the first row, $a_{11}/a_{1.}$, is compared with the observed proportion in the second row, $a_{21}/a_{2.}$; the difference $a_{11}/a_{1.} - a_{21}/a_{2.}$ is divided by its standard error to obtain an asymptotically standardized normal variable,

$$(4.5) \qquad \chi = \left(\frac{a_{11}}{a_{1.}} - \frac{a_{21}}{a_{2.}}\right) \bigg/ \left(\frac{pq}{a_{1.}} + \frac{pq}{a_{2.}}\right)^{\frac{1}{2}}.$$

However p has to be estimated from the data and the relevant information is that $a_{.1}$ observations have fallen into the first column. We estimate

(4.6) $\hat{p} = a_{.1}/N.$

Substituting (4.6) in the denominator of (4.5) and simplifying the numerator, we have

(4.7) $$\chi^2 = \frac{(a_{11}a_{22} - a_{12}a_{21})^2 N}{a_{1.}a_{2.}a_{.1}a_{.2}},$$

which is algebraically equal to the Pearson χ^2 of the form,

(4.8) $$\chi^2 = \sum_{i,j} (a_{ij} - \mathscr{E}a_{ij})^2/\mathscr{E}a_{ij},$$

in which $\mathscr{E}a_{ij}$ is to be taken as $a_{i.}a_{.j}/a_{..}$.

This reasoning can be supported by the observation that the difference is independent of the true value of p and the fact that for neither p nor q very small, the product pq varies only slowly for variations in p.

Model 3 is appropriate in applications more rarely than Models 1 and 2 but it is appropriate in some engineering applications according to Barnard (1947). The test is non-parametric.

Example. Suppose the 52 cards of a bridge pack are dealt into two heaps of 26. There are 26 red cards and 26 black cards, which may be taken as the total numbers in first and second rows. The first heap may be taken as the first column of the table. Considering the whole pack, the 26 cards in the first heap can be obtained in $\binom{52}{26}$ ways. For z cards in the position $(1, 1)$ there are $\binom{26}{z}$ ways of obtaining the z cards. There are $\binom{26}{26 - z}$ ways of obtaining the $(26 - z)$ cards in the position $(2, 1)$. The ratio of favorable cases to all cases is thus

$$\binom{26}{z}\binom{26}{26 - z}\bigg/\binom{52}{26}.$$

In the general case, there are $a_{1.}$ individuals in the first row and $a_{2.}$ in the second row, $a_{1.} + a_{2.} = N$. $a_{.1}$ individuals can appear in the first column in $\binom{N}{a_{.1}}$ ways, which are considered to be equally likely. z can appear in the $(1, 1)$ cell in $\binom{a_{1.}}{z}$ ways and $(a_{.1} - z)$ in the $(2, 1)$ cell in $\binom{a_{2.}}{a_{.1} - z}$ ways. The probability that the number, Z, in position $(1, 1)$ should take the value, z, is thus

(4.9) $\mathscr{P}(Z = z) = \mathscr{P}(\{a_{ij}\} \mid \{a_{i.}\}, \{a_{.j}\}, H_0) = \binom{a_{1.}}{z}\binom{a_{2.}}{a_{.1} - z}\bigg/\binom{N}{a_{.1}}$

and this expression is identical with (4.3) after replacing z by a_{11}. It is remarkable that the three models yield the same expression for the conditional probability, namely, the coefficient of $t^{a_{11}}$ in the expansion for the hypergeometric series,

$$(4.10) \qquad \frac{a_{2.}!a_{.2}!}{N!} F(-a_{.1}, -a_{1.}; N - a_{1.} - a_{.1} + 1; t)$$

where

$$(4.11) \quad F(a, b; c; t) = 1 + \frac{a \cdot b}{1 \cdot c} t + \frac{a(a + 1)b(b + 1)}{1 \cdot 2c(c + 1)} t^2$$

$$+ \frac{a(a + 1)(a + 2)b(b + 1)(b + 2)}{1 \cdot 2 \cdot 3c(c + 1)(c + 2)} t^3 + \dots$$

in the notation of page 281 of Whittaker and Watson (1935). It is always possible to transpose the table and to rearrange the rows so that the inequality holds

$$(4.12) \qquad a_{1.} \leqslant a_{.1} \leqslant a_{.2} \leqslant a_{2.}.$$

Without loss of generality, $a_{1.}$ can thus be taken to be the smallest marginal total. In the cell at position $(1, 1)$, the random variable Z can take values $0, 1, 2, \dots, a_{1.}$. Since (4.3) or (4.9) is a probability, there follows the identity,

$$(4.13) \qquad \sum_{z=0}^{a_{1.}} \frac{a_{1.}!a_{2.}!a_{.1}!a_{.2}!}{N!z!(a_{1.} - z)!(a_{.1} - z)!(N - a_{1.} - a_{.1} + z)!} = 1.$$

This identity is true not only for a single set of marginal totals but for all such sets subject to the inequality (4.12). This remark enables factorial moments of Z of all orders, $r \leqslant a_{1.}$, to be computed in the manner of Romanovsky (1925). For $r > a_{1.}$, the factorial moments are all zero.

$$(4.14) \qquad \mathscr{E}Z^{(r)} = \sum_{z=0}^{a_{1.}} z^{(r)} \mathscr{P}(Z = z)$$

$$= \sum_{z=r}^{a_{1.}} z^{(r)} \mathscr{P}(Z = z)$$

Now a common factor $a_{1.}^{(r)}a_{.1}^{(r)}/N^{(r)}$ can be taken out of each term in the summation; $(z - r)$, $(a_{1.} - r)$, $(a_{.1} - r)$ and $(N - r)$ can be replaced by z', $a_{1.}'$, $a_{.1}'$ and $a_{..}'$; $a_{2.}$ and $a_{.2}$ can be written $a_{2.}'$ and $a_{.2}'$. With these notational conventions, (4.14) becomes

$$(4.15) \quad \mathscr{E}Z^{(r)} = \frac{a_{1.}^{(r)}a_{.1}^{(r)}}{a_{..}^{(r)}} \sum_{z'=0}^{a_{1.}'} \frac{a_{1.}'!a_{2.}'!a_{.1}'!a_{.2}'!}{a_{..}'!z'!(a_{1.}' - z')(a_{.1}' - z')!(a_{..}' - a_{1.}' - a_{.1}' + z')!}$$

$$= a_{1.}^{(r)}a_{.1}^{(r)}/N^{(r)},$$

since the expression in the summation is of the same form as the expression

(4.13). From (4.15) there follow

(4.16) $$\mathscr{E}Z = a_{1.}a_{.1}/N,$$
and

(4.17) $$\operatorname{var} Z = a_1.a_2.a_{.1}a_{.2}/\{N^2(N-1)\}.$$

Further computations show that the higher moments of Z approximate to those of the normal if $a_{1.}$, the smallest marginal total, becomes sufficiently large. $(Z - \mathscr{E}Z)$ divided by its standard deviation is thus asymptotically the square of a standardized normal variable and hence χ^2 with one degree of freedom. Therefore

(4.18) $$\frac{(Z - \mathscr{E}Z)^2}{\operatorname{var} Z} = \frac{(a_{11} - a_{1.}a_{.1}/N)^2}{\operatorname{var} Z} = \frac{(a_{11}a_{22} - a_{12}a_{21})^2(N-1)}{a_{1.}a_{2.}a_{.1}a_{.2}}$$

is approximately χ^2 with one degree of freedom. It may be noted that (4.18) has $(a_{..} - 1)$ in place of $a_{..} = N$ in (4.7). This is not surprising since the results are asymptotic. However, the expectation of the expression on the left of (4.18) is rigorously unity. For some work, perhaps in all cases, (4.18) is preferable to (4.7) as an expression for χ^2. Alternatively, it is evident that the expectation of the usual χ^2 calculated from (4.7) or (4.8) is $N/(N-1)$. (4.7) or (4.8) can be obtained from the factorial expression (4.3) by the application of Stirling's approximation. The asymptotic results in the four-fold table can be obtained in another way using theoretically given or estimated parameters in Model 1. Suppose that ξ is the orthonormal function on the row marginal distribution and η is the orthonormal function on the column distribution. Under the hypothesis of independence, 1, ξ, η and $\xi\eta$ form an orthonormal basis on the distribution given by $p_{ij} = p_{i.}p_{.j}$. The Bernstein generalized central limit theorem can now be invoked. The variables $N^{-\frac{1}{2}}\mathscr{S}\xi$, $N^{-\frac{1}{2}}\mathscr{S}\eta$ and $N^{-\frac{1}{2}}\mathscr{S}\xi\eta$ are all standardized asymptotically normal variables, mutually independent. It is evident from the definitions of ξ and η that the first two variables are the standardized (binomial) variables on the rows and columns and can be exhibited in the form

(4.19) $$N^{-\frac{1}{2}}\mathscr{S}\xi = \frac{a_{1.} - Np_{1.}}{\sqrt{(Np_{1.}p_{2.})}}.$$

Similarly $N^{-\frac{1}{2}}\mathscr{S}\eta$ is the corresponding function on the columns. $N^{-\frac{1}{2}}\mathscr{S}\xi$ and $N^{-\frac{1}{2}}\mathscr{S}\eta$ measure what may be termed "main effects" on analogy with the notation in the analysis of variance. $N^{-\frac{1}{2}}\mathscr{S}\xi\eta$ may similarly be given the title of the "first order interaction". In the notation of (3.1), there follows

(4.20) $$\begin{aligned} \mathscr{S}\eta &= a_{11}\eta_1 + a_{12}\eta_2 + a_{21}\eta_1 + a_{22}\eta_2 = a_{.1}\eta_1 + a_{.2}\eta_2 \\ \mathscr{S}\xi &= a_{11}\xi_1 + a_{12}\xi_1 + a_{21}\xi_2 + a_{22}\xi_2 = a_{1.}\xi_1 + a_{2.}\xi_2 \\ \mathscr{S}\xi\eta &= a_{11}\xi_1\eta_1 + a_{12}\xi_1\eta_2 + a_{21}\xi_2\eta_1 + a_{22}\xi_2\eta_2. \end{aligned}$$

Let now $\{q_{ij}\}$ be written as the elements of a vector in dictionary order, where

(4.21) $$q_{ij} = (a_{ij} - Np_{i.}p_{.j})/(Np_{i.}p_{.j})^{\frac{1}{2}}.$$

Note that $\sum_{i,j} q_{ij}^2$ is the Pearson χ^2 of a multinomial with four cells, when the cell expectations are given by hypothesis.

$$(4.22) \quad \begin{bmatrix} 0 \\ N^{-\frac{1}{2}}\mathscr{S}\eta \\ N^{-\frac{1}{2}}\mathscr{S}\xi \\ N^{-\frac{1}{2}}\mathscr{S}\xi\eta \end{bmatrix} = N^{-\frac{1}{2}} \begin{bmatrix} 1 & 1 & 1 & 1 \\ \eta_1 & \eta_2 & \eta_1 & \eta_2 \\ \xi_1 & \xi_1 & \xi_2 & \xi_2 \\ \xi_1\eta_1 & \xi_1\eta_2 & \xi_2\eta_1 & \xi_2\eta_2 \end{bmatrix} \begin{bmatrix} a_{11} - Np_{11} \\ a_{12} - Np_{12} \\ a_{21} - Np_{21} \\ a_{22} - Np_{22} \end{bmatrix}$$

$$= N^{-\frac{1}{2}} \begin{bmatrix} 1 & 1 \\ \xi_1 & \xi_2 \end{bmatrix} \times \begin{bmatrix} 1 & 1 \\ \eta_1 & \eta_2 \end{bmatrix} \cdot \mathbf{a}$$

$$= \begin{bmatrix} p_{1.}^{\frac{1}{2}} & p_{2.}^{\frac{1}{2}} \\ p_{1.}^{\frac{1}{2}}\xi_1 & p_{2.}^{\frac{1}{2}}\xi_2 \end{bmatrix} \times \begin{bmatrix} p_{.1}^{\frac{1}{2}} & p_{.2}^{\frac{1}{2}} \\ p_{.1}^{\frac{1}{2}}\eta_1 & p_{.2}^{\frac{1}{2}}\eta_2 \end{bmatrix} \mathbf{q}$$

$$= \mathbf{R} \times \mathbf{Cq}$$

where in passing from the second to the third line, the direct product has been post-multiplied by diag $(p_{1.}^{\frac{1}{2}}, p_{2.}^{\frac{1}{2}}) \times$ diag $(p_{.1}^{\frac{1}{2}}, p_{.2}^{\frac{1}{2}})$ and the $\mathbf{a}$ vector has been premultiplied by the inverse of the same direct product of diagonal matrices, but

$$N^{-\frac{1}{2}} \text{ diag } (p_{1.}^{-\frac{1}{2}}, p_{2.}^{-\frac{1}{2}}) \times \text{ diag } (p_{.1}^{-\frac{1}{2}}, p_{.2}^{-\frac{1}{2}})\mathbf{a} = \mathbf{q}.$$

The $\{q_{ij}\}$ could also have been written as the elements of a matrix, $\mathbf{Q}$, and then

$$(4.23) \quad N^{-\frac{1}{2}} \begin{bmatrix} 0 & \mathscr{S}\eta \\ \mathscr{S}\xi & \mathscr{S}\xi\eta \end{bmatrix} = \mathbf{RQC}^T.$$

In either case, the Pearson χ^2 for three degrees of freedom, $\mathbf{q}^T\mathbf{q}$ or tr $\mathbf{QQ}^T$, has been exhibited as the sum of three asymptotically independent components, each asymptotically the square of a standardized normal variable.

Example. The genetic example given earlier in the section may be treated as an example of the method. $\xi_1 = \eta_1 = -1, \xi_2 = \eta_2 = 1.$

$$(3734)^{-\frac{1}{2}} \begin{bmatrix} 0 \\ \mathscr{S}\eta \\ \mathscr{S}\xi \\ \mathscr{S}\xi\eta \end{bmatrix} = (3734)^{-\frac{1}{2}} \begin{bmatrix} 1 & 1 & 1 & 1 \\ -1 & 1 & -1 & 1 \\ -1 & -1 & 1 & 1 \\ 1 & -1 & -1 & 1 \end{bmatrix} \begin{bmatrix} 921 - 933\cdot5 \\ 971 - 933\cdot5 \\ 907 - 933\cdot5 \\ 935 - 933\cdot5 \end{bmatrix}$$

$$= (3734)^{-\frac{1}{2}} \begin{bmatrix} 0 \\ 78 \\ -50 \\ -22 \end{bmatrix} = \begin{bmatrix} 0 \\ 1\cdot276 \\ -0\cdot818 \\ -0\cdot360 \end{bmatrix} = \begin{bmatrix} \text{zero identically} \\ \text{column effect } (B) \\ \text{row effect } (A) \\ \text{interaction } (A \times B) \end{bmatrix}$$

5. COMBINATORIAL THEORY OF THE TWO-WAY TABLES

The two-way tables with fixed marginal totals can be considered combinatorially. This is especially appropriate for the fourfold tables, for which Irwin (1935) and Yates (1934) computed an exact probability. Let us consider a random variable Z taking values z, $z = 0, 1, \ldots, a_{1.}$ with the convention that $a_{1.}$ is the smallest marginal total. The computation of $\mathscr{P}(Z = z)$ does not require the knowledge or estimate of any unknown parameter. In the fourfold table, these conditional probabilities $\mathscr{P}(Z = z)$ are simply the terms of the hypergeometric distribution. They can be computed with the aid of a table of factorials. Alternatively a central term can be computed and then every other term can be obtained by the use of the formula,

$$(5.1) \qquad \frac{\mathscr{P}(Z = z + 1)}{\mathscr{P}(Z = z)} = \frac{(a_{1.} - z)(a_{.1} - z)}{(z + 1)(N - a_{1.} - a_{.1} + z + 1)}.$$

A variation on this computation is to multiply the central term by an arbitrary constant to obtain a convenient number, say unity, to which are applied the ratios to obtain all the terms of appreciable size multiplied by the same constant. The arbitrary constant is then eliminated by dividing through by the sum of the calculated terms.

Example. In the fourfold table with marginal totals, each equal to 10, the probabilities $\mathscr{P}(Z = z)$ are as follows—0 and 10, 0·00295; 1 and 9, 0·00652; 2 and 8, 0·04151; 3 and 7, 0·09640; 4 and 6, 0·20501; 5, 0·29521.

Such computations are practicable in fourfold tables with lowest marginal total of less than one hundred but the arithmetic becomes unduly heavy with the larger tables so that the asymptotic theory becomes more in order. Such computations can be extended in principle to the general two-way contingency tables. For assessing the cumulative probabilities in the fourfold tables, tables have been constructed by Armsen (1955), Finney (1948), Finney, Latscha, Bennett, Hsu and Pearson, with supplement by Bennett and Horst (1963), Garside (1958 and 1961), Swaroop (1938) and Swineford (1948). Bennett and Nakamura (1963) have given similar tables for the 2×3 contingency tables. General routines for the computation of the exact test in the two-way contingency tables have been given by Freeman and Halton (1951) and Robertson (1960). In the fourfold table, the exact probability can be well approximated by the χ or χ^2 corrected for continuity after the manner of Yates (1934); tables to improve the approximation are given in

Fisher and Yates (1963). This normal approximation can be justified even with Model 3 by the application of Stirling's approximation.

A classical problem is to determine the moments of χ^2 in the contingency tables, or the Lexis index of dispersion. A lemma may be stated without proof.

Lemma 5.1. *In the $m \times n$ contingency tables, the distribution of the entries at a given position (I, J) is the same as in the 2×2 contingency tables formed by pooling all other rows $i \neq I$ and all other columns, $j \neq J$. The joint distribution of the entries at positions (I, J) and (I, J') can similarly be determined by considering the 3×2 table with pooling by rows as above and pooling all columns, j such that $j \neq J$, $j \neq J'$. The joint distribution of the entries at (I, J) and (I', J'), $I \neq I'$, $J \neq J'$ can be determined by the consideration of the 3×3 table with rows pooled, $i \neq I$, $i \neq I'$ and columns pooled, $j \neq J$, $j \neq J'$.*

Let now Z_{ij} be the random variable, the entry at position (i, j) in the contingency table with fixed marginal totals $\{a_{i.}\}$ and $\{a_{.j}\}$. Then the following may be written down using the results of Section 4. Let $Z_{ij}^{(r)} = Z_{ij}(Z_{ij} - 1) \ldots (Z_{ij} - r + 1)$

$$(5.2) \qquad \mathscr{E}Z_{ij}^{(r)} = a_{i.}^{(r)} a_{.j}^{(r)} / N^{(r)}$$

$$(5.3) \qquad \mathscr{E}Z_{ij}^{(r)} Z_{ij'}^{(r')} = a_{.j}^{(r)} a_{.j'}^{(r')} a_{i.}^{(r+r')} / N^{(r+r')}$$

$$(5.4) \qquad \mathscr{E}Z_{ij}^{(r)} Z_{i'j'}^{(r')} = a_{i.}^{(r)} a_{i'.}^{(r')} a_{.j}^{(r)} a_{.j'}^{(r')} / N^{(r+r')}$$

from which follow

$$(5.5) \qquad \mathscr{E}Z_{ij} = a_{i.} a_{.j} / N,$$

$$(5.6) \qquad \operatorname{var} Z_{ij} = a_{i.} a_{.j} (N - a_{i.})(N - a_{.j}) / \{N^2(N - 1)\},$$

$$(5.7) \qquad \operatorname{cov}(Z_{ij}, Z_{ij'}) = -a_{.j} a_{.j'} a_{i.} (N - a_{i.}) / \{N^2(N - 1)\},$$

$$(5.8) \qquad \operatorname{corr}(Z_{ij}, Z_{ij'}) = -(a_{.j} a_{.j'} / (N - a_{.j})(N - a_{.j'}))^{\frac{1}{2}},$$

$$(5.9) \qquad \operatorname{cov}(Z_{ij}, Z_{i'j'}) = +a_{i.} a_{i'.} a_{.j} a_{.j'} / \{N^2(N - 1)\}$$

and

$$(5.10) \qquad \mathscr{E}Z_{ij}^2 = a_{i.}^{(2)} a_{.j}^{(2)} / N^{(2)} + a_{i.} a_{.j} / N$$

From (5.10) the expectation of χ^2 can be computed.

$$(5.11) \qquad [N\mathscr{E}Z_{ij}^2 / (a_{i.} a_{.j})] = (a_{i.} - 1)(a_{.j} - 1)/(N - 1) + 1.$$

Hence

$$
(5.12) \quad \mathscr{E}\chi^2 = N \sum_{i,j} \mathscr{E}Z_{ij}^2/(a_{i.}a_{.j}) - N
$$

$$
= \frac{1}{N-1} \sum_{i,j} (a_{i.} - 1)(a_{.j} - 1) + mn - N
$$

$$
= \frac{1}{N-1} \sum_{i,j} (a_{i.}a_{.j} - a_{i.} - a_{.j} + 1) + mn - N
$$

$$
= \frac{1}{N-1} (N^2 - nN - mN + mn) + mn - N
$$

$$
= \frac{1}{N-1} (N^2 - nN - mN + mnN - N^2 + N)
$$

$$
= \frac{N}{N-1} (m-1)(n-1).
$$

This result shows that it would be appropriate to compute

$$
(5.13) \quad \chi^2 = (N-1) \sum a_{ij}^2/(a_{i.}a_{.j}) - (N-1)
$$

in contingency tables as was done in Lancaster (1949b). The higher moments of χ^2 in contingency tables with fixed marginal totals are of less interest than the mean and variance because as numbers become larger it can be assumed that the variables, $(a_{ij} - \mathscr{E}a_{ij})/\sqrt{\mathscr{E}a_{ij}}$, are all approaching normality. Comparatively little work has been done on the enumeration of cases in the contingency tables to determine the approach to normality.

6. ASYMPTOTIC THEORY OF THE TWO-WAY TABLES

Exact probabilities in the sense of Fisher can be computed in the fourfold tables readily if the smallest marginal total is not too large. With the aid of high speed computers, the probabilities associated with all relevant sets, $\{a_{ij}\}$, can be computed and ordered in some specified manner, for example, in the order of decreasing values of $\mathscr{P}(\{a_{ij}\} \mid \{a_{i.}\}, \{a_{.j}\})$. However, this is a time consuming procedure and it can be shown that for expectations in the cells not too small, asymptotic results are adequate; that is, the χ^2 method gives sizes to the tests approximating to the theoretical. If it be assumed that the χ^2 approximation is adequate in the 2×2 tables, then a simple proof can be given for the general table after the manner of Lancaster (1949a).

Let there be defined partial sums of the rows, columns and submatrices as follows:

$$
(6.1) \quad R_{ik} = \sum_{j=1}^{k} a_{ij}, \quad C_{kj} = \sum_{i=1}^{k} a_{ij}, \quad T_{lk} = \sum_{i=1}^{l} \sum_{j=1}^{k} a_{ij}
$$

Then under the hypothesis of independence we have for a 2×3 table,

$$(6.2) \quad \mathscr{P}\{\{a_{ij}\} \mid \{a_{i.}\}, \{a_{.j}\}\} = a_{1.}!a_{2.}!a_{.1}!a_{.2}!a_{.3}! \Big/ \Big(N! \prod_{i,j} a_{ij}! \Big)$$

$$= \frac{R_{12}!R_{22}!a_{.1}!a_{.2}!}{T_{22}!a_{11}!a_{12}!a_{21}!a_{22}!} \frac{T_{22}!a_{.3}!a_{1.}!a_{2.}!}{N!a_{13}!a_{23}!R_{12}!R_{22}!}$$

But the last two expressions refer to two fourfold tables,

a_{11}	a_{12}	R_{12}		R_{12}	a_{13}	$a_{1.}$
a_{21}	a_{22}	R_{22}		R_{22}	a_{23}	$a_{2.}$
$a_{.1}$	$a_{.2}$	T_{22}		T_{22}	$a_{.3}$	$a_{..}$

the first expression being the expectation of the first four entries, conditional on the value of the first two column totals and the first two partial sums R_{12} and R_{22}, of the rows, the second expression being the probability of R_{12}, R_{22}, a_{13} and a_{23} conditional on the marginal totals. It can be shown by induction that the probability of an $m \times n$ table can be factored into a chain of such conditional probabilities, $(m-1)(n-1)$ in number.

Theorem 6.1. *Any $m \times n$ table can be reduced to a set of $(m-1)(n-1)$ fourfold tables, in such a way that the probability of the whole contingency table can be expressed as a product of the probabilities of the fourfold tables. To each fourfold table can be associated a χ_t^2. Then $\sum\limits_t^{(m-1)(n-1)} \chi_t^2$ is asymptotically χ^2 with $(m-1)(n-1)$ degrees of freedom. Also it is asymptotically equal to the usually calculated χ^2, $\sum (observed\text{-}expected)^2/(expected)$, summed for every cell of the table, and is thus asymptotically unique.*

Proof. The proof is by induction for the first statement. Stirling's approximation is then applied to each fourfold table and a χ_t^2 associated with it. Stirling's approximation is also applied to the probability of the complete table. We then obtain an equation of the form

$$(6.3) \qquad K_1 \exp\left(-\tfrac{1}{2} \sum \chi_t^2\right) \simeq K_2 \exp\left(-\tfrac{1}{2}\chi_P^2\right),$$

where on the right, $\chi_P^2 = \sum (observed\text{-}expected)^2/(expected)$ with summation over the mn cells. Now K_1 and K_2 are each of dimension, $-\tfrac{1}{2}$, in the a_{ij}. But (6.3) would also be true for a table with elements ka_{ij} with $k = 2, 3, \dots$. But χ_t^2 and χ^2 are both of dimension unity in the elements a_{ij}. The relation can only be true if $\chi^2 \simeq \sum \chi_t^2$. This shows that $\sum \chi_t^2$ is independent asymptotically of the order of forming the $(m-1)(n-1)$ fourfold tables. χ^2 is asymptotically equal to $\sum \chi_t^2$, which has $(m-1)(n-1)$ degrees of freedom. Define

$$(6.4) \qquad (q_{ij}) = \mathbf{Q} = (\{a_{ij} - a_{..}p_{ij}\}/\sqrt{(a_{..}p_{ij})})$$

and consider an $m \times n$ table with elements, a_{ij}. Let two Helmert matrices be associated with the estimated row and column probabilities and call these **R** and **C**.

Theorem 6.2. *Under the hypothesis, H_0, of independence, the χ^2 of a contingency table of $m \times n$ rows can be partitioned into constituent χ^2, the χ's of which are the elements of*

$$(6.5) \qquad\qquad \mathbf{E} = \mathbf{RQC}^T,$$

e_{11} *being identically zero. If further the row (and/or column) parameters are estimated, then the elements of the first row and/or first column of* **E** *are identically zero. In any case, the elements of* **E** *not identically zero form a set of asymptotically normal and mutually independent variables.*

Proof. The probability of the table can be written

$$(6.6) \quad \mathscr{P}\{\{a_{ij}\} \mid \{a_{i.}\}, \{a_{.j}\}\}$$
$$= \mathscr{P}\{\{a_{ij}\} \mid \{p_{i.}\}, \{p_{.j}\}, a_{..}\}/[\mathscr{P}\{\{a_{i.}\} \mid \{p_{i.}\}, a_{..}\}\mathscr{P}\{\{a_{.j}\} \mid \{p_{.j}\}a_{..}\}]$$
$$= \mathscr{P}\{\{a_{ij}\} \mid \{\lambda p_{i.}p_{.j}\}, H_0\}/$$
$$[\mathscr{P}\{a_{..} \mid \lambda, H_0\}\mathscr{P}\{\{a_{i.}\} \mid \{p_{i.}\}, a_{..}\}\mathscr{P}\{\{a_{.j}\} \mid \{p_{.j}\}, a_{..}\}]$$

The left side is independent of λ and of the true values of $\{p_{i.}\}$ and $\{p_{.j}\}$. Arbitrary values can therefore be given to λ and $p_{i.}$ and $p_{.j}$, namely $\lambda = N$, $\hat{p}_{i.} = a_{i.}/N$, $\hat{p}_{.j} = a_{.j}/N$. Under H_0, the a_{ij} are mutually independent Poisson variates and the variables, $q_{ij} = (a_{ij} - \lambda p_{i.}p_{.j})/\sqrt{(\lambda p_{i.}p_{.j})}$, are asymptotically normal. They thus form a set of mutually independent asymptotically normal variates. If an orthogonal transformation be made on the variables of the form (6.5) the distribution of the variates, e_{ij}, $i = 2, 3, \ldots, r$, $j = 2, 3, \ldots, s$ is jointly independent of the set $\{e_{11}, \{e_{i1}\}, \{e_{1j}\}\}$ $i > 1, j > 1$. Writing in for λ, $p_{i.}$ and $p_{.j}$ the values $\hat{\lambda} = N$, $\hat{p}_{i.} = a_{i.}/N$ and $\hat{p}_{.j} = a_{.j}/N$, we make zero linear forms in the variates, a_{ij}, namely the elements of the first row and column of **E**. The remaining $(m-1)(n-1)$ variables still form an asymptotically independent and normal set with mean zero and unit variance. The sum of their squares is thus χ^2 with $(m-1)(n-1)$ degrees of freedom. Multiplying out the matrix, $\mathbf{QC}^T$ has elements of the form $(a_{i.} - Np_{i.})/\sqrt{(Np_{i.})}$, in the first column $\mathbf{RQC}^T$ transforms these elements into the (asymptotically) independently distributed comparisons of the multinomial. Irwin (1949) showed that the factorization proof was equivalent to a partition of χ^2. A general proof of the asymptotic χ^2 distribution can be given with the aid of Bernstein's generalized central limit theorem. This use is especially appropriate to Model 1 with theoretical parameters or with parameters estimated from the data. It is also appropriate to Model 2, whether the column parameters are given theoretically or are to be estimated from the data. First, it must be noted that, if the distribution of the entries in the cells is considered conditional on the marginal values, the true values of the parameters in the null case of independence are irrelevant; they can therefore be chosen arbitrarily. Define

(6.7) $$\hat{p}_{i.} = a_{i.}/N; \qquad \hat{p}_{.j} = a_{.j}/N.$$

Now consider a probability distribution on the mn points with measure, $\hat{p}_{i.}\hat{p}_{.j}$ and marginal measures, $\hat{p}_{i.}$ and $\hat{p}_{.j}$ respectively on rows and columns.

Let $x^{(0)} = y^{(0)} = 1$ and suppose that $\{x^{(i)}\}$, $i = 0, 1, \ldots, m - 1$ is an orthonormal basis on the row space so that

(6.8) $$\mathscr{E} x^{(i)} x^{(i')} = \delta_{ii'}$$

and similarly $\{y^{(j)}\}$, $j = 0, 1, 2, \ldots, n - 1$ is an orthonormal basis on the column space. It easily verified that

(6.9) $$\{x^{(i)} y^{(j)}\} = \{x^{(i)}\} \times \{y^{(j)}\}$$

is an orthonormal basis on the product probability distribution.

We suppose now that a sample of N independent observations has been made on the probability distribution, $\{\hat{p}_{i.}\hat{p}_{.j}\}$. The standardized sums, if $i + j > 0$, $N^{-\frac{1}{2}}\mathscr{S} x^{(i)} y^{(j)}$, are then asymptotically standardized normal, uncorrelated and so asymptotically independent, in the unconditional case by the generalized Bernstein theorem. Now let us consider the asymptotic joint distribution of the variables $N^{-\frac{1}{2}}\mathscr{S} x^{(i)} y^{(j)}$, $i \neq 0$, $j \neq 0$ conditional on $N^{-\frac{1}{2}}\mathscr{S} x^{(i)} = 0$, $N^{-\frac{1}{2}}\mathscr{S} y^{(j)} = 0$, $i = 1, 2, \ldots, m - 1$, $j = 1, 2, \ldots, n - 1$. It is, by the asymptotic independence of the variables, the same as the unconditional distribution. We have thus $(m - 1)(n - 1)$ standardized asymptotically normal, mutually independent variables. The sum of the squares is thus χ^2 with $(m - 1)(n - 1)$ degrees of freedom. The conditions $N^{-\frac{1}{2}}\mathscr{S} x^{(i)} = 0$ are equivalent to fixing the row totals, for if a typical $x^{(i)}$ takes values $\xi_1, \xi_2, \ldots, \xi_m$

(6.10) $$0 = \mathscr{S} x^{(i)} = \sum a_{i.} \xi_i$$
$$N = \mathscr{S} x^{(0)} = \sum a_{i.} = N \sum \hat{p}_{i.}.$$

and this is the necessary and sufficient condition that the observed $\{a_{i.}\}$ is the same as $\{N\hat{p}_{i.}\}$.

7. PARAMETERS OF NON-CENTRALITY

It is required to determine the parameters of non-centrality of the test of independence in two dimensions. We suppose that H_0 specifies independence, so that

(7.1) $$F(x, y) = G(x)H(y),$$

in the notation already used. If $\{x^{(i)}\}$ and $\{y^{(j)}\}$ are complete orthonormal sets (or orthonormal bases) on the marginal distributions, then (7.1) is equivalent to

$$(7.2) \qquad \mathscr{E}_0 x^{(i)} y^{(j)} = 0, \qquad i = 1, 2, \ldots ; \qquad j = 1, 2, \ldots$$

by Theorem VI.4.1. H_1 will specify that

$$(7.3) \qquad c_{ij} = \mathscr{E}_1 x^{(i)} y^{(j)} \neq 0$$

for at least one pair (i, j), $ij \neq 0$. $\mathscr{E}_0$ and $\mathscr{E}_1$ are the expectation operators under H_0 and H_1 respectively.

Under H_0, $\{x^{(i)} y^{(j)}\}$, $i = 0, 1, 2, \ldots, j = 0, 1, 2, \ldots$ is an orthonormal set with $x^{(0)} y^{(0)} = 1$. With sample size N, the set of functions, $\{X_{ij}\}$, given by

$$(7.4) \qquad X_{ij} = N^{-\frac{1}{2}} \mathscr{S} x^{(i)} y^{(j)}, \qquad i = 0, 1, 2, \ldots, \qquad j = 0, 1, 2, \ldots$$

is also an orthonormal set by the generalized central limit theorem and its elements are asymptotically normal and mutually independent, if the constant function be excluded. Any sum of squares, $\sum X_{ij}^2$, is thus distributed approximately as χ^2 under H_0.

Example. Let H_0 specify that X and Y are mutually independent and normal $(0, 1)$. Let $\{x^{(i)}\}$ and $\{y^{(i)}\}$ be the standardized Hermite polynomials. Let H_1 specify that X and Y are both normal $(0, 1)$ and jointly normal with correlation, ρ. Then if $ij \neq 0$, $\mathscr{E}_0 x^{(i)} y^{(j)} = 0$ and $\mathscr{E}_1 x^{(i)} y^{(j)} = \delta_{ij} \rho^j$. The standardized variables $N^{-\frac{1}{2}} \mathscr{S} x^{(i)} y^{(j)}$ are approximately normal $(0, 1)$ under H_0 but normal $(N^{\frac{1}{2}} \delta_{ij} \rho^i, v_{ij})$ under H_1. The v_{ij} can be obtained by a consideration of the function, $\mathscr{E}_1(\exp \{tX - \frac{1}{2}t^2 + uY - \frac{1}{2}u^2\})$. The variables, X_{ij}, under H_1 will thus not be central variables, nor will they be necessarily uncorrelated. However, if ρ is small, the variances will not differ greatly from unity. Hamdan (1968b) may be consulted for the values of the variances and covariances in this special case. Hamdan (1968a) has shown how the theory can be applied to obtain the power of the test of independence in the bivariate normal distribution using the χ^2 test in a contingency table.

More generally, let us suppose that sets of functions, $\{u^{(i)}\}$ and $\{v^{(i)}\}$, are defined on the marginal distributions of X and Y, respectively. Let them be written as elements of vectors, $\mathbf{u}$ and $\mathbf{v}$. Suppose that $\mathscr{E}_0 \mathbf{u} = \mathbf{0}$, $\mathscr{E}_0 \mathbf{v} = \mathbf{0}$ and that $\mathscr{E}_0 \mathbf{u}\mathbf{u}^T = \mathbf{1}_p$ and $\mathscr{E}_0 \mathbf{v}\mathbf{v}^T = \mathbf{1}_q$. For example, $\{u^{(i)}\}$ may be the set of functions obtained by orthonormalizing the indicator variables of the $(p + 1) = m$ row classes of a contingency table and $\{v^{(i)}\}$, the set of functions obtained by orthonormalizing the column variables. Alternatively the functions might be a sine or cosine series or appropriate orthonormal polynomials; p and/or q may be infinite. In any case, there are the relations,

$$(7.5) \qquad \mathbf{u} = \mathbf{A}\mathbf{x}, \qquad \mathbf{v} = \mathbf{B}\mathbf{y}, \qquad \text{where} \qquad \mathbf{A}\mathbf{A}^T = \mathbf{1}_p, \qquad \mathbf{B}\mathbf{B}^T = \mathbf{1}_q.$$

Further $\mathbf{A}^T\mathbf{A}$ and $\mathbf{B}^T\mathbf{B}$ are idempotents. The expectations under H_1 of the products $u^{(i)}v^{(j)}$ can be obtained,

$$(7.6) \qquad \mathscr{E}_1\mathbf{uv}^T = \mathscr{E}_1\mathbf{Axy}^T\mathbf{B}^T = \mathbf{A}\mathscr{E}_1\mathbf{xy}^T\mathbf{B}^T = \mathbf{ARB}^T,$$

where $r_{ij} = \mathscr{E}_1 x^{(i)}y^{(j)}$. Corresponding to the products on a single distribution, there are also the expectations of the standardized sums, or χ-analogues,

$$(7.7) \qquad \mathscr{E}_1 N^{-\frac{1}{2}}\mathscr{S}(u^{(i)}v^{(j)}) = N^{\frac{1}{2}}(\mathbf{ARB}^T)_{ij}.$$

The expectation of the test function,

$$(7.8) \qquad Q_{pq} = N^{-1}\sum_{i=1}^{p}\sum_{j=1}^{q}[\mathscr{S}(u^{(i)}v^{(j)})]^2,$$

is not less than the sum of the squares of terms of the form given in (7.7).

$$(7.9) \qquad \mathscr{E}_1 Q_{pq} \geqslant N^{-1}\mathscr{E}_1\sum_{i,j}[\mathscr{S}(u^{(i)}v^{(j)})]^2 = N^{-1}\sum_{i,j}[N(\mathbf{ARB}^T)_{ij}]^2$$

$$= N\,\mathrm{tr}\,\mathbf{ARB}^T\mathbf{BR}^T\mathbf{A}^T = N\,\mathrm{tr}\,\mathbf{A}^T\mathbf{ARB}^T\mathbf{BR}^T.$$

The choice of variables $\mathbf{u}$ and $\mathbf{v}$ has not yet been considered. We are still supposing that we know the true values of the marginal parameters and only $\mathbf{u}$ and $\mathbf{v}$ and consequently $c_{ij} = \mathscr{E}_1(u_i v_j)$ are varying. If we choose as the elements of $\mathbf{u}$ and $\mathbf{v}$ the canonical variables under H_1, then

$$(7.10) \qquad \mathscr{E}_1\mathbf{uv}^T = \mathrm{diag}\,(\rho_k), \mathbf{0}_{p,q-p},$$

where we have supposed that $m \leqslant n$. If the aim is to maximize the expectation under the alternative hypothesis, then p should be chosen equal to q. The choice of the number of variables will depend on the properties of the theoretical canonical correlations. Let us specialise the problem and suppose that $\mathbf{u}$ and $\mathbf{v}$ have as elements the sets of orthonormal functions on the partitions of the marginal spaces into $(p + 1)$ and $(q + 1)$ classes. Let us denote by Q_{pq} the resulting χ^2. Then if the marginal frequencies are given by hypotheses, H_0 and H_1

$$(7.11) \qquad \mathscr{E}_0 Q_{pq} = pq, \qquad \mathscr{E}_1 Q_{pq} > N\sum_{1}^{p}\sum_{1}^{q}c_{ij}^2,$$

where $c_{ij} = \mathscr{E}_1 u_i v_j$. If c_{ij} is small for each (i,j), then it may be expected that Q_{pq} can be regarded as non-central χ^2 with pq degrees of freedom and parameter of non-centrality, $N\sum c_{ij}^2$. The variances and covariances of the components of Q_{pq} can always be computed, at least in principle, and are independent of the sample size, N; so approximate distributions can always be computed. As a rule, this is unnecessary. In any case, we can assume that the 50% point of the distribution of Q_{pq} is greater than the parameter of non-centrality. Using this rough rule, one could choose the sample size,

N, so that $N \sum c_{ij}^2$ is greater than the value of χ^2 with pq degrees of freedom at the α level of significance, $t_{pq}^{(\alpha)}$ say. This choice of N will give a size approximately α under H_0 and a power of 50% under all alternative hypotheses such that $N \sum c_{ij}^2 > t_{pq}^{(\alpha)}$.

If $\sum c_{ij}^2$ is not small, practically all observations made with N large will yield a large value for Q_{pq} and questions of power hardly rise. So far, we have considered the marginal distributions as given and have chosen Q_{pq} as the sum of squares of functions orthogonal to the marginal orthonormal systems. It is not known how far this theory has to be modified if the parameters of the marginal distributions have to be estimated. The examples of Lancaster, (1949a) suggest that the value of Q_{pq}, calculated when the marginal parameters are estimated, does not differ greatly from the value calculated when they are given.

In closing this Section, it may be asked why the test functions are chosen in the form Q_{pq}, the sum of squares of a $p \times q$ array. This is only appropriate when there is no definite alternative hypothesis. Q_{pq} tests all possible departures of c_{ij} from the null values. It may well be that if a precise alternative hypothesis is given, there are many better choices of the test function, for example, in the bivariate normal if a particular value of ρ is specified, the standardized sum of the standardized likelihood ratio will be asymptotically normal with parameter of non-centrality equal to $N\phi^2$, $\phi^2 = \rho^2(1 - \rho^2)^{-1}$; in other words, let

(7.12) $$\Omega(x, y) = f(x, y; \rho)/g(x)h(y)$$

and set

(7.13) $$L = [\Omega(x, y) - 1)]/\phi.$$

Then $\mathscr{E}_0 L = 0$, $\mathscr{E}_0 L^2 = 1$, $\mathscr{E}_1 L = \phi$, $\mathscr{E}_1 L^2 = 1 + 6\rho^2 + 20\rho^4 +$ terms of higher order, the integration being valid for $|\rho| < \frac{1}{2}$. The variance of L under H_1 is thus approximately $1 + 5\rho^2 + 19\rho^4$. The test function $N^{-\frac{1}{2}}\mathscr{S}L$ is thus asymptotically normal $(0, 1)$ under H_0 and if ρ be small normal $(N^{\frac{1}{2}}\phi, 1)$ under H_1. Its square thus has a parameter of non-centrality not less than any other χ^2-analogue and is thus the most powerful member in the class.

8. SYMMETRY AND EXCHANGEABILITY IN TWO-WAY TABLES

Symmetry of distribution can be defined in univariate distributions and hence in the marginal distributions as being present if and only if

(8.1) $$F(-x) = 1 - F(x)$$

at all points of continuity, for symmetry about the origin and if and only if

(8.2) $$F(a - x) = 1 - F(x - a)$$

at all points of continuity for symmetry about the point, a.

In multivariate distributions, various forms of symmetry can be defined, of which exchangeability of random variables is an important special case. Random variables, $X_1, X_2, \ldots, X_n$ are said to be exchangeable if and only if

(8.3) $$F(x_{i_1}, x_{i_2}, \ldots, x_{i_n}) = F(x_1, x_2, \ldots, x_n)$$

for all choices of the permutation $\{i_1, i_2, \ldots, i_n\}$ of the set of indices $\{1, 2, \ldots, n\}$.

Distributions possessing some features of asymmetry can be symmetrized. Thus,

(8.4) $$G(x) = \tfrac{1}{2}(F(a - x) + 1 - F(x - a)),$$

at all points of continuity, defines a distribution symmetrical about the point, a. A symmetrical distribution can also be obtained by the convolution of the distributions of X_1 and $-X_2$, where X_1 and X_2 have the same distribution but are mutually independent. Multivariate distributions can be symmetrized by defining

(8.5) $$\bar{F}(x_1, x_2, \ldots, x_n) = (n!)^{-1} \sum_{k \in K} F_k(x_1, x_2, \ldots, x_n)$$

where the notation implies that F_k at the point $(x_1, x_2, \ldots, x_n)$ is to be set equal to the value of $\mathscr{P}(X_{i_1} \leqslant x_1; X_{i_2} \leqslant x_2; \ldots; X_{i_n} \leqslant x_n)$ where $\{i_1, i_2, \ldots, i_n\}$ is a permutation of $\{1, 2, \ldots, n\}$ and k runs through the $n!$ elements of K, the set of all permutations of n indices.

Example. A symmetrized distribution arises naturally in genetics. In matings of the form, $Aa \times Aa$, the offspring can be separated into four classes of which two cannot as a rule be distinguished; for, in the Aa class it will not be known from which parent the A has been received. There are thus three distinguishable classes. If the original observations are set up in the form of a 2×2 table, the contents of the cells at positions $(1, 2)$ and $(2, 1)$ are combined. It is possible to form a symmetrized 2×2 table from the original (a_{ij}) by setting $a'_{11} = a_{11}$, $a'_{22} = a_{22}$, $a'_{12} = a'_{21} = \tfrac{1}{2}(a_{12} + a_{21})$. It is easy to prove that this may be treated as a 2×2 table in the usual way with parameters estimated from the data,

(8.6) $$\hat{p}_{1.} = \hat{p}_{.1} = a'_{1.}/N$$

or with theoretical parameters

(8.7) $$p_{1.} = p_{.1} = \tfrac{1}{2},$$

in which case the expected values are $Np_{i.}p_{.j}$ and the degrees of freedom due to binomial sampling can be obtained by regarding $a'_{1.} + a'_{.1}$ as a binomial variable from a population with parameters, $2N$ and $\frac{1}{2}$.

From Chapter 6, bivariate distributions are characterized by the marginal distributions and by the sets of correlations of the form,

$$\rho_{ij} = \int x^{(i)} y^{(j)} \, dF(x, y),$$

so that a necessary and sufficient condition for symmetry in a two-way table is

(8.8)
$$p_{i.} = p_{.i}, \qquad i = 1, 2, \ldots, m$$
$$\rho_{ij} = \rho_{ji}, \qquad i \neq j.$$

If H_0 specifies the independence of the two variables, $\{p_{i.}\}$ and $p_{i.} = p_{.i}$, the the sampling problem is simple. We can define identical sets of orthogonal functions on the two marginal distributions and thus obtain a set of asymptotically normal and mutually independent standardized sums, $N^{-\frac{1}{2}}\mathscr{S}x^{(i)}$, $N^{-\frac{1}{2}}\mathscr{S}y^{(j)}$, $i, j = 1, 2, \ldots, m - 1$, $N^{-\frac{1}{2}}\mathscr{S}x^{(i)}y^{(j)}$, $i, j = 1, 2, \ldots, m - 1$. The symmetry comparisons can then be made using the variables, $2^{-\frac{1}{2}}N^{-\frac{1}{2}} \times (\mathscr{S}x^{(i)} - \mathscr{S}y^{(i)})$ and $2^{-\frac{1}{2}}N^{-\frac{1}{2}}(\mathscr{S}x^{(i)}y^{(j)} - \mathscr{S}x^{(j)}y^{(i)})$ $i > 0, j > 0$. These variables under the hypotheses made are all asymptotically normal $(0, 1)$ and mutually independent. An overall test of symmetry would then be a sum of the squares of these variables

(8.9) $\quad \chi^2 = (2N)^{-1}\left\{\sum_i (\mathscr{S}x^{(i)} - \mathscr{S}y^{(i)})^2 + \sum_{i > j}(\mathscr{S}x^{(i)}y^{(j)} - \mathscr{S}x^{(j)}y^{(i)})^2\right\}$

A test, which is of more general application and calls for less general theory is due to Bowker (1947). Suppose that H_0 specifies $\{p_{ij}\}$ and also that $p_{ij} = p_{ji}$ for $i \neq j$. Then in a sample of N, a_{ij}, given $a_{ij} + a_{ji}$, is a binomial variable with parameters, $a_{ij} + a_{ji}$ and $\frac{1}{2}$. Further all these conditional probabilities for fixed $\{a_{ij} + a_{ji}\}$ are mutually independent so that

(8.10)
$$X^2_{ij} = \frac{(a_{ij} - a_{ji})^2}{a_{ij} + a_{ji}},$$

is distributed asymptotically as χ^2 with one degree of freedom and

(8.11)
$$\sum_{i < j} X^2_{ij} = \sum_{i < j}(a_{ij} - a_{ji})^2/(a_{ij} + a_{ji})$$

is distributed approximately as χ^2 with $\frac{1}{2}m(m - 1)$ degrees of freedom. The justification for the above is

(8.12) $\quad \mathscr{P}(\{a_{ij}\} \mid \{p_{ij}\}, N) = N! \prod_{i, j} p_{ij}^{a_{ij}}/a_{ij}!$

$$= N! \prod_{i > j}(2p_{ij})^{a_{ij}+a_{ji}}/(a_{ij} + a_{ji})! \prod_i p_{ii}^{a_{ii}}/a_{ii}!$$

$$\times \prod_{i > j}(\tfrac{1}{2})^{a_{ij}+a_{ji}}(a_{ij} + a_{ji})!/a_{ij}!a_{ji}!$$

and for fixed values of $(a_{ij} + a_{ji})$, the binomial expressions in the last factors are those of mutually independent random variables and each factor can be approximated by an asymptotically normal variable.

More difficulties arise if in this general model it is desired to test the homogeneity of the margins, since p_{ij} may not be equal to $p_i.p_{.j}$. However, the necessary variances and covariances can be computed with the aid of the combinatorial theory above and then the covariance matrix of the variables,

$$(8.13) \qquad d_i = a_i. - a_{.i}, \qquad i = 1, 2, \ldots, m$$

can be determined. From general multivariate theory this covariance matrix is the maximum likelihood estimate. However, in (8.13) the variables d_i are linearly dependent and so a subset of $(m - 1)$ of them can be chosen and the covariance V estimated by $\hat{V}$. $d^T\hat{V}^{-1}d$ is then approximately distributed as χ^2 with $(m - 1)$ degrees of freedom where d is a vector, the elements of which are the $(m - 1)$ linearly independent d_i. The solution of this problem is suggested in the exercises.

9. REPARAMETRIZATION

Often bivariate tables are given in which there is no quantitative information on the meaning of the classes. Thus two variables may be given in which there is no quantitative scale, and the observations are given only by rank. If it be assumed, as is often done, that the underlying distribution is jointly normal, then for such ungrouped data, Tables XX and XXI of the *Statistical Tables*, Fisher and Yates (1963) may be used. These tables give the order statistics in samples of size n drawn from a standardized normal population.

With grouped data, if there is a natural ranking of the classes, values can be given for the boundaries between the classes, $d_1, d_2, \ldots, d_{m-1}$ as the solution of the equation

$$(9.1) \qquad w_i = \int_{-\infty}^{d_i} f(u)\, du,$$

where w_i is the cumulative frequency and where $f(u)$ is the density of the standardized normal variable. The value of the variable within the class is usually taken as the mid-point. However, the point corresponding to the mean value of the variable over the class has the highest possible correlation with the normal deviate and possibly this is the best value to assign to the marginal variable.

Maung (1942) showed that in the theoretical bivariate normal distribution, the first pair of canonical variables are the marginal variables themselves. R. A. Fisher suggested that the values of the canonical variables would not be greatly different from the true values and this proposition is frequently used to reparametrise the marginal distributions.

10. MEASURES OF ASSOCIATION

The early workers such as Galton and Pearson considered correlation between the variables themselves rather than the correlation between functions of the variables, and this notion long dominated thoughts on dependence between random variables. In Chapter VI, we have mentioned how satisfactory the Pearson coefficient of correlation is for the description of dependence in the bivariate normal distribution or in those distributions which can be brought to this form by an appropriate transformation.

The normal distributions form a class of additive variables. Other classes with an additive property include the binomial, the Poisson and the Γ-distributions. If bivariate distributions are formed by transformations,

$$(10.1) \qquad \begin{aligned} X &= W_1 + W_2 \\ Y &= W_2 + W_3 \end{aligned}$$

where W_1, W_2 and W_3 are mutually independent but belong to the same additive class, then X and Y also belong to the same additive class and the joint distribution of X and Y can be determined by integrating out the variable, W_2. Some simple results are possible if the variables possess a distribution, belonging to the "Meixner class". In this class a generalization of Runge's identity for the Hermite polynomial system holds and the expectations of the cross products of the polynomials on the marginal distributions are readily evaluated with the aid of the polynomial generating function or more directly by the use of Runge's identity. The canonical form (VI.3.17) for the bivariate distribution now holds and the canonical variables are the orthonormal polynomials on the marginal distributions. In particular, the two canonical variables corresponding to the maximum correlation, ρ_1, are the standardised variables, $[X - \mathscr{E}(X)]/(\text{var } X)^{\frac{1}{2}} = x^{(1)}$ and $[Y - \mathscr{E}(Y)]/(\text{var } Y)^{\frac{1}{2}} = y^{(1)}$ and the maximum correlation is $\mathscr{E}x^{(1)}y^{(1)} = \text{corr}(X, Y)$. This correlation bears a simple relation to the variances of W_1, W_2 and W_3, for

$$(10.2) \qquad \rho_1 = \text{var } W_2[\text{var } X \text{ var } Y]^{-\frac{1}{2}},$$

and ρ_1 can be given an interpretation. If W_1, W_2 and W_3 are sums of n_1, n_2 and n_3 identically distributed summands, then $\rho_1 = n_2(n_1 n_3)^{-\frac{1}{2}}$ and this takes the form $\rho_1 = n_2/n$, if $n_1 = n_3 = n$; in this last form ρ_1 is the proportion of the summands held in common by X and Y, which have in this special case a common distribution. A generalization is possible to this class of Pearson's methods for estimating the correlation in the normal distribution. If it has been given that the joint distribution of X and Y has been obtained by an additive process as given above and that some random variables are held in common, then ρ_1 could be calculated with the parameters n_1, n_2 and

n_3. Conversely, if ρ_1, $(n_1 + n_2)$ and $(n_2 + n_3)$ were given, n_2 could be computed; if only the additive class were given, the parameters $(n_1 + n_2)$ and $(n_2 + n_3)$ could be estimated from the observed marginal distributions and then ρ_1 estimated from the correlation of X and Y. An analogue of Pearson's tetrachoric correlation arises if the marginal parameters are known but only a dichotomy of the marginal spaces is available. ρ_1 can then be estimated by a method generalizing Pearson's procedure; Gonin (1966) has done this for the bivariate binomial and Poisson distributions.

For a given additive Meixner class, each ρ_i is determined by the variances of W_1, W_2 and W_3 so that $\phi^2 = \sum \rho_i^2$ is also determined. If an $m \times n$ contingency table is constructed from N independent observations from the joint distribution of $X = W_1 + W_2$ and $Y = W_2 + W_3$, then χ^2 will have an expectation, $N\phi^2 + (m - 1)(n - 1)$, approximately if the partitions of the marginal spaces are sufficiently fine. In such cases, we could generalize Pearson's procedure to give the estimate,

(10.3) $$\hat{\phi}^2 = [\chi^2 - (m - 1)(n - 1)]/N.$$

If the marginal partitions were not sufficiently fine, a generalization of the method of Lancaster and Hamdan (1964) would be possible. For joint distributions of variables not in some special class, it cannot be expected that the theory will be simple.

Rényi (1959b) has given general conditions on measures of association $\delta(\xi, \eta)$:

(A) $\delta(\xi, \eta)$ is defined for any pair of random variables ξ and η, neither of them being constant with probability, 1.

(B) $\delta(\eta, \xi) = \delta(\xi, \eta)$.

(C) $0 \leqslant \delta(\xi, \eta) \leqslant 1$.

(D) $\delta(\xi, \eta) = 0$ if and only if ξ and η are independent.

(E) $\delta(\xi, \eta) = 1$ if there is a strict dependence between ξ and η, i.e. either $\xi = g(\eta)$ or $\eta = f(\xi)$ where $g(x)$ and $f(x)$ are Borel-measurable functions.

(F) If the Borel-measurable functions $f(x)$ and $g(x)$ map the real axis in a one-to-one way onto itself $\delta(f(\xi), g(\eta)) = \delta(\xi, \eta)$.

(G) If the joint distribution of ξ and η is normal, then $\delta(\xi, \eta) = |R(\xi, \eta)|$ where $R(\xi, \eta)$ is the correlation coefficient of ξ and η.

No measure $\delta(\xi, \eta)$ yet devised, which purports to measure all possible forms of departure, appears to be satisfactory. For example, none of them would treat the artificial population of Exercise VI.36 in a manner that was in agreement with intuition, namely that the measure $\delta(\xi, \eta)$ should increase monotonically from 0 to 1 with β. The reason for this is evident. A joint distribution by Theorem VI.4.1 is determined by its marginal distributions and the possibly infinite matrix, $\mathbf{R}$, of correlations between the members of orthonormal sets on the two marginal distributions. Any measure of dependence thus is a statement about the infinite matrix. General and precise statements can be made in two cases; if $\mathbf{R}$ is a matrix of zeroes, then the marginal variables are stochastically independent (Corollary 2 of Theorem VI.4.1) due to Sarmanov (1958b), Rényi (1959b) and Lancaster (1959 and 1961b). If $\mathbf{R}$ is orthogonal, then the marginal variables are completely mutually dependent (Lancaster, 1963c). In the Meixner class, $\mathbf{R}$ is a diagonal matrix and there is a simple interpretation as we have shown above. Outside these restricted classes, there possibly may be no general solution to the problem of finding a $\delta(\xi, \eta)$.

11. THE HOMOGENEITY OF SEVERAL POPULATIONS

Tests of significance for the homogeneity of several populations are very common. The theory has been discussed above. For further details, the reader is referred to the bibliography.

12. CONTINGENCY TABLES. MISCELLANEOUS TOPICS.

This section is included for bibliographical purposes only.

EXERCISES AND COMPLEMENTS

1. Suppose that of a population of M individuals Mp possess the property B and that Mq do not. Let now a sample of $a_{1.}$ individuals be selected from this population by a random process assigning equal probability to each individual remaining in the population at each selection. Prove that Z, the number of selected individuals having the property B, has the binomial distribution with parameters, $a_{1.}$ and p, if sampling is with replacement. Prove that, if the sampling or selection is without replacement, Z has the hypergeometric distribution. Generalize this to an n-fold classification in which individuals possess precisely one of the properties $B_1, B_2, \ldots, B_n$, and show that the distributions are the multinomial and that of the $2 \times n$ contingency table.

2. Derive the factorial moments of the hypergeometric distribution. With the notation of Section 2, prove that if $a_{1.}$ is large, the results are approximately

$$\mu_2 = a_{1.}a_{2.}a_{.1}a_{.2}/N^3.$$
$$\mu_3 = a_{1.}a_{2.}a_{.1}a_{.2}(a_{1.} - a_{2.})(a_{.1} - a_{.2})/N^5$$
$$\mu_4 = a_{1.}a_{2.}a_{.1}a_{.2}(N^3 + 3a_{1.}a_{2.}a_{.1}a_{.2} - 6a_{1.}a_{2.}N)/N^6$$

If $a_{1.}$ is not negligible with respect to N, this last equation is equivalent to $\mu_4 = 3\mu_2^2$. (Pearson, 1899 and 1906c; Greenwood, 1913).

3. Suppose in the hypergeometric distribution that $a_{1.}$ is small relative to N but $a_{.1}/N = p$ is not negligible. Show that the factorial moments of Z are equal to $a_{1.}^{(r)}a_{.1}^{(r)}/N^{(r)} \simeq a_{1.}^{(r)}p^r$. Hence Z has approximately the binomial distribution.

Suppose now that $a_1.p$ is not negligible although p is. The factorial moments are now approximately $\lambda^r \equiv (a_1.p)^r \simeq a_1^{(r)}p^r$. Z has thus approximately the distribution of a Poisson variable.

4. Obtain the results of Exercise 3 by a consideration of the variable, $U = \Delta^2/\mathscr{E}a_{11} + \Delta^2/\mathscr{E}a_{12} + \Delta^2/\mathscr{E}a_{21} + \Delta^2/\mathscr{E}a_{22}$, where $\Delta = a_{11} - a_1.a_{.1}/N$. Suppose first that $a_1.$ but not $a_{.1}$ is negligible with respect to N; prove that then only the contributions from the first two terms are non-negligible and that the sum is approximately the χ^2 of a binomial distribution with parameters, $a_1.$ and $a_{.1}/N$. Suppose secondly that $a_1.$ and $a_{.1}$ are both negligible with respect to N but that $\lambda = a_1.a_{.1}/N$ is finite; now only the first term contributes to U, which is now equal to $\Delta^2/\lambda = (a_{11} - \lambda)^2/\lambda$, the square of a standardized Poisson variable.

5. Suppose that X and Y take values $1, 2, \ldots, m$ and $1, 2, \ldots, n$ respectively with probabilities $\{f_{i.}\}$ and $\{f_{.j}\}$. Suppose further that X is distributed independently of Y so that $f_{ij} = \mathscr{P}(X = i, Y = j) = \mathscr{P}(X = i)\mathscr{P}(Y = j) = f_{i.}f_{.j}$. If N independent observations are made on the pair (X, Y), prove that the sum $\mathscr{S}X$ is independent of $\mathscr{S}Y$ and that if $\xi(X)$ and $\eta(Y)$ be arbitrary functions of X and Y, then $\mathscr{S}\xi(X)$ is independent of $\mathscr{S}\eta(Y)$.

6. Generalize Exercise 5 to prove that if X and Y have arbitrary distributions and X is independent of Y, then every $\xi(X)$ is independent of every $\eta(Y)$. If these functions have finite variance then $N^{-\frac{1}{2}}\mathscr{S}\xi(X)$ and $N^{-\frac{1}{2}}\mathscr{S}\eta(Y)$ are mutually independent, where $\mathscr{S}$ is the sum of the values of ξ and η over N independent observations on (X, Y) and $N \to \infty$.

7. Prove that if every step function in X is uncorrelated with every step function in Y, then every simple function in X is uncorrelated with every simple function in Y. Hence prove that every square summable function in X is uncorrelated with every square summable function in Y.

8. Prove that if X is independent of Y and if the spaces of X and/or Y are condensed, then the new variables, X^* and Y^* say, are mutually stochastically independent.

9. The differences between the analyses of the three models of the fourfold table are not as absolute as at first seems. Model 1 becomes Model 2 if sampling is by the following rule. Sample the infinite population until a prescribed number $a_1.$ possessing property A or $a_2.$ not possessing property A have been obtained. If $a_2.$ of the not-A's have been obtained, continue sampling but reject all further not-A's until $a_1.$ A's have been obtained. Model 2 can be converted into Model 1 by the artificial device of giving $a_1.$ a binomial distribution with parameters, $f_1.$ and N. Similar procedures convert Model 2 into Model 3 and conversely.

10. With Model 2 in a fourfold table, suppose that $a_1.$ and $a_2.$ have been prescribed and that H_0 specifies some unknown p. Then $\hat{p} = a_{.1}/N$, where $N = a_1. + a_2.$, is the overall mean of a variable Y taking values, unity in the first column and zero in the second column. $\hat{\sigma}^2 = a_1.a_2./N^2(N - 1)$ is the variance of the estimate $\hat{p}$ of the mean. The difference between the means in the two rows is $a_{11}/a_1. - a_{21}/a_2.$, which can be compared with its standard error $\hat{\sigma}\sqrt{(a_1^{-1} + a_2^{-1})}$. Assuming that

Student's t-test is robust, then the usual χ can be set equal to a Student's t with $N - 2$ degrees of freedom (Welch, 1938; Barnard, 1947; Pearson, 1947).

11. Deaths from cancer in Edinburgh and Dundee for 1891–1900 were as follows:

| | Edinburgh | | Dundee | |
| | | | | |
Age Group	Population	Cancer Deaths	Population	Cancer Deaths
0–4	149,763*	5	89,775	6
5–14	280,655	8	168,510	2
15–24	274,343	18	142,917	7
25–34	214,063	47	98,953	10
35	158,133	117	76,132	47
45	115,206	295	59,066	114
55	69,954	356	37,337	151
65	32,966	266	16,958	109
75	10,311	93	4,625	30
85 and over	1,045	8	535	3
Totals.	1,306,439	1,213	694,808	479

* Years of life experienced at these ages 1891–1900.

Compare the mortality rates in the two cities at the various ages by means of fourfold tables. Give a combined result. (Neyman and Pearson, 1928; K. Pearson and Tocher, 1915–17.)

12. Verify that the square of χ in (4.5) is the χ^2 of (4.7).

13. "When it is desired to determine the probability of a given divergence from the independence-values calculated from the total of rows and columns the number of algebraically independent values of δ is only $(r - 1)(c - 1)$ and n' should be taken as given by $n' = (r - 1)(c - 1) + 1$." [$n' - 1$ was, in the notation introduced by K. Pearson, the number of degrees of freedom.] Yule (1922).

14. Prove that the additional necessary conditions for independence in Yule's remark in Exercise 13, are provided by the conditions, $\mathscr{E}x^{(i)}y^{(j)} = 0, i = 1, 2, \ldots,$ $r - 1; \ j = 1, 2, \ldots, c - 1$, where $\{x^{(i)}\}$ and $\{y^{(j)}\}$ are orthonormal bases on the marginal distributions. Alternatively the additional conditions are provided by

$$\mathbf{X}^T\mathbf{F}\mathbf{Y} = \begin{pmatrix} 1 & \mathbf{0}^T \\ \mathbf{0} & \mathbf{0} \end{pmatrix}$$

where $\mathbf{X}$ is a matrix such that $x_{i1} = 1$, x_{ij} is the value of the $(j - 1)^{\text{th}}$ orthonormal function at the point $X = i$ and $\mathbf{Y}$ is similarly defined; $\mathbf{F} = (f_{ij})$.

15. Prove that given the marginal probabilities $\{f_{i.}\}$ and $\{f_{.j}\}$ and an appropriate set of $(r - 1)(c - 1)$ ratios, $\psi(i, i'; j, j') = f_{ij}f_{i'j'}/(f_{ij'}f_{i'j})$, the cell probabilities are determined. If each ψ is unity, then X is independent of Y. Show by examples

that the term "appropriate" cannot be omitted from the above statement; in other words, the ratios must be algebraically independent and not deducible from a subset.

16. Under a hypothesis of mutual independence of the marginal variables, prove that the expectation of χ^2 in an $m \times n$ table is $N(m - 1)(n - 1)/(N - 1)$. (Bartlett, 1937; Cochran 1936a, 1937; Geary, 1940; Haldane, 1939c; Uspensky, 1937; Stevens, 1938, 1951.)

17. Given a 3×3 contingency table with H_0 specifying mutual independence of the marginal variables and row and column parameters r_i and c_j, prove that the covariance matrix, divided by the factor $(N - 1)$, is given by

	a_{11}	a_{12}	a_{22}
a_{11}	$r_1(1 - r_1)c_1(1 - c_1)$	$-r_1(1 - r_1)c_1c_2$	$r_1r_2c_1c_2$
a_{12}	$-r_1(1 - r_1)c_1c_2$	$r_1(1 - r_1)c_2(1 - c_2)$	$-r_1r_2c_2(1 - c_2)$
a_{22}	$r_1r_2c_1c_2$	$-r_1r_2c_2(1 - c_2)$	$r_2(1 - r_2)c_2(1 - c_2)$

(Blakeman and Pearson, 1906; Sheppard, 1929.)

18. For two-way contingency tables with fixed marginal totals, determine the mean and variance of the entry in any particular cell. Determine also the correlation between the entries in cells in the same row or the same column or in neither. Determine the distribution of the sum of diagonal elements or of elements along a diagonal. (Stevens, 1938).

19. Compare the coefficients of $\prod_i t_i^{a_i}$ on the two sides of

$$(t_1 + t_2 + \ldots + t_n)^{a_1.}(t_1 + t_2 + \ldots + t_n)^{a_2.} \ldots (t_1 + t_2 + \ldots t_n)^{a_m.}$$
$$= (t_1 + t_2 + \ldots + t_n)^{a_{..}}$$

to obtain the identity (4.13). By setting $t_1 = u_1$, $t_2 = t_3 = \ldots = t_n = u_2$ and $b_{1.} = a_{1.}, b_{2.} = a_2. + a_3. + \ldots a_{m.}$, show that the distribution of a_{11} in the $m \times n$ table is the same as its distribution in the 2×2 formed by pooling the last $(m - 1)$ rows and $(n - 1)$ columns.

20. "In a $m \times n$ contingency table an alternative criterion is given by

$$\mu = \prod_{i=1}^{m} \prod_{j=1}^{n} a'_{ij}!/a_{ij}!$$

where $a'_{ij}!$ are the values of a_{ij} maximising $\mathscr{P}(\{a_{ij}\} \mid \{a_{i.}\}, \{a_{.j}\})$. For large samples $-2 \log \mu$ will, like $-2 \log \lambda$, be distributed as χ^2, the three tests becoming equivalent. For medium to small samples it is doubtful whether the usual χ^2 test can be bettered. The distribution of μ is known only in so far as it approximates to the form $\exp(-\frac{1}{2}\chi^2)$". (Bartlett, 1937b.)

21. With the convention, $a_{1.} \leqslant a_{.1} \leqslant a_{.2} \leqslant a_{2.}$, there are $a_{1.} + 1$ possible values of the entry, a_{11}. If $a_{1.}$ is not small give an estimate of the number of observations falling in the interval $\mathscr{E}a_{11} \pm 2\sqrt{\operatorname{var} a_{11}}$. Give also an estimate of the magnitude of the probability of the observation, $a_{11} = A$ say, such that $\mathscr{P}(A) \geqslant \alpha$, $\mathscr{P}(A + 1) < \alpha$ using the asymptotic theory. (Lancaster, 1961a.)

22. In the 2×3 table, suppose that $a_1 < a_2$ and that $a_1 < a_{.1} < a_{.2} < a_{.3} < a_2$. Suppose also that $a_{.1} < a_2 - a_1$. Evaluate the number of possible tables which can be constructed subject to the marginal constraints. [Write the table as

u	v	x	a_1
x	x	x	a_2
$a_{.1}$	$a_{.2}$	$a_{.3}$	N

in which x means the value that can be written in to satisfy the constraints $\sum_j a_{ij} = a_i$, $\sum_i a_{ij} = a_{.j}$.] u can assume $a_1 + 1$ values under the hypotheses made. For a fixed u, v can assume $a_1 - u + 1$ distinct values. The required sum is thus

$$\sum_{u=0}^{a_1} (a_1 - u + 1) = \tfrac{1}{2}(a_1^2 + 3a_1 + 2)$$

23. The following table gives the sex of subsequent births, given that the sex of the first birth in the sibship is known.

Sex of First-Born of Sibship	Subsequent Births Males	Females	Totals
Male	17,341	16,518	33,859
Female	16,797	16,232	33,029
Total	34,138	32,750	66,888

Test the null hypothesis that p, the probability of a male birth is the same whether the first birth be male or female. [$\chi^2 = 0.867$ for 1 d.f.]. (Lancaster, 1950d.)

24. "Now in recent work on such things as temper in man, eye colour in man and hair colour in man or other animals, I have proceeded to arrange my groups in two or three different orders, and to calculate the correlation on the basis of these different orders. The results for the different orders came out in rather striking agreement . . .". Comment on this passage from Pearson and consider it as a statement of the invariance of ϕ^2. (Pearson, 1904.)

25. Pearson writes $N + \chi^2 = N \sum a_{ij}^2/(a_i . a_{.j})$ for the χ^2 in a contingency table. He comments that if we sample a parent population by taking individuals out one at a time and recording their characters, we obtain samples in which there is no fixing of the marginal totals.

On the other hand, if we draw two independent samples such as a number of boys and then a number of girls, a different sampling procedure is being used. Verify that these are Models 1 and 2 sampling methods. (Pearson, 1932b.)

26. Of a list of school-teachers who passed a certain examination, a classification has been made.

First Letter of Name

List	A–D	E–J	K–R	S–Z	Total
Men	166	174	180	164	684
Women, 1st year	427	379	411	366	1,583
Women, 2nd year	549	493	577	492	2,111
	1,142	1,046	1,168	1,022	4,378

Calculate $a_{ij} - a_i a_{.j}/a_{..}$ for each square and standardise these differences by dividing by the probable error of the number namely $0.67448975 \sqrt{\{a_i a_{.j}/a_{..}\}}$ to obtain the deviations in terms of probable error. [The standardized ratios are

$$
\begin{array}{cccc}
-1.7 & +1.5 & -0.3 & +0.6 \\
+1.5 & +0.1 & -1.2 & -0.4 \\
-0.2 & -1.2 & +1.4 & -0.1]
\end{array}
$$

Comment on evidence for independence. Analyse the contingency table by the methods of this chapter. (Sheppard, 1898a.)

27. Nine carriers of a parasite were examined repeatedly by a standard procedure with the following results.

Carrier	Positive	Negative	Total
1	5	17	22
2	12	1	13
3	8	0	8
4	3	17	20
5	1	22	23
6	3	11	14
7	11	4	15
8	6	15	21
9	22	21	43
Total	71	108	179

Verify that if the probability of detecting the parasite at an examination is regarded as a constant peculiar to the carrier, this table can be regarded as an example of supernormal dispersion in the Lexis sense. Verify that $\chi^2 = 59.44$ for 8 d.f. and $P < 0.001$. The probability of detection cannot be regarded as constant between carriers. (Lancaster, 1950c.)

28. Given the marginal frequencies, $a_{1.} = 17$, $a_{2.} = 13$, $a_{.1} = 13$, $a_{.2} = 11$, $a_{.3} = 6$, let a_{22} and a_{23} run over all their possible values. Evaluate the values of $\mathscr{P}(\{a_{ij}\} \mid$ marginal totals). Verify that if these sets of values of "$\mathscr{P}$" are arranged in order of magnitude, the cumulative probability is fairly well approximated by $\mathscr{P}(\chi^2)$. (Yates, 1934.) [Note. Tables for testing the significance of 2×3 tables are available in Bennett and Nakamura (1963) and Chakravarti and Rao (1959); the power function of the exact test of the 2×3 tables is given by Bennett and Nakamura (1964).]

29. In a 4×2 table with $a_{i.} = 8$ and $a_{.1} = 8$, show that there are only 15 distinct configurations owing to the equality of the row totals. In such cases, it is possible for several distinct configurations to give the same value of χ^2, whereas in general, no two distinct configurations give the same value of L, the likelihood function. (Cochran, 1936a.)

30. Prove that the χ^2 of a $2 \times n$ contingency table can be divided into components, one of which corresponds to a t-test between variables defined arbitrarily on the columns. Extend this by showing that χ^2 is the sum of squares of $(n-1)$ standardized orthogonal functions of the form

$$N^{-\frac{1}{2}}\mathscr{S}y^{(1)}, N^{-\frac{1}{2}}\mathscr{S}y^{(2)}, \ldots, N^{-\frac{1}{2}}\mathscr{S}y^{(n-1)},$$

where $1, y^{(1)}, \ldots, y^{(n-1)}$ is an orthonormal basis on the probability distribution with weights, $p_{.j}$, at the point, j. This corresponds to a reconciliation of χ^2, considered as a sum of squares of either enumerative or metrical variables. (Lancaster, 1953a.)

31. Show that, if row and column parameters have been estimated from the data, the pooling of two adjacent cells and recalculation of χ^2 does not remove the whole of χ^2 of one degree of freedom. Prove that if a_{31} and a_{32} are the numbers pooled in a $3 \times n$ table, the χ^2 subtracted should be that of the fourfold table,

$$\begin{bmatrix} a_{11} + a_{21} & a_{12} + a_{22} \\ a_{31} & a_{32} \end{bmatrix}$$

which is the only degree of freedom in which a_{31} is compared with a_{32}. (Lancaster, 1949a.)

32. Consider the 3×3 contingency table with numbers in the nine cells of a matrix, **A**.

3009	2832	3008
3047	3051	2997
2974	3038	3018

Calculate χ^2 on the hypothesis that $p_{ij} = \frac{1}{9}$, every (i, j). Next compute the χ^2 for the row totals, using the hypothesis that $p_{i.} = \frac{1}{3}$. Similarly compute χ^2 for the column totals. Now form a table, analysing the total χ^2 into the parts due to rows, columns and remainder. Verify this arithmetic by forming the matrix, **Q**,

$$q_{ij} = (a_{ij} - \mathscr{E}a_{ij})/(\mathscr{E}a_{ij})^{\frac{1}{2}}$$

and the matrix

$$\mathbf{R} = \mathbf{C} = \begin{bmatrix} 3^{-\frac{1}{2}} & 3^{-\frac{1}{2}} & 3^{-\frac{1}{2}} \\ 2^{-\frac{1}{2}} & -2^{-\frac{1}{2}} & 0 \\ 6^{-\frac{1}{2}} & 6^{-\frac{1}{2}} & -2 \cdot 6^{-\frac{1}{2}} \end{bmatrix}$$

Identify the individual components of χ^2 in the matrix,

$$(e_{ij}) = \mathbf{RQC}^T,$$

and note that $e_{11} = 0$ identically. Analyse the above table with parameters estimated from the data i.e. $\hat{p}_{i.} = a_{i.}/a_{..}$, $\hat{p}_{.j} = a_{.j}/a_{..}$, calculating the total χ^2 and form Helmert matrices $\mathbf{R}$ and $\mathbf{C}$ with $r_{1j} = \hat{p}_{j.}^{\frac{1}{2}}$ and $c_{1j} = \hat{p}_{.j}^{\frac{1}{2}}$ respectively. Verify that the interaction standardized normal variables with theoretical and fitted parameters respectively are

$$\begin{bmatrix} 1 \cdot 6531 & -1 \cdot 4712 \\ 1 \cdot 5872 & -0 \cdot 0700 \end{bmatrix} \quad \text{and} \quad \begin{bmatrix} 1 \cdot 6834 & -1 \cdot 4767 \\ 1 \cdot 5896 & -0 \cdot 0711 \end{bmatrix}$$

(Lancaster 1949a).

33. Show that the partition of the interaction χ^2 given above can be approximately obtained by computing the χ^2 of the tables

3009	2832	5841	3008
3047	3051	6098	2997
6056	5883	11939	6005
2974	3038	6012	3018

of which the χ^2's are

$$\begin{bmatrix} 2 \cdot 86004 & 2 \cdot 18027 \\ 2 \cdot 52637 & 0 \cdot 00506 \end{bmatrix} \quad \text{and the } \chi\text{'s are} \quad \begin{bmatrix} 1 \cdot 691 & -1 \cdot 477 \\ 1 \cdot 589 & -0 \cdot 071 \end{bmatrix}$$

(Lancaster, 1949a).

34. Let (A, μ) be a probability measure space. Prove that if $\zeta_1 = \zeta_2$ a.e. $[\mu]$, then their correlation is unity; the converse also holds. Prove that the normalized functions having unit correlation with a given function form an equivalence class so that $\zeta_i = \zeta_j$ a.e. $[\mu]$.

35. "A true coefficient of association should not necessarily become perfect when one of the four quadrants of the fourfold table becomes zero". (Page 261 of Pearson and Heron, 1913.) Use Exercise 34 to comment on this.

36. Let X and Y be the row and column variables of an $m \times n$ contingency table. Let $x^{(0)} = y^{(0)} = 1$. Suppose $\{x^{(i)}\}$ and $\{y^{(j)}\}$ are orthonormal bases. The $(m - 1)(n - 1)$ asymptotically normal $(0, 1)$ variables of the contingency table theory can be represented as the standardized sums of the form, $N^{-\frac{1}{2}}\mathscr{S}(x^{(i)}y^{(j)})$, $i = 1, 2, \ldots, m - 1$; $j = 1, 2, \ldots, n - 1$, which are estimates $N^{\frac{1}{2}}\hat{\rho}_{ij}$ of the $N^{\frac{1}{2}}\rho_{ij}$. The standardized sums $N^{-\frac{1}{2}}\mathscr{S}x^{(i)}$ and $N^{-\frac{1}{2}}\mathscr{S}y^{(j)}$ can be written as elements

of vectors, $\mathbf{X}$ and $\mathbf{Y}$ with covariance matrices,

$$\mathscr{E}\mathbf{X}\mathbf{X}^T = \mathbf{V}_{11} = \mathbf{1}_{m-1}, \qquad \mathscr{E}\mathbf{Y}\mathbf{Y}^T = \mathbf{V}_{22} = \mathbf{1}_{n-1}.$$

These covariance matrices are estimated from the data since the contingency table is being studied with fixed marginal totals. $\mathbf{V}_{12}$ is estimated by setting the (i,j) element of $\mathbf{V}_{12}$ equal to $N^{-\frac{1}{2}}\mathscr{S}x^{(i)}y^{(j)}$. The Fisher canonical theory thus gives estimates, $\hat{\rho}_i$, of the canonical correlations, ρ_i say, by setting $\hat{\rho}_i$ equal to the ith diagonal element of $\mathbf{C} = \mathbf{P}^T\hat{\mathbf{V}}_{12}\mathbf{Q}$, where $\mathbf{P}$ and $\mathbf{Q}$ are orthogonal matrices chosen so that $\mathbf{C}$ is in the canonical form of (VI.2.5). (Fisher, 1940; Hotelling, 1936; Lancaster, 1958; Maung, 1942.)

37. It is required to determine $\xi^{(1)}$, $\eta^{(1)}$ and ρ_1, the first pair of canonical variables and the maximal correlation in an $m \times n$ bivariate distribution with notation as in Exercise 36. Prove the feasibility of the following iteration. Let ξ_1 be a standardized variable taking values ξ_{1i} at the points $X = i$. Now determine the expectation of ξ_1 for $Y = j$, $\mathscr{E}(\xi_1 \mid Y = j)$. Define η_1 as the standardized form of $\mathscr{E}(\xi_1 \mid Y = j)$. Define ξ_2 by standardizing the function $\mathscr{E}(\eta_1 \mid X = i)$ and proceed to obtain η_2, ξ_3, η_3, ξ_4, ... iteratively. Continue until the cycle is repeated or two successive ξ's are sufficiently close. Show that this process will yield the first pair of canonical variables and the maximal correlation. (Fisher, 1940; Lancaster, 1958.)

38. Prove that NR_1^2 is stochastically greater than a χ^2 variable with $m + n - 3$ degrees of freedom. (Kendall and Stuart, 1961; Lancaster 1963c; Williams, 1952.)

39. For the First World War, an experience of inoculation against typhoid fever is given as follows:

	Not Attacked	Attacked	Total
Inoculated	10,322	56	10,378
Not Inoculated	8,664	272	8,936
Total	18,986	328	19,314

$\chi^2 = 180 \cdot 38$ for 1 d.f.; $P < 0 \cdot 0001$. The question arises as the authors point out whether this table can be regarded as suitable for the analysis. Let us write down the experience of two areas (hypothetical) in the same form as the table above

50	50	950	50
500	500	95	5.

Let us now pool the experience and by summing the two tables to form a single table,

1000	100
595	505,

for which $\chi^2 = 373 \cdot 95$.

In the area with a high attack rate, the proportion of the inoculated is small. In the area with the low attack rate, the proportion of the inoculated is high. In other words the comparison between inoculated and not inoculated is not fair in the combined table since the inoculated have enjoyed more favorable conditions than the not inoculated. We might think of the first table as the experience of front-line troops and of the second table as the experience of base troops in an army.

40. Generalize the preceding example to show that if $p_{ij} = p_{i.}p_{.j}$ and $p'_{ij} = p'_{i.}p'_{.j}$ and $f_{ij} = \theta p_{ij} + (1 - \theta)p'_{ij}$ then $f_{ij} = f_{i.}f_{.j}$ only for special values of the parameters. Hence mixing of distributions or pooling of heterogeneous experience will lead to spurious correlation. Note that $p'_{11} = 1$ is a special case of the above. Set up a design for testing H_0, that inoculation is ineffective by observing the results in regiments, whose members may be considered to have had similar experiences, and suggest an appropriate way of combining the results from the different regiments.

41. Suppose that $0 < p_1 < 1$, $0 < p_2 < 1$. Let $b(a_{11}, p_1)$ and $b(a_{21}, p_2)$ be the probabilities that binomial variables, Z_1 and Z_2, take values a_{11} and a_{21} respectively. Let H_0 specify that $p_1 = p_2$. Derive the likelihood ratio test,

$$\lambda = \frac{a^{a_{.1}}_{.1} a^{a_{.2}}_{.2} a_{11}^{-a_{11}} a_{12}^{-a_{12}} a_{21}^{-a_{21}} a_{22}^{-a_{22}}}{N^N a_{1.}^{-a_{1.}} \cdot a_{2.}^{-a_{2.}}}$$

and show that, as $a_{1.} \to \infty$, $a_{2.} \to \infty$, the limiting distribution of $-2 \log \lambda$ is that of χ^2 for 1 d.f. under H_0. Generalise this to $m \times 2$ and $m \times n$ tables.

42. Let $\{p_{ij}\}$ be the probabilities in an $m \times n$ probability distribution and suppose that p_{ij} is positive for every (i, j). Let Ω be the space of all possible sets $\{p_{ij}\}$ subject to the above conditions. Let H_0 specify $p_{ij} = p_{i.}p_{.j}$. Derive the likelihood ratio test,

$$\lambda = \frac{\prod_{i=1}^{m} \left(\frac{a_{i.}}{N}\right)^{a_{i.}} \cdot \prod_{j=1}^{n} \left(\frac{a_{.j}}{N}\right)^{a_{.j}}}{\prod_i \prod_j \left(\frac{a_{ij}}{N}\right)^{a_{ij}}}$$

and show that as $N \to \infty$, the limiting distribution of $-2 \log \lambda$ is χ^2 with $(m - 1)(n - 1)$ d.f. Define $Q = \sum \sum (a_{ij} - \bar{a}_{ij})^2/\bar{a}_{ij}$, where $\bar{a}_{ij} = a_{i.}a_{.j}/N$. Show that under a true null hypothesis, H_0, Q and $-2 \log \lambda$ converge together in distribution to χ^2 with $(m - 1)(n - 1)$ d.f.

43. Suppose an entry, x, at position $(1, 1)$ in a contingency table is missing. Let R_i be the row totals and C_j the column totals, $i = 1, 2, \ldots, r$; $j = 1, 2, \ldots, c$. Estimate the unknown marginal class probabilities by setting

$$x/N = R_1 C_1/(N - R_1 - C_1)$$

and then $\hat{p}_1 = (R_1 + x)/(N + x)$, $\hat{p}_k = R_k/(N + x)$, $k \neq 1$; $\hat{q}_1 = (C_1 + x)/(N + x)$, $\hat{q}_k = C_k/(N + x)$, $k \neq 1$. The resulting χ^2 may be written as

$$Q = \sum_{(i,j) \neq (1,1)} f_{ij}^2/[(N + x)\hat{p}_i \hat{q}_j] - N;$$

Q has $(c - 1)(r - 1) - 1$ d.f. Verify that this is equivalent to dropping the interaction χ corresponding to the fourfold table consisting of the first row and all other rows pooled and the first column and all other columns pooled. (Watson, 1956.)

44. Suppose the entries at positions, a_{11} and a_{12}, $(1, 1)$ and $(1, 2)$ in an $r \times c$

contingency table are mixed up. The row proportions can be estimated by $\hat{p}_i = R_i/N$, $i = 1(1)r$. The column probabilities are $\hat{q}_j = C_j/N$, $j = 3(1)c$,

$$\hat{q}_1 = (C_1 - a_{11})(C_1 + C_2)/\{(C_1 + C_2 - a_{11} - a_{12})N\},$$
$$\hat{q}_2 = (C_2 - a_{12})(C_1 + C_2)/\{(C_1 + C_2 - a_{11} - a_{12})N\}.$$

In other words the estimated entries, $\hat{a}_{11}$ and $\hat{a}_{12}$, are distributed in the cells $(1, 1)$ and $(1, 2)$ so that the ratio, $\hat{a}_{11}/\hat{a}_{12}$, is the same as $(C_1 - a_{11})/(C_2 - a_{12})$. One component χ^2 of the usual χ^2 of the contingency table is thus annihilated. (Watson, 1956.)

45. Prove that if a bivariate distribution is partitioned into $p + 1$ rows and $q + 1$ columns, $p \leqslant q$, then the parameter of non-centrality cannot exceed

$$N(\rho_1^2 + \rho_2^2 + \ldots + \rho_p^2),$$

where $\{\rho_i\}$ is the set of canonical correlations.

46. Suppose the alternative hypothesis specifies p in the first row and p' in the second row. How ought $a_{1.}$ and $a_{2.}$ be chosen subject to $a_{1.} + a_{2.} = N$ so that the test shall have power as large as possible? (Lehmann, 1959, p. 146.)

47. In a bivariate distribution, there can be asymmetry although the variable $X - Y$ has a symmetrical distribution. Let $|\delta| < 1$. Suppose that $\mathscr{P}\{X = i, Y = j\} = p_{ij}$ and that $16p_{ij}$ is displayed as a matrix

X \ Y	-3	-1	$+1$	$+3$
-3	1	$1 + \delta$	$1 - \delta$	1
-1	$1 - \delta$	1	1	$1 + \delta$
$+1$	$1 + \delta$	1	1	$1 - \delta$
$+3$	1	$1 - \delta$	$1 + \delta$	1

Show that the distributions of $X + Y$ and $X - Y$ are the same as that of the sum or difference of two independently distributed random variables with the same distributions. (cf. Feller, 1966, p. 98.)

48. With sampling under Model 1, suppose (X, Y) is a pair of symmetrical variables such as heights of brothers or efficiency of eyesight in the two eyes. Suppose that the same classification (i.e. partition of the space) is made on both variables and that there are m marginal classes with probabilities, $p_{1.}, p_{2.}, \ldots, p_{m.}$ for rows $p_{.1}, \ldots$ for columns and p_{ij} for cells. Calculate the mean and expectation of a cell content, a row total and a column total. Calculate the correlations between cell contents noting that marginal totals are not fixed. Calculate cov $(n_{i.}, n_{.j})$ by forming a 2×2 table by pooling all row classes $i' \neq i$ and all column classes $j' \neq j$. Defining $d_i = n_{i.} - n_{.i}$, calculate the expectation and variance of d_i and cov (d_i, d_j) if $i \neq j$. (Stuart, 1955.)

49. Using the result of the previous exercise, namely

$$\text{var}(d_i) = N\{p_{i.} + p_{.i} - 2p_{ii} - (p_{i.} - p_{.i})^2\}$$
$$\text{cov}(d_i, d_j) = -N\{(p_{ij} + p_{ji}) + (p_{i.} - p_{.i})(p_{j.} - p_{.j})\}.$$

Show that the covariance matrix of the d_i is of rank $(m - 1)$. Give the simplified form of the above equations when H_0 specifies $d_i = 0$. Write down the maximum likelihood estimates for the covariance matrix of the first $(m - 1)$ of the d_i as $\hat{V} = (\hat{v}_{ij})$. Give a justification of $d^T\hat{V}^{-1}d$ as χ^2 with $(m - 1)$ degrees of freedom and show that this is a test of the homogeneity of the marginal distributions. (Stuart, 1955.)

50. Consider the following table from Stuart (1955) with the sight in the two eyes classified in 7477 persons.

Right \ Left	1	2	3	4	Total
1	1520	266	124	66	1976
2	234	1512	432	78	2256
3	117	362	1772	205	2456
4	36	82	179	492	789
Total	1907	2222	2507	841	7477

Carry through the computations suggested by the analysis of the two preceding exercises. (Stuart, 1955.)

51. Show that the table of the previous exercise could be analysed for symmetry by considering variables of the form $X_{ij} = (a_{ij} - a_{ji})/\sqrt{(a_{ij} + a_{ji})}$ as asymptotically normal (0, 1). Since most of the distribution is concentrated along the main diagonal, it may be expected that the covariance matrix of the variables, X_{ij}, $i < j$, will be close to the unit matrix and its inverse will also be close to the unit matrix so that $\sum X_{ij}^2$ is, to a fair approximation, χ^2 with $\frac{1}{2}m(m - 1)$ d.f. $m = 4$ in the example of the exercises above. [Modified test of Bowker, 1947.]

52. The rankings by $N = 123$ consumers of three varieties of snap beans are given in a 3 × 3 table,

$$\begin{matrix} 42 & 64 & 17 \\ 31 & 16 & 76 \\ 50 & 43 & 30, \end{matrix}$$

where the rankings of 123 persons (columns) are given of 3 varieties (rows). If H_0 specifies that the three varieties are of equal consumer rating, the table may be regarded as generated by the summation of 123 random permutation matrices. Work out the variance-covariance ratio of the entries, a_{ij}. Hence prove that if $\bar{a}_{ij} = a_{i.}/r = a_{.j}/r = N/r$, r being the number of rows, the test criterion $\sum a_{ij}^2/\bar{a}_{ij} - N$ is distributed as $k\chi^2$, where $k = r/(r - 1)$. Prove that, if $\mathbf{R}$ is the matrix

$$(r_1 + \delta_{ij}(1 - r_1)), \qquad r_1 = -(r - 1)^{-1},$$

the covariance matrix of single entries into the table is given by the Kronecker product, $\mathbf{R} \times \mathbf{R}$. (Anderson, 1959.)

CHAPTER XII

Contingency Tables of Higher Dimensions

1. INTRODUCTORY AND HISTORICAL

The first attempts to give a general theory of the association or mutual dependence of random variables were due to Yule and Pearson. Pearson (1904) treated his contingency table as if it had been generated by the partition of a multivariate normal distribution. We may summarize his computation of ϕ^2 as follows.

Let $f(\mathbf{X})$ be the density function of n jointly normal variables and $g_i(x_i)$ the density function of the ith marginal variable. Let $f^*(\mathbf{X})$ be the product of the marginal densities, $g_i(x_i)$. Then

(1.1)
$$\phi^2 + 1 = \int \cdots \int f^2/f^* \, dx_1 \, dx_2 \ldots dx_n$$
$$= |1 - \mathbf{P}|^{-\frac{1}{2}} |1 + \mathbf{P}|^{-\frac{1}{2}},$$

in which $1 + \mathbf{P}$ is the correlation matrix of the $\{X_i\}$. A special case is provided by $n = 2$, when we have the bivariate normal distribution and

(1.2)
$$\phi^2 = \rho^2/(1 - \rho^2)$$

Pearson concluded that he had generalized the notions of Yule (1900) who had confined his attention to variables taking precisely two values; that he had developed a theory of dependent random variables or generalized theory of association which is a generalization of the independent random variables of the elementary texts; he had found a relation between contingency and correlation in the normal case. The tests devised were independent of the scaling or ordering of the marginal classifications. The theory could be extended to non-normal distributions; the marginal distributions should not be too fine if ϕ^2 were to be used to estimate ρ^2.

As we have noted in the general historical introduction in Chapter I, Pearson (1916b) returned to the theory of multiple contingency. He imagined a population, effectively infinite, from which a sample of M individuals were chosen. He defined

(1.3) $$\chi^2 = \sum_{i,j,\dots k} (a_{ij\dots k} - \mathscr{E} a_{ij\dots k})^2 / \mathscr{E} a_{ij\dots k},$$

in which $a_{ij\dots k}$ are the observed numbers in the class with co-ordinates $(i, j, \dots, k)$ and the expected numbers are computed on the hypothesis of complete independence. Pearson (1916b) suggested variations on the hypothesis of independence. The cell expectations may be given by an hypothesis in which independence is not assumed. Pearson then showed that if q linear restrictions are placed on the cell frequencies, the degrees of freedom will be reduced from $n' - 1$ to $n' - q - 1$. He noticed the close analogy with the multivariate normal theory of partial association.

If we except the work on partial association in the multivariate normal distribution and its applications to various fields, such as sociology and public health, there was comparatively little interest in the tables of higher dimensions for many years. The reason is possibly that only for the multivariate normal distribution was it possible to obtain neat analytic results. Moreover, the study of the multivariate normal distribution appears to have led to a general belief that pairwise independence implied complete independence. Wilks (1935) applied the likelihood ratio methods, which are asymptotically equivalent to χ^2, to the problem of multidimensional contingency tables and derived the likelihood ratio,

(1.4) $$-2 \log \lambda'_c = 2 \sum_{i,j,k} (n_{ijk} \log n_{ijk}) + 4N \log N - 2 \sum_i n_{i..} \log n_{i..}$$

and two similar terms. He stated that $-2 \log \lambda'$ is distributed approximately as χ^2 with degrees of freedom equal to $rst - r - s - t + 2 \equiv rst - 1 - (r - 1) - (s - 1) - (t - 1)$. Wilks thus combined tests of all sources of variation from the null hypothesis of complete independence into one likelihood ratio test criterion or χ^2.

Bartlett (1935), at the suggestion of R. A. Fisher, sought a test of second order interaction by generalising from the two dimensional condition for independence, $p_{ij}p_{i'j'} = p_{ij'}p_{i'j}$ to

(1.5) $$p_1 p_4 p_6 p_7 = p_2 p_3 p_5 p_8.$$

Elaborations on this procedure have been made by Kullback (1959), Lindley (1964), Norton (1945), Roy (1957), Roy and Kastenbaum (1956) and Kastenbaum and Lamphiear (1959).

An alternative approach also came from R. A. Fisher. In various editions of his *Statistical Methods*, he had considered the overall χ^2 for 7 degrees of freedom in a breeding experiment with three pairs of allelomorphic genes and hence $2^3 = 8$ classes of offspring. He separated out the contributions of the three main effects to the total χ^2 and the three interaction χ^2's in the three two dimensional tables and then noted that a seventh contribution was left which was included "to complete the analysis".

Haldane (1939a) commenting on the analysis of similar genetic experiments, analysed the χ^2 of $31 = 2^5 - 1$ degrees of freedom into the 5 main effects and 10 interactions of the first order, 10 interactions of the second order, 5 interactions of the third order and 1 of the fourth order.

Lancaster (1951) generalized the models that had been developed in two dimensions to higher dimensions, mentioning three important models in which H_0 specified (i) $p_{ijk} = p_{i..}p_{.j.}p_{..k}$ or (ii) $p_{ijk} = a_{i..}p_{.j.}p_{..k}/N$ or (iii) $p_{ijk} = a_{ij.}p_{..k}/N$; (i) is a generalization of Model 1 and (ii) and (iii) are generalizations of Model 2 of the two-way tables. In this paper and in Lancaster (1960a) much attention is paid to the partition of χ^2, which is the analogue of the familiar analysis of variance technique. On the computational side, this analogy was stressed also by Kimball (1954) and Claringbold (1961).

2. INTERACTIONS AND GENERALIZED CORRELATIONS

Two definitions of interaction are now given and proved equivalent; the first is measure theoretical and the second depends on notions of regression and correlation. Some unsatisfactory alternative definitions are discussed in the exercises. There are close analogies between our second definition and the definitions of interactions in factorial experiments. To simplify the notation, the joint distribution of three random variables, X, Y and Z, is considered in detail. The joint distribution function is denoted by $F(x, y, z)$; the one dimensional marginal distribution functions by $G(x)$, $H(y)$ and $K(z)$; the two dimensional marginal distribution functions, by $L(y, z)$, $M(x, z)$ and $Q(x, y)$. The two dimensional distribution functions are possibly not symmetrical with respect to the two arguments. Where no ambiguity is likely to result, the arguments of these distribution functions will be omitted; the corresponding density functions, when they exist, will be denoted by the corresponding lower case letter; measures assigned to individual points in the discrete distributions will also be denoted by lower case letters. Ortho-normal systems are defined on the margins and written $\{x^{(i)}\}$, $\{y^{(i)}\}$ and $\{z^{(i)}\}$; $x^{(0)} = y^{(0)} = z^{(0)} \equiv 1$.

When the theoretical parameters of the marginal distributions have been specified, the *main effects* are defined as the set of expectations of any complete set (or orthonormal basis) of orthonormal functions. The estimators of the main effects we shall take to be functions of the form $N^{-1}\mathscr{S}x^{(i)}$, where N is the sample size and $\mathscr{S}$ is the summation operator over the sample of N. Under H_0, the main effects are all zero. The test criterion for the presence of a main effect is of the form $N^{-\frac{1}{2}}\mathscr{S}(x^{(i)})$. Under H_1, let a_i be the Fourier

coefficient in the expansion of the Radon-Nikodym derivative of the likelihood ratio, dG_1/dG, so that

$$(2.1) \qquad a_i = \mathscr{E}_1 x^{(i)} \equiv \int x^{(i)} \, dG_1(x),$$

where $\mathscr{E}_1$ is the expectation operator and G_1 the distribution function under
H_1. The measure theoretical definition of main effects would be that main
effects are present if and only if $G_1(x) \not\equiv G(x)$ or equivalently $a_i \neq 0$ for at
least one element of a complete set $\{x^{(i)}\}$, $i > 0$.

First order interactions have already been defined in Chapter XI. They are
said to be present in a multidimensional distribution if and only if they are
present in some of the two dimensional marginal distributions. In the
notation introduced above, the absence of first order interactions is equivalent
to the equations

$$(2.2) \qquad \begin{array}{lll} L = HK, & M = GK, & Q = GH, \quad \text{or} \\ l = hk, & m = gk \quad \text{and} & q = gh. \end{array}$$

Considering a single marginal distribution, Q say, we can give two equivalent
definitions of first order interaction,

(i) Measure-Theoretical Definition

Interaction beween two random variables is present if and only if

$$(2.3) \qquad Q(x, y) \neq G(x)H(y), \quad \text{at some point } (x, y).$$

This is the classical probabilistic definition. All other definitions must be
shown to be equivalent. For ease of generalization let us write $F_{12}(x, y) \equiv$
F_{12} in place of $Q(x, y)$, and $F_1(x) \equiv F_1$ and $F_2(y) \equiv F_2$ in place of $G(x)$ and
$H(y)$. Then (2.3) can be rewritten as

$$(2.4) \qquad (F_1^* - F_1)(F_2^* - F_2) \neq 0,$$

provided that we interpret $F_1^* F_2^*$ as F_{12}, $F_i^* = F_i$, $F_i^* F_j = F_j F_i^*$. In other
words the multiplication of a number of symbols $F_1^*, F_2^*, F_3^* \ldots$ is to be
interpreted symbolically and the product is to be set equal to the multivariate distribution function of the variables so that for any set of increasing
indices,

$$(2.5) \qquad F_{i_1}^* F_{i_2}^* \ldots = F_{i_1 i_2 \ldots}(x_{i_1}, x_{i_2}, \ldots).$$

It is assumed that the F_i^* commute with the F_i. No new identity results from
(2.4), as after appropriate cancellations it becomes simply

$$(2.6) \qquad F_{ij} = F_i F_j, \qquad i < j,$$

which is (2.2) in different notation. With three variables (2.2) becomes

(2.7) $F_{12} = F_1 F_2;$ $F_{13} = F_1 F_3$ and $F_{23} = F_2 F_3$

and we take these as the conditions for no first order interactions.

This definition can be extended to higher dimensions. Let F_{i_1}, $F_{i_1 i_2}$, $F_{i_1 i_2 i_3}, \ldots$, denote the one-, two-, three- ... dimensional distribution functions, where $i_1 < i_2 < i_3 \ldots$. Let F' denote the $(n-1)$-dimensional distribution function of the set complementary to the particular random variable chosen, F'' as the $(n-2)$-dimensional distribution function of the $(n-2)$ variables complementary to the particular pair (i_1, i_2) chosen, and so on. With this convention, interactions of the $(n-1)^{\text{th}}$ order are said to exist if and only if F cannot be displayed as a sum

(2.8) $F = \sum F_{i_1} F' - \sum F_{i_1} F_{i_2} F'' + \sum F_{i_1} F_{i_2} F_{i_3} F''' - \ldots$
$$+ (-1)^n \sum F_{i_1} F_{i_2} \ldots F_{i_{n-1}} F^{(n-1)} + (-1)^{n-1} F_{i_1} F_{i_2} \ldots F_{i_n}$$

where the summation is over all combinations of indices $(i_1, i_2, \ldots, i_k)$ $k = 1, 2, \ldots, n - 1$. We have written $F^{(n-1)}$ to mean F with $(n-1)$ primes as superscripts. The last two terms are of the same form and may be consolidated as

$$(-1)^n (n - 1) F_{i_1} F_{i_2} \ldots F_{i_n}, \qquad n > 2.$$

A general definition of no interaction can now be given by means of the symbolic identities,

(2.9) $$\prod_{k \in A} (F_k^* - F_k) = 0,$$

where A is a subset of the integers $1, 2, \ldots, n$. Interaction of the $(s - 1)^{\text{th}}$ order holds between the s variables with indices $\{i_1, i_2, \ldots, i_s\} \equiv A$ if and only if (2.9) does not hold.

Interaction of the $(s - 1)^{\text{th}}$ order in a set of n random variables holds if and only if (2.9) is not true for a set A of s random variables. In particular, interaction of the $(n - 1)^{\text{th}}$ order is present if and only if (2.9) is not true when A is the set of the n variables.

Example. In a three-dimensional distribution, second-order interaction is present if and only if

(2.10) $F \neq KQ + HM + GL - 2GHK.$

(ii) **The Regression or Correlation Theory Definition**

First order interaction is said to be present if there exists a non-null square summable function (i.e. a function of finite variance) on the product space of the two variables, which is orthogonal, with respect to the product measure, to all square summable functions on each of the two marginal spaces.

An analogous definition occurs in the two-way analysis of variance, where the estimators of the interactions are the linear forms in the observations orthogonal to the linear forms giving the grand total and the "main effects". Let $\psi(x, y)$ be such a function, taken to be orthonormal on the product distribution. Then we are given that $\int \psi(x, y)\xi(x)\, dG(x)\, dH(y) = \int \psi(x, y)\eta(y)\, dG(x)\, dH(y) = 0$. Hence, $\psi(x, y)$ can be expanded as a series,

$$(2.11) \qquad \psi(x, y) = \sum_{ij \neq 0} a_{ij}x^{(i)}y^{(j)},$$

for no terms will appear in which either $i = 0$ or $j = 0$ because of the definition of ψ. Interaction is present if and only if

$$(2.12) \qquad \iint \psi(x, y)\, dQ(x, y) \neq 0$$

for some ψ. From (2.11) and (2.12) it follows that

$$(2.13) \qquad \iint x^{(i)}y^{(j)}\, dQ(x, y) \neq 0$$

for some pair (i, j) with $ij \neq 0$. The definition can be reworded so that first order interaction is present if and only if there is some pair (i, j) with $ij \neq 0$ such that (2.13) holds. By Corollary (i) of Theorem VI.4.1, this definition of no interaction by means of the correlations (2.13) is equivalent to the measure-theoretical definition.

The definition of this section can be conveniently rephrased. *First order interaction is present between the random variables, X and Y, if and only if for some $\xi_X(x)$ and $\eta_Y(y)$ the correlation is not zero.*

If interactions are absent, a square summable function ξ_X must have zero linear regression on every orthonormal function in Y.

$$(2.14) \qquad \int (\xi_X(x) - a)y^{(j)}\, dQ = 0, \qquad \text{where } \iint \xi_X(x)\, dQ = a.$$

Second order interactions can be defined similarly. Let $\psi(x, y, z)$ be a representative of the class of square summable functions which are orthogonal to the constant term, functions in X, Y or Z alone and functions in (X, Y), (X, Z) or (Y, Z). Then second order interaction is present if and only if a $\psi(x, y, z)$ can be found in this class such that

$$(2.15) \qquad \iiint \psi(x, y, z)\, dF(x, y, z) \neq 0.$$

As before $\psi(x, y, z)$ can be expanded as a series

(2.16)
$$\psi(x, y, z) = \sum_{ijk \neq 0} a_{ijk} x^{(i)} y^{(j)} z^{(k)}$$

and the definition is equivalent to interaction being present if and only if

(2.17)
$$\iiint x^{(i)} y^{(j)} z^{(k)} \, dF(x, y, z) \equiv \rho_{ijk} \neq 0 \qquad \text{for some } ijk \neq 0.$$

This means that an interaction is present if and only if at least one of the generalized coefficients of correlation, ρ_{ijk}, $ijk \neq 0$, is non-zero for complete sets or orthonormal bases $\{x^{(i)}\}$, $\{y^{(j)}\}$ and $\{z^{(k)}\}$.

If now we consider the regression equation of a square summable function $\xi_X(x)$ in terms of functions $\eta_Y(y)$, $\zeta_Z(z)$ and $\xi_{YZ}(y, z)$, interaction is present if the coefficient of $\xi_{YZ}(y, z)$ is non-zero for at least one choice of $\xi_{YZ}(y, z)$ from the class of all functions which are orthogonal to functions in Y alone and functions in Z alone.

Higher order interactions can be defined recursively. The example of equation (VI.5.5) and obvious generalizations to higher dimensions show that the higher order interactions can be present without the lower order being present and conversely.

Example (i) Let X and Y be jointly normal with correlation, ρ. Then there are first order interactions by both definitions if and only if $\rho \neq 0$.

Example (ii) Let X, Y and Z be standardized jointly normal variables with positive correlation coefficients ρ_{XY}, ρ_{XZ} and ρ_{YZ}. Then the variables have first order interactions as in Example (i). By the first definition, we find from (2.10) interpreted as a relation between densities, that there must be second order interaction between X, Y and Z, since (2.10) would otherwise give an identity between sums of distinct exponentials with variable exponents, which is impossible.

Theorem 2.1. *The two definitions of interaction are equivalent.*

Proof. With ϕ^2-bounded distributions, the proof is easy. Without such an assumption we may obtain a proof by considering orthogonal functions of the form

(2.18)
$$\xi_0 = -[1 - G(x_0)]^{+\frac{1}{2}} [G(x_0)]^{-\frac{1}{2}}, \qquad \text{for } x \leqslant x_0$$
$$= [1 - G(x_0)]^{-\frac{1}{2}} [G(x_0)]^{+\frac{1}{2}}, \qquad \text{for } x > x_0$$

at each point x_0 of a dense enumerable set. If the product of such functions, $\xi_0 \eta_0 \zeta_0$, corresponding to partitions at the point (x_0, y_0, z_0) has a zero expectation, then the value of $F(x_0, y_0, z_0)$ can be obtained as an expansion, for we now have a finite distribution of $2^3 = 8$ points and the amount of distribution in one octant is given by

(2.19)

$$
\begin{aligned}
F(x_0, y_0, z_0) &= (1 + c\xi_0\eta_0 + b\xi_0\zeta_0 + a\eta_0\zeta_0)G(x_0)H(y_0)K(z_0) \\
&= [(1 + c\xi_0\eta_0) + (1 + b\xi_0\zeta_0) + (1 + a\eta_0\zeta_0) - 2]G(x_0)H(y_0)K(z_0) \\
&= Q(x_0, y_0)K(z_0) + M(x_0, z_0)H(y_0) + L(y_0, z_0)G(x_0) \\
&\qquad\qquad - 2G(x_0)H(y_0)K(z_0),
\end{aligned}
$$

which is of the form (2.10). The equations can be followed through in the reverse directions to prove the equivalence of the two definitions. If ϕ^2 is unbounded we can always consider condensed distributions for which all the relevant distributions are ϕ^2-bounded with respect to the product distributions.

Example (*iii*) Let $q_{XY}(x, y)$ be the joint normal density function where X and Y are standardized and the coefficient of correlation is ρ_{XY}. Let q_{XZ} and q_{YZ} be defined similarly with coefficients of correlation ρ_{XZ} and ρ_{YZ} respectively. Now suppose that a_X, a_Y and a_Z are positive constants and $a_X + a_Y + a_Z = 1$. Suppose further that f_X, f_Y and f_Z are the marginal density functions of standardized normal variables. Then

(2.20) $$a_X f_X(x)q_{YZ} + a_Y f_Y(y)q_{XZ} + a_Z f_Z(z)q_{XY}$$

is a density function in three dimensions. The two dimensional densities can be obtained by integrating out any one of the variables.

(2.21) $$l(y, z) = a_X q_{YZ} + a_Y f_Y f_Z + a_Z f_Z f_Y.$$

It is easily verified that

(2.22) $$f(x, y, z) = \sum_{X, Y, Z} f_X[a_X q_{YZ} + a_Y f_Y f_Z + a_Z f_Z f_Y] - 2f_X f_Y f_Z,$$

so that (2.3) is satisfied and there is no second order interaction. The correlation between X and Y in this mixture is $a_Z \rho_{XY}$ and this is also the regression coefficient of X on Y in the mixture (2.21). First order interactions are present by the first and second definitions. However, all expectations of the product of standardized functions, $\xi_X \eta_Y \zeta_Z$ are zero. In particular if $x^{(i)}$, $y^{(j)}$ and $z^{(k)}$ are normalized Hermite functions, the integral of every $x^{(i)}y^{(j)}z^{(k)}$, for $ijk \neq 0$, is zero. There is thus no second order interaction in the sense of the first two definitions.

Example (*iv*) Let X, Y and Z take values ± 1 with probabilities, $\frac{1}{2}$. Let the joint distribution be defined on the eight points $(\pm 1, \pm 1, \pm 1)$ by

(2.23) $$f(x, y, z) = \frac{1}{8}[1 + \rho_{XY}xy + \rho_{YZ}yz + \rho_{XZ}xz + \rho_{XYZ}xyz],$$

subject to $|\rho_{XY}| + |\rho_{YZ}| + |\rho_{XZ}| + |\rho_{XYZ}| \leqslant 1$. Then the marginal distributions are of the form, $\frac{1}{4}(1 + \rho_{XY}xy)$ and two similar expressions. There is first order interaction by both definitions if $|\rho_{XY}| + |\rho_{YZ}| + |\rho_{XZ}| > 0$ and second order interaction by both if $\rho_{XYZ} \neq 0$. By the first definition

(2.24) $f(x, y, z) =$

$\frac{1}{2} \cdot \frac{1}{4}(1 + \rho_{XY}xy) + \frac{1}{2} \cdot \frac{1}{4}(1 + \rho_{YZ}yz) + \frac{1}{2} \cdot \frac{1}{4}(1 + \rho_{XZ}xz) - 2 \cdot \frac{1}{2} \cdot \frac{1}{2} \cdot \frac{1}{2}$

if and only if $\rho_{XYZ} = 0$. Example (iv) suggests a third definition for second order interaction, namely second order interaction is present if and only if for at least one non-trivial dichotomy of each of the three marginal spaces, the correlation coefficient of the second order is non-zero. In terms of Example (iv), this is ρ_{XYZ}. In words, we could say if the Boas coefficient of correlation of the second order is non-zero for at least one non-trivial dichotomy of the marginal spaces, then second order interaction is present. A non-trivial dichotomy means a partition of a marginal space into two sets, neither of which is associated with zero probability.

It is evident that all these definitions set limits on the complexity of the multivariate distributions. In the bivariate case absence of first order interactions means that the bivariate measure is a simple product. In the case of three variables, absence of second order but presence of the first order interaction means that the joint distribution cannot be written as a simple product but it can be written as a sum of products as in (2.10).

Example (v) Let X, Y and Z be standardized jointly normal variables with $\rho_{XZ} = \rho_{XY} = \rho$, $\rho_{YZ} = \rho^2$ as coefficients of correlation. Then it can be verified that first and second order interactions are present by the definitions given above. For fixed values of X, Y is independent of Z; however, for fixed values of Y, X is not independent of Z.

A fourth definition could be given using ϕ^2. If ϕ^2 of each of the three two-dimensional distributions is finite, we can say that second order interaction is present if and only if

(2.25) $$\phi^2 > \phi^2_{XY} + \phi^2_{XZ} + \phi^2_{YZ}.$$

If any ϕ^2 of the two-dimensional distributions is infinite, we can apply the criterion (2.9) to all finite distributions made by finite partitions of the marginal spaces; second order interaction by this definition is present if (2.25) holds for any such partition.

3. MODELS

It has been convenient for notational purposes to treat the distributions as though the spaces of the random variables were subsets of the real line. In what follows we assume that generalizations to abstract spaces, in which the points may be qualities, are valid. In this section, some fundamental distinctions arise because under some conditions of sampling, all the observed marginal numbers or frequencies are random variables or randomly produced, whereas under other models certain of the marginal numbers have been chosen arbitrarily in planning the experiments. For contingency tables of more than two dimensions, this distinction was first pointed out by Lancaster (1951), who noticed three important classes of sampling procedures in the three-way tables and generalized the three models of the two-way tables to the three-way tables. These models are now discussed and numbered in such a way as to bring out the analogies with the models of Section 11.2.

Model 1. *Unrestricted Sampling.* N independent observations are made on the parent population.

Model 2A. *Comparative Trial.* $a_{i..}$ observations are chosen arbitrarily from the ith class of the row classification but no selection is made with respect to columns or layers. $\sum a_{i..} = N$

Model 2B. *Comparative Trial.* $a_{ij.}$ observations are chosen arbitrarily from the intersection of the ith class of the row space and the jth class of the column space, but no selection is made with respect to layers. $\sum \sum a_{ij.} = N$.

Model 3. *Permutation model.* This model is rather artificial and is included only for completeness. The one dimensional frequencies are given, namely $\{a_{i..}\}$, $\{a_{.j.}\}$ and $\{a_{..k}\}$. It is desired to test hypotheses about interactions.

Example (*i*) An example of Model 3 could be provided by bridge deals. Let $a_{i..} = 13$, the number of cards in the ith suit; let $a_{.j.}$ be the number of cards dealt to the jth player. By the rules of the game $a_{i..} = a_{.j.} = 13$ for each i and each j. Suppose now that $a_{..1}$ are clean cards and $a_{..2}$ are marked or perhaps merely dirty cards. The various hypotheses could be tested about whether the cards marked belong to particular suits or have been dealt to particular players or whether marked cards of a particular suit tend to go to a particular player.

Example (*ii*) Model 1 is often appropriate in the analysis of genetic experiments. In Roberts, Dawson and Madden (1939), the following table is given for a cross of the form $AaBbCc \times aabbcc$.

Factors Present	A^-B^-	A^-B^+	A^+B^-	A^+B^+
D^-	427	440	509	460
D^+	494	467	462	475

The theoretical parameters are $p(A^-) = p(A^+) = p(B^-) = \ldots = \frac{1}{2}$. This is an example of unrestricted sampling from a parent population on a space of $2 \times 2 \times 2$ points. Haldane (1939) gives a systematic analysis of the data of these authors using Model 1.

Example (*iii*) Bartlett's (1936) example is of Model 2B. Each $a_{ij.} = 240$.

	Layer 1		Layer 2	
	156	84	84	156
	107	31	133	209

Under the sampling rules of Model 1, an overall χ^2 is defined by

$$(3.1) \qquad \chi^2 = \sum_{i,j,k} q_{ijk}^2 = \sum_{i,j,k} (a_{ijk} - Np_{i..}p_{.j.}p_{..k})^2/(Np_{i..}p_{.j.}p_{..k}),$$

and this is asymptotically distributed as χ^2 with $mnl - 1$ degrees of freedom, where m, n and l are the numbers of rows, columns and layers. From this overall χ^2, the main effects of the form

$$(3.2) \qquad \sum_i (a_{i..} - Np_{i..})^2/(Np_{i..}),$$

with $m - 1$ degrees of freedom and two similar expressions for the columns and layers with $(n - 1)$ and $(l - 1)$ degrees of freedom can be subtracted. A residual χ^2 with $mnl - m - n - l + 2$ degrees of freedom results. From this the χ^2 of the first order interactions can be subtracted, leaving a residual χ^2 with $(m - 1)(n - 1)(l - 1)$ degrees of freedom. It is sometimes objected that this method of obtaining the second order interaction might conceivably lead to negative values of χ^2. This fear can be shown to be unfounded by making orthogonal transformations on the array of variables, q_{ijk} in (3.1).

Let the q_{ijk} be written as the elements of a vector with the mnl elements in dictionary order. Let **M**, **N** and **L** be orthogonal matrices of sizes, m, n and l. Let the first rows of these matrices be determined by $m_{1j} = \sqrt{p_{j..}}$, $n_{1j} = \sqrt{p_{.j.}}$ and $l_{1j} = \sqrt{p_{..j}}$. The elements of

$$(3.3) \qquad \mathbf{u} = \mathbf{M} \times \mathbf{N} \times \mathbf{L}\mathbf{q},$$

can then be identified as the interactions of various orders. For any s, $1 \leqslant s \leqslant mnl$,

$$(3.4) \qquad s = (s_1 - 1)nl + (s_2 - 1)l + (s_3 - 1) + 1,$$

where s_1, s_2 and s_3 run through the indices $1(1)m$, $1(1)n$ and $1(1)l$. The elements u_s of **u** can be identified as follows. If $s_1 = 1$, there is no row comparison involved, whether main effect or interaction. Similarly $s_2 = 1$ or $s_3 = 1$ show that the particular u_s does not involve the column or layer main effects or interactions. q_1 is identically zero; $s_2 = s_3 = 1$, $s_1 \neq 1$ gives the main effects of X; $s_1 = s_3 = 1, s_2 \neq 1$ gives those of Y; $s_1 = s_2 = 1, s_3 \neq 1$ gives those of Z. $s_1 = 1, s_2 > 1, s_3 > 1$ gives the interactions of Y and Z; $s_2 = 1$, $s_1 > 1$, $s_3 > 1$ gives the interactions of X and Z; $s_3 = 1$, $s_1 > 1$, $s_2 > 1$ gives the interactions of X and Y. $s_1 > 1$, $s_2 > 1$, $s_3 > 1$ gives the second order interactions of X, Y and Z.

These results can be obtained after the manner of Section XI.2. Let ξ, η and ζ be typical orthonormal functions on the marginal distributions of X, Y and Z, so that $\mathscr{E}\xi = 0$, $\mathscr{E}\xi^2 = 1$ and similarly for η and ζ. Under the null hypothesis of independence $\{1, \xi, \eta, \zeta, \xi\eta, \xi\zeta, \eta\zeta, \xi\eta\zeta\}$ is an ortho-normal set on the joint distribution. In general, ξ, η and ζ are representative of the orthonormal sets but for notational convenience superscripts are not given to them. In the finite case these superscripts would run through the indices $1, 2, \ldots, m - 1$; $1, 2, \ldots, n - 1$; and $1, 2, \ldots, l - 1$ respec-tively. ξ takes values $\xi_1, \xi_2, \ldots, \xi_m$ or more generally $\xi = \xi(x)$. Now let us suppose that N independent observations have been made on the joint distri-bution of X, Y and Z. Under the hypothesis of independence, $N^{-\frac{1}{2}}\mathscr{S}\xi$, $N^{-\frac{1}{2}}\mathscr{S}\eta, \ldots, N^{-\frac{1}{2}}\mathscr{S}\eta\zeta$ and $N^{-\frac{1}{2}}\mathscr{S}\xi\eta\zeta$ are again mutually uncorrelated standardized variables, asymptotically normal and mutually independent by the generalized central limit theorem. Since variables of the forms, $N^{-\frac{1}{2}}\mathscr{S}\xi$ and $N^{-\frac{1}{2}}\mathscr{S}\xi\eta$, have already been considered in Section XI.2, the only new function is $N^{-\frac{1}{2}}\mathscr{S}\xi\eta\zeta$ and this is $N^{-\frac{1}{2}} \sum_{i,j,k} \xi_i\eta_j\zeta_k a_{ijk}$. The exhibition of these three classes of functions, asymptotically normal, yields a partition of χ^2. Each such expression can be located as an element of the mnl-vector,

(3.5) $$\mathbf{w} = \mathbf{X} \times \mathbf{Y} \times \mathbf{Z}\mathbf{a} - (N, 0, 0, \ldots, 0)^T$$
$$= \mathbf{M} \times \mathbf{N} \times \mathbf{L}\mathbf{q}$$

where the matrices $\mathbf{X}$, $\mathbf{Y}$ and $\mathbf{Z}$ are defined by such equations as $x_{ij} = x_j^{(i-1)}$, the value of the $(i - 1)^{\text{th}}$ orthonormal function at position, j. (3.5) is easily justified in the manner of Section XI.2. The transformation (3.5) from $\mathbf{q}$ to $\mathbf{w}$ is orthogonal, so that $\chi^2 = \mathbf{q}^T\mathbf{q} = \mathbf{w}^T\mathbf{w}$, and we have already seen that the elements of $\mathbf{w}$ can be identified as identically zero, main effects, first order interactions and second order interactions. The sum of squares due to the second order interactions can be obtained by subtraction. We set out the transformation as a Table 3.1. The standardized variables can easily be produced by a modern computer (Claringbold, 1961); however, inspection of Table 3.1 will suggest appropriate methods, so that the calculation of the standardized variables is not difficult even with a hand machine.

So far the hypothesis of complete independence with unrestricted sampling has been considered. In this case all the linear forms may be considered simultaneously. If it is decided to test the first order interactions and then

Table 3.1

THE STANDARDISED SUMS AS ESTIMATES OF INTERACTIONS

$$
\begin{bmatrix}
1 & 1 & 1 & 1 & 1 & 1 & 1 & 1 \\
\zeta_1 & \zeta_2 & \zeta_1 & \zeta_2 & \zeta_1 & \zeta_2 & \zeta_1 & \zeta_2 \\
\eta_1 & \eta_1 & \eta_2 & \eta_2 & \eta_1 & \eta_1 & \eta_2 & \eta_2 \\
\eta_1\zeta_1 & \eta_1\zeta_2 & \eta_2\zeta_1 & \eta_2\zeta_2 & \eta_1\zeta_1 & \eta_1\zeta_2 & \eta_2\zeta_1 & \eta_2\zeta_2 \\
\xi_1 & \xi_1 & \xi_1 & \xi_1 & \xi_2 & \xi_2 & \xi_2 & \xi_2 \\
\xi_1\zeta_1 & \xi_1\zeta_2 & \xi_1\zeta_1 & \xi_1\zeta_2 & \xi_2\zeta_1 & \xi_2\zeta_2 & \xi_2\zeta_1 & \xi_2\zeta_2 \\
\xi_1\eta_1 & \xi_1\eta_1 & \xi_1\eta_2 & \xi_1\eta_2 & \xi_2\eta_1 & \xi_2\eta_1 & \xi_2\eta_2 & \xi_2\eta_2 \\
\xi_1\eta_1\zeta_1 & \xi_1\eta_1\zeta_2 & \xi_1\eta_2\zeta_1 & \xi_1\eta_2\zeta_2 & \xi_2\eta_1\zeta_1 & \xi_2\eta_1\zeta_2 & \xi_2\eta_2\zeta_1 & \xi_2\eta_2\zeta_2
\end{bmatrix}
\begin{bmatrix}
a_{111} \\
a_{112} \\
a_{121} \\
a_{122} \\
a_{211} \\
a_{212} \\
a_{221} \\
a_{222}
\end{bmatrix}
$$

$$
=
\begin{bmatrix}
a_{...} \\
a_{..1}\zeta_1 + a_{..2}\zeta_2 \\
a_{.1.}\eta_1 + a_{.2.}\eta_2 \\
a_{.11}\eta_1\zeta_1 + a_{.12}\eta_1\zeta_2 + a_{.21}\eta_2\zeta_1 + a_{.22}\eta_2\zeta_2 \\
a_{1..}\xi_1 + a_{2..}\xi_2 \\
a_{1.1}\xi_1\zeta_1 + a_{1.2}\xi_1\zeta_2 + a_{2.1}\xi_2\zeta_1 + a_{2.2}\xi_2\zeta_2 \\
a_{11.}\xi_1\eta_1 + a_{12.}\xi_1\eta_2 + a_{21.}\xi_2\eta_1 + a_{22.}\xi_2\eta_2 \\
a_{111}\xi_1\eta_1\zeta_1 + a_{112}\xi_1\eta_1\zeta_2 + a_{121}\xi_1\eta_2\zeta_1 + a_{122}\xi_1\eta_2\zeta_2 \\
\quad + a_{211}\xi_2\eta_1\zeta_1 + a_{212}\xi_2\eta_1\zeta_2 + a_{221}\xi_2\eta_2\zeta_1 + a_{222}\xi_2\eta_2\zeta_2
\end{bmatrix}
\simeq
\begin{bmatrix}
1 \\
L \\
C \\
C \times L \\
R \\
R \times L \\
R \times C \\
R \times C \times L
\end{bmatrix}
$$

the second order interactions, the hypothesis may be modified so that the first order interaction is assumed to be present. H_0 now specifies that $\mathscr{E}x^{(i)}y^{(i)}z^{(k)} = 0$ for all orthonormal functions for which $ijk \neq 0$. Test functions of the form, $N^{-\frac{1}{2}}\mathscr{S}\xi\eta\zeta$, will now have zero expectation but the variances may not be unity and between any two such functions, the correlation may no longer be zero. In specific cases, these means and variances can be computed. It may be said that as a rule these considerations are not important for we are usually only interested in small departures of the various standardized functions from their mean. The mean will have an expectation of $0(N^{\frac{1}{2}})$ whereas the variance will still remain $0(1)$. So that, if the existence or the estimate of the second order interaction is required, only its variance and not its mean value will be altered by the existence of lower order interactions.

4. COMBINATORIAL THEORY

Bartlett (1935) was the first to consider the combinatorial aspects of the $2 \times 2 \times 2$ table along the following lines. The unconditional probability of the entries, given the theoretical probabilities, $\{p_j\}$, is

$$(4.1) \qquad \mathscr{P}(\{n_j\}) = N! \prod p_j^{n_j} / \prod n_j!$$

It is assumed now that the $\{p_j\}$ have to be estimated jointly by $\{\hat{p}_j\}$ subject to the condition,

$$(4.2) \qquad \hat{p}_1 \hat{p}_4 \hat{p}_6 \hat{p}_7 = \hat{p}_2 \hat{p}_3 \hat{p}_5 \hat{p}_8.$$

If the estimated values or the theoretical values of the p_j are substituted in (4.1), the value of $\prod p_j^{n_j}$ is constant over the admissible sets of values of n_j, namely over those sets of n_j which yield the given marginal totals, so that

$$(4.3) \qquad \mathscr{P}(\{n_j\} \mid \text{marginal totals}) = C / \prod n_j!$$

where the constant C can be calculated because the sum of the probabilities in (4.3) is unity. The conditional distribution of the number in the first cell can be computed and an "exact" test is available. A similar procedure could be adopted in the case of $2 \times 2 \times 3$ tables or $r \times s \times l$ tables, but computing difficulties would soon arise.

Care must be taken not to generalize the usual treatment of the $r \times s$ tables by the assertion that the joint probability of the marginal contingency tables for given row, column and layer totals is equal to their product. It is not true that

$$(4.4) \quad \mathscr{P}(\{a_{ijk}\} \mid \{a_{ij.}\}, \{a_{i.k}\}, \{a_{.jk}\})$$
$$= \mathscr{P}(\{a_{ijk}\} \mid \{\hat{p}_{ijk}\}) / [\mathscr{P}(\{a_{ij.}\} \mid \{a_{i..}\}, \{a_{.j.}\})$$
$$\times \mathscr{P}(\{a_{i.k}\} \mid \{a_{i..}\}, \{a_{..k}\}) \mathscr{P}(\{a_{.jk}\} \mid \{a_{.j.}\}, \{a_{..k}\})]$$

even under the null hypothesis, $p_{i..} p_{.j.} p_{..k} = p_{ijk}$, complete independence.

5. THE FISHER-BARTLETT METHODS

The $2 \times 2 \times 2$ contingency table can be considered after the manner of Bartlett (1935). The cell contents can be written out in the form,

First Layer		Second Layer	
n_1	n_2	n_5	n_6
n_3	n_4	n_7	n_8.

As a generalization of the treatment of the 2×2 tables, the marginal totals formed by combining corresponding cells in the two rows, columns or layers can be regarded as fixed. With these marginal totals fixed, there is only one choice possible in filling the cells of the table. If n_j is the observed number and a_j is the expected number in the jth cell, then an analogy with the 2×2 table suggests that $\sum (n_j - a_j)^2/a_j$ is distributed as χ^2 with 1 d.f., for it is a quadratic form in the 8 random variables, n_j, which are subject to 7 algebraically independent linear restrictions. These restrictions ensure that every absolute difference, $|n_j - a_j|$, has the same value, so that

(5.1) $$n_j = a_j \pm x.$$

The expected values, a_j, for the cells can be computed under the null hypothesis of no second order interaction, defined by

(5.2) $$a_1 a_4 a_6 a_7 = a_2 a_3 a_5 a_8, \quad \text{or equivalently}$$

(5.3) $$p_1 p_4 p_6 p_7 = p_2 p_3 p_5 p_8,$$

where p_j is the unconditional probability that a single observation should fall into the jth cell. Combining (5.1) and (5.2), estimates $\hat{p}_j$ and expected values $a_j = N\hat{p}_j$, can be obtained by solving the cubic equation in x

(5.4) $$(n_1 + x)(n_4 + x)(n_6 + x)(n_7 + x) = (n_2 - x)(n_3 - x)(n_5 - x)(n_8 - x).$$

This equation is readily solved by giving x trial values and proceeding by successive approximations; finally a linear interpolation can be used to obtain a root of the cubic equation to any level of approximation. It is obvious that the only admissible solutions are those which give non-negative values to the expected numbers. The root must lie in the interval $[-\min (n_1, n_4, n_6, n_7), \min (n_2, n_3, n_5, n_8)]$ or $[-l, r]$ say. The expressions on the left and right of (5.4) may be written as L and R; and the equation can be written $L - R = 0$. $L - R$, the difference between two biquadratic expressions is a cubic function of x. At $x = -l$, L is zero and R is positive. As x increases from $-l$ to r, L increases and R decreases steadily so that $L - R$ increases steadily and $L - R$ is positive at $x = r$, where $R = 0$. There is thus precisely one real root in the interval $[-l, r]$. With x_0, the root so calculated, the χ^2 of the second order interaction can now be computed

(5.5) $$\sum (n_j - a_j)^2/a_j = x_0^2 \sum a_j^{-1}.$$

The method is readily generalized to the higher $2 \times 2 \times \ldots \times 2$ tables, but computational difficulties arise, when say a $2 \times 3 \times 5$ table is treated.

Some theoretical criticism of the criterion of no second-order interaction of (5.2) or (5.3) is possible. For a given partition of the marginal spaces, (5.3) may yield a criterion of no second order interaction but after condensation of one or more of the marginal spaces, the criterion may no longer hold. The multivariate normal distribution is a convenient example. Suppose at least two of the correlations differ from zero. Then first order interaction is present between two of the variables according to the criterion corresponding to (5.3) in two dimensions; but the second order interaction is always absent if the p_j of (5.3) are interpreted as infinitesimal probabilities. However, with a finite dichotomy of each marginal space, the criterion will declare that second order interaction is present, except for certain special choices of the points of dichotomy.

6. ASYMPTOTIC THEORY

The analysis of contingency tables of several dimensions will depend on the hypotheses made about the underlying probability distribution and the model under which sampling has been carried out. It is supposed that N independent observations have been made.

Model 1. In an important class, the hypothesis is of the form

$$(6.1) \qquad p_{ijk...} = p_{i...}p_{.j...}p_{..k...}\cdots$$

The hypothesis gives the expected numbers $\{Np_{ijk...}\}$ and an overall test function is given by

$$(6.2) \qquad \chi^2 = \sum (a_{ijk...} - Np_{ijk...})^2/(Np_{ijk...}),$$

distributed approximately as χ^2 with $(rcl \ldots - 1)$ d.f. Some genetic examples can be appropriately analysed in this manner with $p_{ijk...} = 2^{-n}$, where n is the number of gene loci in a cross of the form, $AaBbCc \ldots \times aabbcc \ldots$ and $\{a_{ijk...}\}$ is the set of observed frequencies in the 2^n possible classes of the N offspring. All comparisons of the $a_{ijk...}$ are included in a single test. However, in a number of physical or biological applications, it may be desirable to examine more closely certain comparisons. In this genetical example, it may be desirable to establish that the ratio of the number of heterozygotes Aa to the number of homozygotes aa is unity. If this is not so, it will be appropriate to carry out the test of all interactions by substituting estimated values in (6.1) and (6.2). The degrees of freedom, with all parameters estimated from the data will be $2^n - n - 1$ in the genetic example and $rcl \ldots t - 1 - (r - 1) - (c - 1) - \ldots - (t - 1)$ in the general case.

If the multidimensional distribution is of the form, (VI.5.2), then the transformation (3.5) yields a set of estimators of the coefficients of correlation, since $\mathscr{E}x^{(i)}y^{(j)}z^{(k)} = \rho_{ijk}$ and similarly for other coefficients of correlation, ordinary and generalized. However, if the ordinary or first order correlations are not small, the usual distribution theory may not be applicable to such variables as $N^{-\frac{1}{2}}\mathscr{S}x^{(i)}y^{(j)}z^{(k)}$, although $N^{-1}\mathscr{S}x^{(i)}y^{(j)}z^{(k)}$ is an unbiased estimator of ρ_{ijk}; so that the test function $N^{-\frac{1}{2}}\mathscr{S}x^{(i)}y^{(j)}z^{(k)}$ will test the null hypothesis that $\rho_{ijk} = 0$, but the computations of power based on an assumption of unit variance of the test function may be misleading.

In many cases, it will be possible to test the functions of the form $N^{-\frac{1}{2}}\mathscr{S}(\xi\eta\zeta)$, using the asymptotic theory and assuming unit variances, where ξ, η and ζ are orthonormal on the respective marginal distributions.

Let us consider a procedure for the cases in a three-way contingency table for which the first order interactions have been judged significant. There are $(r-1)(c-1)(l-1)$ comparisons testing the second order interactions. Let us suppose (which is not always true) that there is a three dimensional distribution in which the second order interactions are zero but which has the same first order interactions as the marginal two dimensional distributions. If this distribution exists, it has the form

$$(6.3) \quad \hat{f}(x, y, z) = g(x)l(y, z) + h(y)m(x, z) + k(z)q(x, y) - 2g(x)h(y)k(z).$$

The covariance matrix, $\mathbf{V}$, of the $(r-1)(c-1)(l-1)$ variables $\{\xi\eta\zeta\}$ can be determined in principle since if we have two such functions $\xi_1\eta_1\zeta_1$ and $\xi_2\eta_2\zeta_2$, then their product is the product of certain linear forms in the members of orthonormal sets on the marginal distributions. Thus a variable of the form $\mathbf{u}^T\mathbf{V}^{-1}\mathbf{u}$ can be computed where the elements of $\mathbf{u}$ are the $(r-1)(c-1)(l-1)$ standardized sums of the form $N^{-\frac{1}{2}}\mathscr{S}(\xi\eta\zeta)$. These functions are linearly independent on the distribution (6.3) and so $\mathbf{V}$ will be positive definite and the standardized sum,

$$(6.4) \qquad\qquad N^{-1}\mathscr{S}\mathbf{u}^T\mathbf{V}^{-1}\mathbf{u},$$

will have asymptotically the distribution of χ^2 with $(r-1)(c-1)(l-1)$ degrees of freedom. This procedure is related to a suggestion of Plackett (1962). If $\mathbf{u}^T\mathbf{V}^{-1}\mathbf{u}$ is significantly large, the null hypothesis of zero interaction in the table can be rejected. Then the ρ_{ijk} could be estimated by means of

$$(6.5) \qquad\qquad \hat{\rho}_{ijk} = N^{-1}\mathscr{S}x^{(i)}y^{(j)}z^{(k)}.$$

Model 2A. The numbers of observations in one marginal variable are chosen arbitrarily without loss of generality; $a_{i..}$, $i = 1, 2, \ldots, m$ are chosen arbitrarily so that $\sum a_{i..} = N$. The distribution of the a_{ijk} for a fixed i is multinomial in this model so that

$$(6.6) \qquad \mathscr{P}(\{a_{ijk}\} \mid a_{i..}, \{p_{ijk}\}) = a_{i..}! \prod_{j,k} (p_{ijk}^{a_{ijk}}/a_{ijk}!)$$

and sampling for a fixed i is of the unrestricted bivariate type, the Model 1 of Section XI.2. The observations are assumed to be made independently so that

$$(6.7) \qquad \mathscr{P}(\{a_{ijk}\} \mid \{a_{i..}\}, \{p_{ijk}\}) = \prod_i \mathscr{P}(\{a_{ijk}\} \mid a_{i..}, \{p_{ijk}\}).$$

The p_{ijk} are the probabilities of a frequency distribution so they are non-negative and

$$(6.8) \qquad\qquad \sum_{j,k} p_{ijk} = 1, \qquad i = 1, 2, \ldots, r.$$

H_0 will specify $p_{ijk} = p_{0jk}$, namely that p_{ijk} is independent of the value of i. Various alternative hypotheses are possible.

The probability of obtaining a marginal set $\{a_{.jk}\}$ is readily obtained as

$$(6.9) \qquad \mathscr{P}(\{a_{.jk}\} \mid H_0, \{p_{0jk}\}) = N! \prod_{j,k} (p_{0jk}^{a_{.jk}}/a_{.jk}!).$$

The probability of $\{a_{ijk}\}$ conditional on $\{a_{i..}\}$ and $\{a_{.jk}\}$ is obtained by dividing each side of (6.7) by the corresponding side of (6.9). This conditional distribution is independent of the unknown parameters, p_{0jk}, which may be estimated by

$$(6.10) \qquad \hat{p}_{0jk} = a_{.jk}/N.$$

This estimation procedure is equivalent to specifying the marginal distributions of Y and Z and their first order interactions.

The Pearson χ^2 can be computed in the form,

$$(6.11) \qquad Q = \sum (a_{ijk} - a_{i..}\hat{p}_{0jk})^2/(a_{i..}\hat{p}_{0jk}).$$

Q of (6.11) can be analysed into its components, the interactions between X and Y, between X and Z and the second order interactions between X, Y and Z. If p_{0jk} are given theoretically and the theoretical values are used in (6.11), then there will be additional comparisons to be made—the main effects for Y and Z and the interactions between Y and Z. The interactions between the fixed variable X and the random variable Y are the variables $N^{-\frac{1}{2}}\mathscr{S}x^{(i)}y^{(j)}$, where $x^{(i)}$ are fixed variables in the sense that $N^{-\frac{1}{2}}\mathscr{S}x^{(i)} = 0$. Such variables occur in ordinary regression analysis where the values of the independent variable may be chosen arbitrarily before making the observations.

Model 2B. One two-dimensional marginal set of frequencies, say $\{a_{.jk}\}$ is chosen arbitrarily and it follows that for a fixed (j, k), the frequencies a_{ijk} are distributed as a multinomial distribution. For fixed (j, k),

$$(6.12) \qquad \mathscr{P}(\{a_{ijk}\} \mid a_{.jk}, \{p_{i..}\}) = a_{.jk}! \prod_i (p_{i..}^{a_{ijk}}/a_{ijk}!).$$

Since the observations are assumed to be mutually independent

$$(6.13) \qquad \mathscr{P}(\{a_{ijk}\} \mid \{a_{.jk}\}, \{p_{i..}\}) = \prod_{j,k} a_{.jk}! \prod_i (p_{i..}^{a_{ijk}}/a_{ijk}!).$$

However, the $a_{i..}$ are also jointly multinomial variables, so that

$$(6.14) \qquad \mathscr{P}(\{a_{i..}\} \mid N, \{p_{i..}\}) = \prod_i (p_{i..}^{a_{i..}}/a_{i..}!)$$

and a probability of $\{a_{ijk}\}$ conditional on $\{a_{i..}\}$ and $\{a_{.jk}\}$ can be obtained from (6.13) and (6.14). This conditional probability is independent of the true value of $\{p_{i..}\}$. The estimates of the row probabilities are given by

$$(6.15) \qquad \hat{p}_{i..} = a_{i..}/N.$$

In this form of analysis, the test of the first order interactions between X and Y and between X and Z are combined with the test of the second order interactions between X, Y and Z. If the $a_{.jk}$ have been chosen so that $a_{.jk} = a_{.j.}a_{..k}/N$, an orthogonal transformation on the variables is available which will separate the components of the χ^2.

We give now a table which summarizes some of the possible partitions of χ^2 under Models 1, 2A and 2B.

THE ANALYSIS OF χ^2 FOR THE THREE-WAY CONTINGENCY TABLES

Model Estimates	1 $N\hat{p}_{i..}\hat{p}_{.j.}\hat{p}_{..k}$	2A $a_{i..}\hat{p}_{0jk}$	2B $a_{.jk}\hat{p}_{i..}$
Interactions and Main Effects			
R	—*	—	—*
C	—*	—*	—
L	—*	—*	—
RC	$(r-1)(c-1)$	$(r-1)(c-1)$	$(r-1)(c-1)$
RL	$(r-1)(l-1)$	$(r-1)(l-1)$	$(r-1)(l-1)$
CL	$(c-1)(l-1)$	—*	—
RCL	$(r-1)(c-1)(l-1)$	$(r-1)(c-1)(l-1)$	$(r-1)(c-1)(l-1)$

* These degrees of freedom are eliminated by the process of estimation.

The contributions due to the various main effects and interactions may be computed by the direct product of (3.3) or by forming the appropriate tables for the first order interactions and obtaining the second order interaction χ^2 by subtraction.

7.   CANONICAL VARIABLES

The notions of canonical correlation and canonical variables cannot be expected to be as important in the theory of several dimensions as in the theory of two dimensions. Indeed, Exercise VI.64 shows that if the first pair of canonical variables in X and Y is (ξ_1, η) and if the first pair of canonical variables in X and Z is (ξ_2, ζ), then ξ_1 need not be identical with ξ_2. Nor can it be expected that a series of the form $\sum \rho_i x^{(i)} y^{(i)} z^{(i)}$ will be available to give a generalization of (VI.3.17). The discussion is therefore limited to a few general remarks.

Let us consider variables of the form, $N^{-\frac{1}{2}}\mathscr{S}x^{(i)}y^{(j)}z^{(k)} = T_{ijk}$, say, where

$\{x^{(i)}\}$, $\{y^{(i)}\}$ and $\{z^{(i)}\}$ are orthonormal sets on the respective marginal distributions. Suppose that H_0 specifies the mutual independence of X, Y and Z. Then for all quadratic forms in the T_{ijk}, having the χ^2 distribution under H_0, the parameter of non-centrality is bounded by $N\phi^2$, when H_1 specifies a distribution function, $F(x, y, z)$ for which ϕ^2 is defined by (VI.5.1). The reasoning of Section XI.7 could be used to derive the parameter of non-centrality for the χ^2 of an $r \times c \times l$ contingency table, introducing sets of orthonormal functions $\{u^{(i)}\}$, $\{v^{(i)}\}$ and $\{w^{(i)}\}$ on the three marginal distributions, which would correspond to the orthonormal functions on the marginal distributions partitioned into a finite number of parts. The expectations of $u^{(i)}v^{(j)}$, $u^{(i)}w^{(k)}$, $v^{(j)}w^{(k)}$ and $u^{(i)}v^{(j)}w^{(k)}$ under H_1 could be computed and the parameters of non-centrality for the tests of the first order interactions and the second order interactions calculated. For the test of departure from H_0, specifying independence, there will be a quadratic form having the χ^2 distribution under H_0, which has the largest parameter of centrality. Thus suppose that X, Y and Z are three standard normal distributions and that H_0 specifies independence, but H_1 specifies joint normality. Then if the data had been presented in the form of a three-dimensional contingency table, each of the three component χ^2's corresponding to the first order interactions would have a positive parameter of non-centrality; so also would the component χ^2 corresponding to the second-order interactions. If, however, H_1 specified that the distribution was a mixture of three joint normal distributions with correlations $(\rho_{12}, 0, 0)$, $(0, \rho_{13}, 0)$ and $(0, 0, \rho_{23})$, each $|\rho_{ij}| > 0$, the χ^2 corresponding to the second order interactions would have zero parameter of non-centrality.

8. EXCHANGEABLE RANDOM VARIABLES AND SYMMETRY

Let $F_{12\ldots n}(x_1, x_2, \ldots, x_n)$ be the joint distribution function of random variables $X_1, X_2, \ldots, X_n$. The random variables are said to be *exchangeable* if

$$(8.1) \qquad F_{12\ldots n}(x_1, x_2, \ldots, x_n) = F_{12\ldots n}(x_{i_1}, x_{i_2}, \ldots, x_{i_n})$$

for every permutation $\{i_1, i_2, \ldots, i_n\}$ of the indices $1, 2, \ldots, n$. Exchangeability of random variables so defined corresponds to symmetry under permutation of the values of the n arguments. Further, random variables can be exchangeable without having symmetrical univariate distributions. A consequence of (8.1) is that the variables are identically distributed whether considered singly, or in pairs, or in triples, and so on.

Example (i) Let $X_1, X_2, \ldots, X_n$ be independent and identically distributed, then they are exchangeable.

Example (*ii*) Let $X_1, X_2, \ldots, X_n$ be jointly normal with unit variances and zero means. Let $\mathscr{E}X_iX_j = \rho$ for $i \neq j$. The variables are exchangeable.

Examples can be chosen to show that exchangeability of the variables in sets of k does not imply exchangeability of the variables in sets of $(k + 1)$.

A distribution F^* is said to be the *symmetrized* form of F if

$$(8.2) \qquad F^*_{12\ldots n}(x_1, x_2, \ldots, x_n) = (n!)^{-1} \sum F_{12\ldots n}(x_{i_1}, x_{i_2}, \ldots, x_{i_n})$$

where summation is over all permutations $(i_1, i_2, \ldots, i_n)$ of $1, 2, \ldots, n$. Tests of exchangeability of random variables can be based on the following.

Theorem 8.1. *In the class of distributions, ϕ^2-bounded with respect to the product distribution, a necessary and sufficient condition for exchangeability is that for every complete orthonormal set, the generalized coefficients of correlation should be unaffected by a permutation of the indices $\{i_1, i_2, \ldots, i_n\}$ for every choice of the n indices $i_k = 0, 1, 2, \ldots, k = 1, 2, \ldots, n$.*

Proof. *Necessity.* If the random variables are exchangeable,

$$\mathscr{E}(x_1^{(i_1)}x_2^{(i_2)} \ldots x_n^{(i_n)}) = \rho^{\{i_k\}} = \rho^{\{j_k\}}$$

where $\{j_k\}$ is a permutation of the n-set $\{i_k\}$, some or all the i_k being possibly zeroes. Hence $\rho^{\{i_k\}}$ is the average value of the generalized correlation for all possible permutations of $\{i_k\}$ and this average value is also the corresponding value $\rho^{*\{i_k\}}$ in the symmetrized distribution.

Sufficiency. If $\rho^{\{i_k\}}$ is invariant under permutation of the indices it must be equal to the $\rho^{*\{i_k\}}$ of the symmetrized distribution, which is ϕ^2-bounded; hence the $\rho^{*\{i_k\}}$ determine the symmetrized distribution. If the marginal distributions have identical partitions into, say, r classes, a symmetrized distribution is readily computed. Suppose that a_{ijk} are the observed numbers with $i, j, k = 1, 2, \ldots, r$ and that for i, j, k distinct we write $a^*_{ijk} = (3!)^{-1} \sum a_{i'j'k'}$ with summation over all permutations $(i'j'k')$ of $\{ijk\}$, where i, j and k are distinct; $a^*_{iik} = \sum a_{i'j'k'}/3$, where summation is over the three permutations of $\{i, i, k\}$, $i \neq k$; and $a^*_{iii} = a_{iii}$. Then

$$(8.3) \qquad Q = \sum (a_{ijk} - a^*_{ijk})^2/a^*_{ijk}$$

is asymptotically distributed as χ^2 with $5\binom{r}{3} + 2.2\binom{r}{2} = r(r - 1)(5r + 2)/6$ degrees of freedom. The right side of (8.3) will contain the test of exchangeability between pairs of variables and triples of variables. It is possible to test these separately. As in Section XI.8 the exchangeability in two dimensions can be tested by the standardized difference $2^{-\frac{1}{2}}N^{-\frac{1}{2}}\mathscr{S}(x^{(i)}y^{(j)} - x^{(j)}y^{(i)})$ or its square; this test is of the variance of the two functions about the common mean. In three dimensions, the corresponding test function for distinct non-zero i, j, k would be

$$\sum [\mathscr{S}(x^{(i)}y^{(j)}z^{(k)})]^2/N - [\sum \mathscr{S}(x^{(i)}y^{(j)}z^{(k)})]^2/(6N),$$

and for distinct i and k

$$\sum [\mathscr{S}(x^{(i)}y^{(i)}z^{(k)})]^2/N - [\sum \mathscr{S}(x^{(i)}y^{(i)}z^{(k)})]^2/(3N)$$

with 5 d.f. in the first case for every choice of sets of distinct elements i, j and k; and with 2 d.f. in the second case for every choice of distinct i and k.

For certain classes of marginal distributions such as the binomial distribution on two points, the symmetrized distribution becomes the distribution of the sum of the marginal random variables (Bahadur, 1961a; Lancaster, 1965a).

Example (*iii*) Suppose a fourfold table is generated by sampling from a distribution with probabilities p^2, pq, pq and q^2 respectively. If the distribution is symmetrized by averaging the entries in cells at the positions, 12 and 21, the interaction χ^2 of the symmetrized table is identical with the test function of the standardized sum of the second orthonormal polynomial on a binomial variable taking values 0, 1 and 2 with probabilities q^2, $2pq$ and p^2.

EXERCISES AND COMPLEMENTS

1. Evaluate the integral of (1.1). (Pearson, 1904; Lancaster, 1957.)

2. If X, Y and Z are jointly normal, show that a necessary and sufficient condition for complete independence is that the coefficients of correlation should vanish, $\rho_{XY} = \rho_{XZ} = \rho_{YZ} = 0$. In the joint normal distribution the vanishing of the partial correlations, $\rho_{XY.Z}$, $\rho_{XZ.Y}$ and $\rho_{YZ.X}$, is an equivalent condition.

3. With distributions obtained by mixing jointly normal distributions, prove that the vanishing of $\mathscr{E}(XY)$, $\mathscr{E}(XZ)$ and $\mathscr{E}(YZ)$ is not sufficient for independence even though each one dimensional marginal distribution is normal.

4. Let $q(u, v; \rho)$ be the bivariate joint normal density with correlation ρ and with marginal density functions, $g(u)$ and $g(v)$, so that the marginal variables are standardized. Let

$$f(x, y, z) = a_1 g(x)q(y, z; \rho_1) + a_2 g(y)q(x, z; \rho_2)$$
$$+ a_3 g(z)q(x, y; \rho_3) + a_4 g(x)g(y)g(z),$$

with $a_i \geqslant 0$ and $\sum a_i = 1$. Determine ϕ^2 of $f(x, y, z)$ with respect to the product density $g(x)g(y)g(z)$ in the following cases

(i) $a_2 = a_3 = a_4 = 0$
(ii) $a_2 = a_3 = 0$
(iii) $a_3 = 0$
(iv) $a_3 = a_4 = 0$
(v) $a_4 = 0$
(vi) no coefficients zero.

5. (Continuation of Exercise 4.)

Determine the regressions of the Hermite polynomials in X on the Hermite polynomials in Y and Z and products of them. Show that, in terms of either definition of Section 2, there is no second order interaction in any of the mixtures of Exercise 4; nor is there second order interaction in any condensed distributions formed from them.

6. Let X, Y and Z be distributed jointly with probabilities, $p_{xyz} = (1 + \rho_{xyz})/8$, where each variable takes the values ± 1 and $|\rho| \leqslant 1$ and $\rho \neq 0$. Show that the correlations of the form, $\rho_{XY} = \mathscr{E}(XY)$ are all zero. Hence, if the partial correlations are defined in the usual way, $\rho_{XY.Z} = \rho_{XZ.Y} = \rho_{YZ.X} = 0$. However, there is not complete independence. (cf. Lombard and Doering, 1947.)

7. Let X, Y and Z be jointly normal standardized variables. Suppose that $\rho_{XZ} = \rho$, $\rho_{YZ} = \rho$ and $\rho_{XY} = \rho^2$. Then $\rho_{XY.Z}$ is zero and X is independent of Y in every Z-plane, $Z = z$ but $\rho_{XZ.Y} = \rho_{YZ.X} = \rho(1 + \rho^2)^{-\frac{1}{2}} \neq 0$. X is not independent of Z in every Y-plane; nor is Y independent of X in every Z-plane. (cf. Simpson, 1951.)

8. Prove that three or more jointly normal variables obey the criterion of no second order interaction if we interpret as a relation between the contents of infinitesimal cubes the criterion given by some authors as "no second-order interaction", namely

$$\frac{p_{ijk} p_{i'j'k}}{p_{i'jk} p_{i'j'k'}} = \frac{p_{ij'k} p_{ijk'}}{p_{i'j'k} p_{i'jk'}}.$$

9. Let a $2 \times 2 \times 3$ distribution be defined on 12 points with measures proportional to

1	3	1	7	1	11
5	15	13	91	13	143.

There is according to the likelihood ratio rule no second order interaction. If the first two layers are pooled, a new distribution is obtained

2	10	1	11
18	106	13	143.

There is now second order interaction by the likelihood rule. With less information about the variables, one can detect an interaction not present before pooling.

10. Let ξ and η be orthonormal functions on a measure space. Then a NASC for ξ to have non-zero linear regression on η is $\mathscr{E}(\xi - \eta)^2 \neq 2$. Show that the definition is symmetrical in ξ and η. Use this to show that the definition of interaction by regression or correlation theory is symmetrical. Give examples from the normal theory.

11. Prove that a NASC for there to be no second order interaction in the sense

of the definitions given in 3 is that no set of dichotomies of the three marginal distributions yields a $2 \times 2 \times 2$ table with non-zero second order interaction.

12. Let the pair of random variables (X_1, X_2) be independent of the pair of random variables (X_3, X_4). Let (X_1, X_2) and also (X_3, X_4) have first order interactions. Show that the joint distribution of (X_1, X_2, X_3, X_4) has interaction of the third order in terms of the definitions of Section 3.

13. In a theoretically given $2 \times 2 \times 2$ distribution, complete independence is present if

$$(*) \qquad\qquad p_{ijk}/(p_{i..}p_{.j.}p_{..k}) = 1, \qquad i, j, k = 1, 2.$$

Suppose that the ratios (*) are written in dictionary order as elements of a vector, $\mathbf{v}$. Let $\mathbf{R}$ be the matrix defined by $r_{11} = r_{12} = 1$, $r_{21} = p_{1..}^{-1}$, $r_{22} = -p_{2..}^{-1}$ and let matrices $\mathbf{C}$ and $\mathbf{L}$ be similarly defined with $p_{.j.}$ and $p_{..k}$ replacing $p_{i..}$. The editorial comment on Simpson's paper was equivalent to the statement that the final element of

$$(**) \qquad\qquad \mathbf{R} \times \mathbf{C} \times \mathbf{L}\mathbf{v} = \mathbf{u} \quad \text{is zero.}$$

Show that the first element of $\mathbf{u}$ is unity. There are three elements identically zero since the multivariate distribution and the product distributions have the same margins. The equation of any one of the remaining elements to zero is the condition for no first order interaction. (Editorial comment to Simpson, 1951; Plackett, 1962.)

14. Plackett (1962) gives a $2 \times 2 \times 2$ table of the following form with probabilities multiplied by 24,

1	2		3	2
2	6		4	4.

Verify that $1 . 6 . 2 . 4 = 2 . 2 . 3 . 4 = 48$. Second order interaction is not present according to Bartlett's criterion. Verify that the marginal tables are

4	4		3	5		3	8
6	10		8	8		7	6.

Verify that the Boas correlations in these tables obtained by the rule,

$$r = (a_{11}a_{22} - a_{21}a_{12})/(a_{1.}a_{2.}a_{.1}a_{.2})^{\frac{1}{2}}$$

are $70^{-\frac{1}{2}}$, $2 \cdot 286^{-\frac{1}{2}}$ and $38/\sqrt{20020} \simeq 0 \cdot 27$.

15. Analyse the χ^2 in the following table estimating the row and column parameters and assuming that the layer totals are chosen as in Model 2A.

	Row 1		Row 2	
Layer	Column 1	Column 2	Column 1	Column 2
1	17	40	46	6
2	14	44	44	5
3	19	42	48	5
4	21	33	41	4
5	9	39	68	8
6	21	38	70	5
7	19	40	56	4
8	15	32	51	8
9	20	35	73	9
10	15	29	78	5
11	12	19	69	2
12	12	29	75	3

Norton gave as the interaction χ^2 7·59 for 11 d.f. Suggest a computing routine for testing second order interaction as defined in Section 3. (Norton, 1945.)

16. Calculate the χ^2 of the second order interaction of the following table.

	Layer 1			Layer 2		
	Col. 1	Col. 2		Col. 1	Col. 2	
Row 1	156	84	240	84	156	240
Row 2	107	31	138	133	209	342
	263	115	378	217	365	582

All totals formed by summing over layers are equal to 240. Compute Bartlett's χ^2 and also the second-order interaction of the orthonormal theory. (Bartlett, 1935; Lancaster, 1951 and 1960a.)

17. Show that it is immaterial in Models 2A and 2B whether the space of the column variable is condensed before or after sampling.

18. A genetic experiment with 155 matings of the form $AaBbCc \times aabbcc$ has resulted in the following frequencies, ABC 24 ABc 23 AbC 19 Abc 9 aBC 30 aBc 16 abC 11 abc 23. Compute the χ's or χ^2's corresponding to the comparisons A, B and C, the first order interactions AB, AC and BC and the second order interaction ABC. Compare Fisher's commentary "the 7th degree of freedom has no simple biological meaning but is necessary to complete the analysis", with the interpretation of this Chapter that it measures an interaction. [Table 22 of Fisher, 1944; Haldane, 1939a.]

19. The observed frequencies in a $5 \times 2 \times 3$ table can be given by the following

rule. The marginal ordinates are "layers", r_1 taking values 0, 1 and 2; "columns", r_2 taking values 0 and 1 and "rows", r_3 taking values 0, 1, 2, 3 and 4. Write

$$r = 6r_3 + 3r_2 + r_1;$$

then r runs through the values 0(1)29, and the following table defines a contingency table

r	f_r	r	f_r	r	f_r	r	f_r	r	f_r
0	58	6	49	12	33	18	15	24	4
1	11	7	14	13	18	19	13	25	12
2	5	8	10	14	15	20	15	26	17
3	75	9	58	15	45	21	39	27	5
4	19	10	17	16	22	22	22	28	15
5	7	11	18	17	10	23	18	29	8

Calculate the χ^2, assuming that the 5×2 layers by columns marginal distribution has been chosen arbitrarily and estimating the new parameters. (Claringbold, 1961.)

20. Verify the calculation of Bartlett's χ^2 and the following partitions of χ^2 in the analysis of the $2 \times 2 \times 2$ table,

3009	2832	3008	2974
3047	3051	2997	3038

Comparison	χ^2 obtained using Theoretical Parameters, $p_{i..}$ etc. $= \frac{1}{2}$	χ^2 obtained using Parameters Estimated from Data	Bartlett's χ^2
R	4·0115	0	—
C	1·1503	0	—
L	0·2540	0	—
RC	2·7357	2·7824	—
RL	1·7372	1·7547	—
CL	1·3524	1·3607	—
RCL	0·4690	0·5107	0·5116
Total	11·7101	6·4085·	—

(Lancaster, 1951)

21. Discuss the following table as evidence for the assertion that in many distributions, the second order interaction χ^2 as defined in Section 3 and Bartlett's χ^2 approximate.

	First Series			Second Series			Third Series		
Drawing	(1)	(2)	(3)	(1)	(2)	(3)	(1)	(2)	(3)
1	+3·67	+3·07	+3·50	−0·23	−0·60	−0·31	+3·06	+3·06	+3·09
2	−0·12	−0·26	−0·24	+0·28	+0·11	+0·45	+0·30	+0·34	+0·36
3	+1·65	+1·40	+1·65	+0·22	+0·39	+0·53	−1·74	−1·75	−1·75
4	−0·52	−0·39	−0·45	+0·04	−0·27	+0·10	−0·95	−0·90	−0·89
5	+1·40	+1·40	+1·50	−0·66	−0·96	−0·81	+0·69	+0·40	+0·61
6	+0·44	+0·38	+0·44	+1·53	+1·64	+1·82	−1·52	−1·25	−1·53
7	−0·75	−1·01	−0·99	−0·17	−0·29	−0·06	−0·88	−0·85	−0·85
8	+0·76	+0·04	+0·26	−1·19	−1·11	−0·97	+0·84	+0·81	+0·86
9	−0·44	−0·51	−0·52	+0·29	−0·35	0·00	+0·76	+0·80	+0·74
10	−1·17	−1·41	−1·39	−0·21	−0·32	−0·11	−1·49	−1·56	−1·51

First Series $\rho_{110} = \rho_{101} = \rho_{011} = 0.25$, $\rho_{111} = 0$
Second Series $\rho_{110} = 0.25 = \rho_{111}$
Third Series $\rho_{111} = \frac{1}{3}$. All other ρ's zero.
(1) Second order interaction χ less $\rho_{111}N^{\frac{1}{2}}$, theoretical parameters.
(2) Second order interaction χ less $\rho_{111}N^{\frac{1}{2}}$, estimated parameters.
(3) Second order interaction χ less $\rho_{111}N^{\frac{1}{2}}$, by Bartlett's method. (Lancaster, 1960a.)

22. The result of a breeding experiment of the form $AaBbCcDdEe \times aabbccddee$ is given as a matrix **M** in the first table below where the rows and columns give the dominants present in the offspring and the numbers in the table are the frequencies and the superscripts are the numbers of dominants present.

DOMINANTS

		0	1	2	3	4	5	6	7
		1	E	D	DE	C	CE	CD	CDE
0	1	17^0	13^1	17^1	15^2	16^1	17^2	15^2	15^3
8	B	19^1	14^2	22^2	16^3	15^2	14^3	24^3	25^4
16	A	16^1	9^2	15^2	20^3	14^2	19^3	16^3	12^4
24	AB	16^2	23^3	17^3	24^4	16^3	20^4	23^4	17^5

Make orthogonal transformations of the form,

(*) $$Q = (A \times A)M(A \times A \times A)^T,$$

where

(**) $$A = \begin{bmatrix} 1 & 1 \\ -1 & 1 \end{bmatrix}.$$

Verify that the matrix $\mathbf{Q}$ gives the component χ's multiplied by $N^{\frac{1}{2}} = \sqrt{551}$, except that $q_{11} = 551$, the grand total.

TABLE OF INTERACTIONS

		0	1	2	3	4	5	6	7
		1	E	D	DE	C	CE	CD	CDE
0	1	551^{-1}	-5^0	35^0	-5^1	5^0	5^1	-3^1	-31^2
8	B	59^0	7^1	27^1	-13^2	1^1	-15^2	37^2	17^3
16	A	3^0	27^1	-13^1	-9^2	-11^1	-31^2	-23^2	-31^3
24	AB	11^1	19^2	-25^2	-13^3	-11^2	-23^3	-3^3	5^4

In the first table 15 opposite 1 in the first row and under DE is to be interpreted as 15 offspring have DE but no other dominant. In the second table, 7 under EB is the sum $(1) - (E) - (B) + (EB)$ where (1) is the total, (E) and (B) are the numbers possessing dominants E and B respectively and (EB) is the number possessing both. (Haldane, 1939a; Lancaster 1965a; Roberts, Dawson and Madden, 1939.)

23. Verify that the numbers $\{f_k\}$ in the table below are the sums of the frequencies of those offspring possessing precisely 0, 1, 2, 3, 4, 5 dominants of the first table of Exercise 22.

OBSERVED VALUES OF THE ORTHOGONAL POLYNOMIALS

Number of Dominants Present	Frequency f_k	Unstandardized Polynomials, $\{P_k^*\}$					
		P_0^*	P_1^*	P_2^*	P_3^*	P_4^*	P_5^*
0	17	1	-5	10	-10	5	-1
1	81	1	-3	2	2	-3	1
2	152	1	-1	-2	2	1	-1
3	180	1	1	-2	-2	1	1
4	104	1	3	2	-2	-3	-1
5	17	1	5	10	10	5	1
$\sum bP_k^*$	—	551	97	46	-102	-53	5

Calculate χ^2 for 5 d.f. $= \sum \left(f_k - \frac{551}{32}\binom{5}{k}\right)^2 \bigg/ \left(\frac{551}{32}\binom{5}{k}\right)$. Partition this sum into the components corresponding to the standardized sums $551^{-\frac{1}{2}}\mathscr{S}x^{(i)}$ where

$x^{(i)}$ is the orthonormal polynomial of order i on the binomial distribution, $(\frac{1}{2} + \frac{1}{2})^5$. (Lancaster, 1965a.)

24. Consider the $2 \times 2 \times 2$ table with sampling in accordance with Model $2B$. Suppose that the a_{ij}, $i, j = 1, 2$, are selected with $\sum a_{ij} = N$. Suppose that H_0 specifies a probability p that an observation falls into the first layer and q that it falls into the second, $p + q = 1$; further, p is independent of (i, j). The distribution of entries into position $(i, j, 1)$ in the $2 \times 2 \times 2$ table is then binomial with parameters, p and a_{ij}. Moreover, the entries in the four cells $(i, j, 1)$ are mutually independent, since the model has specified independence between successive trials. The probability of the eight entries can now be written

(*) $$\mathscr{P}(\{a_{ijk}\} \mid p, \{a_{ij.}\}) = \prod_{i,j} [a_{ij.}! p^{a_{ij1}} q^{a_{ij2}} / (a_{ij1}! a_{ij2}!)]$$

$$= \frac{N! p^{a_{..1}} q^{a_{..2}}}{a_{..1}! a_{..2}!} (N!)^{-1} a_{..1}! a_{..2}! \prod_{i,j} \frac{a_{ij.}!}{a_{ij1}! a_{ij2}!}$$

The expression for the probability is thus factorized into a binomial probability and the probability of a 2×4 table, the margins of which are $a_{..1}$, $a_{..2}$ and $a_{11.}$, $a_{12.}, a_{21.}, a_{22.}$. This table could be treated as a 2×4 table as in Model 2 of Section XI.2 and may be written

$$
\begin{array}{ccccc}
a_{111} & a_{121} & a_{211} & a_{221} & a_{..1} \\
a_{112} & a_{122} & a_{212} & a_{222} & a_{..2} \\
a_{11.} & a_{12.} & a_{21.} & a_{22.} & a_{...} = N.
\end{array}
$$

It may be noted that two marginal 2×2 tables may be constructed by consolidating columns, 1 and 2, 3 and 4 or by consolidating columns 1 and 3, and 2 and 4. In the combinatorial treatment, test marginal 2×2 tables are to be taken as given. So the problem is to obtain the distribution of the entries conditional on fixed values of $a_{111} + a_{121}$ and $a_{111} + a_{211}$. It is sometimes argued as in Freeman and Halton (1951) that the probability of a given value of a_{111} can be obtained by means of the formula

(**) $\mathscr{P}(a_{111} \mid a_{111} + a_{211}, a_{111} + a_{121}, \{a_{..k}\})$

$$= \mathscr{P}(a_{111} \mid \{a_{..k}\}) / [\mathscr{P}(a_{111} + a_{211} \mid \{a_{..k}\}) \mathscr{P}(a_{111} + a_{121} \mid \{a_{..k}\})].$$

However, the two expressions in the denominator of (**) cannot legitimately be multiplied since the variables $a_{111} + a_{211}$ and $a_{111} + a_{121}$ are not mutually independent. In (**) the conditions in each probability expression are supposed to imply fixed values of $\{a_{ij.}\}$ by the statement of the model, and consequently fixed values of $\{a_{i..}\}$ and $\{a_{.j.}\}$.

Use (XI.5.6) and (XI.5.7) to compute the variances and covariances of the a_{ij1}, $i, j = 1, 2$ and verify the result

(***) $\operatorname{cov}(a_{111} + a_{121}, a_{111} + a_{211}) = (a_{11.} a_{22.} - a_{12.} a_{21.})(a_{1..} a_{2..} a_{.1.} a_{.2.})^{\frac{1}{2}}$

and this expression does not vanish unless there are special relations between the $a_{ij.}$. A warning must therefore be given that there cannot be a justification of the statement of Freeman and Halton (1951), who write "in addition, should the choice of the sample considered, although otherwise random be restricted by a number of conditions, $c_1, c_2, \ldots, c_s$, whose *a priori* probabilities are respectively $P_{c_1}, P_{c_2}, \ldots, P_{c_s}$, the expression (1) must be divided by their product". The results dependent on their Sections II and III are invalid because of this false assumption. However, their Section IV deals with two-way tables in which the multiplicative rule does hold under Model 1 of our Section 2 since the marginal multinomial distributions are mutually independent and in Models 2 and 3, the question of multiplication does not arise. Some of their other computations do not depend on their Sections II and III. In fact, the method of Bartlett (1935) of obtaining the "exact" probabilities subject to the fixed marginal totals can be followed as in Section 4.

25. Under some conditions, a relationship can be demonstrated between the interaction χ^2 of Section 3 and Bartlett's χ^2. If the first order interactions are small, this relationship holds approximately. In the procedure for the test, quantities are added or subtracted to the weights at the eight points of the $2 \times 2 \times 2$ table to obtain the multiplicative relation (5.2) or more generally (5.3). In this more general case, the distribution function is to be modified by adding a signed measure to it in such a way that the marginal two-dimensional measures are unchanged but the multiplicative property is obtained. Let us write the signed measure as a function $U(x, y, z)$, such that $U(\infty, y, z)$, $U(x, \infty, z)$ and $U(x, y, \infty)$ are all identically zero. Prove the following

Theorem. *The measure assigned by the signed measure U to any set is of the form*

(*) $$dU(x, y, z) = \sum \tau_{ijk} x^{(i)} y^{(j)} z^{(k)} \, dG \, dH \, dK.$$

(Lancaster, 1960a.)

26. Show that for fixed marginal totals χ_B, the square root of Bartlett's χ^2, is a linear function of the observed frequency in any given cell. Prove that the same is true of the interaction χ_{RCL} considered over the same set of marginal totals. Hence $\chi_B = bn_1 + b_0$, $\chi_{RCL} = ln_1 + l_0$ for some constants b, b_0, l and l_0.

$$\text{corr}\,(\chi_B, \chi_{RCL}) = 1.$$

b and l may differ. They are identical if the estimated a_i are equal to their values under complete independence with the same row, column and layer totals.

Bibliography

All the articles cited of E.S. Pearson and J. Neyman are available in the volumes of their collected works. Those of R.A. Fisher and K. Pearson, which are available in the collected works, are distinguished by an asterisk.

Abdel-Aty, S. H. (1954) "Approximate formulae for the percentage points and the probability integral of the non-central χ^2 distribution," *Biometrika* **41**, 538–540.

Achieser, N. I. (1956) *The theory of approximation;* trans. by J. Hyman, Ungar, New York, x + 307.

Adler, F. (1951) "Yates' correction and the statisticians," *J. Amer. Statist. Ass.* **46**, 490–501.

Aitchison, J. (1962) "Large-sample restricted parametric tests," *J. Roy. Statist. Soc.* **B24**, 234–250.

Aitken, A. C. (1931) "Some applications of generating functions to normal frequency," *Quart. J. Math.* **2**, 130–135.

Aitken, A. C. (1933*a*) "On the graduation of data by the orthogonal polynomials of least squares," *Proc. Roy. Soc. Edinburgh* **53**, 54–78.

Aitken, A. C. (1933*b*) "On fitting polynomials to weighted data by least squares," *Proc. Roy. Soc. Edinburgh* **54**, 1–11.

Aitken, A. C. (1933*c*) "On fitting polynomials to data with weighted and correlated errors," *Proc. Roy. Soc. Edinburgh* **54**, 12–16.

Aitken, A. C. (1935) "On least squares and linear combination of observations," *Proc. Roy. Soc. Edinburgh* **55**, 42–47.

Aitken, A. C. (1939) "Note on the derivation and distribution of Pearson's χ^2," *Proc. Edinburgh Math. Soc.* **6**, 57–60.

Aitken, A. C. (1940) "On the independence of linear and quadratic forms in samples of normally distributed variables," *Proc. Roy. Soc. Edinburgh* **60**, 40–46. Corrig. **A62**, 277.

Aitken, A. C. (1948) "On a problem in correlated errors," *Proc. Roy. Soc. Edinburgh* **A62**, 273–277.

Aitken, A. C. (1949) "On the Wishart distribution in statistics," *Biometrika* **36**, 59–62.

Aitken, A. C. (1950) "On the statistical independence of quadratic forms in normal variables," *Biometrika* **37**, 93–96.

Aitken, A. C. and Gonin, H. T. (1935) "On fourfold sampling with and without replacement," *Proc. Roy. Soc. Edinburgh* **55**, 114–125.

Aitken, A. C. and Oppenheim, A. (1931) "On Charlier's new form of the frequency function," *Proc. Roy. Soc. Edinburgh* **51**, 35–41.

Åkesson, O. A. (1916) "On the dissection of correlation surfaces," *Ark. f. mat. Astr. och Fysik* Bd. **11** (16) 18 pp.

Alexits, G. (1961) *Convergence problem for orthogonal series*, trans. by L. Foldes, Akadémiai Kiadó, Budapest. ix + 350.

Ali, S. M. and Silvey, S. D. (1965a) "Association between random variables and the dispersion of a Radon-Nikodym derivative," *J. Roy. Statist. Soc.* **B27**, 100–107.

Ali, S. M. and Silvey, S. D. (1965b) "A further result on the relevance of the dispersion of a Radon–Nikodym derivative to the problem of measuring association," *J. Roy. Statist. Soc.* **B27**, 108–110.

Ali, S. M. and Silvey, S. D. (1966) "A general class of coefficients of divergence of one distribution from another," *J. Roy. Statist. Soc.* **B28**, 131–142.

Almond, J. (1954) "A note on χ^2 applied to epidemic chains," *Biometrics* **10**, 459–477.

Anderson, R. L. (1959) "Use of contingency tables in the analysis of consumer preference studies," *Biometrics* **15**, 582–590.

Anderson, T. W. (1958) *An introduction to multivariate statistical analysis.* John Wiley, New York. xii + 374.

Anderson, T. W. and Darling, D. A. (1952) "Asymptotic theory of certain 'goodness of fit' criteria based on stochastic processes," *Ann. Math. Statist.* **23**, 193–212.

Andersson, W. (1941) "The binomial type of Gram's series," *Skand. Aktuarietidskr.* **24**, 203–213.

Andersson, W. (1942) "On the Gram series on Pearson's system of frequency functions," *Skand. Aktuarietidskr.* **25**, 141–149.

Anscombe, F. J. (1950) "Sampling theory of the negative binomial and logarithmic series distributions," *Biometrika* **37**, 358–382.

Anscombe, F. J. (1963) "Tests of goodness of fit," *J. Roy. Statist. Soc.* **B25**, 81–94.

Aoyama, H. (1953) "On the chi-square test for weighted samples," *Ann. Inst. Statist. Math. Tokyo* **5**, 25–28.

Armsen, P. (1955) "Tables for significance tests of 2 × 2 contingency tables," *Biometrika* **42**, 494–511.

Arnold, B. C. (1967) "A note on multivariate distributions with specified marginals," *J. Amer. Statist. Assoc.* **62**, 1460–1461.

Aroian, L. A. (1943) "A new approximation to the levels of significance of the chi-square distribution," *Ann. Math. Statist.* **14**, 93–95.

Bahadur, R. R. (1961a) "A representation of the joint distribution of responses to n dichotomous items," pp. 158–168 of Solomon (1961).

Bahadur, R. R. (1961b) "On classification based on responses to n dichotomous items," pp. 169–176 of Solomon (1961).

Baillie, D. C. (1946) "On testing the significance of mortality ratios by the use of χ^2," *Trans. Actuar. Soc. Amer.* **47**, 326–344.

Baker, G. A. (1941) "Tests of homogeneity for normal populations," *Ann. Math. Statist.* **12**, 233–236.

Baker, P. C. (1952) "Combining tests of significance in cross-validation," *Educ. Psych. Measurement* **12**, 300–306.

Barnard, G. A. (1945) "A new test for 2 × 2 tables," *Nature* **156**, 177 and 783.

Barnard, G. A. (1947a) "Significance tests for the 2 × 2 tables," *Biometrika* **34**, 123–138.

Barnard, G. A. (1947b) "2 × 2 tables. A note on E. S. Pearson's paper," *Biometrika* **34**, 168–169.

Barnard, M. M. (1936) "An enumeration of the confounded arrangements in the 2 × 2 × 2.... factorial designs," *J. Roy. Statist. Soc. Suppl.* **3**, 195–202.

Barrett, J. F. and Lampard, D. G. (1955) "An expansion for some second-order probability distributions and its application to noise problems." *IRE Trans. Profess. Group on Information Theory.* IT-1, **1**, 10–15.

Bartholomew, D. J. (1959) "A test of homogeneity for ordered alternatives," *Biometrika*, **46**, 36–48.

Bartholomew, D. J. (1967) "Hypothesis testing when the random size is treated as a random variable," *J. Roy. Statist. Soc.* **B28**, 53–70.

Bartko, J. J., Greenhouse, S. W. and Patlak, C. S. (1968) "On expectations of some functions of Poisson variates," *Biometrics* **24**, 97–102.

Bartlett, M. S. (1935) "Contingency table interactions," *J. Roy. Statist. Soc. Suppl.* **2**, 248–252.

Bartlett, (1937a) "Sub-sampling for attributes," *J. Roy. Statist. Soc. Suppl.* **4**, 131–135.

Bartlett, M. S. (1937b) "Properties of sufficiency and statistical tests," *Proc. Roy. Soc.*, A, **160**, 268–282.

Bartlett, M. S. (1947) "The general canonical correlation distribution," *Ann. Math. Statist.* **18**, 1–17.

Bartlett, M. S. (1951a) "The goodness of fit of a single hypothetical discriminant function in the case of several groups," *Ann. Eugen. Camb.* **16**, 199–214.

Bartlett, M. S. (1951b) "The effect of standardization on a χ^2 approximation in factor analysis," *Biometrika* **38**, 337–344.

Bartlett, M. S. (1951c) "The frequency goodness of fit test for probability chains," *Proc. Camb. Philos. Soc.* **47**, 86–95.

Bartlett, M. S. (1952a) "A sampling test of the χ^2 theory for probability chains," *Biometrika* **39**, 118–121.

Bartlett, M. S. (1952b) "The statistical significance of odd bits of information," *Biometrika* **39**, 228–237.

Bartlett, M. S. (1954) "A note on the multiplying factors for various χ^2 approximations," *J. Roy. Statist. Soc.* **B16**, 296–298.

Barton, D. E. (1953) "The probability distribution function of a sum of squares," *Trabajos Estadist.* **4**, 199–207.

Barton, D. E. (1954) "Neyman's Ψ_k^2 test of goodness of fit when the null hypothesis is composite," *Skand. Aktuarietidskr.* **37**, 216–245.

Barton, D. E. (1955) "A form of Neyman's Ψ_k^2 test of goodness of fit applicable to grouped and discrete data," *Skand. Aktuarietidskr.* **38**, 1–16.

Barton, D. E. (1956) "Newman's Ψ_k^2 test of goodness of fit when the null hypothesis is composite," *Skand. Aktuarietidskr.* **39**, 216–245.

Barton, D. E. and David, F. N. (1959) "Haemocytometer counts and occupancy theory, *Trabajos Estadist.* **10**, 13–18.

Basharin, G. P. (1957) "On using the test of goodness of fit (χ^2 agreement as) a test of independence of trials," *Teor. Veroyatnost. i Primenen.* **2**, 141–2.

Basu, D. (1956) "A note on the multivariate extensions of some theorems related to the univariate normal distributions," *Sankhyā* **17**, 221–224.

Bateman, G. I. (1949) "The characteristic function of a weighted sum of non-central squares of normal variates subject to s linear restraints," *Biometrika* **36**, 460–462.

Bateman, G. I. (1950) "The power of the χ^2 index of dispersion test, when Neyman's contagious distribution is the alternate hypothesis," *Biometrika* **37**, 59–63.

Batschelet, E. (1960) "Ueber eine Kontingenztafel mit fehlenden Daten," *Biometrische Zeit.* **2**, 236–243.

Bejar, J. (1958) "Tablas de contingencia," *Trabajos Estadist.* **9**, 85–101.

Bellman, R. (1960) *Introduction to matrix analysis.* McGraw-Hill Book Company, New York, xx + 328.

Bennett, B. M. (1955/56) "Note on the moments of the logarithmic non-central χ^2 and z distributions," *Ann. Inst. Statist. Math., Tokyo,* **7**, 57–61.

Bennett, B. M. (1959) "Note on the power function of the X_n test in genetics," *Skand. Aktuarietidskr.* **42**, 1–5.

Bennett, B. M. (1962) "On a heuristic treatment of the 'indices of dispersion'." *Ann. Inst. Statist. Math. Tokyo*, **14**, 151–157.

Bennett, B. M. (1965) "Note on a χ^2 approximation for the multivariate sign test," *J. Roy. Statist. Soc.* **B27**, 82–85.

Bennett, B. M. and Hsu, P. (1960) "On the power function of the exact test for the 2 × 2 contingency test," *Biometrika* **47**, 393–398.

Bennett, B. M. and Hsu, P. (1961) "A sampling study of the power function of the binomial χ^2 'index of dispersion' test," *J. Hyg. (Camb.)* **59**, 449–455.

Bennett, B. M. and Nakamura, E. (1963) "Tables for testing significance in a 2 × 3 contingency table," *Technometrics* **5**, 501–511.

Bennett, B. M. and Nakamura, E. (1964) "The power function of the exact test for the 2 × 3 contingency table," *Technometrics* **6**, 439–458.

Bennett, R. W. (1962) "Size of the χ^2 test in the multinomial distribution," *Aust. J. Statist.* **4**, 86–88.

Berger, A. (1961) "On comparing intensities of association between two binary characteristics in two different populations," *J. Amer. Statist. Assoc.* **56**, 889–908.

Berkson, J. (1938) "Some difficulties of interpretation encountered in the application of the chi-square table," *J. Amer. Statist. Assoc.* **33**, 526–536.

Berkson, J. (1940) "A note on the chi-square test, the Poisson and the binomial," *J. Amer. Statist. Assoc.* **35**, 362–367.

Berkson, J. (1946a) "Approximation of chi-square by "probits" and by "logits"," *J. Amer. Statist. Assoc.* **41**, 70–74.

Berkson, J. (1946b) "Limitation of the application of fourfold table analysis to hospital data," *Biometrics* **2**, 47–53.

Berkson, J. (1949) "Minimum χ^2 and maximum likelihood solution in terms of a linear transform with particular reference to bio-assay," *J. Amer. Statist. Assoc.* **44**, 273–278.

Berkson, J. (1951) "Relative precision of minimum chi-square and maximum likelihood estimates of regression coefficients," *2nd Berkeley Symposium* 471–479.

Berkson, J. (1968) "Application of minimum logit χ^2 estimate to a problem of Grizzle with a notation on the problem of 'no interaction'," *Biometrics* **24**, 75–95.

Berkson, J. and Geary, R. C. (1941) "Comments on Dr. Madow's 'Note on tests of departure from normality' with some remarks concerning tests of significance," *J. Amer. Statist. Assoc.* **36**, 539–543.

Bernstein, S. (1926) "Sur l'extension du théorème limite du calcul des probabilités aux sommes de quantités dépendantes," *Math. Annalen* **97**, 1–59.

Bernstein, S. (1927) "Fondements géométriques de la théorie des corrélations," *Metron*, **7**, Pt 2, 3–27.

Bernstein, S. (1934) *Theory of probability* (second edition). Goz. Izdat. Moscow–Leningrad. 556 pp.

Bernstein, S. (1943) "Retour au problème de l'évaluation de l'approximation de la formule limite de Laplace," *Izvestia Akad. Nauk SSSR. Ser. Mat. Mathematical* series **7**, 3–16.

Bernstein, S. N. (1911) "Sur le calcul approché des probabilités par la formule de Laplace," *Proc. Kharkov Math. Soc.* (2) **12**, 106–110.

Bernt, G. D. (1958) "Power functions of the gamma distribution," *Ann. Math. Statist.* **29**, 302–306.

Berry, A. C. (1941) "The accuracy of the Gaussian approximation to the sum of independent variates," *Trans. Amer. Math. Soc.* **49**, 122–136.

Bhapkar, V. P. (1961) "Some tests for categorical data," *Ann. Math. Statist.* **32**, 72–83.

Bhat, B. R. (1961) "On the asymptotic distribution of the "psi-squared" goodness of fit criteria for Markov chains and Markov sequences," *Ann. Math. Statist.* **32**, 49–58.

Bhattacharya, S. K. (1966) "Confluent hypergeometric distributions of discrete and continuous type with applications to accident proneness," *Calcutta Statist. Assoc. Bull.* **15**, 20–31.

Bhattacharyya, A. (1943) "On a measure of divergence between two statistical populations defined by their probability distributions," *Bull. Calcutta Math. Soc.* **35**, 99–109.

Bhattacharyya, A. (1945) "A note on the distribution of the sum of chi-squares," *Sankhyā* **7**, 27–28.

Bhattacharyya, A. (1946) "On a measure of divergence between two multinomial populations," *Sankhyā* **7**, 401–406.

Bienaymé, J. (1838) "Sur la probabilité des résultats moyens des observations; démonstration directe de la règle de Laplace," *Mémor. Sav. Étrangers Acad. Sci. Paris* **5**, 513–558.

Bienaymé, J. (1852) "Sur la probabilité des erreurs d'après la méthode des moindres carrés," *J. Math. Pures Appl.* **17**, 33–78.

Billingsley, P. (1956) "Asymptotic distribution of two goodness of fit criteria," *Ann. Math. Statist.* **27**, 1123–1129.

Birch, M. W. (1963) "Maximum likelihood in three-way contingency tables," *J. Roy. Statist. Soc.* **B25**, 220–233.

Birch, M. W. (1964) "The detection of partial association, I: the 2×2 case," *J. Roy. Statist. Soc.* **B26**, 313–324.

Birch, M. W. (1965) "The detection of partial association, II: the general case," *J. Roy. Statist. Soc.* **B27**, 111–124.

Birnbaum, A. (1954) "Combining independent tests of significance," *J. Amer. Statist. Assoc.*, **49**, 559–574.

Blakeman, J. (1905) "On tests of linearity of regression in frequency distributions," *Biometrika* **4**, 332–350.

Blakeman, J. and Pearson, K. (1906) "On the probable error of the mean-square contingency," *Biometrika* **5**, 191–196.

Blalock, H. M. (1958) "Probabilistic interpretations of the mean-square contingency," *J. Amer. Statist. Assoc.* **53**, 102–105.

Bliss, C. I. (1944) "A chart of the chi-square distribution," *J. Amer. Statist. Assoc.* **39**, 246–248.

Blom, G. (1954) "Transformation of the binomial, negative binomial, Poisson and χ distribution," *Biometrika* **41**, 302–316. Corrig. **43**, 235.

Blomqvist, N. (1950) "On a measure of dependence between two random variables," *Ann. Math. Statist.* **21**, 593–600.

Blomqvist, N. (1951) "Some tests based on dichotomization," *Ann. Math. Statist.* **22**, 362–371.

Boas, F. (1909) "Determination of the coefficient of correlation," *Science* **29**, 823–4.

Boas, F. (1922) "The measurement of differences between variable quantities," *J. Amer. Statist. Assoc.* **18**, 425–445.

Bodmer, W. F. (1959) "A significantly extreme deviate in data with a non-significant heterogeneity chi-square," *Biometrics* **15**, 538–542.

Bodmer, W. F. and Parson, P. A. (1959) "The analogy between factorial experimentation and balanced multi-point linkage tests," *Heredity* **13**, 145–156.

Bofinger, E. and Bofinger, V. J. (1961a) "A runs test for sequences of random digits," *Austral. J. Statist.* **3**, 37–41.

Bofinger, E. and Bofinger, V. J. (1961b) "The gap test for random sequences," *Ann. Math. Statist.* **32**, 524–534.

Bol'šev, L. N. (1965) "On a characterization of the Poisson distribution and its statistical applications," (Russian, English summary) *Teor. Verojatnost. i Primenen.* **10**, 488–499.

Bonnier, G. (1942) "The fourfold table and the heterogeneity test," *Science* (N.S.), **96**, 13–14.

Boole, G. (1854) *An investigation of the laws of thought.* Walton and Maberly, London. xvi + 448.

Bowker, A. H. (1948) "A test for symmetry in contingency tables," *J. Amer. Statist. Assoc.* **43**, 572–574.

Bowley, A. L. (1920) *Elements of Statistics*, P. S. King and Sons, London, xi + 459.

Bowley, A. L. and Connor, L. R. (1923) "Test of correspondence between statistical grouping and formulae," *Economica* **3**, 1–9.

Boyd, W. C. (1965) "A nomogram for χ^2," *J. Amer. Statist. Assoc.* **60**, 344–346.

Brambilla, F. (1956) *Statistica.* Vol. I *La variabilita strutturale*—1955. Vol. II *La teoria della stima*—1956. La Goliardica, Milan.

Brauer, R. (1929) "Die stetigen Darstellungen der Komplexen orthogonalen Gruppe," *Akad. d. Wissenschaft. zu Berlin, Berlin. Berichte*, **1929**, 626–638.

Bravais, A. (1846) "Analyse mathématique sur les probabilités des erreurs de situation d'un point," *Mém de l'Instit. de France* **9**, 255–332.

Bresciani, C. (1909) "Sui metodi per la misura della correlazioni," *Giorn. d. Economisti* **38**, 401–414, 491–522.

Broadbent, S. R. (1956) "A measure of dispersion applied to cosmic-ray and other problems," *Proc. Cambridge. Philos. Soc.* **52**, 499–513.

Brogden, H. E. (1949) "A new coefficient: application to biserial correlation and to estimation of selective efficiency," *Psychometrika* **14**, 169–182.

Bross, I. (1954) "Misclassification in 2×2 tables," *Biometrics* **10**, 478–486.

Bross, I. D. J. (1964) "Taking a covariable into account," *J. Amer. Statist. Assoc.* **59**, 725–736.

Bross, I. J. and Delaney, M. M. (1952) "Significance tests for certain 2×2 tables," *Amer. J. Hyg.* **55**, 357–362.

Bross, I. D. J. and Kasten, E. L. (1957) "Rapid analysis of 2×2 tables," *J. Amer. Statist. Assoc.* **52**, 18–28.

Brown, J. L. Jr. (1958) "A criterion for the diagonal probability distribution in orthogonal polynomials," *IRE Trans.* **IT4**, 172.

Brown, J. L. Jr. (1960) "Mean square truncation error in series expansions of random functions," *J. Soc. Indust. Appl. Math.* **8**, 28–32.

Brownlee, J. (1924) "Experiments to test the theory of goodness of fit," *J. Roy. Statist. Soc.* **87**, 76–82.

Bruen, C. (1938) "Methods for the combination of observations," *Metron* **13(2)**, 61–140.

Bui Trong Lieu and Carton, D. A. (1964) "Sur la loi du χ^2 dans certains processus de Markov," *Studia Math.* **24**, 25–35.

Burkhardt, F. (1957) "Uber die inversion von Korrelationen." *Bull. Inst. Internat. Statist.* **35**, 3–6.

Burnside, W. (1925) "On the approximate sum of selected terms from the multinomial expansion," *Messenger Math.* **54**, 189–192.

Burnside, W. (1928) *Theory of Probability.* Cambridge Univ. Press, xxx + 106.

Byrne, L. (1935) "A theory of validation for derivative specification and check lists," *Ann. Math. Statist.* **6**, 146–157.

Cadwell, J. H. (1953) "Approximating to the distributions of measures of dispersion by a power of χ^2," *Biometrika* **40**, 336–346.

Camp, B. H. (1924) "Probability integrals for the point binomial," *Biometrika* **16**, 163–171.

Camp, B. H. (1925) "Probability integrals for a hypergeometric series," *Biometrika* **17**, 61–67.

Camp, B. H. (1929) "The multinomial solid and the chi-test," *Trans. Amer. Math. Soc.* **31**, 133–144. Corrig. **44** (1938) 151.

Camp, B. H. (1938) "Further interpretations of the chi-square test," *J. Amer. Statist. Assoc.* **33**, 537–542.

Camp, B. H. (1940) "Further comments on Berkson's problem," *J. Amer. Statist. Assoc.* **35**, 368–376.

Camp, B. H. (1951) "Approximation to the point binomial," *Ann. Math. Statist.* **22**, 130–131.

Campbell, J. T. (1932a) "Factorial moments and frequencies of Charliers type B," *Proc. Edinburgh Math. Soc.* **(2)3**, 99–106.

Campbell, J. T. (1932b) "The Poisson correlation function," *Proc. Edinburgh Math. Soc.* **(2)4**, 18–26.

Carroll, J. B. (with C. C. Bennett) (1950) "Machine short-cuts in the computation of chi-square and the contingency coefficient," *Psychometrika* **15**, 441–447.

Castellan, N. J. Jr. (1966) "On the estimation of the tetrachoric correlation coefficient," *Psychometrika* **31**, 67–73.

Cauchy, A. L. (1829) "Sur l'équation à l'aide de laquelle on détermine les inégalités séculaires des mouvements des planètes," *Ex. de Math.* **4**, 140–160.

Caussinus, H. (1962) "Sur certaines généralisations de l'emploi du test du χ^2," *C. R. Acad. Sci. Paris* **254**, 3306–3308.

Caussinus, H. (1966) "Contribution à l'analyse statistique des tableaux de corrélation." *Ann. Fac. Sci. Univ. Toulouse* **29**, 77–183.

Cayley, A. (1869) See Todhunter (1869).

Chacko, V. J. (1963) "Testing homogeneity against ordered alternatives," *Ann. Math. Statist.* **34**, 945–956.

Chakrabarti, M. C. (1949) "On the moments of the non-central χ^2," *Bull. Calcutta Math. Soc.* **41**, 208–210.

Chakravarti, I. M. and Rao, C. R. (1959) "Tables for some small sample tests of significance for Poisson distributions and 2 × 3 contingency tables," *Sankhyā* **21**, 315–326.

Chapelin, J. (1932) "On a method of proceeding from partial cell frequencies to ordinates and to total cell frequencies in the case of a bivariate frequency surface," *Biometrika* **22**, 495–497.

Chapman, D. G. (1951) "Some properties of the hypergeometric distribution with applications to zoological sample censuses," *Univ. California Publ. Statist.* **1**, 131–159.

Chapman, D. G. (1958) "A comparative study of several one-sided goodness of fit tests," *Ann. Math. Statist.* **29**, 655–674.

Cherian, K. C. (1941) "A bivariate correlated gamma-type distribution function," *J. Indian. Math. Soc.* (N.S.) **5**, 133–144.

Chernoff, H. and Lehmann, E. L. (1954) "The use of the maximum likelihood estimates in χ^2 tests for goodness of fit," *Ann. Math. Statist.* **25**, 579–586.

Chernoff, H. and Teicher, H. (1958) "A central limit theorem for sums of interchangeable random variables," *Ann. Math. Statist.* **29**, 118–130.

Cheshire, L. Saffir, M. and Thurstone, L. L. (1933) *Computing Diagrams for the tetrachoric correlation coefficient*. Univ. Chicago Press, Chicago.

Chiang, C. L. (1951) "On the design of mass medical surveys," *Hum. Biol.* **23**, 242–271.

Chu, J. T. (1956) "Errors in normal approximations to the t, τ, and similar types of distribution," *Ann. Math. Statist.* **27**, 780–789.

Chung, J. H. and de Lury, D. B. (1950) *Confidence limits for the hypergeometric distribution*. Univ. Toronto Press, Toronto, 158 pp.

290 BIBLIOGRAPHY

Čibisov, D. M. (1965) "An investigation of the asymptotic power to tests of fit," *Teor. Veroyatnost. i Primenen.* **10**, 460–478.

Claringbold, P. J. (1961) "The use of orthogonal polynomials in the partition of chi-square," *Austral. J. Statist.* **3**, 48–63.

Clark, A. and Leonard, W. H. (1939) "The analysis of variance with special reference to data expressed as percentages," *J. Amer. Soc. Agronom.* **31**, 55–66.

Clopper, C. J. and Pearson, E. S. (1934) "The use of confidence or fiducial limits illustrated in the case of the binomial," *Biometrika* **26**, 404–413.

Cochran, W. G. (1934) "The distribution of quadratic forms in a normal system, with applications to the analysis of covariance," *Proc. Camb. Philos. Soc.* **30**, 178–191.

Cochran, W. G. (1936a) "The χ^2 distribution for the binomial and Poisson series, with small expectations," *Ann. Eugen. London* **7**, 207–217.

Cochran, W. G. (1936b) "Statistical analysis of field counts of diseased plants," *J. Roy. Statist. Soc. Suppl.* **3**, 49–67.

Cochran, W. G. (1937) "Note on J. B. S. Haldane's paper 'The exact value of the moments of the distribution of χ^2,' *Biometrika* **29**, 133–143," *Biometrika* **29**, 407.

Cochran, W. G. (1940) "The analysis of variance when experimental errors follow the Poisson or binomial laws," *Ann. Math. Statist.* **11**, 335–347.

Cochran, W. G. (1942) "The χ^2 correction for continuity," *Iowa State College J. Sci.* **16**, 421–436.

Cochran, W. G. (1950) "The comparison of percentages in matched samples," *Biometrika* **37**, 256–266.

Cochran, W. G. (1952) "The test of goodness of fit," *Ann. Math. Statist.* **23**, 315–345.

Cochran, W. G. (1954) "Some methods for strengthening the common χ^2 tests," *Biometrics* **10**, 417–451.

Cochran, W. G. (1955) "A test of a linear function of the deviations between observed and expected numbers," *J. Amer. Statist. Assoc.* **50**, 377–397.

Cornish, E. A. and Fisher, R. A. (1937) "Moments and cumulants in the specification of distributions," *Rev. Institut Internat. Statistique.* **4**, 1–14.

Corsten, L. C. A. (1957) "Partition of experimental vectors connected with multinomial distributions," *Biometrics* **13**, 451–484.

Corsten, L. C. A. (1958) "Vectors, a tool in statistical regression theory," *Meded. Landbouwhogeschool Wageningen* **58**, 1–92.

Cournot, A. (1843) *Exposition de la théorie des chances et des probabilités.* Paris, L. Hachette et cie. viii + 448.

Cox, D. R. (1957) "Note on grouping," *J. Amer. Statist. Assoc.* **52**, 543–547.

Craig, A. T. (1936) "Note on a certain bilinear form that occurs in statistics." *Amer. J. Math.* **58**, 864–866.

Craig, A. T. (1943) "Note on the independence of certain quadratic forms," *Ann. Math. Statist.* **14**, 195–197.

Craig, C. C. (1929) "Sampling when the parent population is of Pearson's Type III," *Biometrika* **21**, 287–293.

Craig, C. C. (1953) "Combination of neighboring cells in contingency tables," *J. Amer. Statist. Assoc.* **48**, 104–112.

Cramér, H. (1924) "Remarks on correlation," *Skand. Aktuarietidskr.* **7**, 220–240.

Cramér, H. (1928) "On the composition of elementary errors. First paper: mathematical deductions," "Second paper: statistical applications," *Skand. Aktuarietidskr.* **11**, 13–74 and 141–180.

Cramér, H. (1937) *Random variables and probability distributions.* Cambridge tracts on Mathematics, No. 36. pp. 119.

Cramér, H. (1946) *Mathematical methods of statistics*. Princeton University Press, Princeton. xvi + 575.

Cramér, H. and Wold, H. (1936) "Some theorems on distribution functions," *J. Lond. Math. Soc.* **11**, 290–294.

Crathorne, A. R. (1922) "Calculation of the correlation ratio," *Quart. Publ. Amer. Statist. Assoc.* **18**, 394–396.

Crow, E. L. (1952) "Some cases in which Yates' correction should not be applied," *J. Amer. Statist. Assoc.* **47**, 303–304.

Csaki, P. and Fischer, J. (1960a) "Contributions to the problem of maximum correlation," *Magyar Tud. Akad. Mat. Kutato Int. Közl.* **5**, 325–332.

Csaki, P. and Fischer, J. (1960b) "On bivariate stochastic connection," *Magyar Tud. Akad. Mat. Kutato Int. Közl.* **5**, 331–323.

Crow, J. F. (1945) "A chart of the χ^2 and t distributions," *J. Amer. Statist. Assoc.* **40**, 376.

Czuber, E. (1898) *Theorie der Beobachtungsfehler*. Teubner, Leipzig, xii + 418.

Czuber, E. (1920) "Über Funktionen von Variablen, zwischen welchen Korrelationen bestehen," *Metron* **1(1)**, 53–63.

Daboni, L. (1959) "Una proprietà delle distribuzioni poissoniane," *Boll. Un. Mat. Ital.* **(3)14**, 318–320.

Dagnelie, P. (1962) "L'application de l'analyse multi-variable à l'étude des communautés végétales," *Bull. Internat. Statist. Inst.* **39(2)**, 265–275.

Dall'Aglio, G. (1959) "Sulla compatibilità delle funzioni di ripartizione doppia," *Rend. Mat. e Appl.* (5). **18**, 385–413.

Daly, C. (1962) "A simple test for trends in a contingency table," *Biometrics* **18**, 114–119.

Dalzell, D. P. (1945) "On the completeness of a series of normal orthogonal functions," *J. London Math. Soc.* **20**, 87–93.

Daniels, H. E. (1952) "The covering circle of a sample from a circular normal distribution," *Biometrika* **39**, 137–143.

Darroch, J. N. (1962) "Interactions in multi-factor contingency tables," *J. Roy. Statist. Soc.* **B24**, 251–263.

Darroch, J. N. and Silvey, S. D. (1963) "On testing more than one hypothesis," *Ann. Math. Statist.* **34**, 555–567.

Darwin, J. H. (1957) "The power of the Poisson index of dispersion," *Biometrika* **44**, 286–289.

Darwin, J. H. (1958) "On corrections to the chi-squared distribution," *J. Roy. Statist. Soc.* **B20**, 387–392.

Darwin, J. H. (1959) "Note on a three-decision test for comparing two binomial populations," *Biometrika* **46**, 106–113.

Das Gupta, P. (1964) "On the estimation of the total number of events and of the probabilities of detecting an event from information supplied by several agencies," *Calcutta Statist. Assoc. Bull.* **13**, 89–100.

David, F. N. (1934) "On the $P\lambda_n$ Test for randomness: remarks, further illustration, and table for $P\lambda_n$," *Biometrika* **26**, 1–11.

David, F. N. (1939) "On Neyman's 'Smooth' test for goodness of fit. I. Distribution of the criterion ψ^2 when the hypothesis tested is true," *Biometrika* **31**, 191–199.

David, F. N. (1947) "A 'smooth' test of goodness of fit," *Biometrika* **34**, 299–310.

David, F. N. (1948) "Correlations between χ^2 cells," *Biometrika* **35**, 418–422.

David, F. N. (1950) "An alternative form of χ^2," *Biometrika* **37**, 448–451.

David, F. N. and Johnson, N. L. (1948) "The probability integral transformation when parameters are estimated from the data," *Biometrika* **35**, 182–190.

David, F. N. and Johnson, N. L. (1950) "The probability integral transformation when the variable is discontinuous," *Biometrika* **37**, 42–49.

Davies, O. L. (1933) "On asymptotic formulae for the hypergeometric series I. Hypergeometric series in which the fourth element, x, is unity," *Biometrika* **24,** 295–322.

Davies, O. L. (1934) "On asymptotic formulae for the hypergeometric series II. Hypergeometric series in which the fourth element, x, is not necessarily unity," *Biometrika* **26,** 59–107.

Davis, P. J. (1961) "Advances in orthonormalizing computation," in *Advances in Computers.* F. L. Alt. (editor) 1961. Academic Press. New York, **2,** 55–133.

Davis, P. J. (1962) "Orthonormalizing codes in numerical analysis," *Survey of numerical analysis.* J. A. Todd, McGraw-Hill, New York. 347–379.

Davis, P. J. and Rabinowitz, P. (1961) "Advances in orthonormalizing computation," *Advances in computers.* Academic Press, New York. **2,** 55–133.

Dawson, R. B. (1954) "A simplified expression for the variance of the χ^2-function on a contingency table," *Biometrika* **41,** 280.

Deemer, W. L. and Olkin, I. (1951) "The Jacobians of certain matrix transformations useful in multivariate analysis, based on lectures of P. L. Hsu at the University of North Carolina, 1947," *Biometrika* **38,** 345–367.

Deming, L. S. (1960) "Selected bibliography of statistical literature, 1930 to 1957. I. Correlation and regression theory," *J. Res. Nat. Bur. Standards* **64B,** 55–68.

Deming, W. E. (1938) "Some thoughts on curve fitting and the chi-square test," *J. Amer. Statist. Assoc.* **33,** 543–551.

Deming, W. E. and Stephan, F. F. (1940) "On a least squares adjustment of a sampled frequency table when the expected marginal totals are known," *Ann. Math. Statist.* **11,** 427–444.

De Movire, A. (1733) "Approximatio ad summam terminorum binomii $(a + b)^n$ in seriem expansi," being a supplement to *Miscellanea Analytica*, London.

De Morgan, A. (1847) "On the structure of the syllogism and on the application of the theory of probabilities to questions of argument and authority," *Trans. Camb. Philos. Soc.* **8,** 379–408.

Denny, J. L. (1967) "Sufficient conditions for a family of probabilities to be exponential," *Proc. Nat. Acad. Sci. U.S.A.* **57,** 1184–1187.

Diamond, E. L. (1963) "The limiting power of categorical data chi-square tests analogous to normal analysis of variance," *Ann. Math. Statist.* **34,** 1432–1441.

Diamond, E. L. and Lilienfeld, A. M. (1962*a*) "Effects of errors in classification and diagnosis in various types of epidemiological studies," *Amer. J. Publ. Hlth.* **52,** 1137–44.

Diamond, E. L. and Lilienfeld, A. M. (1962*b*) "Misclassification errors in 2 × 2 tables with one margin fixed: some further comments," *Amer. J. Publ. Hlth.* **52,** 2106–10.

Diamond, E. L., Mitra, S. K., and Roy, S. N. (1960) "Asymptotic power and asymptotic independence in the statistical analysis of categorical data," *Bull. Inst. Internat. Statist.* **37,** 309–329.

Dickson, J. D. Hamilton (1886) "Appendix," *Proc. Roy. Soc. London* **40,** 63–66. See Galton (1886).

DiDonato, A. R. and Jarnagin, M. P. (1962) "A method for computing the circular coverage function," *Math. Comp.* **16,** 347–355.

Dingman, H. F. (1954) "A computing chart for the point biserial correlation coefficient," *Psychometrika* **19,** 257–259.

Ditlevsen, O. (1964) "A simple chi-square test," *Nordisk Mat. Tidskr.* **12,** 157–167.

Dixon, W. J. (1940) "A criterion for testing the hypothesis that two samples are from the same population," *Ann. Math. Statist.* **11,** 199–204.

Dodd, E. L. (1942) "Certain tests for randomness applied to data grouped into small sets," *Econometrica* **10,** 249–257.

Doodson, A. T. (1917) "Relation of the mode, median and mean in frequency curves," *Biometrika* **11**, 425–429.

Doolittle, M. H. (1888) "Association ratios," *Bull. Philos. Soc. Wash.* **10**, 83–87 and 94–96.

Dubois, P. H. (1939) "Note on the calculation of the chi-square test for goodness of fit," *Psychometrika* **4**, 173–174.

Dubois, P. H. (1942) "A note on the computation of biserial *r* in item validation," *Psychometrika* **7**, 143–146.

Duncan, A. J. (1955) "Chi-square tests of independence and the comparison of percentages," *Industr. Qual. Contr.* **11**, (9), 9–14.

Dunlap, J. W. (1936) "Note on computation of biserial correlations in item evaluation," *Psychometrika* **1**, No. 2, 51–60.

Dunlap, J. W. (1940) "Note on the computation of tetrachoric correlation," *Psychometrika* **5**, 137–140.

Dyke, G. V. and Patterson, H. D. (1952) "Analysis of factorial arrangements when the data are proportions," *Biometrics* **8**, 1–12.

Eagleson, G. K. (1964) "Polynomial expansions of bivariate distributions," *Ann. Math. Statist.* **35**, 1208–1215.

Eagleson, G. K. and Lancaster, H. O. (1967) "The regression system of sums with random elements in common," *Austral. J. Statist.* **9**, 119–125.

Edgeworth, F. Y. (1892) "The law of error and correlated averages," *Phil. Mag.* (5) **34**, 429–438 and 518–526.

Edgeworth, F. Y. (1896) "The compound law of error," *Phil. Mag.* (5) **41**, 207–215.

Edgeworth, F. Y. (1905) "The law of error," *Trans. Camb. Philos. Soc.* **20**, 36–65 and 113–141.

Edgeworth, F. Y. (1916) "On the mathematical representation of statistical data," *J. Roy. Statist. Soc.* **79**, 455–500. **80**, (1917) 65–83, 266–288 and 411–437.

Editorial (1917) "The probable error of a Mendelian class frequency," *Biometrika* **11**, 429–432.

Editorial (1930) "Tables for determining the volumes of a bi-variate normal surface. (i) Introduction to the tables," *Biometrika* **22**, 1–11.

Editorial (1955) "The normal probability function: tables of certain area—ordinate ratios and of their reciprocals," *Biometrika* **42**, 217–222.

Edwards, A. L. (1948) "Note on the 'Correction for continuity' in testing the significance of the difference between correlated proportions," *Psychometrika* **13**, 185–187.

Edwards, A. L. (1950) "On the use and misuse of the chi-square test—the case of the 2 × 2 contingency table," *Psych. Bull.* **47**, 341–346.

Edwards, A. W. F. (1960) "The meaning of binomial distribution," *Nature* **186**, 1074.

Edwards, A. W. F. (1963) "The measure of association in a 2 × 2 table," *J. Roy. Statist. Soc.* **A126**, 109–114.

Edwards, J. H. (1957) "A note on the practical interpretation of 2 × 2 tables," *British J. Prev. Soc. Med.* **11**, 73–78.

Efron, B. (1967) "The power of the likelihood ratio tests," *Ann. Math. Statist.* **38**, 802–806.

Egg, K., Rust, H. and van der Waerden, B. L. (1965) "Die Irrtumswahrscheinlichkeit des χ^2-testes im Grenzfall der Poissonverteilung," *Z. Wahrscheinlichkeitstheorie* **4**, 260–264.

Eggenberger, J. (1893) "Beiträge zur Darstellung des bernoullischen Theorems der Gamma Function und des laplaceschen Integrals," *K. J. Wyss, Diss. Bern* 75 pp. 2nd Edition (1906), Fischer, Jena 75 pp.

Eidel'nant, M. I. (1958) "Approximate formulas for hypergeometric distribution," (Russ.) *Izv. Akad. Nauk UzSSR. Ser. Fiz-Mat.* **5**, 79–92.

Eisenhart, C. (1938) "The power function of the χ^2 test," *Bull. Ann. Math. Soc.* **44**, 32.

294   BIBLIOGRAPHY

Eisenhart, C. (Editor). (1949) *Tables of the binomial probability distribution.* U.S. Govt. Printing Office Washington. x + 387.

Eisenhart, C., Haystay, M. W. and Wallis, W. A. (Editors). (1947) *Selected techniques of statistical analysis.* McGraw-Hill. New York. 274–275.

Eisenhart, C. and Wilson, P. W. (1943) "Statistical methods and control in bacteriology," *Bact. Rev.* **7,** 57–137.

Elashoff, R. M. and Afifi, A. (1966) "Missing values in multivariate statistics—I review of the literature," *J. Amer. Statist. Assoc.* **61,** 595–604.

El Badry, M. A. and Stephan, F. F. (1955) "On adjusting sample tabulations to census counts," *J. Amer. Statist. Assoc.* **50,** 738–762.

Elderton, W. P. (1902) "Tables for testing the goodness of fit of theory to observation," *Biometrika* **1,** 155–163.

Elderton, W. P. (1907) *Frequency curves and correlation.* xiii + 172. 2nd edition (1927) vii + 239. Charles and Edward Layton, London.

Elderton, W. P. and Hansmann, G. H. (1934) "Improvement of curves fitted by the method of moments," *J. Roy. Statist. Soc.* **97,** 331–333.

Ellis, R. L. (1844a) "On a question in the theory of probabilities," (Reprinted in 'Mathematical and other writings,' 1863). *Camb. Math. Journ.* **4,** 127–132.

Ellis, R. L. (1844b) "On the method of least squares," *Trans. Camb. Philos. Soc.* **8,** 204–219.

el Shanawany, M. R. (1936) "An illustration of the accuracy of the χ^2 approximation," *Biometrika* **28,** 179–187.

Esary, J. D., Proschan, F. and Walkup, D. W. (1967) "Association of random variables, with applications," *Ann. Math. Statist.* **38,** 1466–1474.

Eudey, M. W. (1949) *Technical report* No. 13 Statistical Laboratory, University of California.

Everitt, P. F. (1910) "Tables of the tetrachoric functions for fourfold correlation tables," *Biometrika* **7,** 437–451.

Everitt, P. F. (1912) "Supplementary tables for finding the correlation coefficient from tetrachoric groupings," *Biometrika* **8,** 385–395.

Federighi, E. (1950) "The use of chi-square in small samples, methodological note on overestimation of relationships in 2 × 2 chi-square tables, where N is fewer than or equal to 40 and one or more of the expected cell frequencies is fewer than 10," *Amer. Sociolog. Rev.* **15,** 777–779.

Feldt, L. S. (1961) "The use of extreme groups to test for the presence of a relationship," *Psychometrika* **26,** 307–316.

Feller, W. (1945) "On the normal approximation to the binomial distribution," *Ann. Math. Statist.* **16,** 319–329. Corrig. **21** (1950) 301–302.

Feller, W. (1966) *Introduction to probability and statistics.* J. Wiley and Sons, New York. **2,** xix + 626.

Feller, W. (1967) *An introduction to probability theory and its applications.* Vol. 1 3rd edition. John Wiley, New York. xviii + 509.

Feller, W. K. (1940) "Statistical aspects of E.S.P.," *J. Parapsychol.* **4,** 271–298.

Fernandez Baños, O. (1946) "Contribution to the study of Pearson's χ^2," *Revista. Mat. Hisp.-Amer.* **(4)6,** 66–83.

Fertig, J. W. (1936) "On a method of testing the hypothesis that an observed sample of n variables and of size N has been drawn from a specified population of the same number of variables," *Ann. Math. Statist.* **7,** 113–121.

Festinger, L. (1946) "The significance of difference between means without reference to the frequency distribution function," *Psychometrika* **11,** 97–105.

Feuell, A. J. and Rybicka, S. M. (1951) "Quality control chart based on the goodness of fit test," *Nature* **167,** 194–5.

Finney, D. J. (1948) "The Fisher-Yates test of significance in 2 × 2 contingency tables," *Biometrika*. **35**, 145–156.

Finney, D. J., Latscha, R., Bennett, B. M. Hsu, P. and Pearson, E. S. (1963) *Tables for testing significance in a 2 × 2 contingency table*. (Supplement by B. M. Bennett and C. Horst i + 28.) Camb. Univ. Press. 103 pp.

Fisher, R. A. (1915) "Frequency distribution of the values of the correlation coefficient in samples from an indefinitely large population," *Biometrika* **10**, 507–521.

Fisher, R. A.* (1922a) "On the interpretation of χ^2 from contingency tables and the calculation of *P*," *J. Roy. Statist. Soc.* **85**, 87–94.

Fisher, R. A.* (1922b) "The goodness of fit of regression formulae and the distribution of regression coefficients," *J. Roy. Statist. Soc.* **85**, 597–612.

Fisher, R. A.* (1922c) "On the mathematical foundations of theoretical statistics," *Philos. Trans. Roy. Soc.* **A122**, 309–368.

Fisher, R. A.* (1923) "Statistical tests of agreement between observations and hypothesis," *Economica* **3**, 139–147.

Fisher, R. A.* (1924) "The conditions under which χ^2 measures the discrepancy between observation and hypothesis," *J. Roy. Statist. Soc.* **87**, 442–450.

Fisher, R. A.* (1925a) "Theory of statistical estimation," *Proc. Camb. Philos. Soc.* **22**, 700–725.

Fisher, R. A. (1925b) *Statistical methods and research workers*. 13th edition (1958) xv + 358. Oliver and Boyd Limited. Edinburgh.

Fisher, R. A. (1926) "Bayes' theorem and the fourfold table," *Eugenics Review* **18**, 32–33.

Fisher, R. A.* (1928a) "On a property connecting the χ^2 measure of discrepancy with the method of maximum likelihood," *Bologna Atti del Congresso Internazionale dei matematici* **6**, 3–10.

Fisher, R. A.* (1928b) "On a distribution yielding the error function of several well known statistics," *Proc. Internat. Math. Cong. Toronto* **2**, 805–813.

Fisher, R. A.* (1928c) "The general sampling distribution of the multiple correlation coefficient," *Proc. Roy. Soc. A.* **121**, 654–673.

Fisher, R. A. (1934) "Two new properties of mathematical likelihood," *Proc. Roy. Soc.* **A144**, 285–307.

Fisher, R. A.* (1935) "The mathematical distributions used in the common tests of significance," *Econometrica* **3**, 353–365.

Fisher, R. A.* (1936) "The use of multiple measurements in taxonomic problems," *Ann. Eugen. London* **7**, 179–188.

Fisher, R. A.* (1937) "Professor Karl Pearson and the method of moments," *Ann. Eugen. London* **7**, 303–318.

Fisher, R. A.* (1938a) "The statistical utilisation of multiple measurements," *Ann. Eugen. London* **8**, 376–386.

Fisher, R. A. (1938b) *Statistical theory of estimation, Calcutta University readership lectures.* Univ. of Calcutta, Calcutta, 45 pp.

Fisher, R. A.* (1939) "The sampling distribution of some statistics obtained from non-linear equations," *Ann. Eugen. London* **9**, 238–249.

Fisher, R. A.* (1940) "The precision of discriminant functions," *Ann. Eugen. London* **10**, 422–429.

Fisher, R. A. (1941a) "The interpretation of experimental fourfold tables," *Science* (N.S.) **94**, 210–211.

Fisher, R. A.* (1941b) "The negative binomial distribution," *Ann. Eugen. London* **11**, 182–187.

Fisher, R. A. (1945a) "The logical inversion of the notion of the random variable," *Sankhyā* **7**, 129–132.

Fisher, R. A. (1945*b*) "A new test for 2 × 2 tables," *Nature* **156**, 388.

Fisher, R. A. (1950*a*) *Contributions to mathematical statistics*. J. Wiley and Sons, Inc. New York.

Fisher, R. A. (1950*b*) "The significance of deviations from expectation in a Poisson series," *Biometrics* **6**, 17–24.

Fisher, R. A. (1962) "Confidence limits for a cross-product ration," *Austral. J. Statist.* **4**, 41. Correction **5**, 125–126.

Fisher, R. A. and Thornton, H. G. and MacKenzie, W. A.* (1922). "The accuracy of the plating method of estimating the density of bacterial populations," *Ann. Appl. Biology* **9**, 325–359.

Fisher, R. A. and Yates, F. (1938) *Statistical tables for biological, agricultural and medical research*. Oliver and Boyd, Edinburgh. 6th Edit. (1963). x + 146.

Fisher, W. D. (1953) "On a pooling problem from the statistical decision viewpoint," *Econometrica* **21**, 567–585.

Fiske, D. W. and Dunlop, J. W. (1945) "A graphical test for the significance of differences between frequencies from different samples," *Psychometrika* **10**, 225–229.

Fix, E. (1949) "Tables of noncentral χ^2," *Univ. California Publ. Statist.* **1**(2), 15–19.

Fix, E. Hodges, J. L. and Lehmann, E. L. (1959) "The restricted chi-square test," 92–102, in Grenander, U. (editor) (1959). *Probability and statistics: The Harold Cramer volume*. 434. Almquist and Wiksell, Stockholm.

Fog, D. (1948) "The geometrical method in the theory of sampling," *Biometrika* **35**, 46–54.

Fog, D. (1953) "Contingency tables and approximate χ^2 distributions," *Math. Scand.* **1**, 93–103.

Fraser, D. A. S. (1950) "Note on the χ^2 smooth test," *Biometrika* **37**, 447–8.

Fraser, D. A. S. (1963) "On the sufficiency and likelihood principles," *J. Amer. Statist. Assoc.* **58**, 641–647.

Freeman, G. H. and Halton, J. H. (1951) "Note on the exact treatment of contingency, goodness of fit and other problems of significance," *Biometrika* **38**, 141–149.

Friede, G. and Münzner, H. (1948) "Zur Maximalkorrelation," *Zeit. Angew. Math. Mech.* **28**, 158–160.

Friedlander, D. (1961) "A technique for estimating a contingency table, given the marginal totals and some supplementary data," *J. Roy. Statist. Soc.* **A124**, 412–420.

Fry, T. C. (1938) "The χ^2 test of significance," *J. Amer. Statist. Assoc.* **33**, 513–525.

Fulcher, J. S. and Zubin, J. (1942) "The item analyser; a mechanical device for treating the fourfold table in large samples," *J. Appl. Psychol.* **26**, 511–522.

Gabriel, K. R. (1962) "Ante-dependence analysis of an ordered set of variables," *Ann. Math. Statist.* **33**, 201–212.

Gabriel, K. R. (1963) "Analysis of variance of proportions with unequal frequencies," *J. Amer. Statist. Assoc.* **58**, 1133–1157.

Gabriel, K. R. (1966) "Simultaneous test procedures for multiple comparisons on categorical data," *J. Amer. Statist. Assoc.* **61**, 1081–1096.

Galton, F. (1886) "Family likeness in stature," *Proc. Roy. Soc. London* **40**, 42–63.

Garner, W. R. and McGill, W. J. (1956) "The relation between information and variance analyses," *Psychometrika* **21**, 219–228.

Garside, R. F. (1958) "Tables for ascertaining whether differences between percentages are statistically significant," *British Med. J.* **1**, 1459–1461.

Garside, R. F. (1961) "Tables for ascertaining whether differences between percentages are statistically significant at the 1% level," *British. Med. J.* **1**, 874–876.

Gart, J. J. (1966) "Alternative analyses of contingency tables," *J. Roy. Statist. Soc.* **B28**, 164–179.

Garwood, F. (1936) "Fiducial limits for the Poisson distribution," *Biometrika* **28**, 437–442.

Gates, C. E. and Beard, B. H. (1961) "Rapid chi-square test of significance for three-part ratios," *Biometrics* **17**, 261–266.

Geary, R. C. (1927) "Some properties of correlation and regression in a limited universe," *Metron* **7** (1), 83–119.

Geary, R. C. (1940) "The mathematical expectation of the mean square contingency when the attributes are mutually independent," *J. Roy. Statist. Soc.* **103**, 90–91.

Geary, R. C. (1942) "Inherent relations between random variables," *Proc. Roy. Irish, Acad. Sect. A.* **47**, 63–76.

Geary, R. C. (1947) "Testing for normality," *Biometrika* **34**, 209–242.

Geary, R. C. (1956) "Tests de la normalité," *Ann. Instit. Henri. Poincaré* **15**, 35–65.

Geary, R. C. and Pearson, E. S. (1938) *Tests of normality.* Biometrika office, London. 15.

Gebelein, H. (1941) "Das statistische Problem der Korrelation als Variations—und Eigenwertproblem und sein Zusammenhang mit der Ausgleichsrechnung," *Z. Angew. Math. Mech.* **21**, 364–379.

Gebelein, H. (1942a) "Bermerkung über ein von W. Hoeffding vorgeschlagenes massstabinvariantes Korrelationsmasz," *Z. Angew. Math. Mech.* **22**, 171–173.

Gebelein, H. (1942b) "Verfahren zur Beurteilung einer sehr geringer Korrelation zwischen zwei statistischen Merkmalsreihen," *Z. Angew. Math. Mech.* **22**, 286–298 and 553–592.

Gebelein, H. (1952) "Maximalkorrelation und Korrelationsspektrum," *Z. Angew. Math. Mech.* **32**, 9–19.

Geiringer, H. (1933) "Korrelationsmessung auf Grund der Summenfunktion," *Zeit. Angew. Math. Mech.* **13**, 121–124.

Geiringer, H. (1942) "A new explanation of non-normal dispersion in the Lexis theory," *Econometrica* **10**, 53–60.

Geppert, M. P. (1939) "Su una classe di distribuzions in due variabili casuali," *Giorn. Ist. Ital. Attuari.* **10**, 225–228.

Geppert, M. P. (1944) "Über den Vergleich zweier beobachteter Häufigkeiten," *Deutsche Math.* **7**, 553–592.

Geppert, M. P. (1961) "Erwartungstreue plausibelste Schätzer aus dreieckig gestutzten Kontingenztafeln," *Biom. Zeit.* **3**, 54–67.

Gilbert, N. E. G. (1956) "Likelihood function for capture-recapture samples," *Biometrika* **43**, 488–489.

Gildemeister, M. and van der Waerden, B. L. (1944) "Die Zulässigkeit des χ^2-Kriteriums für kleine Versuchszahlen," *Ber. Verh. Sächs. Akad. Wiss. Leipzig. Math.-Nat. Kl.* **95**, 145–150.

Gilliland, D. C. (1962) "Integral of the bivariate normal distribution over an offset circle," *J. Amer. Statist. Assoc.* **57**, 758–768.

Gilliland, D. C. (1964) "A note on the maximization of a non-central chi-square probability," *Ann. Math. Statist.* **35**, 441–442.

Gini, C. (1951) "Combinations and sequences of sexes in human families and mammal litters," *Act. Gen. Stat. Med.* **2**, 220–244.

Girshick, M. A. (1939) "On the sampling theory of the roots of determinantal equations," *Ann. Math. Statist.* **10**, 203–224.

Gjeddeback, N. F. (1949) "Contributions to the study of grouped observations. Application of the method of maximum likelihood in case of normally distributed observation," *Skand. Aktuarietidskr.* **32**, 135–159.

Gjeddeback, N. F. (1956/7) "Contribution to the study of grouped observations. ii. Loss of information caused by the grouping of normally distributed observations. iii. The distribution of estimates of the mean," *Skand. Aktuarietidskr.* **39**, 154–9 and **40**, 20–5.

Gjeddeback, N. F. (1959) "Contributions to the study of grouped observations. iv. Some comments on simple estimates," *Biometrics* **15**, 433–439.

Gjeddeback, N. F. (1960) "Contribution to the study of grouped observations. v. Three-class grouping of normal observations," *Skand. Aktuarietidskr.* **42,** 194–207.

Gjeddeback, N. F. (1961) "Contribution to the study of grouped observations. vi." *Skand. Aktuarietidskr.* **44,** 55–73.

Glasser, G. J. (1962) "A distribution-free test of independence with a sample of paired observations," *J. Amer. Statist. Assoc.* **57,** 116–133.

Gnedenko, B. V. and Kolmogorov, A. N. (1954) *Limit distribution for sums of independent random variables* . . . translated from Russian by K. L. Chung, with appendix by J. L. Doob. vii + 264 Cambridge, Mass. Addison-Wesley, Publishing Co.

Goheen, H. W. and Davidoff, M. D. (1951) "A graphical method for the rapid calculation of biserial and point biserial correlation in test research," *Psychometrika* **16,** 239–242. corr. **19,** 163–4.

Gold, R. Z. (1962) "On comparing multinomial probabilities," School of Aerospace Medicine USAF. Aerospace medical division (A.F.S.C.) Brooks Air Force Base, Texas. 13 pp.

Gold, R. Z. (1963) "Tests auxiliary to χ^2 tests in a Markov chain," *Ann. Math. Statist.* **34,** 56–74.

Goldberg, H. and Levine, H. (1946) "Approximate formulas for the percentage points and normalization of t and χ^2," *Ann. Math. Statist.* **17,** 216–225.

Gonin, H. T. (1936) "The use of factorial moments in the treatment of the hypergeometric distribution and in tests for regression," *Philos. Mag.* **(7)21,** 215–226.

Gonin, H. T. (1944) "Curve fitting by means of the orthogonal polynomials in binomial statistical distributions," *Trans. Roy. Soc. South Africa* **30,** 207–215.

Gonin, H. T. (1961) "The use of orthogonal polynomials of the positive and negative binomial frequency functions in curve fitting by Aitken's method," *Biometrika* **48,** 115–123.

Gonin, H. T. (1966) "Poisson and binomial frequency surfaces," *Biometrika* **53,** 617–619.

Gonin, H. T. (1967) "A study of the matrix of fitting of a series of discrete frequency functions analogous to the type A series," *South African Statist. J.* **1,** 55–58.

Gontcharoff, W. (1943) "On the succession of events in a series of independent trials in the scheme of Bernouilli," (Russ.) *Doklady SSSR* **38,** 283–285.

Good, I. J. (1955) "On the weighted combination of significance tests," *J. Roy. Statist. Soc.* **B17,** 264–265.

Good, I. J. (1956) "On the estimation of small frequencies in contingency tables," *J. Roy. Statist. Soc.* **B18,** 113–124.

Good, I. J. (1958a) "The interaction algorithm and practical Fourier analysis," *J. Roy. Statist. Soc.* **B20,** 361–372.

Good, I. J. (1958b) "Significance tests in parallel and in series," *J. Amer. Statist. Assoc.* **53,** 799–813.

Good, I. J. (1960) "The interaction algorithm and practical Fourier analysis: an addendum," *J. Roy. Statist. Soc.* **B22,** 372–375.

Good, I. J. (1961) "The multivariate saddlepoint method and chi-square for the multinomial distribution," *Ann. Math. Statist.* **32,** 535–548.

Good, I. J. (1963) "Maximum entropy for hypothesis formulation especially for multidimension contingency tables," *Ann. Math. Statist.* **34,** 911–934.

Good, I. J. (1965) *The estimation of probabilities: An essay on modern Bayesian methods.* Res. Mon. *No. 3* xii + 109. The M.I.T. Press, Cambridge, Mass.

Good, I. J. (1967) "A Bayesian significance test for multinomial distribution (with discussion)," *J. Roy. Statist. Soc.* **B29,** 399–431.

Goodman, L. A. (1958) "Asymptotic distributions of psi-squared goodness of fit criteria for mth order Markov chains," *Ann. Math. Statist.* **29,** 1123–1133.

Goodman, L. A. (1963a) "On methods for comparing contingency tables," *J. Roy. Statist. Soc.* **A126**, 94–108.

Goodman, L. A. (1963b) "On Plackett's test for contingency table interactions," *J. Roy. Statist. Soc.* **B25**, 179–188.

Goodman, L. A. (1964a) "Interactions in multidimensional contingency tables," *Ann. Math. Statist.* **35**, 632–646.

Goodman, L. A. (1964b) "Simultaneous confidence intervals for contrasts among multinomial populations," *Ann. Math. Statist.* **35**, 716–725.

Goodman, L. A. (1964c) "Simultaneous confidence limits for cross-product ratios in contingency tables," *J. Roy. Statist. Soc.* **B26**, 86–102.

Goodman, L. A. (1964d) "The analysis of persistence in a chain of multiple events," *Biometrika* **51**, 405–411.

Goodman, L. A. (1964e) "Simple methods for analyzing three-factor interaction in contingency tables," *J. Amer. Statist. Assoc.* **59**, 319–352.

Goodman, L. A. and Kruskal, W. H. (1954) "Measures of association for cross classifications," *J. Amer. Statist. Assoc.* **49**, 732–764. Corrig. **52**, 578.

Goodman, L. A. and Kruskal, W. H. (1959) "Measures of association for cross classifications. ii. Further discussion and references," *J. Amer. Statist. Assoc.* **54**, 123–163.

Goodman, L. A. and Kruskal, W. H. (1963) "Measures of association for cross classifications iii. Approximate sampling theory," *J. Amer. Statist. Assoc.* **58**, 310–364.

Gordon, M. H., Loveland, E. H. and Cureton, E. E. (1952) "An extended table of chi-square for two degrees of freedom, for use in combining probabilities from independent samples," *Psychometrika* **17**, 311–316.

Gordon, R. D. (1939) "Estimating bacterial populations by the dilution method," *Biometrika* **31**, 167–180.

Govindarajulu, Z. (1965) "Normal approximations to the classical discrete distributions," in Patil (1965). 79–108.

Gower, J. C. (1961/2) "The handling of multiway tables on computers," *Comput. J.* **4**, 280–286.

Gray, J. and Tocher, J. F. (1900) "The physical characteristics of adults and school children in East Aberdeenshire," *J. Anthrop. Instit.* **30**, 104–124.

Gray, P. (1947) "The effect of the death rate in biological experiments on the validity of observations, and on the chi-square test of association," *Arch. Biochem.* **13**, 461–474.

Greenberg, R. A. and White, C. (1965) "The detection of a correlation between the sexes of adjacent sibs in human families," *J. Amer. Statist. Assoc.* **60**, 1035–1045.

Greenhood, E. R. (1940) *A detailed proof of the chi-square test of goodness of fit.* Harvard Univ. Press. xiii + 61.

Greenwood, J. A. and Hartley, H. O. (1962) *Guide to tables in mathematical statistics.* Princeton Univ. Press. Princeton N.J. lxii + 1014.

Greenwood, M. (1913) "On errors of random sampling in certain cases not suitable for the application of a 'normal' curve of frequency," *Biometrika* **9**, 69–90.

Greenwood, M. and Yule, G. V. (1915) "The statistics of anti-typhoid and anti-cholera inoculations and the interpretation of such statistics in general," *Proc. Roy. Soc. Medicine* **8**, 113–194.

Gregory, G. (1961) "Contingency tables with a dependent classification," *Austral. J. Statist.* **3**, 42–47.

Greiner, R. (1909) "Ueber das Fehlersystem der Kollektivmasslehre," *Z. Math. Phys.* **57**, 121–158, 225–260, and 337–373.

Greville, T. N. E. (1941) "The frequency distribution of a general matching problem," *Ann. Math. Statist.* **12**, 350–354.

Gridgeman, N. T. (1963) "Matching trials by double criteria," *Biometrics* **19**, 398–405.

Grizzle, J. E. (1967) "Continuity correction in the χ^2-test for 2×2 tables," *Amer. Statistician* **21(4)**, 28–32.

Grüneberg, H. and Haldane, J. B. S. (1937) "Tests of goodness of fit applied to records of Mendelian segregation in mice," *Biometrika* **29**, 144–153.

Guenther, W. C. (1964a) "A generalization of the integral of the circular coverage function," *Amer. Math. Monthly* **71**, 278–283.

Guenther, W. C. (1964b) "Another derivation of the non-central chi-square distribution," *J. Amer. Statist. Assoc.* **59**, 957–960.

Guenther, W. C. and Terragno, P. J. (1964) "A review of the literature on a class of coverage problems," *Ann. Math. Statist.* **35**, 232–260.

Guenther, W. C. and Thomas, P. O. (1965) "Some graphs useful for statistical inference," *J. Amer. Statist. Assoc.* **60**, 334–343.

Guest, P. G. (1956) "Grouping methods in the fitting of polynomials to unequally spaced observations," *Biometrika* **43**, 149–160.

Guilford, J. P. (1941) "Phi-coefficient and chi-square as indices of item validity," *Psychometrika* **6**, 11–19.

Guilford, J. P. and Lyons, T. C. (1942) "On determining the reliability and significance of a tetrachoric coefficient of correlation," *Psychometrika* **7**, 243–249.

Guilford, J. P. and Perry, N. C. (1951) "Estimation of other coefficients of correlation from the phi-coefficient," *Psychometrika* **16**, 335–346.

Guldberg, A. (1923a) "Zur Theorie der Korrelation," *Metron* **2**, 637–680.

Guldberg, A. (1923b) "Über normale stabile statistische Reihen," *Skand. Aktuarietidskr.* **6**, 90–96.

Guldberg, A. (1934) "On discontinuous frequency functions of two variables," *Skand. Aktuarietidskr.* **17**, 89–117.

Gumbel, E. J. (1942) "Simple tests for given hypotheses," *Biometrika* **32**, 317–333.

Gumbel, E. J. (1943) "On the reliability of the classical χ^2-test," *Ann. Math. Statist.* **14**, 253–263.

Gumbel, E. J. (1954) "Applications of the circular normal distribution," *J. Amer. Statist. Assoc.* **49**, 267–297.

Gumbel, E. J. (1958) *Statistics of Extremes*, Oxford University Press, London. xx + 375.

Gumbel, E. J. (1960) "Bivariate exponential Distributions," *J. Amer. Statist. Assoc.* **55**, 698–707.

Gumbel, E. J. (1961) "Bivariate logistic distributions," *J. Amer. Statist. Assoc.* **56**, 335–349.

Gumbel, E. J. (1962) "Multivariate extremal distributions," *Bull. Internat. Statist. Inst.* **39(2)**, 471–475.

Gumbel, E. J. and Goldstein, N. (1964) "Analysis of empirical bivariate extremal distributions," *J. Amer. Statist. Assoc.* **59**, 794–816.

Gupta S. S. (1963) "Bibliography of the multivariate normal integrals and related topics," *Ann. Math. Statist.* **34**, 829–838.

Gurian, J. M., Cornfield, J. and Mosimann, J. E. (1964) "Comparisons of power for some exact multinomial significance tests," *Psychometrika* **29**, 409–419.

Guttman, I. (1967) "The use of the concept of a future observation in goodness of fit problems," *J. Roy. Statist. Soc.* **29B**, 83–100.

Haar, A. (1910) "Zur Theorie der orthogonalen Funktionsysteme I.," *Math. Annalen* **69**, 331–371.

Hadwiger, H. (1943) "Gruppierung mit Nebenbedingungen," *Mitt. Verein. Schweiz. Versich.-Math.* **43**, 113–122.

Haight, F. A. (1967) *Handbook of the Poisson distribution.* John Wiley and Sons, Inc. New. York, xi + 168.

Hald, A. (1952) *Statistical tables and formulas*. J. Wiley and Sons, Inc., New York
Chapman and Hall, London. 97 pp.

Hald, A. and Sinbaek, S. (1950) "A table of percentage points of the χ^2 distribution,"
Skand. Aktuarietidskr. **33**, 168–175.

Haldane, J. B. S. (1937a) "The approximate normalization of a class of frequency functions," *Biometrika* **29**, 392–404 (Corr. **31**, 220).

Haldane, J. B. S. (1937b) "The exact value of the moments of the distribution of χ^2 used as a test of goodness of fit when expectations are small," *Biometrika* **29**, 133–143. (Corr. **31**, 220).

Haldane, J. B. S. (1937c) "The first six moments of χ^2 for the n-fold table with n degrees of freedom when some expectations are small," *Biometrika* **29**, 389–391. (Corr. **31**, 220).

Haldane, J. B. S. (1939a) "Note on the preceding analysis of Mendelian segregations," *Biometrika* **31**, 67–71.

Haldane, J. B. S. (1939b) "Corrections to formulae in papers on the moments of χ^2," *Biometrika* **31**, 220.

Haldane, J. B. S. (1939c) "The mean and variance of χ^2 when used as a test of homogeneity when samples are small," *Biometrika* **31**, 346–355.

Haldane, J. B. S. (1939d) "The cumulants and moments of the binomial distribution and the cumulants of χ^2 for a $(n \times 2)$-fold table," *Biometrika* **31**, 392–396.

Haldane, J. B. S. (1941) "The fitting of binomial distributions," *Ann. Eugen. London* **11**, 179–181.

Haldane, J. B. S. (1942a) "Moments of the distributions of power and products of normal variates," *Biometrika* **32**, 226–242.

Haldane, J. B. S. (1942b) "The mode and median of a nearly normal distribution with given cumulants," *Biometrika* **32**, 294–299.

Haldane, J. B. S. (1945a) "Moments of r and χ^2 for a fourfold table in the absence of association," *Biometrika* **33**, 231–233.

Haldane, J. B. (1945b) "The use of χ^2 as a test of homogeneity in a $(n \times 2)$-fold table when expectations are small," *Biometrika* **33**, 234–238.

Haldane, J. B. S. (1955a) "Substitutes for χ^2," *Biometrika* **42**, 265–266.

Haldane, J. B. S. (1955b) "A problem in the significance of small numbers," *Biometrika* **42**, 266–267.

Haldane, J. B. S. (1955c) "The rapid calculation of χ^2 as a test of homogeneity from a $2 \times n$ table," *Biometrika* **42**, 519–520.

Halikov, M. K. (1959) "An exact estimate for a multidimensional local theorem," (Russ.). *Izv. Akad. Nauk UzSSR Ser. Fiz.-Mat.* **2**, 3–6.

Halmos, P. R. (1954) *Measure Theory*. New York, London: Van Nostrand. xi + 304.

Halmos, P. R. and Savage, L. J. (1949) "Application of the Radon-Nikodym theorem to the theory of sufficient statistics," *Ann. Math. Statist.* **20**, 225–241.

Halphen, E. (1957) "L'analyse intrinséque des distributions de probabilité," *Publ. Inst. Statist. Paris Univ. Paris* **6(2)**, 79–159.

Hamdan, M. A. (1962) "The powers of certain smooth tests of goodness of fit," *Austral. J. Statist.* **4**, 25–40.

Hamdan, M. A. (1963) "The number and width of classes in the chi-square test," *J. Amer. Statist. Assoc.* **58**, 678–689.

Hamdan, M. A. (1964) "A smooth test of goodness of fit based on the Walsh functions," *Austral. J. Statist.* **6**, 130–136.

Hamdan, M. A. (1968a) "Optimum choice of classes for contingency tables," *J. Amer. Statist. Assoc.* **63**, 291–7.

Hamdan, M. A. (1968b) "On the structure of the tetrachoric series," *Biometrika* **55**, 261–262.

Hamdan, M. A. (1968c) "The variance of the components of chi-square in contingency tables," *Austral. J. Statist.* 10, 49–55.

Hannan, E. J. (1961) "The general theory of canonical correlation and its relation to functional analysis," *J. Austral. Math. Soc.* 2, 229–242.

Hannan, J. and Harkness, W. (1963) "Normal approximation to the distribution of two independent binomials, conditional on fixed sum," *Ann. Math. Statist.* 34, 1593–1595.

Hannan, J. F. and Tate, R. F. (1965) "Estimation of the parameters for a multivariate normal distribution when one variable is dichotomized," *Biometrika* 52, 664–668.

Hansmann, G. H. (1934) "On certain non-normal symmetrical frequency distributions," *Biometrika* 26, 129–195.

Hardy, G. H. (1933) "A theorem concerning Fourier transforms," *J. London Math. Soc.* 8, 227–231.

Harkness, W. L. and Katz, L. (1964) "Comparison of the power functions for the test of independence in 2 × 2 contingency tables," *Ann. Math. Statist.* 35, 1115–1127.

Harris, A. J. (1957) "A maximum-minimum problem related to statistical distributions in two dimensions," *Biometrika* 44, 384–398.

Harris, J. A. and Treloar, A. E. (1927) "On a limitation in the applicability of the contingency coefficient," *J. Amer. Statist. Assoc.* 22, 460–472.

Harris, J. A., Treloar, A. E. and Wilder, M. (1930) "On the theory of contingency II. Professor Pearson's note on our paper on contingency," *J. Amer. Statist. Assoc.* 25, 323–327.

Harris, J. A. and Tu, C. (1929) "A second category of limitation in the applicability of the contingency coefficient," *J. Amer. Statist. Assoc.* 24, 367–375.

Harris, J. A., Tu, C. and Wilder, M. (1930) "The biological significance of certain differences between the values of the correlation coefficient, correlation ratio and contingency coefficient," *Amer. J. Bot.* 17, 175–183.

Harris, W. A. Jr. and Helvig, T. N. (1965/66) "Marginal and conditional distributions of singular distributions," *Publ. Res. Inst. Math. Sci. Ser. A.* 1, 199–204.

Harter, H. L. (1964a) "A new table of percentage points of the chi-square distribution," *Biometrika* 51, 231–239.

Harter, H. L. (1964b) *New tables of the incomplete gamma function ratio and of percentage points of the chi-square and beta distributions.* U.S. Government Printing Office, Washington, D.C. xv + 245.

Hartley, H. O. and Fitch, E. R. (1951) "A chart for the incomplete beta-function and the cumulative binomial distribution," *Biometrika* 38, 423–426.

Hartley, H. O. and Pearson, E. S. (1950) "Tables of the χ^2 integral and of the cumulative Poisson distribution," *Biometrika* 37, 313–325.

Hayashi, Takakura, Makita, Saito, (1958, 1960) "One method of quantification of attitude II Analysis of political attitude," (Japanese, English summary.) *Proc. Inst. Statist. Math.* 6, 8. 1–39 and 45–46.

Hayes, S. P. (1943) "Tables of the standard error of tetrachoric correlation coefficient," *Psychometrika*, 8, 193–203.

Hayes, S. P. (1946) "Diagrams for computing tetrachoric correlation coefficients from percentage differences," *Psychometrika* 11, 163–172.

Haynam, G. E. and Leone, F. C. (1965) "Analysis of categorical data," *Biometrika* 52, 654–660.

Heck, D. L. (1960) "Charts of some upper percentage points of the distribution of the largest characteristic root," *Ann. Math. Statist.* 31, 625–642.

Hellinger, E. (1909) "Neue Begruendung der Theorie quadratischer Formen von unendichvielen Veraenderlichen," *J. Reine Angew. Math.* 136, 210–271.

Helmert, F. R. (1875a) "Über die Formeln für den Durchschnittsfehler," *Astron. Nachr.* **85**, 353–366.

Helmert, F. R. (1875b) "Ueber die Berechnung des wahrscheinlichen Fehlers aus einer endlichen Anzahl wahrer Beobachtungsfehler," *Zeit. f. Math. u. Phys.* **20**, 300–303.

Helmert, F. R. (1876a) "Über die Wahrscheinlichkeit der Potenzsummen und über einige damit in Zusammenhange stehende Fragen," *Zeit. f. Math. u. Phys.* **21**, 192–218.

Helmert, F. R. (1876b) "Die Genauigkeit der Formel von Peters zur Berechnung des wahrscheinlichen Beobachtungsfehlers directer Beobachtungen gleicher Genauigkeit," *Astron. Nachr.* **88**, 115–132.

Henderson, J. (1922) "On expansions in tetrachoric functions," *Biometrika* **14**, 157–185.

Hendricks, W. A. (1935) "A problem involving the Lexis theory of dispersion," *Ann. Math. Statist.* **6**, 76–82.

Heron, D. (1910) "The danger of certain formulae suggested as substitutes for the correlation coefficient," *Biometrika* **8**, 109–122.

Herzel, A. (1958) "Influenza del raggruppamento in classi sulla probabilità e sull'intensità di transvariazione," *Metron* **19**, (1–2), 199–242.

Hildebrandt, E. H. (1931) "Systems of polynomials connected with the Charlier expansions and the Pearson differential and difference equations," *Ann. Math. Statist.* **2**, 379–439.

Hirschfeld, H. O. (1935) "A connection between correlation and contingency," *Proc. Camb. Philos. Soc.* **31**, 520–524.

Hobson, E. W. (1919) "On Hellinger's integrals," *Proc. London. Math. Soc.* (2) **18**, 249–265.

Hodges, J. L. (1955) "On the noncentral beta-distribution," *Ann. Math. Statist.* **26**, 648–653.

Hoeffding, W. (1940) "Maszstabinvariante Korrelationstheorie," *Schr. Math. Inst. Angew Math. Univ. Berlin.* **5** (3), 181–233.

Hoeffding, W. (1941) "Maszstabinvariante Korrelationsmasse für diskontinuierliche Verteilungen," *Arch. Math. Wirtsch. und Sozialforschung* **7**(2), 49–70.

Hoeffding, W. (1942) "Stochastische Abhängigkeit und funktionaler Zusammenhang," *Skand. Aktuarieidskr.* **25**, 200–227.

Hoeffding, W. (Russian Gefding, V.) (1964) "On a theorem of V. M. Zolotarev," (Russian. English Summary) *Teor. Verojatnost. i. Primenen* **9**, 96–99.

Hoeffding, W. (1965) "Asymptotically optimal tests for multinomial distributions," *Ann. Math. Statist.* **36**, 369–400.

Hoeffding, W. and Robbins, H. (1948) "The central limit theorem for dependent random variables," *Duke Math. J.* **15**, 773–780.

Hoel, P. G. (1938) "On the chi-square test for small samples," *Ann. Math. Statist.* **9**, 158–165.

Hoel, P. G. (1945) "Testing the homogeneity of Poisson frequencies," *Ann. Math. Statist.* **16**, 362–368.

Hoel, P. G. (1947) "Discriminating between binomial populations," *Ann. Math. Statist.* **18**, 556–564.

Holland, P. W. (1967) "A variation on the minimum chi-square test," *J. Math. Psychol.* **4**, 377–413.

Horst, P. (1961) "Relations among *m* sets of measures," *Psychometrika* **26**, 129–149.

Hostinsky, B. (1937) "Sur les probabilités relatives aux variables aléatoires liées entre elles. Applications diverses," *Ann. Inst. H.-Poincaré* **7**, 69–116.

Hotelling, H. (1930) "The consistency and ultimate distribution of optimum statistics," *Trans. Amer. Math. Soc.*, **32**, 847–859.

Hotelling, H. (1933) "Analysis of a complex of statistical variables into principal components," *J. Educ. Psychol* **24**, 417–441, 498–520.

Hotelling, H. (1936) "Relations between two sets of variables," *Biometrika* **28**, 321–377.

Hotelling, H. (1951) "A generalized *T* test and measure of multivariate dispersion," *Proc. 2nd Berkeley Symp.* 23–41.

Hoyt, C. J., Krishnaiah, P. R. and Torrance, E. P. (1959) "Analysis of complex contingency data," *J. Exper. Educ.* **27**, 187–194.

Hsu, P. L. (1939) "On the distribution of roots of certain determinantal equations," *Ann. Eugen. London* **9**, 250–258.

Hsu, P. L. (1942) "The limiting distribution of a general class of statistics," *Science Record* (Academia Sinica) **1**, 37–41.

Hsu, P. L. (1943) "Some simple facts about the separation of degrees of freedom in factorial experiments," *Sankhyā* **6**, 253–254.

Hsu, P. L. (1949) "The limiting distribution of functions of sample means and application to testing hypotheses," *Proc. 1st. Berkeley Symp.* **1**, 359–402.

Hussain, Q. M. (1943) "A note on interaction," *Sankhyā* **6**, 321–322.

Huttley, N. H. (1951) "A survey of the historical development and application of the statistic χ^2," *Ph. D. Thesis, Univ. London.*

Ikeda, S. (1959) "A note on the normal approximation to the sum of independent random variables," *Ann. Inst. Statist. Math.—Tokyo* **11**, 121–130.

Ireland, C. T. and Kullback, S. (1968) "Contingency tables with given marginals," *Biometrika* **55**, 179–188.

Irwin, J. O. (1929) "Note on the χ^2-test of goodness of fit," *J. Roy. Statist. Soc.* **92**, 264–266.

Irwin, J. O. (1931) "Mathematical theorems involved in the analysis of variance," *J. Roy. Statist. Soc.* **94**, 284–300.

Irwin, J. O. (1934) "On the independence of the constituent items in the analysis of variance," *J. Roy. Statist. Soc. Suppl 1.* 236–251.

Irwin, J. O. (1935) "Tests of significance for differences between percentages based on small numbers," *Metron* **12** (2), 83–94.

Irwin, J. O. (1937) "The frequency-distribution of the difference between two independent variates, following the same Poisson distribution," *J. Roy. Statist. Soc.* **100**, 415–416.

Irwin, J. O. (1942) "On the distribution of a weighted estimate of variance and on analysis of variance in certain cases of unequal weighting." *J. Roy. Statist. Soc.* **105**, 115–118.

Irwin, J. O. (1949) "A note on the subdivision of χ^2 into components," *Biometrika* **36**, 130–134.

Ishii, G. (1960) "Intraclass contingency tables," *Ann. Inst. Statist. Math.* **12**, 161–207, Corrig. 279.

Isserlis, L. (1914) "The application of solid hypergeometric series to frequency distributions in space," *Phil. Mag.* (6) **28**, 379–403.

Isserlis, L. (1917) "On the representation of statistical data," *Biometrika* **11**, 418–425.

Isserlis, L. (1931) "On the moment distributions of moments in the case of samples drawn from a limited universe," *Proc. Roy. Soc.* **A132**, 586–605.

Jackson, J. E. and Bradley, R. A. (1961a) "Sequential χ^2 and T^2 tests and their application to an acceptance sampling problem," *Technometrics* **3**, 519–534.

Jackson, J. E. and Bradley, R. A. (1961b) "Sequential χ^2 and T^2 tests," *Ann. Math. Statist.* **32**, 1063–1077.

Jackson, R. W. (1936) "Tests of statistical hypotheses in the case when the set of alternatives is discontinuous, illustrated on some genetical problems," *Statist. Res. Mem.* **1**, 138–161.

Jambunathan, M. V. (1954) "Some properties of beta and gamma distributions," *Ann. Math. Statist.* **25**, 401–405.

James, A. T. (1964) "Distributions of matrix variates and latent roots derived from normal samples," *Ann. Math. Statist.* **35**, 475–501.

James, G. S. (1952) "Note on a theorem of Cochran," *Proc. Camb. Philos. Soc.* **48**, 443–446.

Jamnik, R. (1962) "Über vollständige Systeme von paarweise unabhängigen zufälligen Grössen," *Bull. Soc. Math. Physics* (Serbie) **14**, 161–164.

Jamnik, R. (1964) "Über vollständige orthonormierte Systeme von paarweise unabhängigen zufälligen Grössen," *Publ. Dept. Math.* (Ljubljana) **1**, 23–41.

Jarrett, F. F. (1946) "Normal approximations to the chi-square distribution," *Amer. Psychologist* **1**, 454.

Jeffreys, H. (1936) "Further significance tests," *Proc. Camb. Philos. Soc.* **32**, 416–445.

Jeffreys, H. (1937a) "Maximum likelihood, inverse probability and the method of moments," *Ann. Eugen.* **8**, 146–151.

Jeffreys, H. (1937b) "The tests for sampling differences and contingency," *Proc. Roy. Soc. London* **A162**, 479–495.

Jeffreys, H. (1937c) "The law of errors and the combination of observations," *Phil. Trans. A.* **237**, 231–271.

Jeffreys, H. (1938a) "The use of minimum χ^2 as an approximation to the method of maximum likelihood," *Proc. Camb. Philos. Soc.*, **34**, 156–57.

Jeffreys, H. (1938b) "Significance tests for continuous departures from suggested distributions of chance," *Proc. Roy. Soc. London* **A164**, 307–315.

Jeffreys, H. (1939a) "The minimum χ^2 approximation," *Proc. Camb. Philos. Soc.* **35**, 520.

Jeffreys, H. (1941) "Some applications of the method of minimum χ'^2," *Ann. Eugen.* **11**, 108–114.

Jevons, W. S. (1870) "On a general system of numerically definite reasoning," *Mem. Manchester Lit. Phil. Soc.* Reprinted in *Pure Logic and other Minor Works*, Macmillan, London.

Joffe, A. D. (1967) "Minimum chi-squared estimation using independent statistics," *Ann. Math. Statist.* **38**, 267–270.

Johnson, H. M. (1945) "Maximum selectivity, correctivity and correlation obtainable in a 2 × 2 contingency table," *Amer. J. Phychol.* **58**, 65–68.

Johnson, N. L. (1959) "On the extension of the connection between Poisson and χ^2 distributions," *Biometrika* **46**, 352–363.

Johnson, N. L. and Welch, B. L. (1939) "On the calculation of the cumulants of the χ^2 distribution," *Biometrika* **31**, 216–218.

Jordan, C. (1937) "On approximation and on test criteria by the χ^2 test and by Bayes' theorem," *J. Soc. Hongroise de Statist.* **15**, 101–128.

Jurgensen, C. E. (1947). "Table for determining phi coefficients," *Psychometrika* **12**, 17–29.

Kac, M. (1959) *Statistical independence in probability analysis and number theory.* Math. Assoc. America. (Carus. Math. Mon. No. 12) John Wiley and Sons, New York. 93 + xiv.

Kaczmarz, S. (1929) "Über ein Orthogonalysystem," *C.R. du .1ᵉCong. des Math. des Payes Slaves* 189–192.

Kale, B. K. (1966) "Approximations to the maximum-likelihood estimator using grouped data," *Biometrika* **53**, 282–285.

Kamat, A. R. (1962) "Some more estimates of circular probable error," *J. Amer. Statist. Assoc.* **57**, 191–195.

Karhunen K. (1947) "Über lineare Methoden in der Wahrscheinlichkeitsrechnung," *Ann. Acad. Sci. Fenn. A.I., Math.-Phys.* **37**, 3–79.

Karhunen, K. (1950) "Über die Struktur stätionarer zufälliger Funktioner," *Ark. Mat.* **1**, 141–160.

Karhunen, K. (1952) "Methodos linealies en el calculo de probabilidades," *Trab. Estadist.* **3**, 59–136.

Karon, B. P. and Alexander, I. E. (1958) "A modification of Kendall's tau for measuring association in contingency tables," *Psychometrika* **23**, 379–383.

306   BIBLIOGRAPHY

Kastenbaum, M. A. (1960) "A note on the additive partitioning of chi-square in contingency tables," *Biometrics* **16**, 416–422.

Kastenbaum, M. A. and Lamphiear, D. E. (1959) "Calculation of chi-square to test the no three-factor interaction hypothesis," *Biometrics* **15**, 107–115.

Kathirgamatamby, N. (1953) "Note on the Poisson index of dispersion," *Biometrika* **40**, 225–228.

Katti, S. K. and Sastry, A. N. (1965) "Biological examples of small expected frequencies and the chi-square test," *Biometrics* **21**, 49–54.

Kellerer, H. G. (1964) "Verteilungsfunktionen mit gegebenen Marginalverteilungen," *Z. Wahrscheinlichkeitstheorie* **3**, 247–270.

Kemp, C. D. and A. W. (1956) "Generalized hypergeometric distributions," *J. Roy. Statist. Soc.* **B18**, 202–211.

Kempthorne, O. (1967) "The classical problem of inference—goodness of fit." *Proc. 5th Berkeley Symp.* **1**, 235–249.

Kendall, M. G. (1941) "Proof of relations connected with the tetrachoric series and its generalization," *Biometrika* **32**, 196–198.

Kendall, M. G. and Stuart, A. (1961) *The advanced theory of statistics*, **2**, *Inference and relationship*. Griffin and Co., London. ix + 676.

Kendall, M. G. and Stuart, A. (1963) *The advanced theory of statistics* **1**, C. Griffin and Company. London. xii + 433.

Kermack, W. O. and MacKendrick, A. G. (1940) "The design and interpretation of experiments based on a fourfold table. The statistical assessment of the effect of treatment," *Proc. Roy. Soc. Edin.* **60**, 362–375.

Kerridge, D. (1965) "A probabilistic derivation of the non-central χ^2 distribution," *Austral. J. Statist.* **7**, 37–39. Corrig. 7–114.

Keynes, J. M. (1921) *A treatise on probability.* Macmillan and Company Limited, London. xi + 466.

Khamis, S. H. (1960) "Incomplete gamma functions expansions of statistical distribution function," (French summary). *Bull. Inst. International Statist.* **37** no. 3 385–396.

Khamis, S. H. (1963) "New tables of the chi-squared integral," *Bull. Inter. Statist. Inst.* **40**, 799–822.

Khatri, C. G. (1963) "Some more estimates of circular probable error," *J. Indian Statist. Assoc.* **1**, 40–47.

Khatri, C. G. and Pillai, K. G. S. (1965) "Some results on the non-central multivariate beta distribution and moments of traces of two matrices," *Ann. Math. Statist.* **36**, 1511–1520.

Kibble, W. F. (1941) "A two-variate gamma type distribution," *Sankhyā* **5**, 137–150.

Kibble, W. F. (1945) "An extension of a theorem of Mehler on Hermite polynomials," *Proc. Camb. Philos. Soc.* **41**, 12–15.

Kimball, A. W. (1954) "Short-cut formulas for the exact partition of χ^2 in contingency tables," *Biometrics* **10**, 452–8.

Kincaid, W. M. (1962a) "The combination of 2 × m contingency tables," *Biometrics* **18**, 224–228.

Kincaid, W. M. (1962b) "The combination of tests based on discrete distributions," *J. Amer. Statist. Assoc.* **57**, 10–19.

Kingman, J. F. C. and Taylor, S. J. (1966) *Introduction to measure and probability.* Camb. Univ. Press Cambridge. x + 401.

Koessler, M. (1924) "La formule sommatoire pour la serie, $S = \dfrac{h}{\sqrt{\pi}} \sum\limits_{k=-n}^{n} e^{-h^2 k^2}$," *Casopis pro Pestovani mat. a Fysiky*, Prague **53**, 110–114.

Kondo, T. (1929) "On the standard error of the mean square contingency," *Biometrika* 21, 376–428.

Kondo, T. and Elderton, E. M. (1931) "Table of functions of the normal curve to ten decimal places," *Biometrika* 22, 368–376.

Konijn, H. S. (1956) "On the power of certain tests for independence in bivariate populations," *Ann. Math. Statist.* 27, 300–323, (Corrig. 435–6.).

Kono, K. (1952) "On inefficient statistics for measurement of dependency of normal bivariates," *Mem. Fac. Sci. Kysyu Univ.* A7, 1–12.

Koshal, R. S. (1933) "Application of the method of maximum likelihood to the improvement of curves fitted by the method of moments," *J. Roy. Statist. Soc.* 96, 303–313.

Koshal, R. S. (1935) "Application of the method of maximum likelihood to the derivation of efficient statistics for fitting frequency curves," *J. Roy. Statist. Soc.* 98, 128.

Koshal, R. S. (1939) "Maximal likelihood and minimal χ^2 in relation to frequency curves," *Ann. Eugen. London* 9, 209–31.

Koshal, R. S. and Turner, A. J. (1930) "Studies in the sampling of cotton for the determination of fibre properties," *J. Text. Inst.* 21, T325–370.

Kotz, S., Johnson, N. L. and Boyd, D. W. (1967a) "Series representations of distributions of quadratic forms in normal variables. I. Central case," *Ann. Math. Statist.* 38, 823–837.

Kotz, S., Johnson, N. L. and Boyd, D. W. (1967b) "Series representations of distributions of quadratic forms in normal variables. II. Non central case," *Ann. Math. Statist.* 38, 838–848.

Krawtchouk, W. (1929) "Sur une généralisation des polynomes d'Hermite," *Compt. Rend. Acad. Sci. Paris* 189, 620–622.

Krishnaiah, P. R. (Editor). (1966) *Multivariate analysis: Proceedings of an international symposium held in Dayton, Ohio, June, 14–19, 1965.* Academic Press, New York. xix + 592.

Krishnaiah, P. R. and Armitage, J. V. (1965) "Tables for the distribution of the maximum of correlated chi-square variates with one degree of freedom," (Spanish summary). *Trabajos Estadist.* 16, 91–96.

Krishnaiah, P. R. and Rao, M. M. (1961) "Remarks on a multivariate gamma distribution," *Amer. Math. Monthly* 68, 342–346.

Krishnaiah, P. R., Hagis, P. Jr. and Steinberg, L. (1963) "A note on the bivariate chi distribution," *SIAM Rev.* 5, 140–144.

Krishnamoorthy, A. S. (1951) "Multivariate binomial and Poisson distributions," *Sankhyā* 11, 117–124.

Krishnamoorthy, A. S. and Parthasarathy, M (1951) "A multivariate gamma-type distribution," *Ann. Math. Statist.* 22, 549–557.

Krishnaswami Ayyangar, A. A. (1934a) "Note on the recurrence formulae for the moments of the point binomial," *Biometrika* 26, 262–264.

Krishnaswami Ayyangar, A. A. (1934b) "Note on the incomplete moments of the hypergeometric distribution," *Biometrika* 26, 264–265.

Kruskal, W. H. (1958) "Ordinal measures of association," *J. Amer. Statist. Assoc.* 53, 814–861.

Ku, H. H. (1963) "A note on contingency tables involving zero frequencies and the 2*I* test," *Technometrics* 5, 398–400.

Kuder, G. F. (1937) "Nomograph for point biserial *r*, biserial *r* and fourfold correlations," *Psychometrika* 2, 135–138.

Kudô, A. and Fujisawa, H. (1966) "Some multivariate tests with restricted alternative hypotheses," pp. 73–85 of (*Krishnaiah, P. R. Editor, 1966*) *Multivariate analysis* J. Wiley and Sons Inc., New York.

Kuiper, N. H. (1952) "Variantie-analyse," *Statistica Neerlandica* 6, 149–194.

Kullback, S. (1959) *Information theory and statistics*. J. Wiley and Sons, Inc. New York, xvii + 395.

Kullback, S., Kupperman, M. and Ku, H. H. (1962a) "Tests for contingency tables and Markov chains," *Technometrics* **4**, 573–608.

Kullback, S., Kupperman, M. and Ku, H. H. (1962b) "An application of information theory to the analysis of contingency tables, with a table of $2n \ln n$, $n = 1$ (1) 10,000," *J. Res. Nat. Bur. Standards. Sect. B.* **66B**, 217–243.

Kulldorff, G. (1958) "Maximum likelihood estimation of the standard deviation of a normal random variable when the sample is grouped," *Skand. Aktuarietidskr.* **41**, 18–36.

Kulldorff, G. (1959) "A problem of maximum likelihood estimation from a grouped sample," *Metrika* **2**, 94–99.

Kulldorff, G. (1961) *Contributions to the theory of estimation from grouped and partially grouped samples*. Almqvist-Wicksell, Stockholm, 144 pp.

Kupperman, M. (1960) "On comparing two observed frequency counts," *Appl. Statist.* **9**, 37–42.

Lancaster, H. O. (1949a) "The derivation and partition of χ^2 in certain discrete distributions," *Biometrika* **36**, 117–129.

Lancaster, H. O. (1949b) "The combination of probabilities arising from data in discrete distributions," *Biometrika* **36**, 370–382. Corrig. **37**, 452.

Lancaster, H. O. (1950a) "The exact partition of χ^2 and its application to the problem of the pooling of small expectations," *Biometrika* **37**, 267–270.

Lancaster, H. O. (1950b) "Statistical control in haematology," *J. Hyg. Camb.* **48**, 402–417.

Lancaster, H. O. (1950c) "The theory of amoebic surveys," *J. Hyg. Camb.* **48**, 257–276.

Lancaster, H. O. (1950d) "The sex ratios in sibships with special reference to Geissler's data," *Ann. Eugen. London* **15**, 153–158.

Lancaster, H. O. (1951) "Complex contingency tables treated by the partition of χ^2," *J. Roy. Statist. Soc.* **B13**, 242–249.

Lancaster, H. O. (1952) "Statistical control of counting experiments," *Biometrika* **39**, 419–422.

Lancaster, H. O. (1953a) "A reconciliation of χ^2 considered from metrical and enumerative aspects," *Sankhyā* **13**, 1–10.

Lancaster, H. O. (1953b) "Accuracy of blood cell counting," *Austral. J. Exp. Biol. Med. Sc.* **31**, 603–606.

Lancaster, H. O. (1954) "Traces and cumulants of quadratic forms in normal variables," *J. Roy. Statist. Soc. Ser. B.* **16**, 247–254.

Lancaster, H. O. (1957) "Some properties of the bivariate normal distribution considered in the form of a contingency table," *Biometrika* **44**, 289–292.

Lancaster, H. O. (1958) "The structure of bivariate distributions," *Ann. Math. Statist.* **29**, 719–736. Corrig. **35**, (1964) 1388.

Lancaster, H. O. (1959) "Zero correlation and independence," *Austral. J. Statist.* **1**, 53–56.

Lancaster, H. O. (1960a) "On tests of independence in several dimensions," *J. Austral. Math. Soc.* **1**, 241–254. Corrig. **1**, 496.

Lancaster, H. O. (1960b) "The characterisation of the normal distribution," *J. Austral. Math. Soc.* **1**, 368–383.

Lancaster, H. O. (1960c) "On statistical independence and zero correlation in several dimensions," *J. Austral. Math. Soc.* **1**, 492–496.

Lancaster, H. O. (1961a) "Significance tests in discrete distributions," *J. Amer. Statist. Assoc.* **56**, 223–234.

Lancaster, H. O. (1961b) "The combination of probabilities: an application of orthonormal functions," *Austral. J. Statist.* **3**, 20–33.

Lancaster, H. O. (1963a) "Canonical correlations and partitions of χ^2," *Quart. J. Maths.* **14**, 220–224.

Lancaster, H. O. (1963b) "Correlations and canonical forms of bivariate distributions," *Ann. Math. Statist.* **34**, 532–538.

Lancaster, H. O. (1963c) "Correlation and complete dependence of random variables," *Ann. Math. Statist.* **34**, 1315–1321.

Lancaster, H. O. (1965a) "Symmetry in multivariate distributions," *Austral. J. Statist.* **7**, 115–126.

Lancaster, H. O. (1965b) "The Helmert matrices," *Amer. Math. Monthly* **72**, 4–11.

Lancaster, H. O. (1966a) "Forerunners of the Pearson χ^2," *Austral. J. Statist.* **8**, 117–126.

Lancaster, H. O. (1966b) "Kolmogorov's remark on the Hotelling canonical correlations," *Biometrika* **53**, 585–588.

Lancaster, H. O. (1967) "The combination of probabilities," *Biometrics* **23**, 840–842.

Lancaster, H. O. (1968) "The median of χ^2," *Austral. J. Statist.* **10**, 83.

Lancaster, H. O. and Brown I. A. I. (1965) "Sizes of the χ^2 test in the symmetrical multinomials," *Austral. J. Statist.* **7**, 40–44.

Lancaster, H. O. and Hamdan, M. A. (1964) "Estimation of the correlation coefficient in contingency tables with possibly non-metrical characters," *Psychometrika* **29**, 383–391.

Laplace, P. S. de (1878/1912) *Oeuvres complètes de Laplace publiées sous les auspices de l'Académie des Sciences par M.M. les secrétaires perpétuels.* 14 volumes, (especially volume 7.) Gauthier-Villar Paris.

Larsen, H. D. (1939) "Moments about the arithmetic mean of a hypergeometric frequency distribution," *Ann. Math. Statist.* **10**, 198–201.

Latscha, R. (1953) "Test of significance in a 2 × 2 contingency table: Extension of Finney's table," *Biometrika* **40**, 74–86.

Lauricella, G. (1912). "Sulla chiusura dei sistemi di funzioni ortogonali," *Rend. R. Accad. naz. Lincei.* **(5), 21.** 675–685.

Lawley, D. N. (1956) "A general method for approximating to the distribution of likelihood ratio criteria," *Biometrika* **43**, 295–303.

Lawley, D. N. (1959) "Tests of significance in canonical analysis," *Biometrika* **46**, 59–66.

Lazarsfeld, P. F. (1961) "The algebra of dichotomous systems," (In Solomon 1961, 111–152).

Lee, A. (1917) "Further supplemementary tables for determining high correlations from tetrachoric grouping," *Biometrika* **11**, 284–291.

Lee, A. (1925) "Table of the first twenty tetrachoric functions to seven decimal places," *Biometrika* **17**, 343–354.

Lee, A. (1927) "Supplementary tables for determining correlation from tetrachoric groupings," *Biometrika* **19**, 354–404.

Lehmann, E. L. (1947) "On optimum tests of composite hypotheses with one constraint," *Ann. Math. Statist.* **18**, 473–494.

Lehmann, E. L. (1959) *Testing statistical hypotheses.* J. Wiley and Sons Inc., New York. xiii + 369.

Lehmann, E. L. (1966) "Some concepts of dependence," *Ann. Math. Statist.* **37**, 1137–1153

Leipnik, R. (1961) "When does zero correlation imply independence?" *Amer. Math. Monthly* **68**, 563–564.

Lengyel, B. A. (1939) "On testing the hypothesis that two samples have been drawn from a common normal population," *Ann. Math. Statist.* **10**, 365–375.

Leslie, P. H. (1951) "The calculation of χ^2 for an $r \times c$ contingency table," *Biometrics* **7**, 283–286.

Leslie, P. H. (1955) "A simple method of calculating the exact probability in 2 × 2 contingency tables with small marginal totals," *Biometrika* **42**, 522–523.

310   BIBLIOGRAPHY

Lesser, P. C. V. (1933) "Note on the shrinkage of physical characters in man and woman with age as an illustration of the use of χ^2, P methods," *Biometrika*, **25**, 197–202.

Lev, J. (1949). "The point biserial coefficient of correlation," *Ann. Math. Statist.* **20**, 125–126.

Lewis, B. N. (1962) "On the analysis of interaction in multi-dimensional contingency tables," *J. Roy. Statist. Soc.* **A125**, 88–117.

Lewis, D. and Burke, C. J. (1949) "The use and misuse of the chi-square test," *Psychol Bull.* **46**, 433–489.

Lewis, D. and Burke, C. J. (1950) "Further discussion of the use and misuse of the chi-square test," *Psychol. Bull.* **47**, 347–355.

Lewis, T. (1953) "99·9 and 0·1% points of the χ^2 distribution," *Biometrika* **40**, 421–426.

Lexis, W. (1877) *Zur Theorie der Massenerscheinungen in der menschlichen Gesellschaft.* Freiberg i B, 95 pp.

Lexis, W. (1879) "Über die Theorie der Stabilität statistischer Reihen," *Conrad's Jahrbücher* **(1) 32**, 60–98. (Jahrb. f. nat. Ök. u. Statist.)

Lexis, W. (1903) *Abhandlungen zur Theorie der Bevölkerungs und Moralstatistik.* Fischer, Jena. 253 pp.

Li, S. (1963) "On the asymptotic power of a non-parametric test analogous to the χ^2 statistic," *Dopovidi Akad. Nauk Ukrain. RSR. 1963* 856–861.

Lieberman, G. J. and Owen, D. B. (1961) *Tables of the hypergeometric probability distribution.* vii + 726. Stanford University Press, Stanford, California.

Liebermeister, C. (1877) "Über Wahrscheinlichkeitrechnung in Anwendung auf therapeutische Statistik," *Sammlung Klinischer Vorträge*, edited R. Volkmann, No. 110. (*Innere Medicin* No. 39), 935–962.

Lienau, C. C. (1941) "Discrete bivariate distributions in certain problems of statistical order," *Amer. J. Hyg.* **33**, 65–85.

Lindley, D. V. (1964) "The Bayesian analysis of contingency tables," *Ann. Math. Statist.* **35**, 1622–1643.

Lindley, D. V. (1965) *Introduction to probability and statistics. 1.* xi + 259, *II.* xiii + 292. Cambridge Univ. Press.

Linfoot, E. H. (1957) "An informational measure of association," *Information and Control.* **1**, 85–89.

Linnik, Ju. V. (1961) "On certain connections of information theory of C. Shannon and R. Fisher with the theory of symmetrization of random vectors," (Russian). *Trans. 2nd Prague Conf. information theory. Publ. House Czechoslovak. Acad. Sci. Prague* Academic Press, New York. 313–327.

Lins Martins, O. (1946) "Sôbre o coeficiente de correlação disserial; estudo experimental de distribução de seus valores obtidos em pequenas amostras," (Portuguese) *Rev. Brasil. Estatist.* **7**, 713–62.

Lipmann, O. (1908) "Eine methode zum Vergleich zweier Kollektivgegenstände," *Z.f. Psychol.* **48**, 421–431.

Lipps, G. F. (1905) "Die Bestimmung der Abhängigkeit zwischen den Merkmalen eines Gegenstandes," *Ber. Verhand. König. Sächs Gesells. f. Wissenschaft zu Leipzig. Math. Phy. Klasse.* **57**, 1–32.

Liptak, T. (1958) "On the combination of independent tests," *Magyar Tud. Akad. Mat. Kutato. Int. Közl.* **3**, 171–197.

Loève, M. (1955) *Probability theory foundations, random sequences.* Van Nostrand, New York. xvii + 515. 2nd Edit. 1960.

Lombard, H. L. and Doering, C. R. (1947) "Treatment of the fourfold table by partial associatiion and partial correlation as it relates to public health problems," *Biometrics* **3**, 123–28.

Lord, F. M. (1944) "Alignment chart for calculating the fourfold point correlation coefficient," *Psychometrika* 9, 41–42.

Lord, F. M. (1963) "Biserial estimates of correlation," *Psychometrika* 28, 81–85.

Lorenz, P. (1949) "Herleitung der Näherungsformel von Laplace für die Binomialverteilung, ohne Grenzübergang," *Z. Angew. Math. Mech.* 29, 368–374.

Lüders, R. (1934). "Die Statistik der seltenen Ereignisse," *Biometrika* 26, 108–128.

Lukacs, E. (1959) *Characteristic functions in statistics.* Statist. Mon. Courses. C. Griffin and Co. London. 216.

Lukacs, E. and Laha, R. G. (1963) *Applications of characteristic functions in statistics.* Griffin's Statist. Mon. and Courses. 202 pp.

Lukaszewicz, J. and Sadowski, W. (1960/61) "On comparing several populations with a control population," *Zastos. Mat.* 5, 309–320.

Lurquin, C. (1937) "Sur la loi de Bernouilli à deux variables," *Bull. Classe Sci. Acad. R. Belg.* (5) 23, 857–60.

McGill, W. J. (1954) "Multivariate information transmission," *Psychometrika* 19, 97–116.

MacKinnon, W. J. (1959) "Compact table of twelve probability levels of the symmetric binomial cumulative distribution for sample sizes to 1,000," *J. Amer. Statist. Soc.* 54, 164–172. Errata 811.

McNeil, D. R. (1966) "Consistent statistics for estimating and testing hypotheses from grouped samples," *Biometrika* 53, 545–557.

McNemar, Q. (1947) "Note on the sampling error of the difference between correlated proportions or percentages," *Psychometrika* 12, 153–157.

McNemar, Q. (1952) "Statistical theory and research design," *Ann. Rev. Psychol.* 1, 409–418.

McNolty, F. (1962) "A contour-integral derivation of the non-central chi-square distribution," *Ann. Math. Statist.* 33, 796–800.

Maag, U. R. (1966) "A k-sample analogue of Watson's U^2 statistic," *Biometrika* 53, 579–583.

Mahalanobis, P. G., Bose, R. C. and Roy, S. N. (1937) "Normalisation of statistical variates and the use of rectangular co-ordinates in the theory of sampling distributions," *Sankhyā* 3, 1–40.

Mainland, D. (1948) "Statistical methods in medical research. I. Qualitative statistics (enumeration data)," *Canada J. Res. E* 26, 1–166.

Mainland, D. and Murray, I. M. (1952) "Tables for use in fourfold contingency tables," *Science* 116, 591–594.

Malécot, G. (1948) "Le regroupement des classes d' une table de contingence et ses applications à la génétique," *C.R. Acad. Sci. Paris* 226, 1682–1683.

Mallows, C. L. (1959) "On the probability integral transformation," *Biometrika* 46, 481–486.

Mann, H. B. and Wald, A. (1942) "On the choice of the number of intervals in the application of the chi-square test," *Ann. Math. Statist.* 13, 306–317.

Mantel, N. (1963) "Chi-square tests with one degree of freedom, extensions of the Mantel-Haenszel procedure," *J. Amer. Statist. Assoc.* 58, 690–700.

March, L. (1905) "Comparaison numérique de courbes statistiques," *J. Soc. Statist. Paris.* 46, 255–277, and 306–311.

Mardia, K. V. (1967). "Some contributions to contingency-type bivariate distributions," *Biometrika* 54, 235–249.

Maritz, J. S. (1953) "Estimation of the correlation coefficient in the case of a bivariate normal population when one of the variables is dichotomized," *Psychometrika* 18, 97–110.

Marriott, F. H. C. (1952) "Tests of significance in canonical analysis," *Biometrika* **39**, 58–64.

Masuyama, M. and Yoshimura, I. (1962) "Tables of n, $\log_e n$ and $n \log_e n$ for $n = 0.5$ (0.5) 200.0," *Rep. Statist. App. Res. Un. Jap. Sci. Engrs.* **9**, No. 4, 137–181.

Matérn, B. (1949) "Independence of non-negative quadratic forms in normally correlated variables," *Ann. Math. Statist.* **20**, 119–120.

Mathen, K. K. (1946) "A criterion for testing whether two samples have come from the same population without assuming the nature of the population," *Sankhyā* **7**, 329.

Mathen, K. K. (1954) "Note on the design of experiments and testing the efficiency of drugs having local healing power," *Sankhyā* **14**, 175–179.

Mather, K. (1935) "The combination of data," *Ann. Eugen. London* **6**, 399–410.

Mathur, R. K. (1961) "A note on Wilson-Hilferty transformation of χ^2," *Calcutta Statist. Assoc. Bull.* **10**, 103–105.

Matusita, K. (1955) "Decision rules based on the distance, for problems of fit, two samples and median," *Ann. Math. Statist.* **26**, 631–640.

Maung, K. (1942) "Measurement of association in a contingency table with special reference to the pigmentation of hair and eye colours of Scottish school children," *Ann. Eugen. London* **11**, 189–223.

Maxwell, A. E. (1961) *Analysing qualitative data.* Methuen, London. 163pp.

Mehler, F. G. (1866) "Über die Entwicklung einer Funktion von beliebig vielen Variablen nach Laplaceschen Funktionen hoeherer Ordnung," *J. Reine Angew. Math.* **66**, 161–176.

Meixner, J. (1934) "Orthogonale Polynomsysteme mit einer besonderen Gestalt der erzeugenden Funktion," *J. London Math. Soc.* **9**, 6–13.

Meixner, J. (1938) "Erzeugende Funktionen der Charlierschen Polynome," *Math. Zeit.* **44**, 531–535.

Meng, R. C. and Chapman, D. G. (1966) "The power of chi-square tests for contingency tables," *J. Amer. Statist. Assoc.* **61**, 967–975.

Merrington, M. (1941) "Numerical approximations to the percentage points of the distribution," *Biometrika* **32**, 200–202.

Merrington, M. (1951) "Tables of the 5% and 0.5% points of Pearson curves (with argument β_1 and β_2), expressed in standard measure," *Biometrika* **38**, 4–10.

Meyer, P. L. (1967) "The maximum likelihood estimate of the non-centrality parameter of a non-central χ^2-variate," *J. Amer. Statist. Assoc.*, **62**, 1258–1264.

Michael, W. B. Perry, N. C. and Guilford, J. P. (1952) "The estimation of a point biserial coefficient of correlation from a phi coefficient," *British J. Psychol. Statist. Sect.* **5**, 139–150.

Mirimanoff, D. (1929) "Les épreuves répétées et les formules approchées de Laplace et de Charlier," *Comment. Math. Helvet.* **1**, 15–41 and 156.

Mirimanoff, D. (1930) "Le jeu de pile on face et les formules de Laplace et de J. Eggenberger," *Comment. Math. Helvet.* **2**, 133–168.

Mirimanoff, S. and Dovaz, R. (1927) "Les épreuves répétées et la formule de Laplace," *C.R. Acad. Sci. Paris* **185**, 827–829.

Mitra, S. K. (1955) "Contributions to the statistical analysis of categorical data," *Inst. Statistics. Univ. Nth. Carolina* Mimeo Ser. 142.

Mitra, S. K. (1956) "On Bartlett's test of complex contingency table interaction," *Ann. Math. Statist.* **27**, 214.

Mitra, S. K. (1958) "On the limiting power function of the frequency chi-square test," *Ann. Math. Statist.* **29**, 1221–1233.

Molina, E. C. (1945) *Tables of the Poisson exponential limit.* Van Nostrand and Co., New York. vi + 47.

Monk, D. T. and Owen, D. B. (1957) *Tables of the normal probability integral.* U.S. Department Commerce, Office of Technical Serivces. Washington, D.C. 58 pp.

Monro, C. J. (1874) "Note on the inversion of Bernoulli's theorem in probabilities," *Proc. London. Math. Soc.* **5**, 74–78 and 145–146.

Mood, A. M. (1941) "On the joint distribution of the medians in samples from a multivariate population," *Ann. Math. Statist.* **12**, 268–278.

Mood, A. M. (1949) "Tests of independence in contingency tables as unconditional tests," *Ann. Math. Statist.* **20**, 114–116.

Mood, A. M. (1950) *Introduction to the theory of statistics.* McGraw-Hill Book Company Inc., New York. xiii + 433. 2nd Edit. by A. M. Mood and F. A. Graybill, 1963. xv + 443.

Mood, A. M. (1951) "On the distribution of the characteristic roots of normal second-moment matrices," *Ann. Math. Statist.* **22**, 266–273.

Moore, P. G. (1949) "A test for randomness in a sequence of two alternatives involving a 2 × 2 table," *Biometrika* **36**, 305–316.

Moran, P. A. P. (1948) "Rank correlation and product-moment correlation," *Biometrika*, **35**, 203–206.

Moran, P. A. (1952) "A characteristic property of the Poisson distribution," *Proc. Camb. Philos. Soc.* **48**, 206–207.

Morgenstern, D. (1958) "Elementarer Beweis für die asymptotische χ^2-Verteilung," *Metrika* **1**, 239–242.

Morrell, A. J. (1944) "Note on Wilson and Hilferty's approximation to the χ^2 distribution," *J. Roy. Statist. Soc.* **107**, 59.

Mosteller, F. (1946) "On some useful inefficient statistics," *Ann. Math. Statist.* **17**, 377–408

Mosteller, F. (1948) "On pooling data," *J. Amer. Statist. Assoc.* **43**, 231–242.

Mosteller, F. (1968) "Association and estimation in contingency tables," *J. Amer. Statist. Assoc.* **63**, 1–28.

Mote, V. L. and Anderson, R. L. (1965) "An investigation of the effect of misclassification on the properties of χ^2-tests in the analysis of categorical data," *Biometrika* **52**, 95–109.

Mourier, E. (1953) "Eléments aléatoires dans un espace de Banach," *Ann. Inst. H. Poincaré* **13**, 161–244.

Myers, R. J. (1934) "Note on Koshal's method of improving the parameters of curves by the use of the method of maximum likelihood," *Ann. Math. Statist.* **5**, 320–323.

Nadaraja, E. (1961) "On the χ^2-test," *Trudy Vyčisl. Centra Akad. Nauk. Gruzin. SSR.* **2**, 241–244.

Nair, K. R. (1937) "A note on the exact distribution of λ_n," *Sankhyā* **3**, 171–174.

Nandi, H. K. (1946) "On the average power of test criteria," *Sankhyā* **8**, 67–72.

Nass, C. A. G. (1959) "The χ^2 test for small expectations in contingency tables, with special reference to accidents and absenteeism," *Biometrika* **46**, 365–385.

Neall, P. S. (1952) "A χ^2 test for frequency data of simple Markoff chain processes," Unpublished .

Newbold, E. M. (1925) "Notes on an experimental test of errors in partial correlation coefficients, derived from fourfold and biserial total coefficients," *Biometrika* **17**, 251–265.

Neyman, J. (1930) "Contribution to the theory of certain test criteria," *Bull. Int. Statist. Instit.* **24**, 44–88.

Neyman, J. (1935a) "Sur la vérification des hypothèses statistiques composées," *Bull. Soc. Math. France.* **63(1)** 246–266.

Neyman, J. (1935b) "La vérification de l'hypothèse concernant la loi de probabilité d'une variable aléatoire," *C.R. Acad. Sci. Paris* **203**, 1047–1049.

Neyman, J. (1937a) "Outline of a theory of statistical estimation based on the classical theory of probability," *Phil. Trans. Roy. Soc.* **A236**, 333–380.

Neyman, J. (1937b). "Smooth test of goodness of fit," *Skand. Aktuarietidskr.* **20**, 149–199.

Neyman, J. (1938a) "On statistics the distribution of which is independent of the parameters involved in the original probability law of the observed variables," *Statist. Res. Mem.* **2**, 58–9.

Neyman, J. (1938b) "A historical note on Karl Pearson's deduction of the moments of the binomial," *Biometrika* **30**, 11–15.

Neyman, J. (1940a) "Conception of equivalence in the limit of tests and its application to certain λ- and χ^2-tests," *Ann. Math. Statist.* **11**, 477–478.

Neyman, J. (1940b) "Empirical comparison of the 'smooth' test of goodness of fit with the Pearson's χ^2 test," *Ann. Math. Statist.* **11**, 478.

Neyman, J. (1942) "Basic ideas and some recent results in the theory of testing statistical hypotheses," *J. Roy. Statist. Soc.* **105**, 292–327.

Neyman, J. (1949) "Contribution to the theory of the χ^2 test," *Proc. 1st Berkeley Symp.* **1**, 239–273.

Neyman, J. (1967) *A selection of the early statistical papers of J. Neyman.* Univ. Press, Cambridge. ix + 1429.

Neyman, J., Chernoff, H. and Chapman, D. G. (1965) "Discussion on "Asymptotically optimal tests for multinomial distributions," by W. Hoeffding," *Ann. Math. Statist.* **36**, 401–408.

Neyman, J. and Pearson, E. S. (1928) "On the use and interpretation of certain test criteria for purposes of statistical inference," II *Biometrika* **20A**, 263–294.

Neyman, J. and Pearson, E. S. (1931) "Further notes on the χ^2 distribution," *Biometrika* **22**, 298–305.

Neyman, J. and Pearson, E. S. (1938) "Contributions to the theory of testing statistical hypotheses," *Statist. Res. Mem.* **2**, 25–57.

Neyman, J. and Pearson, E. S. (1967) *Joint statistical papers.* Univ. Press. Cambridge. viii + 299.

Niceforo, A. (1911a) "Contributo allo studio della variabilità di alcuni caratteri anthropologici," *Atti. d. Soc. rom. d. Anthropologia* **1**.

Niceforo, A. (1911b) "Contribution à l'étude des corrélations entre le bien—être économique et quelques faits de la vie démographique," *J. Soc. Statist. Paris.* 20 pp.

Nixon, J. W. (1913) "An experimental test of the normal law of error," *J. Roy. Statist. Soc.* **76**, 702–706.

Norton, H.W. (1945) "Calculation of chi-square for complex contingency tables," *J. Amer. Statist. Assoc.* **40**, 251–258.

Obukhov, A. M. (1938) "Normal correlation of vectors," (Russ.) *Izv. Akad. Nauk (Math and Phys. Ser.)* **3**, 339–370.

Ogawa, J. and Ikeda, S (1964) "On the asymptotic distribution of the likelihood ratio under the regularity conditions due to Doob," *Ann. Instit. Statist. Math.* **16**, 369–385.

Ogawa, J., Moustafa, M. D. and Roy, S. N. (1966) "On the asymptotic distribution of the likelihood ratio in some problems on mixed variate populations, I and II," *Proc. math. phys. Soc. U.A.R. (Egypt)* **26**, (1962), 1–9 and 11–17.

Okamoto, M. (1952) "Unbiasedness in the test of goodness of fit," *Osaka Math. J.* **4**, 211–214.

Okamoto, M. (1955) "Chi-square test of parametric fit," (Japanese). *Osaka Tokeidanwakai Hokoku.* **1**.

Okamoto, M. (1959) "A convergence theorem for discrete probability distributions," *Ann. Inst. Statist. Math. Tokyo* **11**, 107–112. Errata **11**, 220.

Okamoto, M. (1960) "An inequality for the weighted sum of χ^2 variates," *Bull. Math. Statist.* **9**, (2/3), 69–70.

Okamoto, M. (1963) "Chi-square statistic based on the pooled frequencies of several observations," *Biometrika* **50,** 524–528.

Okamoto, M. and Ishii, G. (1961) "Tests of independence in intraclass 2×2 tables," *Biometrika* **48,** 181–190.

Okuno, T. and Takeuchi, K. (1961) "A note on the degrees of freedom of combined chi-square variables," *Rep. Statist. Appl. Res. Un. Jap. Sci. Engrs.* **8,** 182–184.

Onicescu, O. and Sacuiu, I. (1964) "Extensions of the notions of moment and coefficient of correlation to arbitrary variables," *Stud. Cerc. Mat.* **15,** 325–330.

Ottaviani, G. (1939) "Sulla probabilità che una prova su due variabili casuali X e Y verifichi la disuguaglianza $X < Y$ e sul corrispondente scarto quadratico medio," *Giorn. Inst. Ital. Attuari* **10,** 186–192.

Ottestad, P. (1943) "On Bernoullian Lexis, Poisson and Poisson-Lexis series," *Skand. Aktuarietiedskr.* **26,** 15–67.

Owen, D. B. (1956) "Tables for computing bivariate normal probabilities," *Ann. Math. Statist.* **27,** 1075–1090.

Owen, D. B. (1957) *The bivariate normal probability distribution.* Sandia Corp. and Off. Techn. Serv., Dept. Commerce, Washington 25, D.C. pp. 136.

Owen, D. B. (1962) *Handbook of statistical tables.* Addison-Wesley, New York. xiii + 580.

Pagurova, V. I. (1963) *Tables of the incomplete gamma function.* Vyčisl. Centr. Akad. Nauk SSSR, Moscow. xvi + 236.

Pastore, N. (1950) "Some comments on the use and misuse of the chi-square test," *Psychol. Bull.* **47,** 338–40.

Patankar, V. N. (1954) "The goodness of fit of frequency distributions obtained from stochastic processes," *Biometrika* **41,** 450–462.

Patau, K. (1942) "Eine neue χ^2 Tafel," *Zeit. Indukt. Abstamm. u. Vereblehre* **80,** 558–564.

Pathria, R. K. (1961) "A statistical analysis of the first 2,500 decimal places of e and $1/e$," *Proc. Nat. Inst. Sci. India.* A**27,** 270–282.

Pathria, R. K. (1962) "A statistical study of randomness among the first 10,000 digits of π." *Mathematics of Computation* **16,** 188–197.

Pathria, R. K. (1964) "A statistical study of randomness among the first 60,000 digits of e," *Proc. Nat. Inst. Sci. India* A. **30,** 663–674.

Patil, G. P. (Editor). (1965) *Classical and contagious discrete distributions. Statistical Publ. Soc.,* Calcutta and Pergamon Press, London. xiv + 552.

Patnaik, P. B. (1948) "The power function of the test for the difference between two proportions in a 2×2 table," *Biometrika* **35,** 157–175.

Patnaik, P. B. (1949) "The non-central χ^2-and F-distributions and their applications," *Biometrika* **36,** 202–232.

Patnaik, P. B. (1954) "A test of significance of a difference between two sample proportions when the proportions are very small," *Sankhyā* **14,** 187–202.

Paulson, E. and Wallis, W. A. (1947) "Planning and analysing experiments for comparing two percentages," Chapter 7, in Statistical Research Group, Columbia University, *Techniques of Statistical Analysis.* McGraw-Hill Book Co. Inc., New York.

Pawlik, K. (1959) "Der maximale Kontingenzkoeffizient im Falle nicht quadritisches Kontingenztafeln," *Metrika* **2,** 150–166.

Pearson, E. S. (1923) "The probable error of a class index correlation," *Biometrika* **14,** 261–280.

Pearson, E. S. (1931) "Note on test for normality," *Biometrika* **22,** 423–424.

Pearson, E. S. (1936/37) "Karl Pearson: An appreciation of some aspects of his life and work," *Biometrika.* **28,** 193–257 and **29,** 161–248.

Pearson, E. S. (1938) "The probability integral transformation for testing goodness of fit and combining independent tests of significance," *Biometrika* **30**, 134–138.

Pearson, E. S. (1942) "Notes on testing statistical hypotheses," *Biometrika* **32**, 311–316.

Pearson, E. S. (1947) "The choice of statistical tests illustrated on the interpretation of data classed in a 2 × 2 table," *Biometrika* **34**, 139–167.

Pearson, E. S. (1950) "On questions raised by the combination of tests based on discontinuous distributions," *Biometrika* **37**, 383–398. Corrig. **38**, 265.

Pearson, E. S. (1959) "Note on an approximation to the distribution of non-central χ^2," *Biometrika* **46**, 364.

Pearson, E. S. (1966) *The selected papers of E. S. Pearson.* Cambridge Univ. Press, Cambridge. ix + 327.

Pearson, E. S. and Hartley, H. O. (1951) "Charts of the power function for analysis of variance tests, derived from the non-central F-distribution," *Biometrika* **38**, 112–130

Pearson, E. S. and Hartley, H. O. (1958) *Biometrika tables for statisticians.* Camb. Univ. Press. Cambridge. **1**, xiv + 240.

Pearson, E. S. and Merrington, M. (1948) "2 × 2 tables, the power function of the test on a randomized experiment," *Biometrika* **35**, 331–345.

Pearson, E. S. and Stephens, M. A. (1962) "The goodness of fit tests based on W_N^2 and U_N^2," *Biometrika* **49**, 397–402.

Pearson, E. S. and Wilks, S. S. (1933) "Methods of statistical analysis appropriate for k samples of two variables," *Biometrika* **25**, 353–378.

Pearson, K. (1895) "Contributions to the mathematical theory of evolution. II Skew variation in homogeneous material," *Philos. Trans. Roy. Soc.* **A186**, 343–414.

Pearson, K. (1896) "Mathematical contributions to the theory of evolution. III. Regression, heredity and panmixia," *Philos. Trans. Roy. Soc.* **A187**, 253–318.

Pearson, K. (1899) "On certain properties of the hypergeometrical series, and on the fitting of such series to observation polygons in the theory of chance," *Phil. Mag.* **(5)47**, 236–246.

Pearson, K. (1900a) "On a criterion that a given system of deviations from the probable in the case of a correlated system of variables is such that it can be reasonably supposed to have arisen from random sampling," *Philos. Mag.* **(5)50**, 157–175.

Pearson, K. (1900b) "Mathematical contributions to the theory of evolution on the inheritance of characters not capable of exact quantitative measurement, VIII," *Philos. Trans. Roy. Soc.* **A195**, 79–150.

Pearson, K. (1900c) "Mathematical contributions to the theory of evolution. VII. On the correlation of characters not quantitatively measureable," *Philos. Trans. Roy. Soc. London* **A195**, 1–47.

Pearson, K. (1901) "On lines and planes of closest fit to systems of points in space," *Phil. Mag.* **(6)2**, 559–72.

Pearson, K. (Editorial) (1903) "On the probable error of frequency constants," *Biometrika*, **2**, 273–281.

Pearson, K. (1904) "Mathematical contributions to the theory of evolution. XIII. On the theory of contingency and its relation to association and normal correlation," *Draper's Company Research Memoirs, Biometric Series 1.* 35 pp.

Pearson, K. (1905a) "Das Fehlergesetz und eine Verallgemeinerungen durch Fechner und Pearson, A Rejoinder," *Biometrika* **4**, 169–212.

Pearson, K. (1905b) "Mathematical contributions to the theory of evolution. XIV. On the general theory of skew correlation and non-linear regression," *Drapers' Company Research Memoirs, Biometric Ser. II,* 54 pp.

Pearson, K. (1906a) "On the curves which are most suitable for describing the frequency of random samples of population," *Biometrika* **5**, 172–175.

Pearson, K. (1906b) "On certain points connected with scale order in the case of the correlation of two characters which for some arrangement give a linear regression line," *Biometrika* 5, 176–178.

Pearson, K. (1906c) "Note on the significant or non-significant character of a sub-sample drawn from a sample," *Biometrika* 5, 181–183.

Pearson, K. (1906d) "On a coefficient of class heterogeneity or divergence," *Biometrika* 5, 198–203.

Pearson, K. (1908) "On the influence of double selection on the variation and correlation of two characters," *Biometrika* 6, 111–112.

Pearson, *K.* (1909) "On a new method of determining correlation between a measured character *A* and a character *B*, of which only the percentage of cases wherein *B* exceeds (or falls short of) a given intensity is recorded for each grade of *A*," *Biometrika* 7, 96–105.

Pearson, K. (1910) "On a new method of determining correlation when one variable is given by alternative and other by multiple categories," *Biometrika* 7, 248–257.

Pearson, K. (1911a) "On the probability that two independent distributions of frequency are really samples from the same population,"*Biometrika* 8, 250–254.

Pearson, K. (1911b) "On a correction to be made to the correlation ratio η," *Biometrika* 8, 254–256.

Pearson, K. (1912) "Mathematical contributions to the theory of evolution. XVIII. On a novel method of regarding the association of two variates classed solely in alternative cateogries," *Draper' Company Research Memoirs. Biometric Series VII.* 29 pp.

Pearson, K. (1913a) "On the probable error of a coefficient of correlation as found from a fourfold table," *Biometrika* 9, 22–27.

Pearson, K. (1913b) "On the measurement of the influence of 'Broad Categories' on correlation," *Biometrika* 9, 116–139.

Pearson, K. (1913c) "On the surface of constant association $Q = 0.6$," *Biometrika* 9, 534–537.

Pearson, K. (1915) "On the probable error of a coefficient of mean square contingency," *Biometrika* 10, 570–573.

Pearson, K. (1916a) "On some novel properties of partial and multiple correlation coefficients in a universe of manifold characteristics." *Biometrika* 11, 231–238.

Pearson, K. (1916b) "On the general theory of multiple contingency with special reference to partial contingency," *Biometrika* 11, 145–158.

Pearson, K. (1916c) "On a brief proof of the fundamental formula for testing the goodness of fit of frequency-distributions and of the probable error of *P*," *Phil. Mag.* (6)31, 369–378.

Pearson, K. (1916d) "On the application of goodness of fit tables to test regression curves and theoretical curves used to describe observational or experimental data," *Biometrika* 11, 239–261. Corrig. 12, 259–81.

Pearson, K. (1916e) "Second supplement to a memoir on skew variation. Mathematical contributions to the theory of evolution. XIX," *Philos. Trans. Roy. Soc. of London* A216, 429–457. Corrig. in Pearson (1919).

Pearson, K. (1917) "On the probable error of biserial η," *Biometrika*, 11, 292–302.

Pearson, K. (1919) "Peccavimus!," Editorial. *Biometrika* 12, 259–281.

Pearson, K. (1920) "Notes on the history of correlation," *Biometrika* 13, 25–45.

Pearson, K. (1921) "Second note on the coefficient of correlation as determined from the quantitative measurement of one variate and the ranking of a second variate," *Biometrika* 13, 302–305.

Pearson, K. (Editor). (1922a) *Tables of the incomplete Γ-function*, Cambridge University Press. xxxii + 164.

Pearson, K. (1922b) "On the χ^2 test of goodness of fit," *Biometrika* 14, 186–191.

Pearson, K. (1923a) "Further note on the χ^2 test of goodness of fit," *Biometrika* **14**, 418.

Pearson, K. (1923b) "On the correction necessary for the correlation ratio η," *Biometrika* **14**, 412–417.

Pearson, (1924a) "On the moments of the hypergeometrical series," *Biometrika* **16**, 157–162.

Pearson, K. (1924b) "On the difference and the doublet tests for ascertaining whether two samples have been drawn from the same population," *Biometrika* **16**, 249–252.

Pearson, K. (1924c) "Historical note on the origin of the normal curve of errors," *Biometrika* **16**, 402–404.

Pearson, K. (1925) "The fifteen constant bivariate frequency surface," *Biometrika* **17**, 268–313.

Pearson, K. (1927) "Note on the relationship of the (P, χ^2) test of goodness of fit to the distribution of standard deviations in samples from a normal population," *Biometrika* **19**, 215.

Pearson, K. (1928a) "On a method of ascertaining limits to the actual number of marked members in a population of given size from a sample," *Biometrika* **20A**, 149–174.

Pearson, K. (1928b) "The contribution of Giovanni Plana to the normal bivariate frequency surface," *Biometrika* **20A** 295–298.

Pearson, K. (1930) "On the theory of contingency. I. Note on Professor J. Arthur Harris' paper on the limitation in the applicability of the contingency coefficient," *J. Amer. Statist. Assoc.* **25**, 320–323 and 327.

Pearson, K. (Editor). (1931) *Tables for statisticians and biometricians, Part II.* Biometric Lab. Univ. Coll. London. Cambridge Univ. Press. ccl + 262.

Pearson, K. (1932a) "Experimental discussion of the (χ^2, P) test of goodness of fit," *Biometrika* **24**, 351–381.

Pearson, K. (1932b) "On the probability that two independent distributions are really samples from the same parent population," *Biometrika* **24**, 457–470.

Pearson, K. (1933a) "On the parent population with independent variates which gives the minimum value of ϕ^2 for a given sample," *Biometrika* **25**, 134–146.

Pearson, K. (1933b) "On a method of determining whether a sample of size n supposed to have been drawn from a parent population having a known probability integral has probably been drawn at random," *Biometrika* **25**, 379–410.

Pearson, K. (1934a) "On a new method of determining goodness of fit," *Biometrika* **26**, 425–442.

Pearson, K. (1934b) "Remarks on Professor Steffensen's measure of contingency," *Biometrika* **26**, 255–260.

Pearson, K. (1948) *Karl Pearson's early statistical papers.* Edited by E. S. Pearson, Cambridge University Press. viii + 557.

Pearson, K. and Heron, D. (1913) "On theories of association," *Biometrika* **9**, 159–315.

Pearson, K. and Lee, A. (1908) "On the generalised probable error in multiple normal correlation," *Biometrika* **6**, 59–79.

Pearson, K. and Pearson, E. S. (1922) "Polychoric coefficients of correlation," *Biometrika* **14**, 127–156.

Pearson, K. and Tocher, J. F. (1916) "On criteria for the existence of differential death-rates," *Biometrika* **11**, 159–84.

Pearson, K. and Young, A. W. (1918) "On the product-moments of various orders of the normal correlation surface of two variates." *Biometrika* **12**, 86–92.

Peierls, R. S. (1935) "Statistical methods in counting experiments," *Proc. Roy. Soc. (Lond.)* **A149**, 467–86.

Peiser, A. M. (1943) "Asymptotic formulas for significance levels of certain distributions," *Ann. Math. Statist.* **14**, 56–62. Corrig. **20**, 128–29.

Perry, N. C. and Michael, W. B. (1954) "The reliability of a point biserial coefficient of correlation," *Psychometrika* **19**, 313–325.

Peters, C. C. (1950) "The misuse of chi-square—a reply to Lewis and Burke," *Psychol. Bull.* **47**, 331–337.

Picone, M. (1934) "Trattazione elementare dell approssimazione lineare in insiemi non limitati," *Giorn. Inst. Ital. Attuari* **5**, 155–195.

Pillai, K. C. S. (1956) "On the distribution of the largest or the smallest root of a matrix in multivariate analysis," *Biometrika* **43**, 122–127.

Pillai, K. C. S. (1964) "On the distribution of the largest of seven roots of a matrix in multivariate analysis," *Biometrika* **51**, 270–275.

Pillai, K. C. S. (1967) "Upper percentage points of the largest root of a matrix in multivariate analysis," *Biometrika* **54**, 189–194.

Pitman, E. J. G. (1936) "Sufficient statistics and intrinsic accuracy," *Proc. Camb. Philos. Soc.* **32**, 567–579.

Plackett, R. L. (1962) "A note on interactions in contingency tables," *J. Roy. Statist. Soc.* **B24**, 162–166.

Plackett, R. L. (1964) "The continuity correction in 2 × 2 tables," *Biometrika* **51**, 327–337.

Plackett, R. L. (1965) "A class of bivariate distributions," *J. Amer. Statist. Assoc.* **60**, 516–522.

Poisson, S. D. (1827) "Sur la probabilité des résultats moyens des observations," *Connaiss. d. Temps. 1827*, 273–302, also *Bull. Sci. math.* (*Férussac*) **13**, 266–277.

Poisson, S. D. (1832) "Suite du mémoire sur la probabilité de résultant moyen des observations," *Connaiss. d. Temps 1832*, 3–22.

Pollaczek-Geiringer, H. (1933) "Korrelationsmessung auf Grund der Summenfunktion," *Zeit. f. Angew. Math.* **13**, 121–124.

Pólya, G. (1923) "Herleitung des gaussschen Gesetzes aus einer Funktionalgleichung," *Math. Z.* **18**, 96–108.

Pompilj, G. (1950) "Osservazioni sull'omogamia: La trasformazione di Yule e il limite della transformazione r i corrente de Gini," *Univ. Roma. Ist. Naz. Mat. Rend. Mat. e Appl.* **(5)9**, 367–388.

Poti, S. J. (1950) "Power function of chi-square test with special reference to analysis of blood group data," *Sankhyā* **10**, 397–406.

Press. S. J. (1966) "Linear combinations of non-central chi-square variates," *Ann. Math. Statist.* **37**, 480–487.

Pretorius, S. J. (1930) "Skew bivariate frequency surfaces, examined in the light of numerical illustrations," *Biometrika* **22**, 109–223.

Price, G. B. (1946) "Distributions derived from the multinomial expansion," *Amer. Math. Monthly* **53**, 59–74

Przyborowski, J. and Wilenski, H. (1935a) "Statistical principles of routine work in testing clover seed for dodder," *Biometrika* **27**, 273–292.

Przyborowski, J. and Wilenski, H. (1935b). "Sur les erreurs de la première et de la second catégorie dans la vérification des hypothèses concernant la loi de Poisson," *C.R. Acad. Sci. Paris* **200**, 1460–1462.

Przyborowski, J. and Wilenski, H. (1940) "Homogeneity of results in testing samples from Poisson series, with an application to testing clover seed for dodder," *Biometrika* **31**, 313–323.

Putter, J. (1964) "The goodness of fit test for a class of cases of dependent observations," *Biometrika* **51**, 250–252.

Pyke, R. (1958) "On renewal processes related to type I and type II counter models," *Ann. Math. Statist.* **29**, 737–754.

Rademacher, H. (1922) "Einige Sätze über Reihen von allgemeinen Orthogonalfunktionen," *Math. Ann.* **87**, 112–138.

Raj, D. (1953) "On Mill's ratio for the type III population," *Ann. Math. Statist.* **24**, 309–312.

Ram, S. (1954) "A note on the calculation of moments of the two-dimensional hypergeometric distribution," *Ganita* **5**, 97–101.

Ram, S. (1955) "Multidimensional hypergeometric distribution," *Sankhyā* **15**, 391–398.

Ram, S. (1956) "On the calculation of moments of hypergeometric distribution," *Ganita* **7**, 1–5.

Ramabhadran, V. K. (1951) "A multivariate gamma-type distribution," *Sankhyā* **11**, 45–46.

Ramachandran, G. (1951) "On a test whether two samples are from the same population," *J. Madras. Univ.* **21 B** 124–139.

Ramachandran, K. V. (1958) "On the Studentized smallest chi-square," *J. Amer. Statist. Assoc.* **53**, 868–872.

Rao, C. R. (1946) "On the linear combination of observations and the general theory of least squares," *Sankhyā* **7**, 237–56.

Rao, C. R. (1947) "Minimum variance and the estimation of several parameters," *Proc. Camb. Philos. Soc.* **43**, 280–283.

Rao, C. R. (1951a) "A theorem in least squares," *Sankhyā* **11**, 9–12.

Rao, C. R. (1951b) "An asymptotic expansion of the distribution of Wilks' Λ criterion," *Bull. Inst. Internat. Statist.* **33(2)**, 177–180.

Rao, C. R. (1952a) *Advanced statistical methods in biometric research.* John Wiley, New York. xvii + 383.

Rao, C. R. (1952b) "Some theorems on minimum variance estimation," *Sankhyā* **12**, 27–42.

Rao, C. R. (1952c) "Minimum variance estimation in distributions admitting ancillary statistics," *Sankhyā* **12**, 53–56.

Rao, C. R. (1957a) "Maximum likelihood estimation for the multinomial distribution," *Sankhyā* **18**, 139–148.

Rao, C. R. (1957b) "Theory of the method of estimation by minimum chi-square," *Bull. Inst. Internat. Statist.* **35(2)**, 25–32.

Rao, C. R. (1965) *Linear statistical inference and its applications.* John Wiley and Sons, New York xviii + 522.

Rao, C. R. and Chakravarti, I. M. (1956) "Some small sample tests of significance for a Poisson distribution," *Biometrics*, **12**, 264–282.

Reiersøl, O. (1944) "Measures of departure from symmetry," *Skand. Aktuarietidskr.* **27**, 229–234.

Rényi, A. (1959a) "New version of the probabilistic generalization of the large sieve," *Acta. Math. Acad. Sci. Hung* **10**, 218–226.

Rényi, A. (1959b) "On measures of dependence," *Acta. Math. Acad. Sci. Hungar.* **10**, 441–451.

Reuning, H. (1952) "Evaluation of square contingency tables. A simple method of condensation for small samples," *Bull. Nat. Inst. Personnel Res. S.A.C.S.I.R. Johannesburg* **4**, 160–167.

Rhodes, E. C. (1924) "On the problem whether two given samples can be supposed to have been drawn from the same population," *Biometrika* **16**, 239–248.

Rhodes, E. C. (1925) "On a skew correlation surface," *Biometrika* **17**, 314–326.

Rhodes, E. C. (1926) "The comparison of two sets of observations," *J. Roy. Statist. Soc.* **89**, 544–552.

Richter, H. (1949) "Zur Maximalkorrelation," *Z. Angew. Math. Mech.* **29**, 127–128.

Richter, W. (1958) "The limit behaviour of the χ^2 distribution in the case of large deviations," (Russ.) *Dokl. Akad. Nauk SSSR.* **119**, 652–654.

Richter, W. (1964) "Mehrdimensionale Grenzwertsätze für grosse Abweichungen und ihre Anwendung auf die Verteilung von χ^2," (Russian summary). *Teor. Veroyatnost i Primenen.* **9**, 31–42.

Riesz, F. and Sz.-Nagy, B. (1956) *Functional analysis.* (translated from the 2nd French edition by L. F. Boron). Blackie and Son Limited, London. xii + 468.

Rietz, H. L. (1923) "Frequency distributions obtained by certain transformations of normally distributed variables," *Ann. Math.* **(2)23**, 292–300.

Rietz, H. L. (1932) "On the Lexis theory and the analysis of variance," *Bull. Amer. Math. Soc.* **38**, 731–735.

Riordan, J. (1937) "Moment recurrence relations for binomial, Poisson and hypergeometric frequency distributions," *Ann. Math. Statist.* **8**, 103–111.

Riordan, J. (1958) *An Introduction to Combinatorial Analysis.* John Wiley and Sons, Inc. New York. x + 242.

Risser R. (1945) "Sur l'équation caractéristique des surfaces de probabilité," *C.R. Acad. Sci. Paris* **220**, 31–32.

Risser, R. (1948a) "Note relative aux surfaces de probabilités," *Assoc. Actuair. Belges. Bull.* **53**, 5–48.

Risser, R. (1948b) "Essai sur les courbes de distribution statistique," *Assoc. Actuair. Belges. Bull.* **54**, 41–72.

Risser, R. (1948c) "Note relative aux surfaces de probabilités," *J. Soc. Statist. Paris* **89**, 381–409.

Risser, R. (1948d) "Essai sur les courbes de distribution statistique," *J. Soc. Statist. Paris* **89**, 288–306.

Risser, R. (1957) "Essai sur les surfaces de probabilité," *Bull. de l'Institut. Internat. de Statistique.* **35(2)** 105–130.

Ritchie-Scott, A. (1918) "The correlation coefficient of a polychoric table," *Biometrika* **12**, 93–133.

Robbins, H. and Pitman, E. J. G. (1949) "Application of the method of mixtures to quadratic forms in normal variates," *Ann. Math. Statist.* **20**, 552–560.

Roberts, E., Dawson, W. M. and Madden, M. (1939) "Observed and theoretical ratios in Mendelian inheritance," *Biometrika* **31**, 56–66.

Robertson, A. (1951) "The analysis of heterogeneity in the binomial distribution," *Ann. Eugen.* **16**, 1–15.

Robertson, W. H. (1960) "Programming Fisher's exact method of comparing two percentages," *Technometrics* **2**, 103–107.

Robinson, J. (1965) "The distribution of a general quadratic form in normal variates," *Austral. J. Statist.* **7**, 110–114.

Robinson, S. (1933) "An experiment regarding the χ^2 test," *Ann. Math. Statist.* **4**, 285–287.

Robson, D. S. and King, A. J. (1952) "Multiple sampling of attributes," *J. Amer. Statist. Assoc.* **47**, 203–215.

Romanovsky, V. (1923) "Note on the moments of a binomial $(p + q)^n$ about its mean," *Biometrika* **15**, 410–412.

Romanovsky, V. (1924) "Generalization of some types of the frequency curves of Professor Pearson," *Biometrika* **16**, 106–117.

Romanovsky, V. (1925) "On the moments of the hypergeometrical series," *Biometrika* **17**, 57–60.

Rosenblatt, M. (1952) "Remarks on a multivariate transformation," *Ann. Math. Statist.* **23**, 470–472.

322   BIBLIOGRAPHY

Rosenblatt, M. (1961) "Independence and dependence," *Proc. 4th Berkeley Symp.* **2**, 431–443.

Roy, A. R. (1956) "On χ^2 statistics with variable intervals," *Stanford Tech. Reports.* **1**.

Roy, A. R. and Mohanty, S. G. (1958) "Distribution of χ^2 analogue for normal population with class intervals defined in terms of sample median," *J. Indian Soc. Agric. Statist.* **10**, 90–98, xix.

Roy, J. and Mohamad, J. (1964) "An approximation to the non-central chi-square distribution," *Sankhyā* **26**, 81–84.

Roy, J. and Murthy, V. K. (1960) "Percentage points of Wilks' L_{mvc} and L_{vc} criteria," *Psychometrika* **25**, 243–250.

Roy, S. N. (1939) "*p*-statistics or some generalisations on analysis of variance appropriate to multivariate problems," *Sankhyā* **4**, 381–96.

Roy, S. N. (1950) "Univariate and multivariate analysis as problems in testing of composite hypotheses, I," *Sankhyā* **10**, 29–80.

Roy, S. N. (1957) *Some aspects of multivariate analysis.* (Indian Statistical Series No. 1.) John Wiley, New York. viii + 214.

Roy, S. N. and Bargmann, R. E. (1958) "Tests of multiple independence and the associated confidence bounds," *Ann. Math. Statist.* **29**, 491–503.

Roy, S. N. and Bhapkar, V. P. (1960) "Some non-parametric analogs of normal anova, manova and of studies in 'normal' association," in Olkin *et al* (1960). 371–387.

Roy, S. N. and Kastenbaum, M. A. (1955) "A generalization of analysis of variance and multivariate analysis to data based on frequencies in qualitative categories or class intervals," *Inst. Statist. Univ. Nth. Carolina Mimeo Series* **131**. 27 pp.

Roy, S. N. and Kastenbaum, M. A. (1956) "On the hypothesis of "no interaction" in a multi-way contingency table," *Ann. Math. Statist.* **27**, 749–757.

Roy, S. N. and Mitra, S. K. (1955) "An introduction to some nonparametric generalizations of analysis of variance and multivariate analysis," *Inst. Statist. Univ. Nth. Carolina Mimeo Series* **139**,

Roy, S. N. and Mitra, S. K. (1956) "An introduction to some non-parametric generalisations of analysis of variance and multivariate analysis," *Biometrika* **43**, 361–376.

Royer, E. B. (1933) "A simple method for calculating mean square contingency," *Ann. Math. Statist.* **4**, 75–78.

Royer, E. B. (1941) "A machine method for computing the biserial correlation coefficient in item validation," *Psychometrika* **6**, 55–59.

Ruben, H. (1963) "A new result on the distribution of quadratic forms," *Ann. Math. Statist.* **34**, 1582–1584.

Rudolph, G. J. (1967) "A quasi-multinomial type of contingency table," *S. African Statist. J.* **1**, 59–65.

Runge, C. (1914) "Über eine besondere Art von Integralgleichungen," *Math. Ann.* **75**, 130–132.

Rust, H. (1965) "Die Momente der Testgrösse des χ^2 test," *Z. Wahrscheinlichkeitstheorie* **4**, 222–231.

Rybarz, J. (1959) "Ein einfacher Beweis für das dem χ^2 Verfahren zugrundliegende Theorem," *Metrika* **2**, 89–93.

Salvosa, L. R. (1930) "Tables of Pearson's type III function," *Ann. Math. Statist.* **1(2)** 191–198 + appendix 187 pp.

Sankaran, M. (1959) "On the non-central chi-square distribution," *Biometrika* **46**, 235–237.

Sankaran, M. (1963) "Approximation to the non-central chi-square distribution," *Biometrika* **50**, 199–204.

Sansone, G. (1933) "La chiusura dei sistemi ortogonali di Legendre, di Laguerre e di Hermite rispette alle funzioni di quadrati sommabili," *G. Ist. Ital. Attuari.* **4**, 71–82.

Sarkadi, K. (1953) "Choice of intervals for grouping of data," (Hungarian), *Magyar Tud. Akad. Alkalm. Mat. Int. Közl.* **2**, 299–306.

Sarkadi, K. (1960) "On testing for normality," *Magyar. Tud. Akad. Mat. Kutató Int. Közl.* **5**, 269–275.

Sarkadi, K. (1964) "Prüfung des Verteilungstypes," *Abh. Deutsch. Akad. Wiss. Berlin Kl. Math. Phys. Tech.* **4**, 109–111.

Sarmanov, O. W. (1941) "Sur la corrélation isogène," *C.R. (Doklady). Acad. Sci. URSS.* **32**, 28–30.

Sarmanov, O. (1945) "On isogeneous correlation," (Russ.) *Bull. Acad. Sci. URSS Ser. Math.* **9**, 169–200.

Sarmanov, O. (1946) "Sur les solutions monotones des équations intégrales de corrélation," *Dokl. Akad. Nauk SSSR.* **53**, 773–776.

Sarmanov, O. V. (1947a) "On the rectification of a symmetric correlation," (Russ.) *Dokl. Akad. Nauk SSSR.* **58**, 745–747.

Sarmanov, O. V. (1947b) "Generalization of a limit theorem of the theory of probability to sums of almost independent variables satisfying Lindeberg's condition," (Russ.) *Izvest. Acad. Nauk SSSR.* **11**, 569–575.

Sarmanov, O. V. (1948a) "On the rectification of asymmetrical correlation," (Russ.) *Dokl. Akad. Nauk SSSR.* **59**, 861–863.

Sarmanov, O. V. (1948b) "On the rectification of correlation," *Uspehi Mat. Nauk* **3(5)**, 190–192.

Sarmanov, O. V. (1952a) "On functional moments of a symmetric correlation," (Russ.) *Dokl. Akad. Nauk SSSR. (NS.).* **84**, 887–890.

Sarmanov, O. V. (1952b) "On functional moments of a non-symmetric correlation," (Russ.) *Dokl. Akad. Nauk SSSR. (N.S.)* **84**, 1139–43.

Sarmanov, O. V. (1958a) "The maximum correlation coefficient (symmetric case)," (Russ.) *Dokl. Akad. Nauk SSSR.* **120**, 715–718.

Sarmanov, O. V. (1958b) "Maximum correlation coefficient (non-symmetrical case)," (Russ.) *Dokl. Akad. Nauk SSSR.* **121**, 52–55.

Sarmanov, O. V. (1960a) "Pseudonormal correlation and its various generalizations," (Russ.) *Dokl. Akad. Nauk SSSR.* **132**, 299–302.

Sarmanov. O. V. (1960b) "Characteristic correlation functions and their applications in the theory of stationary Markov processes," (Russ.). *Dokl. Akad. Nauk SSSR.* **132**, 769–772.

Sarmanov, O. V. (1961a) "The properties of a two dimensional density defining a stationary Markov process," (Russ.) *Dokl. Akad. Nauk SSSR.* **136**, 1295–1297.

Sarmanov, O. V. (1961b) "Investigation of stationary Markov processes by the method of eigenfunction expansion," (Russian) *Trudy Mat. Inst. Steklov* **60**, 238–261.

Sarmanov, O. V. (1966) "Generalized normal correlation and two dimensional Fréchet classes," (Russ.) *Dokl. Akad. Nauk SSSR.* **168**, 32–35.

Sarmanov, O. V. and Zaharov, V. K. (1959) "Spectra of enlarged stochastic matrices," (Russ.) *Dokl. Akad. Nauk SSSR* **128**, 243–245.

Sarmanov, O. V. and Zaharov, V. K. (1960a) "Measures of dependence between random variables and the spectra of stochastic sequences and matrices," (Russ.) *Mat. Sb.* (N.S.) **52**, 953–990.

Sarmanov, O. V. and Zaharov, V. K. (1960b) "Maximum coefficients of multiple correlation," *Dokl. Akad. Nauk SSSR.* **130**, 269–271.

Sato, S. (1961/62) "A multivariate analogue of pooling of data," *Bull. Math. Statist.* **10**, No. 3/4, 61–76.

324 BIBLIOGRAPHY

Satterthwaite, F. E. (1942) "Linear restrictions on chi-square," *Ann. Math. Statist.* **13**, 326–331.

Savage, I. R. (1953) "Bibliography of nonparametric statistics and related topics," *J. Amer. Statist. Assoc.* **48**, 844–906.

Savage, I. R. (1957) "On the independence of tests of randomness and other hypotheses," *J. Amer. Statist. Assoc.* **52**, 53–57.

Savage, I. R. (1962) *Bibliography of nonparametric statistics.* Harvard Univ. Press., Cambridge, Mass. vii + 284.

Savage, L. J. (1957) "When different pairs of hypotheses have the same family of likelihood ratio test regions," *Ann. Math. Statist.* **28**, 1028–1032.

Sawkins, D. T. (1940) "Elementary presentation of the frequency distributions of certain statistical populations associated with the normal population," *Proc. Roy. Soc. New South Wales* **74**, 209–239.

Sawkins, D. T. (1941) "Remarks on goodness of fit hypotheses and on Pearson's χ^2 test," *Proc. Roy. Soc. New South Wales* **75**, 85–95.

Sawkins, D. T. (1947) "A new method of approximating the binomial and hypergeometric probabilities," *J. Roy. Soc. New South Wales* **81**, 38–47.

Schäffer, K. A. (1957) "Der Likelihood—Anpassungstest," *Mitt. Math. Statist.* **9**, 27–54.

Scheffé, H. (1942) "On the theory of testing composite hypotheses with one constraint," *Ann. Math. Statist.* **13**, 280–293.

Scheffé, H. (1947) "The relation of control charts to analysis of variance and chi-square tests," *J. Amer. Statist. Assoc.* **42**, 425–431. Corrig. **42**, 634.

Scheuer, E. M. (1962) "Moments of the radial error," *J. Amer. Statist. Assoc.* **57**, 187–190. Corrig. **60**, 1251.

Schiff, L. I. (1936) "Statistical analysis of counter data," *Phys. Rev.* (2)**50**, 88–96.

Schilling, W. (1947) "A frequency distribution represented as the sum of two Poisson distributions," *J. Amer. Statist. Assoc.* **42**, 407–424.

Schmidt, E. (1907) "Zur Theorie der linearen und nichtlinearen Integralgleichungen," *Math. Ann.* **63**, 433–476.

Schur, I. (1924) "Neue anwendungen der Integralrechnung auf Probleme der Invariant-theorie," *Sitzungsber. d. Preuss. Akad. d. Wissenschaften* **1924**, 297–321.

Schwerdtfeger, H. (1960) "Direct proof of Lanczos' decomposition theorem," *Amer. Math. Monthly* **67**, 856–860.

Scott, E. L. (1949a) "Distribution of the longitude of periastron of spectroscopic binaries," *Astrophysical J.* **109**, 194–207.

Scott, E. L. (1949b) "Further note on the distribution of the longitude of periastron," *Astrophysical. J.* **109**, 446–451.

Seal, H. L. (1943) "Tests of a mortality table graduation," *J. Inst. Actuar.* **71**, 5–67.

Seal, H. L. (1947) "A historical note on the use of χ^2 to test the adequacy of a mortality table graduation," *J. Instit. Actuar. Stud. Soc.* **6**, 185–187.

Seal, H. L. (1948) "A note on the χ^2 smooth test," *Biometrika* **35**, 202.

Seal, H. L. (1967) "Studies in the history of probability and statistics XV. The historical development of the Gauss linear model," *Biometrika* **54**, 1–24.

Seber, G. A. F. (1963) "The non-central chi-squared and beta distributions," *Biometrika* **50**, 542–544.

Seshadri, V. and Patil, G. P. (1964) "A characterization of a bivariate distribution by the marginal and the conditional distributions of the same component," *Inst. of Statist. Math. Ann. Tokyo* **15**, 215–221.

Severo, N. C. and Zelen, M. (1960) "Normal approximation to the chi-square and non-central F probability functions," *Biometrika* **47**, 411–416.

Shanks, D. and Wrench, J. W. (1962) "Calculation of π to 100,000 decimals," *Math. of Computation* **16**, 76–99.

Shannon, S. (1942) "Comparative aspects of the point binomial polygon and its associated normal curve of error," *Record Amer. Inst. Actuar.* **31**, 208–226.

Sheppard, W. F. (1898*a*) "On the application of the theory of error to cases of normal distribution and normal correlation," *Phil. Trans. Roy. Soc. London* **A192**, 101–167.

Sheppard, W. F. (1898*b*) "On the geometrical treatment of the 'normal curve' of statistics with special reference to correlation and to the theory of errors," *Proc. Roy. Soc.* **62**, 170–173.

Sheppard, W. F. (1903) "New tables of the probability integral," *Biometrika* **2**, 174–190.

Sheppard, W. F. (1907) "Tables of deviates of the normal curve," *Biometrika* **5**, 404–406.

Sheppard, W. F. (1914*a*) "Graduation by reduction of mean square error," *J. Inst. Actuar.* **48**, 171–185. **48**, 390–412 and **49**, 148–157 (1915).

Sheppard, W. F. (1914*b*) "Fitting of polynomials by the method of least squares," *Proc. London Math. Soc.* **(2)13**, 97–108.

Sheppard, W. F. (1929) "The fit of formulae for discrepant observations," *Phil. Trans. Roy. Soc. London* **A228**, 115–150.

Sheps, M. C. (1959) "An examination of some methods of comparing several rates or proportions," *Biometrics* **15**, 87–97.

Shohat, J. A. and Tamarkin, J. D. (1943) *The problem of moments.* Math. Surv. No. 1. Amer. Math. Soc. New York. xiv + 140.

Shohat, J. A. Hille, E. and Walsh, J. L. (1940) *A bibliography of orthogonal polynomials* Bull. National Res. Council No. 103. National Academy of Washington, D.C. x + 204.

Shreider, I. A. (1964) *Method of Statistical testing; Monte Carlo method.* Elsevier, Amsterdam. ix + 303.

Sikorski, R. (1964) *Boolean algebras.* 2nd Edit. Springer-Verlag, Berlin x + 237.

Sillitto, G. P. (1949) "Note on approximations to the power function of the 2 × 2 comparative trial," *Biometrika* **36**, 347–352.

Silvey, S. D. (1959) "The Lagrangian multiplier test," *Ann. Math. Statist.* **30**, 389–407.

Silvey, S. D. (1961) "A note on maximum-likelihood in the case of dependent random variables," *J. Roy. Statist. Soc.* **B23**, 444–452.

Silvey, S. D. (1964) "On a measure of association," *Ann. Math. Statist.* **35**, 1157–1166.

Simpson, E. H. (1951) "The interpretation of interaction in contingency tables," *J. Roy. Statist. Soc.* **B13**, 238–241.

Skellam, J. G. (1946) "The frequency distribution of the difference between two Poisson variates belonging to different populations," *J. Roy. Statist. Soc.* **109**, 296.

Skory, J. (1952) "Automatic machine method of calculating contingency χ^2," *Biometrics* **8**, 380–382.

Slakter, M. J. (1966) "Comparative validity of the chi-square and two modified chi-square goodness of fit tests for small but equal expected frequencies," *Biometrika* **53**, 619–623.

Slater, P. (1947) "The factor analysis of a matrix of 2 × 2 tables," *J. Roy. Statist. Soc. Suppl* **9**, 114–127.

Slutsky, E. (1913) "On the criterion of goodness of fit of the regression lines and on the best method of fitting them to the data," *J. Roy. Statist. Soc.* **77**, 78–84.

Slutsky, E. (1941) "On the table of chi-square (incomplete gamma function)," *Akad. Nauk SSSR. Mat. Ser.* **5**, No. 2, 183–4.

Slutskii, E. E. (1950) *Tables for the calculation of the incomplete Γ-function and the χ^2 probability function.* (Editor: A. Kolmogorov). Moscow and Leningradi: Izdatelstvo Akademii Nauk SSSR. 14 + 55.

Smirnov, N. (1948) "Table for estimating the goodness of fit of empirical distributions," *Ann. Math. Statist.* **19**, 279–281.

Smirnov, N. (1949) *Limit distributions for the terms of a variational series. Amer. Math. Soc.* Trans. No. 67. Providence (1952). 64 pp. from the Russian in *Trudy Mat. Inst. Steklov* **25**, 1–60.

Smirnov, N. V. (1965) *Tables of the normal probability integral, the normal density and its normalized derivatives.* (Trans. by D. E. Brown). Pergamon Press, London. 144pp.

Smirnov, N. V. and Bol'šev, L. N. (1962) *Tables for evaluating a function of a two-dimensional normal distribution,* I'zv. Akad. Nauk SSSR. Moscow. 204pp.

Smirnov, S. V. and Potapov, M. K. (1957) "A nomogram for an incomplete Γ-function and probability function χ^2," *Teor. Veroyatnost. i. Primenen.* **2**, 470–2.

Smirnov, S. V. and Potapov, M. K. (1961) "Nomogram for probability functions," *Teor. Veroyatnost i. Primenen* **6**, 138–140.

Smith, C. A. B. (1951) "A test for heterogeneity of proportions," *Ann. Eugen.* **16**, 16–25.

Smith, C. A. B. (1952) "A simplified heterogeneity test," *Ann. Eugen. London* **17**, 35–36.

Smith, C. D. (1951) "Some probability estimates for contingency tables," *Math. Magazine.* **25**, 59–62.

Smith, D. E. (1929) *A source book in mathematics.* McGraw-Hill Book Company—New York. xvii + 701.

Smith, H. F. (1957) "On comparing contingency tables," *Philippine Statistician* **6(2)**, 71–81.

Smith, J. H. (1947) "Estimation of linear functions of cell proportions," *Ann. Math. Statist.* **18**, 231–254.

Smith, K. (1916) "On the best values of the constants in frequency distributions," *Biometrika* **11**, 262–276.

Smith, K. (1918) "On the standard deviations of adjusted and interpolated values of an observed polynomial function and its constants and the guidance they give towards a proper choice of the distribution of observations," *Biometrika* **12**, 1–85.

Smoliakow, P. T. (1933) "Die Fechnersche Korrelationsformel," *Meteorol. Z.* **50**, 87–93.

Snedecor, G. W. (1958) "Chi-squares of Bartlett, Mood and Lancaster, in a 2^3 contingency table," *Biometrics* **14**, 560–562.

Snedecor, G. W. and Irwin, M. R. (1933) "On the chi-square test for homogeneity," *Iowa State Coll. J. Sci.* **8**, 75–81.

Somers, R. H. (1962) "A new asymmetric measure of association for ordinal variables," *Amer. Soc. Rev.* **27**, 799–811.

Somers, R. H. (1964) "Simple measures of association for the triple dichotomy," *J. Roy. Statist. Soc.* **A127**, 409–415.

Soper, H. E. (1914) "Tables of Poisson's exponential limit," *Biometrika* **10**, 25–35.

Soper, H. E. (1915) "On the probable error of the biserial expression for the correlation coefficient," *Biometrika* **10**, 384–390.

Soper, H. E. (1916) cited by Pearson (1916*b*).

Soper, H. E. (1923) *Frequency arrays,* Cambridge Univ. Press. 48.

Soper, H. E. (1926) "The moments of the hypergeometric series," *J. Roy. Statist. Soc.* **89**, 326–328.

Spearman, C. (1904) "The proof and measurement of the association between two things," *Amer. J. Psychol.* **15**, 72–101.

Spearman, C. (1906) "A footrule for measuring correlation," *British J. Psych.* **2**, 89–108.

Srinivasan, R. (1965) "The bivariate gamma distribution and the random walk problem," *Proc. Indian Acad. Sci. Sect. A,* **62**, 358–366.

Stanley, J. C. (1961) "Analysis of unreplicated three-way classifications with applications to rater bias and trait independence," *Psychometrika* **26**, 205–219.

Steck, G. P. (1957) *Limit theorems for conditional distributions.* Univ. California Publ. Statist. **2 (12)**, 237–284.

Steffensen, J. F. (1934) "On certain measures of dependence between statistical variables," *Biometrika* **26**, 251–255.

Steffensen, J. F. (1941*a*) "On the coefficient of correlation for continuous distributions," *Skand. Aktuarietidskr.* **24**, 1–12. corrig. 232.

Steffensen, J. F. (1941*b*) "On the ω test of dependence between statistical variables," *Skand. Aktuarietidskr.* **24**, 13–33.

Stepanov, V. E. (1957) "Some statistical tests for Markov chains," *Teor. Veroyatnost. i. Primenen.* **2**, 143–144.

Stephan, F. F. (1942) "An iterative method of adjusting sample frequency tables when expected marginal totals are known," *Ann. Math. Statist.* **13**, 166–178.

Stephens, M. A. (1963) "The distribution of the goodness of fit statistic U_N^2," *Biometrika* **50**, 303–313.

Stephens, M. A. (1964) "The distribution of the goodness of fit statistic U_N^2 II," *Biometrika* **51**, 393–397.

Stephens, M. A. (1965) "Significance points for the two sample statistic $U_{M,N}^2$," *Biometrika* **52**, 661–663.

Sterne, S. (1938) "A preliminary note on tests of significance and problems of goodness of fit," *Norske Vid. Selsk. Forh.* **11**, 68–71.

Stevens, W. L. (1938) "The distribution of entries in a contingency table with fixed marginal totals," *Ann. Eugen. London* **8**, 238–244.

Stevens, W. L. (1950) "Fiducial limits of the parameter of a discontinuous distribution," *Biometrika* **37**, 117–129.

Stevens, W. L. (1951) "Mean and variance of an entry in a contingency table," *Biometrika* **38**, 468–470.

Steyn, H. S. (1951) "On discrete multivariate probability functions," *Proc. Acad. Sci. Amst.* **A54**, 23–30.

Steyn, H. S. (1957) "On regression properties of discrete systems of probability functions," *Indagationes Math.* **19**, 119–127.

Steyn, H. (1959) "On χ^2 tests for contingency tables of negative multinomial types," *Statistica Neerlandica* **13**, 433–444.

Steyn, H. S. (1960) "On regression properties of multivariate probability functions of Pearson's types," *Indag. Math.* **22**, 302–311.

Steyn, H. S. (1963) "On approximations for the distributions obtained from multiple events," *Nederl. Akad. Wetensch. Proc. Ser.* **A66-*Indag. Math.* 25**, 85–96.

Stieltjes, T. J. (1889) "Extract d'une lettre addressée à M. Hermite," *Bull. Sci. Math.* **(2) XIII**, 170–172.

Stieltjes, T. J. (1914/18) *Oeuvres complètes de Thomas Jan Stieltjes.* I, vii + 471, II, iv + 603. Soc. Math. d'Amsterdam. P. Noordhoff, Groningen.

Stoneham, R. G. (1965) "A study of 60,000 digits of the transcendental '*e*'," *Amer. Math. Monthly* **72**, 483–500.

Stuart, A. (1953) "The estimation and comparison of strengths of association in contingency tables," *Biometrika* **40**, 105–110.

Stuart, A. (1954) "Too good to be true," *Appl. Statistics* **3**, 29–32.

Stuart, A. (1955) "A test for the homogeneity of the marginal distributions in a two way classification," *Biometrika* **42**, 412–416.

"Student" (W. S. Gosset) (1907) "On the error of counting with a haemocytometer," *Biometrika* **5**, 351–360.

"Student" (W. S. Gosset). (1908) "The probable error of a mean," *Biometrika* **6**, 1–25.

"Student" (1919) "An explanation of deviations from Poisson's law in practice," *Biometrika* **12**, 211–215.

Studer, H. (1966) "Prüfung der Annäherung der exakten χ^2 Verteilung durch die stetige χ^2 Verteilung," *Metrika* **11**, 55–78.

Sukhatme, P. V. (1935) "A contribution to the problem of two samples," *Proc. Indian. Acad. Sci.* **2A**, 584–604.

Sukhatme, P. V. (1937a) "Tests of significance for samples of the χ^2 population with two degrees of freedom," *Ann. Eugen. London* **8**, 52–56.

Sukhatme, P. V. (1937b) "The problem of K samples from a Poisson population," *Proc. Nat. Inst. Sci. India* **3**, 297–305.

Sukhatme, P. V. (1938) "On the distribution of χ^2 in samples of the Poisson series," *J. Roy. Statist. Soc.* suppl. **5**, 75–79.

Sutcliffe, J. P. (1957) "A general method of analysis of frequency data for multiple classification designs," *Psychol. Bull.* **54**, 134–137.

Swaroop, S. (1938) "Tables of the exact values of probabilities for testing the significance of difference between proportions based on pairs of small samples," *Sankhyā* **4**, 73–84.

Swaroop, S. (1950) "Exact significance of difference in response under two treatments," *Indian Med. Res. Mem.* **35**, 1–113, Thacker Spink and Co., Calcutta.

Swineford, F. (1946) "Graphical and tabular aids for determining sample size when planning experiments which involve comparisons or percentages," *Psychometrika* **11**, 43–49.

Swineford, F. (1948) "A table for estimating the significance of the difference between correlated percentages," *Psychometrika* **13**, 23–25.

Swineford, F. (1949) "Further notes on differences between percentages," *Psychometrika* **13**, 183–187.

Szegö, G. (1939) *Orthogonal polynomials*, Amer. Math. Soc. Coll. Ser. No. 23, New York. ix + 401 Revised 1959, ix + 421.

Takács, L. (1958) "On a probability problem in the theory of counters," *Ann. Math. Statist.* **29**, 1257–1263.

Tallis, G. M. (1962) "The maximum likelihood estimation of correlation from contingency tables," *Biometrics* **18**, 342–353.

Tamura, R. (1961/62) "On the efficiency of Sukhatme's test," *Bull. Math. Statist.* **10(3/4)** 31–38.

Tang, P. C. (1938) "The power function of the analysis of variance tests with tables and illustrations of their use," *Statist. Res. Mem.* **2**, 126–157.

Tate, R. F. (1954) "Correlation between a discrete and a continuous variable. Point-biserial correlation," *Ann. Math. Statist.* **25**, 603–607.

Tate, R. F. (1955) "The theory of correlation between two continuous variables when one is dichotomized," *Biometrika* **42**, 205–216.

Taylor, W. F. (1951) "Mathematical statistical studies on the tuberculin and histoplasmin skin test. I Overlapping of negative and positive reactions," *Hum. Biol.* **23**, 1–23.

Tchebichef, P. L. (1864) "Sur l'interpolation," *Zapiski Akad. Nauk* **4**, No. 5 reprinted *Oeuvres* **1**, 539–560 edited by A. A. Markov and N. Sonin, Reprinted by Chelsea Publ. Co. 1961.

Thionet, P. (1959) "L'ajustement des résultats des sondages sur ceux des dénombrements," *Rev. Inst. Internat. Statist.* **27**, 8–25.

Thompson, C. M. (1941) "Tables of the percentage points of the χ^2 distribution," *Biometrika* **32**, 187–191.

Thompson, W. R. (1933) "On the likelihood that one unknown probability exceeds another in view of the evidence of two samples," *Biometrika* **25**, 285–294.

Thomson, D. H. (1947) "Approximate formulae for the percentage points of the incomplete beta function and of the χ^2 distribution," *Biometrika* **34**, 368–372.

Thomson, G. H. (1919) "The criterion of goodness of fit of psychophysical curves," *Biometrika* **12**, 216–230.

Tiku, M. L. (1965a) "Laguerre series forms of non-central χ^2 and F distributions," *Biometrika* **52**, 415–427.

Tiku, M. L. (1965b) "Chi-square approximations for the distributions of goodness of fit statistics U_N^2 and W_N^2," *Biometrika* **52**, 630–633.

Tippett, L. H. C. (1932) "A modified method of counting particles," *Proc. Roy. Soc.* A137, 434–446.

Tocher, J. F. (1908) "Pigmentation survey of school children in Scotland" *Biometrika*, **6**, 130–235.

Tocher, K. D. (1950) "Extension of the Neyman–Pearson theory of tests to discontinuous variates," *Biometrika* **37**, 130–144.

Todhunter, I. (1865) *A history of the mathematical theory of probability from the time of Pascal to that of Laplace.* Macmillan, London. xvi + 624.

Todhunter, I. (1869) "On the method of least squares," *Trans. Camb. Philos. Soc.* **11**, 219–238.

Tricomi, F. G. (1955) *Vorlesungen über Orthogonalreihen*, Die Grundlehren der math. Wissenschaften, Band 76, Springer, Berlin. viii + 264.

Tricomi, F. G. (1957) *Integral equations.* Interscience, New York, viii + 238.

Tschuprow, A. A. (1918a) "On the mathematical expectation of the moments of frequency distributions," *Biometrika* **12**, 140–149 and 185–210.

Tschuprow, A. A. (1918b) "Zur Theorie der Stabilität statistischer Reihen I," *Skand. Aktuarietidskr.* **1**, 199–256.

Tschuprow, A. A. (1919) "Zur Theorie der Stabilität statistischer Reihen II," *Skand. Aktuarietidskr.* **2**, 80–133.

Tschuprow, A. A. (1920) "On the mathematical expectation of the moments of frequency distributions. Part II," *Biometrika* **13**, 283–295.

Tschuprow, A. A. (1923a) "On the mathematical expectation of the moments of frequency distributions in the case of correlated observations," *Metron* **2**, No. 3, 461–493 and No. 4, 646–683.

Tschuprow, A. A. (1923b) "Über normal stabile Korrelation," *Skand. Aktuarietidskr.* **6**, 1–17.

Tschuprow, A. A. (1925) *Grundbegriffe und Grundprobleme der Korrelationstheorie.* Teubner, Leipzig. Translated by M. Kantorowitsch. *Principles of the mathematical theory of correlation* (1939) W. M. Hodge and Co. Limited, London. x + 194.

Tschuprow, A. A. (1934) "The mathematical foundations of the methods to be used in statistical investigation of the dependence between two chance variables," *Nordisk Statist. Tidskr.* **5**, 34.

Tukey, J. W. (1957) "Approximations to the upper 5% points of Fisher's B distribution and non-central χ^2. *Biometrika* **44**, 528–530.

Tumanyan, S. H. (1954) "On the asymptotic distribution of the χ^2 criterion," (Russ.) *Dokl. Akad. Nauk. SSSR. (N.S.).* **94**, 1011–1012.

Tumanyan, S. H. (1956) "Asymptotic distribution of χ^2 criterion when the size of observations and the number of groups simultaneously increase," (Russ.) *Teor. Veroyatnost. i. Primenen.* **1**, 131–145.

Tumanyan, S. H. (1958a) "On the power of the chi-square criterion applied to the problem of two samples relative to 'near' alternatives," (Russ.) *Izv. Akad. Nauk. Armyan SSR. Ser. Fiz-Mat. Nauk* **11**, **(6)**. 31–45.

Tumanyan, S. H. (1958b) "The χ^2 test applied to the problem of two samples," (Russian). *Proc. All-Union Conf. Theory Prob and Math. Statist* (Erevan 1958). (Russian.) Izdat. Akad. Nauk Armyan. SSR. (Erevan 1960). (i) 121–138.

330   BIBLIOGRAPHY

Turnbull, H. W. and Aitken, A. C. (1932) *An introduction to the theory of canonical matrices.* Blackie and Sons, London. xiii + 192.

Uhlmann, W. (1966) "Vergleich der hypergeometrischen mit der Binomial—Verteilung," *Metrika* 10, 145–158.

Uppuluri, V. R. R. and Bowman, K. O. (1966) "Likelihood ratio test criterion for small samples from multinomial distributions," *Oak Ridge Nat. Lab.* –3991, UC-32. vi + 47.

U.S.A. National Bureau of Standards, (1949) *Tables of the binomial probability distribution.* (Applied Mathematics Series 6.) Washington, Government Printing Office, x + 387.

U.S.A. National Bureau of Standards, (1952) *A guide to tables of the normal probability integral.* (Applied Mathematics series 21.) Washington, Supt. of Documents. iv + 16.

U.S.A. National Bureau of Standards, (1953) *Tables of normal probability functions.* (Applied mathematics Series 23.) ix + 343.

U.S.A. National Bureau of Standards, (1959) *Tables of the bivariate normal distribution function and related functions.* (Applied Mathematics Series 50.) Nat. Bureau Standard, Superintendent of Documents, Washington. xlv + 258.

Uspensky, J. V. (1937) *Introduction to mathematical probability.* McGraw-Hill and Co., New York and London, ix + 411.

Vajda, S. (1943) "The algebraic analysis of contingency tables," *J. Roy. Statist. Soc.* 106, 333–342.

van der Waerden, B. L. (1956) "The computation of the X-distribution," *Proc. 3rd. Berkely Symp.* 207–208.

van der Waerden, B. L. (1957) *Mathematische Statistik. Die Grundlehren math. Wiss. in Einzeldarstellungen, Band 87.* Springer-Verlag, Berlin. 360 pp.

van der Waerden, B. L. and Nievergelt, N. (1956) *Tables for comparing two samples by X-test and sign test.* Springer-Verlag, Berlin. 34 pp.

van Eeden, C. (1965) "Conditional limit distributions for the entries in a $2 \times k$ contingency table," in Patil (1965), 123–26.

van Eeden, C. and Runnenburg, J. (1960) "Conditional limit distributions for the entries in a 2×2 table," *Statistica Neerlandica* 14, 111–126.

van Heerden, D. F. I. and Gonin, H. T. (1966) "The orthogonal polynomials of power series probability distributions and their uses," *Biometrika* 53, 121–128.

van Heerden, D. F. I. and Gonin, H. T. (1967) "The orthogonal polynomials of the factorial power series probability distributions," *S. African Statist. J.* 1, 49–53.

van Klinken, J. (1959) "On some estimation problems with regard to the Poisson distribution and the χ^2-minimum method," *Mitt. Verein. Schweiz. Versich.-Math.* 59, 297–306.

van Uven, M. J. (1947/8) "Extension of Pearson's probability distributions to two variables, I. II. III and IV," *Proc. Kon. Ned. Akad. Wetensch.* A50, 1063–70, 1252–64, 51, 41–52, 191–6. (*Indag. Math.* 9, 477–84 578–90 10, 12–23 and 62–67.)

van Zwet, W. R. and Oosterhoff, J. (1967) "On the combination of independent test statistics," *Ann. Math. Statist.* 38, 659–680.

Vaswani, S. (1950) "Assumptions underlying the use of the tetrachoric correlation coefficient," *Sankhyā* 10, 269–276.

Venter, J. H. (1967) "Probability measures on product spaces," *South Afr. Statist. J.* 1, 3–20.

Vernon, P. E. (1936) "A note on the standard error in the contingency matching technique," *J. Educ. Psych.* 27, 704–709.

Vessereau, A. (1958) "Sur les conditions d'application du criterium χ^2 de Pearson," *Bull. Inst. Internat. Statist.* 36, 87–101.

Vietoris, L. (1961) "Eine die Stichprobenverteilung betreffende Abschätzung," *Monatsh. Math.* 65, 287–290.

Vincze, I. (1960) "On the deviation of two-variate empirical distribution function," *Magyar Tud. Akad. Mat. Fiz. Oszt. Kozl.* **10**, 361–372.

Vitali, G. (1921) "Sulla condizione di chuisura di un sistema di funzioni ortogonali," *Rend. R. Accad. Naz. Lincei.* (5) **30**, 498–501.

Volodin, I. N. (1964) "Testing the hypothesis of normality of a distribution by small samples (multivariate case)," (Russian.) *Kazansk. Gos. Univ. Ucen. Zap.* **124**, 21–25.

von Bortkiewicz, L. (1898) *Das Gesetz der kleinen Zahlen.* Teubner, Leipzig. vi + 52.

von Bortkiewicz, L. (1910) "Zur Verteidigung des Gesetzes der kleinen Zahlen," *Jahrb. f. Nationalök u. Stat.* **39** (3) 218–236.

von Bortkiewicz, L. (1915a) "Realismus und Formalismus in der mathematischen Statistik," *Allg. Statist. Arch.* (München). **9**, 225–256.

von Bortkiewicz, L. (1915b) "Über den Präzisionsgrad des Divergenz Koeffizientes," *Mitteil des Verbandes der österr. und ungar. Versicherungstechniker.* **5**,

von Bortkiewicz, L. (1917) *Die Iterationen: ein Beitrag zur Wahrscheinlichkeitstheorie.* Springer-Verlag, Berlin, xii + 205.

von Bortkiewicz, L. (1918) "Homogeneität und Stabilität in der Statistik," *Skand. Aktuarietidskr.* **1**, 1–81.

von Bortkiewicz, L. (1922) "Das Helmertsche Verteilungsgesetz für die Quadratsumme zufälliger Beobachtungsfehler," *Zeit. Angew. Math. Mech.* **2**, 358–375.

von Bortkiewicz, L. (1931) "The relations between stability and heterogeneity," *Ann. Math. Statist.* **2**, 1–22.

von Mises, R. (1919) "Grundlagen Wahrscheinlichkeitsrechnung," *Math. Zeit.*, **5**, 52–99. corr. 7 (1920) 323.

von Mises, R. (1931) *Wahrscheinlichkeitsrechnung.* Deutickes Wien, x + 574.

von Schelling, H. (1940) "Über die exakte Behandlung des Zusammenhänges zwischen biologischen Werkmalreihen," *Arbeit a.d. Staatl. Instit. f exper. Therapie.* **39**, 35–71.

von Schelling, H. (1949) "A formula for the partial sums of some hypergeometric series," *Ann. Math. Statist.* **20**, 120–122.

von Schelling, H. (1950) "A second formula for the partial sum of hypergeometric series having unity as the fourth argument," *Ann. Math. Statist.* **21**, 458–460.

Vora, S. A. (1951) "Bounds on the distribution of chi-square," *Sankhyā* **11**, 365–378.

Wahlund, S. (1935) "A new method of determining correlation from tetrachoric groupings," *Lantbrukshögsk. Ann.* **2**, 181–242.

Waite, H. (1915) "Association of finger prints," *Biometrika.* **10**, 421–478.

Wald, A. (1941a) "Asymptotically most powerful tests of statistical hypothesis," *Ann. Math. Statist.* **12**, 1–19.

Wald, A. (1941b) "Some examples of asymptotically most powerful tests," *Ann. Math. Statist.* **12**, 396–408.

Wald, A. (1943) "Tests of statistical hypotheses concerning several parameters when the number of observations is large," *Trans. Amer. Math. Soc.* **54**, 426–482.

Wald, A. and Wolfowitz, J. (1940) "On a test whether two samples are from the same population," *Ann. Math. Statist.* **11**, 147–162.

Walker, H. M. (1929) *Studies in the history of statistical method.* Williams and Wilkins, Baltimore. viii + 229.

Walker, H. M. (1934) "Abraham de Moivre," *Scripta Math.* **2**, 316–333.

Wallace, D. L. (1958a) "Asymptotic approximations to distributions," *Ann. Math. Statist.* **29**, 635–654.

Wallace, D. L. (1958b) "Bounds for normal approximations to Student's ratio and the chi-square distribution," *Report SRC-80710 Wc 89, Statist. Res. Centre, Univ. Chicago.*

Wallace, D. L. (1959) "Bounds on normal approximations to Student's and the chi-square distributions," *Ann. Math. Statist.* **30**, 1121–1130.

332   BIBLIOGRAPHY

Wallis, W. A. (1942) "Compounding probabilities from independent significance tests," *Econometrica* **10**, 229–248.

Walsh, J. E. (1957) "Validity of approximate normality values for $\mu \pm k\sigma$ areas of practical type continuous populations," *Ann. Inst. Statist. Math. Tokyo* **8**, 79–86.

Walsh, J. E. (1957/58) "Further consideration of normality values for $\mu \pm k\sigma$ areas of continuous populations," *Ann. Inst. Statist. Math. Tokyo* **9**, 127–129.

Walsh, J. E. (1962/65) *Handbook of nonparametric statistics. I Investigations of randomness, moments, percentiles and distribution. II Results for two and several sample problems, symmetry and extremes.* D. Van Nostrand Co. Inc. Princeton, N.J. xxvi + 549, and xxvi + 686.

Walsh, J. E. (1963) "Loss in test efficiency due to misclassification for 2 × 2 tables," *Biometrics* **19**, 158–162.

Walsh, J. L. (1923) "A closed set of normal orthogonal functions," *Amer. J. Math.* **45**, 5–24.

Walter, E. (1954) "Test zur Prüfung der Symmetrie bezüglich Null," *Mitt. Math. Statist.* **6**, 92–104.

Walter, E. (1959) "Einige Eigenschaften von Symmetrietests," *Math. Ann.* **137**, 433–453.

Watanabe, S. (1967) "Karhunen–Loève expansion and factor analysis: theoretical remarks and applications," *Trans. 4th Prague Conf. Inform. Theory*, 635–660.

Watson, G. N. (1929) "Theorems stated by Ramanujan (v) approximations connected with e^x," *Proc. London Math. Soc.* (2) **29**, 293–308.

Watson, G. N. (1933a) "Notes on generating functions of polynomials (2) Hermite polynomials," *J. London Math. Soc.* **8**, 194–199.

Watson, G. N. (1933b) "Notes on generating functions of polynomials (3) Polynomials of Legendre and Gegenbauer," *J. London Math. Soc.* **8**, 289–292.

Watson, G. N. (1934) "Notes on generating functions of polynomials (4) Jacobi polynomials," *J. London Math. Soc.* **9**, 22–28.

Watson, G. S. (1956) "Missing and mixed up frequencies in contingency tables," *Biometrics* **12**, 47–50.

Watson, G. S. (1957) "The goodness of fit test for normal distributions," *Biometrika* **44**, 336–348.

Watson, G. S. (1958) "On chi-square goodness of fit tests for continuous distributions," *J. Roy. Statist. Soc.* **B20**, 44–72.

Watson, G. S. (1959) "Some recent results in chi-square goodness of fit tests," *Biometrics* **15**, 440–468.

Wegner, L. H. (1956) "Properties of some two-sample tests based on a particular measure of discrepancy," *Ann. Math. Statist.* **27**, 1006–1016.

Weichselberger, K. (1959) "Über die Parameterschätzung bei Kontingenztafeln, deren Randsummen vorgegeben sind," *Metrika* **2**, 100–130. II *ibid.*, 198–229.

Weida, F. M. (1934) "On measures of contingency," *Ann. Math. Statistics.* **5**, 308–319.

Weiler, H. (1966) "A coefficient measuring the goodness of fit," *Technometrics* **8**, 327–334.

Weiner, I. B. (1959) "A note on the use of Mood's likelihood ratio test for item analyses involving 2 × 2 tables with small samples," *Psychometrika.* **24**, 371–372.

Welch, B. L. (1936) "Note on an extension of the L_1 test," *Statist. Res Mem.* **1**, 52–56.

Welch, B. L. (1938) "On tests of homogeneity," *Biometrika* **30**, 149–158.

Wells, W. T., Anderson, R. L. and Cell, J. W. (1962) "The distribution of the product of two central or non-central chi-square variates," *Ann. Math. Statist.* **33**, 1016–1020.

Wherry, R. J. (1947) "Multiple biserial and multiple point biserial correlation," *Psychometrika* **12**, 189–195.

Wherry, R. J. and Taylor, E. K. (1946) "The relation of multiserial eta to other measures of correlation," *Psychometrika* **11**, 155–161.

Whitaker, L. (1914) "On the Poisson law of small numbers," *Biometrika* **10**, 36–71.

Whittaker, E. T. and Robinson, G. (1940) *The calculus of observations.* 3rd edit. Blackie and Sons, xvi + 395.

Whittaker, E. T. and Watson, G. N. (1943) *A course of modern analysis.* Camb. Univ. Press. London. iv + 608. 4th Edit.

Whittle, P. (1959/60) "Quadratic forms in Poisson and multinomial variables," *J. Austral. Math. Soc.* **1**, 233–240.

Wicksell, S. D. (1916) "Some theorems in the theory of probability with special reference to their importance in the theory of homograde correlation," *Svenska Aktuarietidskr* **3**, 165–213.

Wicksell, S. D. (1917*a*) "The construction of the curves of equal frequency in case of type *A* correlation," *Svenska Aktuarietidskr* **4**, 122–140.

Wicksell, S. D. (1917*b*) "The application of solid hypergeometric series to frequency distributions in space," *Phil. Mag.* **(6)33**, 389–394.

Wicksell, S. D. (1917*c*) "The correlation function of type *A*," *Kungl. Svenska Vetenskapsakad. Handl. Bd 58; Medd. Lunds Astr. Obs.* Series 2, No. 17.

Wicksell, S. D. (1933) "On correlation functions of type III," *Biometrika* **25**, 121–133.

Widén, L. (1962) "The development of mortality within different groups of causes of death in Sweden," 1951–1960. *Statist. Tidskrift.* **11**, 1–14.

Wiid, J. B. (1957/58) "On the moments and regression equations of the fourfold negative and fourfold negative factorial binomial distributions," *Proc. Roy. Soc. Edinburgh* **A 65**, 29–34.

Wijsman, R. A. (1957) "Random orthogonal transformations and their use in some classical distribution problems in multivariate analysis," *Ann. Math. Statist.* **28**, 415–423.

Wijsman, R. A. (1958) "Contribution to the study of the question of association between two diseases," *Hum. Biol.* **30**, 219–236.

Wilks, S. S. (1932) "Certain generalizations of the analysis of variance," *Biometrika* **24**, 471–494.

Wilks, S. S. (1935) "The likelihood test of independence in contingency tables," *Ann. Math. Statist.* **6**, 190–196.

Wilks, S. S. (1938) "The large sample distribution of the likelihood ratio for testing composite hypotheses," *Ann. Math. Statist.* **9**, 60–62.

Wilks, S. S. (1940) "Confidence limits and critical differences between percentages," *Publ. Opin. Quarterly* **4**, 332–338.

Wilks, S. S. and Daly, J. F. (1939) "An optimum property of confidence regions associated with the likelihood function," *Ann. Math. Statist.* **10**, 225–235.

Williams, C. A. Jnr. (1950) "On the choice of the number and width of classes for the chi-square test of goodness of fit," *J. Amer. Statist. Assoc.* **45**, 77–86.

Williams, E. J. (1952) "Use of scores for the analysis of association in contingency tables," *Biometrika* **39**, 274–289.

Williamson, E. and Bretherton, M. H. (1963) *Tables of the negative binomial distribution.* J. Wiley and Sons, New York. 275 pp.

Wilson, E. B. (1931) "Correlation and association," *J. Amer. Statist. Assoc.* **26**, 250–257.

Wilson, E. B. (1941) "The controlled experiment and the fourfold table," *Science* (N.S.). **93**, 557–560.

Wilson, E. B. (1942) "On contingency tables," *Proc. Nat. Acad. Sci. U.S.A.* **28**, 94–100.

Wilson, E. B. and Hilferty, M. M. (1931) "The distribution of chi-square," *Proc. Nat. Acad. Sci.* **17**, 684–688.

Wilson, E. B., Hilferty, M. M. and Maher, H. C (1931) "Goodness of fit," *J. Amer. Statist. Assoc.* **26**, 443–448.

Wilson, E. B. and Worcester, J. (1942a) "The association of three attributes," *Proc. Nat. Acad. Sci. U.S.A.* **28**, 384–390.

Wilson, E. B. and Worcester, J. (1942b) "Contingency tables," *Proc. Nat. Acad. Sci. U.S.A.* **28**, 378–384.

Winsor, C. P. (1948a) "Factorial analysis of a multiple dichotomy," *Hum. Biol.* **20**, 195–204.

Winsor, C. P. (1948b) "Probability and Listerism," *Hum. Biol.* **20**, 161–169.

Wise, M. E. (1954) "A quickly convergent expansion for cumulative hypergeometric probabilities, direct and inverse," *Biometrika* **41**, 317–329. Corrig. **42**, 277.

Wise, M. E. (1963) "Multinomial probabilities and the χ^2 and X^2 distributions," *Biometrika* **50**, 145–154.

Wise, M. E. (1964) "A complete multinomial compared with approximation and an improvement to it," *Biometrika* **51**, 277–281.

Wishart, J. (1932) "A note on the distribution of the correlation ratio," *Biometrika* **24**, 441–456.

Wishart, J. (1947a) "Proof of the distribution of χ^2 of the estimate of variance and of the variance ratio," *J. Inst. Actuar. Stud. Soc.* **7**, 98–103.

Wishart, J. (1947b) "The cumulants of the z and of the logarithmic χ^2 and t distributions," *Biometrika* **34**, 170–178. Corrig. 374.

Wishart, J. (1948) "Proofs of the distribution law of the second order moment statistics," *Biometrika* **35**, 55–57. note: **35**, 422.

Wishart, J. (1949) "Cumulants of multivariate multinomial distributions," *Biometrika* **36**, 47–58.

Wishart, J. (1956) "χ^2 Probabilities for large numbers of degrees of freedom," *Biometrika* **43**, 92–95.

Witting, H. (1959) "On a χ^2 test with cells determined by order statistics," *Arch. d. Mathematik* **10**, 468–479.

Wolfowitz, J. (1949) "The power of the classical tests associated with the normal distribution," *Ann. Math. Statist.* **20**, 540–551.

Wong, E. and Thomas, J. B. (1962) "On polynomial expansions of second-order distributions," *J. Soc. Indust. Appl. Math.* **10**, 507–516.

Woo, T. L. (1929) "Tables for ascertaining the significance or non-significance of association measured by the correlation ratio," *Biometrika* **21**, 1–66.

Woodworth, R. S. (1912) "Combining the results of several tests. A study in statistical method," *Psychol. Rev.* **19**, 97–123.

Woolf, B. (1957) "The log likelihood ratio test (the *G*-test). Methods and tables for tests of heterogeneity in contingency tables," *Ann. Human. Gen.* **21**, 397–409.

Yang, Chung-How (1965) "Chi-square test on hypergeometric variates," *Bull. Math. Soc. Nanyang Univ., 1965*, 76–80.

Yates, F. (1934) "Contingency tables involving small numbers of the χ^2 test," *J. Roy. Statist. Soc. Suppl. 1.* 217–235.

Yates, F. (1948) "The analysis of contingency tables with groupings based on quantitative characters," *Biometrika* **35**, 176–181. corr. **35**, 424

Yates, F. (1955a) "The use of transformations and maximum likelihood in the analysis of quantal experiments involving two treatments," *Biometrika* **42**, 382–403.

Yates, F. (1955b) "A note on the application of the combination of probabilities test to a set of 2 × 2 tables," *Biometrika* **42**, 404–411.

Yates, F. (1961) "Marginal percentages in multiway tables of quantal data with disproportionate frequencies," *Biometrics* **17**, 1–9.

Yoshimura, I. (1963) "A moment recurrence relation and its application to multinomial distributions and others," *Rep. Statist. Appl. Res. Un. Jap. Sci. Engrs.* **10**, No. 2, 137–150.

Young, A. W. and Pearson, K. (1916) "On the probable error of a coefficient of contingency without approximation," *Biometrika* **11**, 215–230.

Young, D. H. (1962) "Two alternatives to the standard χ^2 test of the hypothesis of equal cell frequencies," *Biometrika*, **49**, 107–116.

Yule, G. U. (1897a) "On the theory of correlation," *J. Roy. Statist. Soc.* **60**, 812–854.

Yule, G. U. (1897b) "On the significance of Bravais' formulae for regression, etc. in the case of skew correlation," *Proc. Roy. Soc.* **A60**, 477–489.

Yule, G. U. (1900) "On the association of attributes in statistics," *Phil. Trans. Roy. Soc.* **A194**, 257–319.

Yule, G. U. (1901) "On the theory of consistence of logical class frequencies, and its geometrical representation," *Phil. Trans. Roy. Soc.* **A197**, 91–134.

Yule, G. U. (1903) "Notes on the theory of association of attributes in statistics," *Biometrika* **2**, 121–134.

Yule, G. U. (1906) "On a property which holds good for all groupings of a normal distribution, etc.," *Proc. Roy. Soc.* **A77**, 324–336.

Yule, G. U. (1907) "On the theory of correlation for any number of variables treated by a new system of notation," *Proc. Roy. Soc.* **A79**, 182–193.

Yule, G. U. (1909) "The applications of the method of correlation to social and economic statistics," *J. Roy. Statist. Soc.* **72**, 721–730.

Yule, G. U. (1911) *An Introduction to the Theory of Statistics.* Charles Griffin and Co. Ltd., London.

Yule, G. U. (1912) "On the methods of measuring the association between two attributes," *J. Roy. Statist. Soc.* **75**, 579–652.

Yule, G. U. (1922) "On the application of the χ^2 method to association and contingency tables, with experimental illustrations," *J. Roy. Statist. Soc.* **85**, 95–104.

Yule, G. U. (1938) "On some properties of normal distributions, univariate and bivariate, based on sums of squares of frequencies," *Biometrika* **30**, 1–10.

Zelen, M. and Joel, L. S. (1959) "The weighted compounding of two independent significance tests," *Ann. Math. Statist.* **30**, 885–895.

Zelen, M. and Severo, N. C. (1960) "Graphs for bivariate normal probabilities," *Ann. Math. Statist.* **31**, 619–624.

Zubin, J. (1939) "Nomographs for determining the significance of the differences between the frequencies of events in two contrasted series or groups," *J. Amer. Statist. Assoc.* **34**, 539–544.

Zygmund, A. (1959) *Trigonometric series.* Camb. Univ. Press. Vols. 1 and 2, xii + 383 and vii + 354.

Index to the Bibliography

CHAPTER I. HISTORICAL SURVEY OF χ^2

Bartlett, 1935; Bernstein, 1926; Bienaymé, 1838, 1852; Bowley, 1920; Bravais, 1846; Brownlee, 1924; Cayley A. (See Todhunter, 1869); de Moivre, 1733; Edgeworth, 1892; Ellis, 1844a; Fisher, 1915, 1922a, b, c, 1923, 1924, 1928b, 1935b, 1950a, *Statistical Methods;* Fisher, Thornton and Mackenzie, 1922; Greenwood and Yule, 1915; Haldane, 1939a; Helmert, 1875b; Huttley, 1951; Lancaster, 1966a; Laplace, *Oeuvres;* Lexis, 1877, Liebermeister, 1877, Maung, 1942; K. Pearson, 1896, 1900a, b, 1901, 1904, 1905a, 1916a, b, c, 1920, 1924c, 1928b; K. Pearson and Heron, 1913; K. Pearson and Lee, 1908; Poisson, 1827, 1832; Seal, 1967; Sheppard, 1898a, b; D. E. Smith, 1929; Soper, 1916; Todhunter, 1865, 1869; Walker, 1929, 1934; Winsor, 1948a; E. S. Pearson, 1936/7.

CHAPTER II. DISTRIBUTION THEORY

Texts: Anderson, 1958; Cramér, 1937, 1946; Eisenhart, 1949; Fisher, 1950a, *Statistical Methods;* Fisher and Yates, 1938; Greenwood and Hartley, 1962; Hald, 1952; Harter, 1964b; Kendall and Stuart, 1961, 1963; Laplace, *Oeuvres;* Lieberman and Owen, 1961; Lukacs, 1959; Lukacs and Laha, 1963; Molina, 1945; Monk and Owen, 1957; Mood, 1950; Owen, 1957, 1962; Pagurova, 1963; Patil, 1965; E. S. Pearson and Hartley, 1958; K. Pearson, 1922a, 1931; Rao, 1952a; Shohat and Tamarkin, 1943; Slutskii, 1950; Smirnov, 1965; Soper, 1923; Turnbull and Aitken, 1932; U.S.A. National Bureau of Standards, 1949, 1952, 1953, 1959; van der Waerden, 1957; Whittaker & Watson, 1943.

1.–3. Properties of the Distributions of the Γ-variable and χ^2

Bhattacharyya, 1945; Bienaymé, 1838; Cornish and Fisher, 1937; Craig, 1943; Doodson, 1917, Ellis, 1844a, b; Fisher, 1922a; 1924, 1928b, 1935; Fog, 1948; Gnedenko and Kolmogorov, 1954; Haldane, 1942a; Helmert, 1875a, b; Isserlis, 1931; Jambunathan, 1954; Jarrett, 1946; Johnson and Welch, 1939; Khamis, 1960; Lancaster, 1954, 1965a, 1968; Laplace, *Oeuvres;* K. Pearson, 1895, 1900a, b, 1901, 1916a, b, 1922a; Raj, 1953; Ramachandran, 1958, Rietz, 1923, Rust, 1965; Rybarz, 1959; Satterthwaite, 1942; Sawkins, 1940; Scheuer, 1962; Shohat and Tamarkin, 1943; Student, 1908; Turnbull and Aitken, 1932; van der Waerden, 1956, 1957; von Bortkiewicz, 1922; Vora, 1951; Wallace, 1958a, b; Watson, 1929; Wells, Anderson and Cell, 1962; Wilson and Hilferty, 1931; Wilson, Hilferty and Maher, 1931; Whittaker and Robinson, 1940; Wishart, 1947a, b, 1956; C. C. Craig, 1929; Hoeffding, 1964.

4. Tables of the χ^2 Distribution. (Note the existence of tables in the texts at the head of the references to this chapter.)

Aroian, 1943; Bienaymé, 1838; Bliss, 1944; Blom, 1954; Boyd, 1965; Chakravarti and Rao, 1959; Chu, 1956; Cornish and Fisher, 1937; Crow, 1945; Darwin, 1958; Editorial, 1955; Fisher, *Statistical Methods;* Goldberg, and Levine, 1946; Greenwood and Hartley, 1962; Guenther and Thomas, 1965; Hald and Sinbaek, 1950; Harter, 1964*a*; Hartley and Fitch, 1951; Hartley and Pearson, 1950; Johnson, 1959; Khamis, 1963; Kondo and Elderton, 1931; Lewis, 1953; Mathur, 1961; Merrington, 1941, 1951; Molina, 1945; Pagurova, 1963; Patau, 1941; E. S. Pearson and Hartley, 1951; K. Pearson, 1900*a*, *b*, 1922*a*; Peiser, 1943; Salvosa, 1930; Severo and Zelen, 1960; Sheppard, 1903, 1907; Slutsky, 1941; Smirnov and Potapov, 1957, 1961; Soper, 1914; Thompson, 1941; Thomson, 1947; Wallace, 1958*a*, *b*, 1959; Zubin, 1939; Elderton, 1902; Krishnaiah and Armitage, 1965.

5. The Distribution of Quadratic Forms in Normal Variables

Aitken, 1931, 1940, 1950; Cochran, 1934; Craig, 1936, 1943; James, 1952; Kotz, Johnson and Boyd, 1967*a*, *b*; Lancaster, 1954; Matérn, 1949; Neyman and Pearson, 1928; Okamoto, 1960; K. Pearson, 1900*a*; Robbins and Pitman, 1949; Robinson, 1965.

CHAPTER III. DISCRETE DISTRIBUTIONS

Texts: Eisenhart, 1949; Feller, 1957; Fisher, 1950*a*, *Statistical Methods;* Fisher and Yates, *Tables;* Greenwood and Hartley, 1962; Kendall and Stuart, 1963; Kingman and Taylor, 1966; Lehmann, 1959; Lieberman and Owen, 1961; Molina, 1945; Mood, 1950; Patil, 1965; K. Pearson, 1931; Savage, 1962; Soper, 1923; Todhunter, 1865; U.S.A. National Bureau of Standards, 1949; Uspensky, 1937; von Bortkiewicz, 1898; Walsh, 1962/1965.

1.–2. Condensation and Randomized Partition. Significance Tests in Discrete Distributions. (This list also includes some references concerned with more than one distribution.)

Adler, 1951; Bartholomew, 1967; Berry, 1941; Blom, 1954; Cochran, 1942; Crow, 1952; Edgeworth, 1916, 1917; Eudey, 1949; Fisher, 1922*c*, *Statistical Methods;* Gordon, 1939; Govindarajulu, 1965; Haldane, 1937*a*, 1955*b*; Jordan, 1937; Lancaster, 1949*b*, 1950*b*, 1961*a*; Lehmann, 1959; Mahalanobis, Bose and Roy, 1937; E. S. Pearson, 1950; K. Pearson, 1905*a*, 1934*a*; Peierls, 1935; Rietz, 1923; Riordan, 1937; Stevens, 1950; Tippett, 1932; Tocher, 1950; Walsh, 1957, 1957/58; Yates, 1934.

3. The Normal Approximation to the Binomial Distribution.

Bernstein, 1911, 1926, 1943; Bienaymé, 1852; Blom, 1954; Camp, 1924, 1929, 1951; Chu, 1956; Clopper and Pearson, 1934; Cochran, 1942, de Moivre, 1733; Edgeworth, 1916, 1917; Ellis, 1944*b*; Feller, 1945, 1967; Hannan and Harkness, 1963; Isserlis, 1917; Krishnaswami Ayyangar, 1934*a*; Laplace, *Oeuvres;* Lorenz, 1949; McKinnon, 1959; Mirimanoff, 1929, 1930; Mirimanoff and Dovaz, 1927; Monro, 1874; Neyman, 1938*b*; K. Pearson, 1895, 1905*a*, 1924*c* (footnote to el Shanawany, 1939); Poisson, 1827, 1832; Riordan, 1937; Romanowsky, 1923; Sawkins, 1947; Shannon, 1942; Todhunter, 1865; Uspensky, 1937; Wallace, 1958*a*; Yates, 1934; Eggenberger, 1893; Koessler, 1924.

4a. The Normal Approximation to the Poisson Distribution. (See also V.10.)

Cochran, 1936a; Feller, 1967; Fisher, 1922c, 1950b; Garwood, 1936; Gordon, 1939; Haight, 1967; Hoel, 1938; Lüders, 1934; Peierls, 1935; Schilling, 1947; Soper, 1914; 1923; Tschuprow, 1918a, b, 1920, 1923a; Tippett, 1932; von Bortkiewicz, 1898, 1910; Wallace, 1958a, b; Whittle, 1959/60; Edwards, 1960.

4b. The Normal Approximation to the Hypergeometric Distribution. (See also XI.4.)

Bhattacharya, 1966; Edwards, 1948; Feller, 1967; Fisher, 1962; Greenwood, 1913; Irwin, 1935; Kemp and Kemp, 1956; Larsen, 1939; Plackett, 1964; Ram, 1954, 1955, 1956; Riordan, 1937; Romanovsky, 1925; Sawkins, 1947; Soper, 1923, 1926; Uhlmann, 1966; von Schelling, 1940, 1949, 1950; Yates, 1934; Eidel'nant, 1958.

5. The Normal or χ^2 Approximation to the Multinomial Distribution. (See also V.10).

Adler, 1951; Bennett, 1962; Burnside, 1925; Cochran, 1936a, 1942; Darwin, 1958; el Shanawany, 1936; Fisher, Thornton and Mackenzie, 1922; Haldane, 1939b, c, d; Lancaster, 1952, 1953b, 1961a; Lancaster and Brown, 1965; Neyman and Pearson, 1931; Plackett, 1964; Studer, 1966; Sukhatme, 1938; Uppuluri and Bowman, 1966; Whittle, 1959/60; Wishart, 1949; Yoshimura, 1963.

CHAPTER IV.   ORTHOGONALITY

Texts: Achieser, 1956; Alexits, 1961; Bellman, 1960; Cramér, 1946; Fisher, 1950a, *Statistical Methods;* Fisher and Yates, *Tables;* Kac, 1959; Riesz and Sz-Nagy, 1956; Roy, 1957; Shohat, Hille and Walsh, 1940; Shohat and Tamarkin, 1943; Smirnov, 1965; Stieltjes, 1914/18; Szegö, 1939; Tricomi, 1955, 1957; Turnbull and Aitken, 1932; van der Waerden, 1957; Whittaker and Watson, 1943; Zygmund, 1959.

1.–4. Orthogonality.

Aitken, 1933a, b, c, 1935, 1949; Andersson, 1941, 1942; Bahadur, 1961a, b; Brauer, 1929; Burnside, 1928; Corsten, 1957, 1958; Cramér, 1924; Czuber, 1920; Dalzell, 1945; Davis, 1961, 1962; Davis and Rabinowitz, 1961; Deemer and Olkin, 1951; Eagleson, 1964; Fisher, 1915; Fog, 1948; Good, 1958a, 1960; Haar, 1910, Hannan, 1961; Helmert, 1876b; Hsu, 1943; Hussain, 1943; Irwin, 1929, 1942, 1949; Kaczmarz, 1929; Krawtchouk, 1929; Lancaster, 1949a, 1965a, b; Lauricella, 1912; Meixner, 1934, 1938; Mirimanoff, 1929; Picone, 1934; Rademacher, 1922; Runge, 1914; Sansone, 1933; Schmidt, 1907; Schur, 1924; Tchebichef, 1864; van Heerden and Gonin, 1966, 1967; Vitali, 1921; Walsh, 1923; Watson, 1933a, b, 1934; Wijsman, 1957.

CHAPTER V.   THE MULTINOMIAL DISTRIBUTION

Texts: Cramér, 1937, 1946; Feller, 1966, 1967; Fisher, 1950a, *Statistical Methods;* Gnedenko and Kolmogorov, 1954; Greenhood, 1940; Kendall and Stuart, 1961, 1963; Kullback, 1959; Laplace, *Oeuvres;* Lexis, 1877, 1903; Loève, 1955; Lukacs, 1959; Lukacs and Laha, 1963; Mood, 1950; Rao, 1952a; Steck, 1957; Todhunter, 1865; Uspensky, 1937; von Mises, 1931.

1.–2. The Multivariate Central Limit Theorem.

Bernstein, 1926, Bienaymé, 1838, 1852;   Chernoff and Teicher, 1958;   David, 1948; Ellis, 1844*b*; Fisher, 1922*a*; Hoeffding and Robbins, 1948; Hsu, 1942; Ikeda, 1959; Laplace, *Oeuvres;* Loève, 1955; Lukacs, 1959; Morgenstern, 1958; Neyman, 1937*b;* Okamoto, 1959; K. Pearson, 1900*a*, 1916*b*; Poisson, 1827, 1832; Pólya, 1923, Richter, 1958, 1964; Sheppard, 1898*a, b*; J. H. Smith, 1947; Tschuprow, 1918*a, b*, 1919; Tumanyan, 1954, 1956; von Mises, 1919.

3. The Proofs of K. Pearson.

Aitken, 1939; Camp, 1929; Cochran, 1952; Fisher, 1922*a*; Lancaster, 1949*a*, 1965*b*: Neyman, 1938*b*; K. Pearson, 1900*a*, 1903, 1916*a, c*, 1932*a*; Sarmanov, 1960*a*.

4. Stirling's Approximation.

Camp, 1929; de Moivre, 1733; Feller, 1967; Fry, 1938.

5. The Proof of H. E. Soper.

Bol'sev, 1965; Daboni, 1959; Fisher, 1922*a*, 1950*a*; Fisher, Thornton and Mackenzie, 1922; Moran, 1952; K. Pearson, 1916*b*, 1927; Rybarz, 1959; Soper, 1916, 1923.

6. The Factorization Proof.

Cramér, 1946; Irwin, 1949; Lancaster, 1949*a*; Morgenstern, 1958; K. Pearson, 1916*c*.

7. The Proof by Curve Fitting (Lexis Theory).

Geiringer, 1942; Gontcharoff, 1943; Guldberg, 1923*b*; Haldane, 1937*b, c*, 1939*c*; Hendricks, 1935; Lancaster, 1950*c*; Ottestad, 1943; K. Pearson, 1932*a*; Rietz, 1932; Rust, 1965; Tschuprow, 1918*a*, 1919, 1923*b*; Vessereau, 1958; von Bortkiewicz, 1915*b*, 1917, 1918, 1931; Yoshimura, 1963.

8. Analogues of the Pearson χ^2. (See also Section VII.3.)

Barton, 1953, 1954, 1955, 1956; David, 1939, 1947; Fraser, 1950; Hamdan, 1962, 1964; Jeffreys, 1938; Lancaster, 1953*a*, 1961*c*; Neyman, 1937*b*; Scott, 1949*a, b*.

9. Empirical Verifications of the Distribution of the Discrete χ^2.

B. M. Bennett and Hsu, 1961; R. W. Bennett, 1962; Bronwlee, 1924; Cochran, 1936*a*, 1942, 1952, 1954; Egg, Rust and van der Waerden, 1965; el Shanawany, 1936; Gildemeistèr and van der Waerden, 1944; Lancaster, 1952, 1961*a, c*; Lancaster and Brown, 1965; Neyman and Pearson, 1931; Nixon, 1913; E. S. Pearson, 1950; E. S. Pearson and Merrington, 1948; Price, 1946; Robinson, 1933; Schäffer, 1957; Sukhatme, 1935, 1937*a, b*, 1938; Uppuluri and Bowman, 1966; Vessereau, 1958; Wilson, Hilferty and Maher, 1931; Wise, 1963, 1964; Yule, 1922.

10. Applications of χ^2 in the Multinomial Distribution. (See also III.4a.)

Bartko, Greenhouse and Patlak, 1968; Barton and David, 1959; Basharin, 1957; Bateman, 1950; Bennett, 1962; Bennett and Hsu, 1961; Berkson, 1940; Bol'sev, 1965; Bowley and Connor, 1923; Broadbent, 1956; Cochran, 1936*a*, 1952, 1955; Daboni, 1959; Darwin, 1957; Egg, Rust and van der Waerden, 1965; Eisenhart and Wilson, 1943;

CHAPTER VI. CANONICAL OR STANDARD FORMS FOR PROBABILITY DISTRIBUTIONS

1.–6. Canonical or Standard Forms. (*See also XI and XII.*)

CHAPTER VII.   NON-CENTRAL χ^2

Texts: Cramér, 1946; Fisher, 1950a, Kendall and Stuart, 1961, 1963; Maxwell, 1961.

1. Distribution Theory.

Abdel-aty, 1954; Bateman, 1949, 1950; Bennett, 1955/56; Bernt, 1958; Cadwell, 1953; Chakrabarti, 1949; Chiang, 1951; Čibisov, 1965; Daniels, 1952; Darwin, 1957; Diamond, 1963; Diamond, Mitra and Roy, 1960; di Donato and Jarnagin, 1962; Eisenhart, 1938; Eisehnart, Hastay and Wallis, 1947; Fix, 1949; Gilliland, 1962, 1964; Guenther, 1964a, b; Guenther and Terragno, 1964; Gumbel, 1954; Gumbel and Goldstein, 1964; Gurian, Cornfield and Mosimann, 1964; Harkness and Katz, 1964; Hodges, 1955; Johnson, 1959; Kamat, 1962; Kermack and McKendrick, 1940; Kerridge, 1965; Khatri, 1963; Khatri and Pillai, 1965; Li, 1963; McNolty, 1962; Meng and Chapman, 1966; Meyer, 1967; Nandi, 1946; Owen, 1962; Patnaik, 1948, 1949; E. S. Pearson, 1959; E. S. Pearson and Hartley, 1951; E. S. Pearson and Merrington, 1948; Poti, 1950; Press, 1966; Przyborowski and Wilenski, 1935b; Roy and Mohamad, 1964; Ruben, 1963; Sankaran, 1959, 1963; Seber, 1963; Severo and Zelen, 1960; Sillitto, 1949; Tang, 1938; Taylor, 1951; Tiku, 1965a; Tukey, 1957; Tumanyan, 1958a, b; Wells, Anderson and Cell, 1962, Wishart,, 1932; Wolfowitz, 1949.

2.–3. Analogues of the Pearson χ^2, the Combination of Probabilities.

Aoyama, 1953; Baker, 1952; Bartlett, 1952b, 1954; Barton, 1953, 1954, 1955, 1956; Bernstein, 1927a; Birnbaum, 1954; Bruen, 1938; Chapman, 1958; Cochran, 1954; David, 1939, 1947; David and Johnson, 1948, 1950; Fisher, *Statistical Methods;* Fraser, 1950; Good, 1955, 1958a; Gordon, Loveland and Cureton, 1952; Halmos and Savage, 1949; Hamdan, 1962, 1964; Kincaid, 1962a, b; Lancaster, 1949b, 1953a, 1961b; Lehmann, 1959; Liptak, 1958; Mallows, 1959; Mather, 1935; Mosteller, 1948; Neyman, 1937b, 1940a, b; Neyman and Pearson, 1938a; Okuno and Takeuchi, 1961; E. S. Pearson, 1938, 1942, 1950; E. S. Pearson and Wilks, 1933; K. Pearson, 1895, 1933b; Scott, 1949a, b; Seal, 1948; van Zwet and Oosterhoff, 1967; Wallis, 1942; Watson, 1957; Wilks and Daly, 1939; Woodworth, 1912; Yates, 1934, 1955b; Zelen and Joel, 1959; Denny, 1967.

CHAPTER VIII.   TESTS OF GOODNESS OF FIT IN THE MULTINOMIAL DISTRIBUTION

Texts: Cramér, 1946; Elderton, 1907; Fisher, 1938b, 1950a, *Statistical Methods;* Kendall and Stuart, 1961, 1963; Kullback, 1959; Kulldorff, 1961; Maxwell, 1961.

1. Introductory.

Fisher 1922a, b, 1924; K. Pearson, 1900a, 1913b; 1916b.

2. Least Squares and Minimum χ^2.

Aitken, 1948, 1933a, b, c, 1935; Bartlett, 1937b, Berkson, 1949, 1951; Bienaymé, 1852; Blakeman, 1905; Chernoff and Lehmann, 1954; Fisher, 1922a, b, c; 1923, 1924, 1928a,

3. The Fitting of Sufficient Statistics.

4. χ^2 in the Multinomial Distribution with Estimated Parameters (Fisher Theory).

5. Estimated Parameters (Cramér Theory).

6.–7. Estimated Parameters and Orthonormal Theory. The Test of Goodness of Fit.

CHAPTER IX.   PROBLEMS OF INFERENCE

1. Introductory.

10. χ^2 and the Sample Size.

Berkson, 1938; Camp, 1938, 1940; Diamond, 1963; Edgeworth, 1916; Hoel, 1938; Pathria, 1961, 1962, 1964; Rao and Chakravarti, 1956; Wald, 1943.

11. Small Class Frequencies.

Adler, 1951, Claringbold, 1961; Fisher, 1922a, 1950b, Gumbel, 1943; Haldane, 1955b, Hoel, 1938; Irwin, 1937; Katti and Sastry, 1965; Nass, 1959; Neyman and Pearson, 1931; Patnaik, 1954; Sawkins, 1941; Slakter, 1966; Yates, 1934.

12. The Partition of χ^2.

Bodmer, 1959; Bodmer and Parsons, 1959; Claringbold, 1961; Cochran, 1952, 1955; Corsten, 1957; Fisher, 1915, Statistical Methods; Fog, 1953; Haldane, 1939a, Irwin, 1942 and 1949; Kastenbaum, 1960; Kimball, 1954; Kuiper, 1953; Lancaster, 1949a, 1950a, 1951, 1953a, 1960a, 1965a, b; Neyman and Pearson, 1931, 1938b; K. Pearson, 1906d; Roberts, Dawson and Madden, 1939.

13. Misclassifications and Missing Values.

Batschelet, 1960; Bross, 1954; Diamond and Lilienfeld, 1962a, b; Elashoff and Afifi, 1966; Mote and Anderson, 1965; Watson, 1956.

14. The Reconciliation of χ^2.

Barnard, 1947b; Cochran, 1940, 1950; Fisher, 1922a, 1923, 1924, Statistical Methods; Geary, 1927, 1940, 1942; Hirschfeld, 1935; K. Pearson, 1904; Yule, 1911 (Text).

15. Miscellaneous Inference.

Alternatives. David, 1950; Haldane, 1955a; Savage, 1953, 1962; Silvey, 1959; Smirnov, 1948, 1949; Wald and Wolfowitz, 1940.

Computation. Carroll and Bennett, 1950; Dubois, 1939; Gower, 1961/62; Haldane, 1955c; Royer, 1933; Skory, 1952.

Design. McNemar, 1952; Mathen, 1954.

Fréchet Classes. Caussinus, 1966.

General. Fernandez Baños, 1946.

Generalizations. Caussinus, 1962.

Genetics. Greenberg and White, 1965; Jackson, 1936.

Goodness of fit. Jeffreys, 1936; Kempthorne, 1967.

Information. Kullback, Kupperman and Ku, 1962b; Linfoot, 1957; McGill, 1954; Masuyama and Yoshimura, 1962; Linnik, 1961.

Markov Chains. Almond, 1956; Bartlett, 1951b, c, 1952a; Bhat, 1961; Bofinger and Bofinger, 1961a, b; Bui Trong Lieu and Carton, 1964; Gold, 1963; Goodman, 1964b; Neall, 1952; Patankar, 1954; Stepanov, 1957.

Optimal Tests. Hoeffding, 1965.

Order Statistics. Witting, 1959.

Power. Wald, 1941a, b.

Probits and Logits. Berkson, 1946a.

Quality Control. Feuell and Rybicka, 1951; Scheffé, 1947.

Sequential. Jackson and Bradley, 1961a, b.

Sign Test. Bennett, 1965.

Simple Test. Ditlevsen, 1964.

Guilford and Perry, 1951; Hamdan, 1968b; Hayes, 1943, 1946; Henderson, 1922; Jurgensen, 1947; Kendall, 1941; Kuder, 1937; Lee, 1917, 1925, 1927; Mardia, 1967; Mehler, 1966; Mood, 1941; Mosteller, 1946; Newbold, 1925; E. S. Pearson and K. Pearson, 1922; K. Pearson, 1900a, b, c, 1912, 1913a, 1931, 1933a; K. Pearson and Heron, 1913; Ritchie-Scott, 1918; Smirnov, 1965; Stieltjes, 1889; Vaswani, 1950; Wahlund, 1935.

9. The Polychoric Series.

Guilford, 1941; Harris and Treloar, 1927; Harris, Treloar and Wilder, 1930; Harris and Tu, 1929; Harris, Tu and Wilder, 1930; Lancaster and Hamdan, 1964; E. S. Pearson and K. Pearson, 1922; K. Pearson, 1913b, 1930, 1931; Ritchie-Scott, 1918.

10. The Correlation Ratio.

Blakeman, 1905; Crathorne, 1922; Fisher, Statistical Methods; Kellerer, 1964; Pearson, 1905b, 1909, 1911b; Woo, 1929.

11. Biserial η.

Brogden, 1949; Dingman, 1954; Dubois, 1942; Dunlap, 1936; Editorial, 1955; Goheen and Davidoff, 1951; Hannan and Tate, 1965; Kuder, 1937; Lev, 1949; Lins Martins, 1946; Lord, 1963; Maritz, 1953; Michael, Perry and Guilford, 1952; Newbold, 1925; K. Pearson, 1909, 1910, 1911b, 1913b, 1917, 1921, 1923a, b; Perry and Michael, 1954; Pompilj, 1950; Ritchie-Scott, 1918; Royer, 1933, 1941; Soper, 1915; Tate, 1954, 1955; Wherry, 1947; Wherry and Taylor, 1946.

12. Tests of Normality.

Berkson and Geary, 1941; Cornish and Fisher, 1937; Fisher, Statistical Methods; Geary, 1947, 1956; Geary and Pearson, 1938; Kendall, 1941.

13. Various Measures of Correlation.

Blalock, 1958; Fisher, 1915; Goodman and Kruskal, 1954, 1959, 1963; Hamdan, 1968b; Hotelling, 1933; Kono, 1952; Kruskal, 1958; E. S. Pearson, 1923; K. Pearson, 1901; Ritchie-Scott, 1918; Wilks, 1932.

CHAPTER XI. TWO-WAY CONTINGENCY TABLES

Texts: Anderson, 1958; Cournot, 1843; Elderton, 1907; Fisher, 1950a, Statistical Methods; Gumbel, 1958; Kendall and Stuart, 1961; Kullback, 1959; Maxwell, 1961; Mood, 1950; Rao, 1952a.

1. Introductory.

Anderson, 1959; Bejar, 1958; Bodmer and Parsons, 1959; Dagnelie, 1962; dall'Aglio, 1959; Lindley, 1964; Malécot, 1948; Stephan, 1942; Steyn, 1959; Thionet, 1959; Wilson, 1942.

2. Probability Models in a Two-way Contingency Table.

Barnard, 1945, 1947a, b; Fisher, 1922a; Goodman, 1964a; Lancaster, 1951; Mood, 1949; E. S. Pearson, 1947; K. Pearson, 1904, 1932b and passim; Roy and Kastenbaum, 1956; Tocher, 1908; Yates, 1961.

3. Tests of Independence. (See also XI.4.)

Bartlett, 1935; Fisher, 1922a, Statistical Methods; Lancaster, 1949a, Lehmann, 1959; K. Pearson, 1904, 1932b, Sheppard, 1898a.

4. The Fourfold Table.

Aitken and Gonin, 1935; Armsen, 1955; Barnard, 1945, 1947a, b; Bartholomew, 1967; Bartlett, 1937a; Bennett and Hsu, 1960, 1961; Berger, 1961; Berkson, 1946b; Birch, 1964; Blomqvist, 1951; Boas, 1909; Bonnier, 1942; Bowley, 1920; Bross, 1954, 1964; Bross and Delaney, 1952; Bross and Kasten, 1957; Camp, 1925, 1938; Chapman, 1951; Chung and de Lury, 1950; Cochran, 1950, 1952; Cournot, 1843; Davies, 1933, 1934; Edwards, 1948, 1950, 1963; Federighi, 1950; Festinger, 1946; Finney, 1948; Finney, Latscha, Bennett, Hsu and Pearson, 1963; Fisher, 1922a, Statistical Methods, 1926, 1941a, 1945b, 1962; Fiske and Dunlop, 1945; Fulcher and Zubin, 1942; Gabriel, 1963; Garside, 1958, 1961; Gates and Beard, 1961; Gonin, 1936; Gray, 1947; Gray and Tocher; 1900; Greenwood, 1913; Greenwood and Yule, 1915; Gridgeman, 1963; Grizzle, 1967, Haldane, 1945a, 1955b; Harkness and Katz, 1964; Heron, 1910; Irwin, 1935; Isserlis, 1914; Jeffreys, 1937b; H. M. Johnson, 1945; Kermack and McKendrick, 1940; Krishnaswami Ayyangar, 1934b; Kullback, Kupperman and Ku, 1926b; Kupperman, 1960; Lancaster, 1950d; Larsen, 1939; Latscha, 1953; Leslie, 1955; Liebermeister, 1877; Lombard and Doering, 1947; Lord, 1944; McNemar, 1947; Mainland, 1948; Mainland and Murray, 1952; Mantel, 1963; Moore, 1949; Newbold, 1925; Niceforo, 1911a, b; Okamoto and Ishii, 1961; Ottaviani, 1939; Patnaik, 1948, 1954; Paulson and Wallace, 1947; E. S. Pearson, 1947; E. S. Pearson and Merrington, 1948; K. Pearson, 1899, 1906c, 1912, 1913a, 1916a, 1922b, 1923a, 1924a, b, 1928a; K. Pearson and Heron, 1913; K. Pearson and Tocher, 1916; Robertson, 1960; Romanovsky, 1925; Sheppard, 1898a; Sheps, 1959; Slater, 1947; C. A. B. Smith, 1951; Swaroop, 1938, 1950; Swineford, 1946, 1948, 1949; Thompson, 1933; van Eeden and Runnenburg, 1960; von Schelling, 1949, 1950; Walsh, 1963; Weiner, 1959; Wiid, 1957/8; Wilks, 1940; Wilson, 1941; Winsor, 1948a; Wise, 1954; Woolf, 1957; Yang, 1965; Yates, 1934, 1955a, b; Yule, 1922; Zubin, 1939.
Doolittle, 1888; J. H. Edwards, 1957; Vietoris, 1961.

5. Combinatorial Theory of the Two-way Tables.

Bartlett, 1937a; Bennett and Nakamura, 1963, 1964; Blakeman and Pearson, 1906; Chakravarti and Rao, 1959; Cochran, 1936a, 1937, 1950, 1952; Dawson, 1954; Freeman and Halton, 1951; Geary, 1927, 1940; Greville, 1941; Haldane, 1939c, d, 1945b; Karon and Alexander, 1958; Kondo, 1929; Lancaster, 1949a; Mood, 1949; Nass, 1959; K. Pearson, 1915, 1919, 1921; Stevens, 1938, 1951; Williams, 1952; Yates, 1934; Yoshimura, 1963; Young and Pearson, 1916; Hamdan, 1968c.

6. Asymptotic Theory of the Two-way Tables.

Bhapkar, 1961; Birch, 1965; Claringbold, 1961; Clark and Leonard, 1939; Cochran, 1936b, 1940; Daly, 1962; Fog, 1953; Good, 1956, 1963; Goodman, 1963a, b; Goodman and Kruskal, 1963; Irwin, 1949; Ishii, 1960; Kastenbaum, 1960; Kimball, 1954; Ku, 1963; Lancaster, 1949a, 1950a, 1957, 1958; Leslie, 1951; Mood, 1949; Pearson, 1904; C. D. Smith, 1951; Steyn, 1963; Vajda, 1943; Waite, 1915; Yates, 1934.

7. Parameters of Non-Centrality.

Bennett, 1959; Bennett and Nakamura, 1964; Blakeman and Pearson, 1906; Blalock, 1958; Diamond, Mitra and Roy, 1960; Fisher, 1940, 1941a; Goodman and Kruskal, 1954, 1963; Harris, 1957; Hotelling, 1936; Jurgensen, 1947; Kendall and Stuart, 1961; Konijn, 1956; Lancaster, 1958, 1963c; Lehmann, 1959; Lipps, 1905; Maung, 1942; Meng and Chapman, 1966; Neyman, Chernoff and Chapman, 1965; K. Pearson, 1904, 1906d; Somers, 1962; Stuart, 1953; Tallis, 1962; Weida, 1934; Wijsman, 1958; Williams, 1952.

8. Symmetry and Exchangeability in Two-way Tables.

Bowker, 1948; Gridgeman, 1963; Kendall and Stuart, 1961; Lancaster, 1965a; Mathen, 1954; Reiersøl, 1944; Richmond, 1963; Stuart, 1955.

9. Reparametrization.

Chapelin, 1932; Daly, 1962; Deming and Stephan, 1940; el Badry and Stephan, 1955; Friedlander, 1961; Geppert, 1961; Good, 1956; Hayashi, Takakura, Makita and Saito, 1958, 1960; K. Pearson, 1900a, b, c, 1904; J. H. Smith, 1947; Thionet, 1959; Weichselberger, 1959; Williams, 1952; Yates, 1948.

10. Measures of Association.

Aitken and Gonin, 1935; Barrett and Lampard, 1955; Blomqvist, 1950; Boas, 1909; Caussinus, 1966; Csaki and Fischer, 1960a, b; Deming, 1960; Eagleson, 1964; Fisher, 1940; Friede and Münzner, 1948; for Fréchet references see Caussinus, 1966; Geary, 1942; Gebelein, 1941, 1942a, b, 1952; Geiringer, 1933; Gonin, 1966; Goodman and Kruskal, 1954, 1959, 1963; Hannan, 1961; Hirschfeld, 1935; Hoeffding, 1940, 1941, 1942; Hotelling, 1936; Kendall and Stuart, 1961; Lancaster, 1958, 1959, 1961b, 1963c; Lehmann, 1966; Maung, 1942; K. Pearson, 1900a, b, c, 1904, 1905b; K. Pearson and Heron, 1913; Pretorius, 1930; Putter, 1964; Rényi, 1959a, b; Richter, 1949; Rosenblatt, 1961; Sarmanov, 1958a, b, 1966; Sarmanov and Zaharov, 1960a, b; Silvey, 1964; Smoliakow, 1933; Spearman, 1904, 1906; Steffensen, 1934, 1941a, b; Venter, 1967; Weiler, 1966; Williams, 1952; Yule, 1900, 1903, 1906, 1912.
Bresciani, 1909; Linfoot, 1957; K. Pearson, 1934b; Somers, 1962; Tschuprow, 1934.

11. The Homogeneity of Several Populations.

Baillie, 1946; Baker, 1941; Bhattacharyya, 1943, 1946; Blomqvist, 1950; Boas, 1922; Chacko, 1963; Chakravarti and Rao, 1959; Darwin, 1959; Dixon, 1940; Duncan, 1955; Gart, 1966; Geppert, 1944; Gold, 1962; Goodman, 1964b; Greville, 1941; Haldane, 1955c; Hayman and Leone, 1965; Hoel, 1947; Lancaster, 1950c, 1953a; Lengyel, 1939; Lesser, 1933; Lipmann, 1908; Lukaszewicz and Sadowski, 1960/61; Mantel, 1963; March, 1905; Mathen, 1946; Neyman and Pearson, 1928, 1931; K. Pearson, 1911a, 1924a, b, 1932b; Przyborowski and Wilenski, 1935a, Ramachrandran, 1951; Rhodes, 1924, 1926; Richter, 1958; Robertson, 1951; Robson and King, 1952; C. A. B. Smith, 1952; H. F. Smith, 1957; Snedecor and Irwin, 1933; Sukhatme, 1935; Thompson, 1933;

Tumanyan, 1958a, b; van Eeden, 1965; Vernon, 1936; Vincze, 1960; von Bortkiewicz, 1931; Wegner, 1956; Welch, 1938; Widen, 1962; Woolf, 1957. Bartholomew, 1959; van der Waerden and Nievergelt, 1956.

12. Contingency Tables. Miscellaneous Topics.

Adjustment. Deming and Stephan, 1940; Harris, 1957; Steyn, 1963.
Class Index. E. S. Pearson, 1923; Stephan, 1942.
Entropy. Good, 1963.
Frèchet Classes. Caussinus, 1966.
Extreme Groups. Feldt, 1961.
Likelihood. Ku, 1963.
Interactions. Barnard, 1945; Good, 1958a, 1960, 1963; Simpson, 1951. Williams, 1952.
Multiple Comparisons. Darroch and Silvey, 1963.

CHAPTER XII. CONTINGENCY TABLES OF HIGHER DIMENSIONS

Texts: Fisher, 1950a, *Statistical Methods*, Kendall and Stuart, 1961; Kullback, 1959; Mood, 1950; Roy, 1957.
Articles.
Bahadur, 1961a, b; G. A. Barnard, 1945; M. M. Barnard, 1936; Bartlett, 1935; Bejar, 1958; Berkson, 1968; Bhapkar, 1961; Birch, 1963; Camp, 1938; Claringbold, 1961; Cochran, 1940; Corsten, 1957; dall'Aglio, 1959; Darroch, 1962; Deming and Stephan, 1940; Diamond, Mitra and Roy, 1960; Dyke and Patterson, 1952; Fog, 1953; Freeman and Halton, 1951; Gabriel, 1963; Garner and McGill, 1956; Geppert, 1961; Good, 1963; Goodman, 1963a, 1964b, c, e; Gregory, 1961; Gruneberg and Haldane, 1937; Haldane, 1939a; Hoyt, Krishnaiah and Torrance, 1959; Kastenbaum, 1960; Kastenbaum and Lamphiear, 1959; Kimball, 1954; Kuiper, 1952; Lancaster, 1951; 1957, 1960a, 1965a; Lazarsfeld, 1961; Lewis, 1962; Lindley, 1964; Lombard and Doering, 1947; Mitra, 1955, 1956, 1958; Mosteller, 1968; Newbold, 1925; Norton, 1945; K. Pearson, 1904, 1916a, b; Plackett, 1962; Roberts, Dawson and Madden, 1939; Rosenblatt, 1952; Roy, 1950; Roy and Bargmann, 1958; Roy and Bhapkar, 1960; Roy and Kastenbaum, 1955, 1956; Roy and Mitra, 1955, 1956; Sarmanov and Zaharov, 1960b; Simpson, 1951; Snedecor, 1958; Somers, 1964; Stanley, 1961; Stephan, 1942; Sutcliffe, 1957; Vaswani, 1950; Wilks, 1935; Wilson and Worcester, 1942a, b; Windsor, 1948b; Yates, 1934, 1955a, b, 1961; Yule, 1900; Byrne, 1935.

Index

This index contains the names of authors (i) of historical interest; (ii) of papers not listed in the *Bibliography* from which examples have been discussed; (iii) of personal communications. It is principally a subject index.